Mathematical Models in the Biosciences 2

Mathematical Models in the Biosciences 2

Michael Frame

Yale UNIVERSITY PRESS/NEW HAVEN & LONDON

Set in Adobe Garamond Pro and The Sans Pro type by Newgen North America

Printed in the United States of America.

Library of Congress Control Number: 2020947408
ISBN 978-0-300-25369-6 (paperback : alk.paper)

A catalogue record for this book is available from the British Library.

This paper meets the requirements of ANSI/NISO Z39.48-1992 (Permanence of Paper).

10 9 8 7 6 5 4 3 2 1

Contents

Preface ix

Ways to use this book xiii

Acknowledgments xvii

Review of Volume 1 xxi

13 Higher-dimensional differential equations 1

13.1 Eigenvalues in higher dimensions 1

13.2 SIR calculations 7

13.3 Michaelis-Menten kinetics 17

13.4 Virus dynamics basics 22

13.5 Immune response dynamics 37

13.6 The Hodgkin-Huxley equations 45

13.7 Chaos in predator-prey systems 54

13.8 The Duffing equation and final state sensitivity 63
13.9 Controlling chaos 73
14 Stochastic models 87
14.1 Markov chains 88
14.2 The Perron-Frobenius theorem 105
14.3 Structured populations 114
14.4 Reproductive value and sensitivity analysis 123
15 A tiny bit of genetics 135
15.1 The Hardy-Weinberg law and selection equations 136
15.2 Price's equation and multilevel selection 146
15.3 Fitness landscapes 150
15.4 The quasispecies equation 158
15.5 Bioinformatics 163
15.6 Some other topics 167
16 Markov chains in biology 179
16.1 Genetic drift 180
16.2 Matrix differential equations 190
16.3 Ion channel dynamics 197
16.4 The Clancy-Rudy cardiac arrhythmia model 206
16.5 Tumor suppressor genes 215
17 Some vector calculus 225
17.1 Gradient fields 226
17.2 Line integrals 231
17.3 Double integrals 243
17.4 Green's theorem 250

17.5 Interesting geometries 256
17.6 Bendixson's criterion 263
17.7 The index of a fixed point 266
17.8 Surface integrals 277
17.9 Stokes' theorem 285
17.10 Triple integrals 292
17.11 Gauss' theorem 296
17.12 Diffusion 301
18 A glimpse of systems biology 304
18.1 Genetic toggle switch 307
18.2 Transcription networks 312
18.3 Some other networks 323
18.4 Why should this work? 328
19 What's next? 329
19.1 Evolutionary medicine 329
19.2 Translational bioinformatics 336
19.3 Topological methods 341
19.4 No, really, what's next? 351
Appendix A Technical Notes 353
A.12 Some more linear algebra 353
A.13 Some Markov chain properties 362
A.14 Cell membrane channels 370
A.15 Viruses and the immune system 373
A.16 The normal density function 389
A.17 A sketch of the Perron-Frobenius proof 390

A.18 The Leslie matrix characteristic equation 393
A.19 Liapunov exponents 395
A.20 Stochastic resonance and the Duffing equation 398
A.21 Proof of Lienard's theorem 402
A.22 A bit of molecular genetics 406
Appendix B Some Mathematica code 416
B.11 Eigenvalues in higher dimensions 416
B.12 SIR calculations 417
B.13 Michaelis-Menten calculations 417
B.14 Virus dynamics 418
B.15 Immune system dynamics 418
B.16 The Hodgkin-Huxley equations 419
B.17 Chaos in predator-prey models 421
B.18 The Duffing equation 421
B.19 Control of chaos 422
B.20 Sensitivity analysis 424
B.21 The Clancy-Rudy model 424
B.22 A genetic toggle switch 425
B.23 Transcription networks 426
B.24 Topological methods 426
Appendix C Some useful integrals and hints 429
References 433
Index 453

Preface

In the previous volume we studied techniques from second semester calculus with applications to the biomedical sciences. Much of this involved one- and two-dimensional differential equations, infinite series, and a bit of probability. While the collection of applications of mathematics to biology and medical practice is immense—far more than I, or perhaps anyone, can catalog now, in part because it is a moving target— the examples of Volume 1 miss some important categories. For example, many systems modeled by differential equations require more than two variables. Some epidemiological systems (Sect. 13.2), virus dynamics (Sect. 13.4), immune response dynamics (Sect. 13.5), nerve impulse transmission (Sect. 13.6), transcription networks (Sect. 18.2), and metastatic prostate cancer dynamics (Sect. 19.1) are a few examples that can be approached by extending the 2-dimensional techniques of Chapter 9 to higher dimensions. This was one reason to develop a second semester course.

That alone would not be adequate reason for another course. And premed education is so crowded that it scarcely could allow calculus 2a and calculus 2b. But some premed majors require calculus 3, and vector calculus provides more techniques to study long-term behavior of trajectories of differential equations. For a decade I was in charge of, and taught sections of, calculus 3 at Yale. Through this much field experience I noticed a few tricks that are included in Chapter 17. The notion of the index of a fixed point (Sect. 17.7), and the proofs of Bendixson's criterion (Sect. 17.6) and Lienard's theorem (Sect. A.21), use vector calculus and have interesting biological applications.

Add in Markov chains (Sect. 14.1) and some applications (Sects. 16.3, 16.4, 16.5), examples from network theory (Sects. 18.2, 18.3), and a topological approach (Sect. 19.3) to sequence pattern visualization. That's the second volume. Throughout I've tried to give adequate background for the biological systems involved. Some of these are folded into the text; others—sketches of the immune system (Sect. A.15) and of molecular genetics (Sect. A.22), for example—are so involved that they inhabit sections of the technical notes.

The main message of these books is that now the conversation between math and biology is a true dialogue. However many biomath books you find, another always could be written. This is fertile ground to cultivate curiosity. Why should we care about curiosity? How's this:

> Curiosity will save us, if anything will.

This is my strongest belief. I hope these two books, along with [120], will show you why I believe this.

There's another point I hope you'll see. Chapter 12 in Volume 1 is my best explanation of this point. Here I'll try another approach.

One hot day in early summer when I was ten and my brother, Steve, was five, our father and our uncle Bill took us to a small beach at Lower Falls on the Coal River in West Virginia. Several families were there that morning. Steve and I looked for interesting rocks, while Dad and Bill swam. A boy about my age waded in the river with his brother, who was about Steve's age. Then the younger brother was gone. The older brother yelled for help. Bill threw an inner tube Steve and I had used as a float. A hand came out of the water, tried to grab the inner tube but slid down the side and was gone. Adults, including Dad and Bill, swam over and began to dive to the river bottom. The boys' mother wailed. Someone scrambled up the steep riverbank to find a house and call the police.

Steve and I sat on the beach. The summer sun suddenly felt cold. The older brother sat on the beach about ten feet from us. He pulled his knees up to his chest, wrapped his skinny little-boy arms around his skinny little-boy legs, and put his head on his knees. Eventually the police arrived by boat. They dropped horrible big hooks into the water and began to drag the river. My father decided it was time for us to leave.

I don't recall much of the discussion with Mom and Dad that night, but they were gentle, thoughtful, and as far as I could tell, honest. This was the first death I'd seen up close. The first time I'd seen the incandescent grief of losing a child, of losing a brother. I could not imagine how the older brother felt. How could I? To do that, I'd need to know so many details of their lives. Were they friends? Did the younger brother annoy the older? To know how he felt, I'd need to know answers to these and a thousand other questions.

The best I could do was wonder how I'd feel if Steve died. That may be impossible, too, but at least I know how to think about this. If you become a physician, you'll face this need for empathy many times in your career. You can't know how your patient feels. But you can see a shadow of how you'd feel if you were in that situation. If you have any uncertainty that you need to think as hard as you can, learn as much as possible, imagine how determined you'd be if the person suffering were someone you love.

And think of this: eventually the person suffering *will* be someone you love. How much are you willing to do now, in order to help then?

Ways to use this book

This book grew out of the second of two math for biosciences courses I developed at Yale. Roughly it covers the content of third semester calculus, with emphasis on concepts and examples for bioscience students, and develops some other mathematics useful in biosciences but not often seen in calculus classes.

After a quick review of the main points of Volume 1, we begin with a sketch of how the simple collection of solution curve types for planar differential equations is far more complicated in 3 dimensions. This is the realm of chaotic solutions of differential equations. In Chapter 13 we sketch just a few simple examples, an invitation to an active, rich area of research that also has touched popular culture—an event that is uncommon for subtle math. We'll see a bit of why this is true.

The evolution of the population distributions can be expressed through discrete dynamical systems, which we see in Chapter 14, or differential equations, which we see in Chapter 16. Examples include structured populations of animals and of pathogens, and ion channel dynamics.

In Chapter 15 we sketch some topics in genetics where mathematical models are of use. These include the Hardy-Weinberg law, fitness landscapes, and the quasispecies equation. For bioinformatics we'll investigate applications of BLAST, the basic local alignment search tool.

Vector calculus, the main component of most calculus 3 courses, is the topic of Chapter 17. In addition to the differential tools gradient, divergence, and curl, we study line and surface integrals along with the fundamental theorem of line integrals, Green's theorem, Stokes' theorem, and Gauss' theorem. Applications include Bendixson's criterion for the existence of closed trajectories and the index of a fixed point.

In Chapter 18 we sketch some developments in systems biology, a challenging and exciting confluence of biology and mathematics. For complex biological systems, gene transcription networks for example, we don't have the right math to approach large-scale questions. The 20th century was a period of wild cross-pollinated growth in mathematics and physics: differential geometry was a key ingredient in Albert Einstein's general theory of relativity, and gauge field theory was central to Simon Donaldson's work on 4-manifolds. For a long time, mathematics has had much to say to biology, but has not listened very well to biology. But now biology is asking many questions that mathematics cannot answer. We expect, or at least we hope, that we are on the brink of a period of mathematical growth, in directions unimagined, informed by clever biologists. This is an exciting time to be on the very fuzzy and quickly moving boundary between mathematics and biology.

Finally, in Chapter 19 we explore a few examples of my guess at promising directions. These include evolutionary medicine, translational bioinformatics, and an application of fractals to study patterns in biological sequences. I expect my guesses are wrong. Not that these directions won't yield important results, but rather that the most exciting advances will be in fields unknown to me, or perhaps completely unknown now. I wish I could see what you will see.

I'll list the chapters I used for the second course, and suggest some alternates.

The course started with Markov chains, the Perron-Frobenius theorem, age-structured populations, and sensitivity analysis, the contents of Chapter 14. Then matrix differential equations (Sect. 16.2), ion channel dynamics (Sect. 16.3), and the Clancy-Rudy model (Sect. 16.4). Then we took a short detour into fractal analysis of sequence data in Sect. 19.3, a topic that reinforces the importance of visual reasoning. Vector calculus, Chapter 17, was a focus of the course. Before Sect. 17.6 we returned to Sect. 9.8. Then we looked at more examples of higher-dimensional systems in Sects. 13.2, 13.4, and 13.5. We concluded with two examples of the importance of evolutionary medicine, the quasispecies equation (Sect. 15.4) and tumor suppressor genes (Sect. 16.5), and the survey of Sect. 19.1.

A more traditional course would expand the coverage of vector calculus at the expense of some of the longer biomedical examples. Certainly, the geometry of vector calculus is one of the most beautiful parts of the undergraduate math curriculum. My colleague Steve Stearns argued, convincingly, that the point of this course is to show biomath students that math is relevant to biology and to medicine. So examples of successful applications of math to biomedical problems, especially if some claims of clinical significance are supported, are more effective ways to show future physicians that they should attend to mathematical models.

We should embrace every tool, *every tool*, that can be useful in medicine. These two courses are my answer to the question "Why should I learn this?" asked by premed students in calculus 2 or 3. The answer "That you'll need it for the MCATs" gives scant satisfaction. Far better is "Because it's led to a deeper understanding of disease dynamics, and to improved clinical outcomes." Physicians need to know a lot. These courses are my attempt to show that math should be part of that lot.

Acknowledgments

Some books owe much to earlier books. In writing this text, I was informed and sometimes inspired by these: [6, 9, 12, 16, 107, 138, 174, 188, 258, 261, 351, 387, 388]. If you are interested in, or become interested in, the conversation between biology and math, I encourage you to read all of them. Treasures await in each.

Many people helped me write this book. Garrett Odell, who taught my undergraduate topology course at RPI in 1972, introduced me to the power of mathematical biology, specifically, just how much of invagination in early gastrulation is due to mechanical forces on neighboring cells. This idea stayed in my head for forty-five years. I intended to thank Gary when I finished this book, a quiet voice from the end of my career to recall the start of his. But he died in May of 2018. I waited too long.

My colleague Ted Bick at Union College talked me into teaching a mathematical biology course in 1989. This was my first exposure to the breadth of the subject, immense possibilities then seen by me only as shadows. I intended to thank Ted when I finished this book, but he died in August 2016. You might think this loss of a dear friend would have made me contact Gary, but no. Sometimes I am just staggeringly blind.

So, if you feel you need to thank someone, don't wait. Be better than I've been. Expressed appreciation always improves the world.

The chapters on differential equations developed from a Union College course I taught, aided by colleagues Arnold Seiken and William Fairchild, enthusiastic and thoughtful members of the audience. Much of the material in Chapters 2 and 6 grew out of discussions, some of the most interesting I've ever had, with David Peak, my coauthor of [271]. Conversations with Yale colleagues Sandy Chang, John Hall, Miki Havlickova, Douglas Kankel, David Pollard, Richard Prum, William Segraves, Stephen Stearns, Andy Szymkowiak, Günter Wagner, and Robert Wyman were instructive and enjoyable. The Howard Hughes Medical Institute grant number 52006963 and the Yale College Dean's Office provided generous support. In addition, I've benefitted from the comments and suggestions of my students. Serious, interested students are a delight for any teacher, and I've been especially fortunate with the students I've gotten. In particular, thanks to Monique Arnold, Christopher Coyne, Mariana Do Carmo, Rafael Fernandez, Candice Gurbatri, Megan Jenkins, Misun Jung, Miriam Lauter, Regina Lief, Jonathan Marquez, Aala Mohamed, Susie Park, Miriam Rock, Ashley Schwarzer, Bijan Stephen, Paschalis Toskas, and Zaina Zayyad. And special thanks to Divyansh Agarwal, Aiyana Bobrownicki, Colleen Clancy, Noelle Driver, Liz Hagan, Fran Harris, Christina Stankey, and Taylor Thomas, who contributed substantial amounts of material—ideas, corrections and clarifications, some sections, and many many exercises. Also, Noelle recommended Lauren Sompayrac's excellent book *How the Immune System Works*[338] when I complained about the complexity of the immune system. This is a far better book because of their efforts.

My experiences on earlier projects with Yale University Press made me expect that working with my editor, Joseph Calamia, would be a pleasure. And it was. Thanks, Joe. When Joe left Yale University Press, I found the same high standards working with Jean Thomson Black, Erica Hanson, Jeffrey Schier, and Elizabeth Sylvia.

Anonymous reviewers made suggestions that led to considerable improvements. These include the addition of Sect. 13.3 of Volume 2 on Michaelis-Menten enzyme kinetics, Chapter 17 of Volume 2 on vector calculus, and Appendix B of both volumes on how to replace Java programs, a moving target in these days of web vulnerabilities, with Mathematica code.

I've taught many premed students near the beginning of their careers. They anchored my grasp of the starting point of medical education. My appreciation of the goal of medical education comes from knowing thoughtful doctors who understood when it was necessary to take a big gamble and when it wasn't. In particular, I thank Dr. Steven Artz, who kept my father alive for many years, Dr. John Bubinak, who has kept my wife alive, Dr. John Byrd, who has kept my brother alive (more on this is recounted in Chapter 12 of Volume 1), Dr. Daniel Geisser, who has kept me alive, and Dr. Richard Magliula for taking such good care of our large collection of formerly stray cats. All of these doctors have been very good at explaining disease and treatment, and when talking with my wife and me, all assumed we are intelligent and interested. With some other doctors we have had less satisfying interactions. If you become a physician, when you talk with your patients please remember they are not just ciphers.

Sonya Bahar, a biophysicist I met at the 1996 Gordon Research Conference on Fractals, helped me understand multilevel selection and introduced me to Price's equation. And her work on stochastic resonance is quite interesting, too.

The artist T. E. Breitenbach asked a wonderful question which led to Chapter 12. His insights always take our conversations in directions unexpected and enjoyable.

Amy Chang, the education director of the American Society for Microbiology, made me aware of John Jungck's interesting work on the place of bioinformatics in the undergraduate biology curriculum. Thanks, Amy.

Mike Donnally, next-door neighbor in my childhood and a dear friend all these years later, has been generous with his botanical knowledge, for instance, pointing out an example of a dwarf Alberta spruce that had reverted to wild type. Mary Laine, also a dear friend for many years, showed a class of fractal structures in plants I had not noticed before. That two such serious people become so animated when they talk about plants is a delight, a contagious delight. Evolution has found intricate, beautiful structures in the plant world, too. The effort to understand these constructs gives a different perspective on the dance of evolution that produced our DNA, our cells, our organs, our selves, and our diseases.

Curiosity is the most important force of the mind. The desire to know what's around the corner, what's over the horizon, was my faithful companion—or maybe I was its companion—through my long exploration of how biology and math dance together. But intellect alone is not enough to carry out a project that is so personally important. For that an emotional pole star is needed. My navigation is provided by my family: my brother Steve and his wife Kim, my sister Linda and her husband David, my nephew Scott and his wife Maureen, my wife Jean, and my late parents Mary and Walter. Thoughts of them have helped me to explore the mechanism of empathy, to sculpt how I see its importance. For this, and for giving more understanding and affection than he deserved to a goofy little kid who "always had his nose buried in some durn book," I do not have words adequate for the thanks owed.

Special thanks to Jean Maatta, my med tech wife, for her limitless patience in explaining complicated, contingent biology to her geometer husband. More than any other experience, learning the biology needed for this book emphasized that

> If biology always worked the same way, it would be a subset of math.

(This formulation occurred in a conversation with Dr. Magliula.) But it doesn't and it isn't. We are so far from an axiomatic basis for biology. It is complicated, many aspects interrelated, probably we still are asking questions that don't lead us in useful directions. This is unfamiliar ground for someone (me, for example) who has spent over half a century thinking very hard about geometry. I am grateful for my wife's patience and good humor in explaining again and again that my attempts to fit biology into mathematical categories were wrong-headed.

Still, biology and math can talk with one another. This book is a part of that conversation.

Review of Volume 1

These are the main points from Volume 1 that we'll expand and elaborate.

- The differential equation $x' = f(x,y)$, $y' = g(x,y)$ has fixed at the intersections of the curves $f(x,y) = 0$ (the x-nullcline) and $g(x,y) = 0$ (the y-nullcline). In most cases, the stability and type of the fixed point is determined by the eigenvalues λ_1 and λ_2 of the derivative matrix

$$D\vec{F} = \begin{bmatrix} \partial f/\partial x & \partial f/\partial y \\ \partial g/\partial x & \partial g/\partial y \end{bmatrix}$$

  This is the content of the Hartman-Grobman theorem 9.2.1. Specifically, the fixed point is a saddle if λ_1 and λ_2 are real with opposite signs, an unstable node if λ_1 and λ_2 are real and positive, an asymptotically stable node if λ_1 and λ_2 are real and negative, an unstable spiral if λ_1 and λ_2 are complex with positive real parts, and an asymptotically stable spiral if λ_1 and λ_2 are complex with negative real parts. Alternately, we can apply the trace-determinant plane analysis of Sect. 8.6.
- In situations where the Hartman-Grobman theorem can't be applied— λ_1 and λ_2 are imaginary or 0—the stability of the fixed point may be determined by a Liapunov function. For example, a function V defined on a neighborhood of the plane is positive definite if $V(0,0) = 0$ and all other $V(x,y) > 0$. If we can find a

positive definite V with

$$V' = (\partial V/\partial x)x' + (\partial V/\partial y)y' = (\partial V/\partial x)f + (\partial V/\partial y)g$$

negative definite then the origin is an asymptotically stable fixed point for the differential equation $x' = f(x,y)$, $y' = g(x,y)$. If both V and V' are positive definite, then the origin is unstable. Other cases are described in Sect. 9.4. Although the motivation is the total energy of a mechanical system, we have complete freedom in our choice of the function V. But this does not make Liapunov functions easy to find, for while positive definite and negative definite V abound, V' involves the functions f and g from the differential equation. This is not always an easy method to use.

- Nonlinear systems can exhibit another behavior, a limit cycle. This is a closed trajectory, so signals periodic behavior of $x(t)$ and $y(t)$, and nearby trajectories spiral in to or away from the limit cycle. Simple examples are described in Sect. 9.6. General conditions that guarantee the existence of a limit cycle are given by the Poincaré-Bendixson theorem of Sect. 9.7.
- For a function f with continuous derivatives of all orders, the Taylor series about the point $x = a$ is

$$f(x) = f(a) + f'(a)(x-a) + \frac{f''(a)}{2!}(x-a)^2 + \frac{f'''(a)}{3!}(x-a)^3 + \cdots$$

This is the topic of Sect. 10.8. For x near a, the first few terms give a good approximation to f, and this can simplify some differential equations to be (approximately) solvable.

- From probability theory presented in Chapter 11 we'll use several results, some for discrete distributions, some for continuous. Suppose a discrete distribution has values $x_1, \dots, x_n$, $P(X = x_i) = p_i$ for $i = 1, \dots, n$, and so $\sum_{i=1}^{n} p_i = 1$. Suppose a continuous distribution has probability density function f so $P(a \leq X \leq b) = \int_a^b f(x)\, dx$. Then the expected value is

$$\mathbb{E}(X) = \sum_{i=1}^{n} x_i p_i \text{ (discrete)} \quad \text{and } \mathbb{E}(X) = \int x f(x)\, dx \text{ (continuous)}$$

In the continuous case the integral is taken over the range of values x can take. The variance is $\sigma^2(X) = \mathbb{E}(X^2) - \mathbb{E}(X)^2$, where the second moment is

$$\mathbb{E}(X^2) = \sum_{i=1}^{n} x_i^2 p_i \text{ (discrete)} \quad \text{and } \mathbb{E}(X) = \int x^2 f(x)\, dx \text{ (continuous)}$$

The conditional probability of observing E_1 given that E_2 has been observed is $P(E_1|E_2) = P(E_1 \cap E_2)/P(E_2)$.

If $E_1, \dots, E_n$ is a partition of the space of all possible events into non-overlapping classes E_i, then the law of conditioned probabilities is an expression of the probability of any event A: $P(A) = \sum_{i=1}^{n} P(A|E_i) \cdot P(E_i)$.

Bayes' theorem relates conditional probabilities: $P(E_2|E_1) = P(E_1|E_2) \cdot P(E_2)/P(E_1)$.

In addition to its use for computation, Bayes' theorem emphasizes that the correlation expressed through conditional probabilities does not suggest a causal relation.

- Recall the principle of mass action: the likelihood of an encounter between members of two species is proportional to the product of the fractions of the population that belong to each species. If S denotes the fraction of the population that is susceptible to an infection, and I denotes the fraction of the population that is infected, then the likelihood that a susceptible will encounter an infected is proportional to SI. This is a simplification, of course. For example, it assumes that members of both populations are uniformly distributed throughout the physical space, and that everyone moves around randomly. But it's a good first step.

The first volume contains many biomedical examples, and we'll refer to some of these in this volume. These we'll mention explicitly as we need them. Most important is the awareness that mathematical models can give rise to clinically important treatment strategies. This volume presents more modeling tools, and more models.

Chapter 13 Higher-dimensional differential equations

With the Hartman-Grobman theorem, Liapunov functions, and the Poincaré-Bendixson theorem we understand the basic the geometry of solutions of planar differential equations. But add a dimension and the possible behaviors of solutions becomes so complex that even now some aspects are not fully understood. In this chapter we'll sketch some of these complications.

We'll begin with a 2-dimensional technique that extends to higher dimensions: eigenvalue analysis. Then examples: SIR, Michaelis-Menten kinetics, virus dynamics, immune response dynamics, the Hodgkin-Huxley equations, predator-prey models with two prey species, the Duffing equation and a different kind of sensitivity to initial conditions called final state sensitivity, and finally attempts to control chaos, that is, to temporarily stabilize unstable periodic trajectories in the chaotic system. Probably we don't know the full extent of these complications yet.

13.1 EIGENVALUES IN HIGHER DIMENSIONS

In Sect. 8.4 we presented the ideas of eigenvectors and eigenvalues for 2×2 matrices. These methods generalize in a straightforward way to $n \times n$ matrices, once we know how to compute determinants of matrices of this size. We'll give a 3×3 example first, then state the general approach.

For a matrix M, the (i,j) *cofactor* is $(-1)^{i+j}$ times the determinant of the matrix obtained by deleting the ith row and the jth column of M. For example, if we take

$$M = \begin{bmatrix} 1 & 0 & 2 \\ 0 & -1 & 1 \\ 0 & -1 & 2 \end{bmatrix} \tag{13.1}$$

then the $(1,3)$ cofactor is

$$M_{1,3} = (-1)^{1+3} \det \begin{bmatrix} 0 & -1 \\ 0 & -1 \end{bmatrix} = 1 \cdot (0 \cdot (-1) - 0 \cdot (-1)) = 0$$

Now we can define the determinant of M:

$$\det(M) = m_{1,1} \cdot M_{1,1} + m_{1,2} \cdot M_{1,2} + m_{1,3} \cdot M_{1,3} = 1 \cdot (-1) + 0 \cdot 0 + 2 \cdot 0 = -1$$

This is called expanding along the first row. In fact, the determinant can be computed by expanding along any row or along any column. Exercise 13.1.1 provides some examples.

A useful observation: if a row or a column has mostly 0s, expanding along that row or column gives the determinant with fewer steps.

Here's the general formula for an $n \times n$ matrix M. For each i and j, $1 \leq i,j \leq n$,

$$\begin{aligned} \det(M) &= m_{i,1} \cdot M_{i,1} + m_{i,2} \cdot M_{i,2} + \cdots + m_{i,n} \cdot M_{i,n} \\ &= m_{1,j} \cdot M_{1,j} + m_{2,j} \cdot M_{2,j} + \cdots + m_{n,j} \cdot M_{n,j} \end{aligned}$$

The first equation is expanding along the ith row; the second is expanding along the jth column.

Eigenvalues still are computed by the characteristic equation, Eq. (8.10), except now M can be an $n \times n$ matrix. As an illustration, we'll compute the eigenvalues of the matrix M of Eq. (13.1), expanding along the first column to find the characteristic equation, $0 = \det(M - \lambda I)$

$$\begin{aligned} &= \det \begin{bmatrix} 1-\lambda & 0 & 2 \\ 0 & -1-\lambda & 1 \\ 0 & -1 & 2-\lambda \end{bmatrix} = (1-\lambda) \det \begin{bmatrix} -1-\lambda & 1 \\ -1 & 2-\lambda \end{bmatrix} \\ &= (1-\lambda) \cdot \big((-1-\lambda) \cdot (2-\lambda) - (-1) \cdot 1\big) = (1-\lambda) \cdot (\lambda^2 - \lambda - 1) \end{aligned}$$

Then the eigenvalues are the roots of the characteristic equation

$$0 = (1-\lambda) \cdot (\lambda^2 - \lambda - 1), \text{ that is, } \lambda = 1, \frac{1 \pm \sqrt{5}}{2}$$

For 2×2 matrices, the characteristic equation is quadratic, so can be solved handily. For 3×3 matrices, the characteristic equation is cubic, and while there is a cubic formula, it's pretty complicated. There is an even more complicated quartic formula, and that's as far as general formulas go. There's no general formula for the roots of quintic or higher order polynomials. It isn't that no one has found such a formula but someday someone might. The math used to prove this is called Galois theory, a branch of field theory, part of abstract algebra. Beautiful stuff, but requiring quite a bit of background.

Sometimes we're lucky, as we were in this calculation. Many 0s in a row or column, or some lucky cancellations, can present a characteristic polynomial (the side of the characteristic equation containing all the λs) in factored form. If this doesn't happen, or if we can't divide out linear factors and be left with a quadratic, we'll use a computer algebra system. The Mathematica code to find the exact eigenvalues, if possible or if they are simple enough, and to find the numerical eigenvalues in any case, is given in Sect. B.11. Often numerical solutions will be enough. Because access to computer algebra systems is not easy for everyone, we'll find exact solutions whenever we can.

Now we'll find the eigenvectors of each eigenvalue of this matrix M. The set-up is the obvious generalization of what we did for 2×2 matrices in Sect. 8.4, but higher-dimensional systems can have some complications. We'll discuss these in Sect. A.12. For now, we'll be content with an example. We'll find the eigenvectors of $\lambda = (1 + \sqrt{5})/2$. The eigenvector equation $M\vec{v} = \lambda\vec{v}$, Eq. (8.6), becomes

$$\begin{bmatrix} 1 & 0 & 2 \\ 0 & -1 & 1 \\ 0 & -1 & 2 \end{bmatrix} \begin{bmatrix} x \\ y \\ z \end{bmatrix} = \frac{1+\sqrt{5}}{2} \begin{bmatrix} x \\ y \\ z \end{bmatrix}$$

The first equation gives $x = (\sqrt{5} + 1)z$, the second and third both give $y = ((3 - \sqrt{5})/2)z$, so we're on familiar ground: one of the equations is redundant. (More than one equation can be redundant. We saw an example in Sect. A.7.) Taking $z = 1$, we obtain the eigenvector

$$\langle x, y, z \rangle = \langle \sqrt{5} + 1, (3 - \sqrt{5})/2, 1 \rangle$$

or if we want the first entry to be 1, we just divide all entries of the eigenvector by $\sqrt{5} + 1$ and simplify

$$\langle x, y, z \rangle = \langle 1, (\sqrt{5} - 2)/4, (\sqrt{5} - 1)/4 \rangle$$

Not surprisingly, differential equations in 3 or more dimensions have types of fixed points that we do not see in 2 dimensions. For example, the fixed point at the origin for

$$x' = -x - y \qquad y' = x - 2y \qquad z' = z/2$$

has an asymptotically stable spiral in the xy-plane, while points off the xy-plane spiral in toward the z-axis and move away from the origin along the z-axis. However, even with all these complications, it's still true that

Proposition 13.1.1. If $\vec{x}' = \vec{F}(\vec{x})$ has a fixed point at $\vec{x}_0$ and if $D\vec{F}(\vec{x}_0)$ has at least one eigenvalue with a positive real part, then that fixed point is unstable.

Practice Problems

$$M = \begin{bmatrix} 1 & 2 & 0 \\ 1 & 1 & 2 \\ -1 & 0 & 1 \end{bmatrix} \quad A = \begin{bmatrix} 1 & 2 \\ 3 & 1 \end{bmatrix} \quad B = \begin{bmatrix} 2 & 1 \\ 3 & 1 \end{bmatrix} \quad C = \begin{bmatrix} 1 & 2 & 0 & 0 \\ 3 & 1 & 0 & 0 \\ 0 & 0 & 2 & 1 \\ 0 & 0 & 3 & 1 \end{bmatrix}$$

13.1.1. Find the eigenvalues and eigenvectors of the matrix M. Scale the eigenvectors so each has its first entry equal to 1.

13.1.2. (a) Find the eigenvalues of the matrices A and B.
(b) Find the eigenvalues of the matrix C.
(c) Explain the relationship between the eigenvalues of (a) and (b).

Practice Problem Solutions

13.1.1. The eigenvalues are the roots of the characteristic equation. We evaluate the determinant by expanding along the first row.

$$0 = \det \begin{bmatrix} 1-\lambda & 2 & 0 \\ 1 & 1-\lambda & 2 \\ -1 & 0 & 1-\lambda \end{bmatrix}$$

$$= (1-\lambda)\cdot((1-\lambda)^2 - 2\cdot 0) - 2\cdot(1\cdot(1-\lambda) - (-1)\cdot 2)$$

$$= (1-\lambda)^3 - 2\cdot(3-\lambda) = -\lambda^3 + 3\lambda^2 - \lambda - 5 = (\lambda+1)\cdot(-\lambda^2 + 4\lambda - 5)$$

where the last equality comes from dividing $\lambda+1$ into $-\lambda^3 + 3\lambda^2 - \lambda - 5$. (The other obvious choices for factors to try are $\lambda - 1$, $\lambda + 5$, and $\lambda - 5$.) Then the eigenvalues are $\lambda = -1, 2 \pm i$.

Now we'll find eigenvectors for each eigenvalue. For $\lambda = -1$ we have

$$\begin{bmatrix} 1 & 2 & 0 \\ 1 & 1 & 2 \\ -1 & 0 & 1 \end{bmatrix} \begin{bmatrix} x \\ y \\ z \end{bmatrix} = -1 \cdot \begin{bmatrix} x \\ y \\ z \end{bmatrix}$$

The equation given by the second row is a consequence of the equations determined by the first and third rows, so we'll use the first and third. It's easier to use equations with at least one 0, if possible. The first equation gives $y = -x$, the third gives $z = x/2$. Then the eigenvector with $x = 1$ is $\langle 1, -1, 1/2 \rangle$.

For the eigenvalue $2+i$ we have

$$\begin{bmatrix} 1 & 2 & 0 \\ 1 & 1 & 2 \\ -1 & 0 & 1 \end{bmatrix} \begin{bmatrix} x \\ y \\ z \end{bmatrix} = (2+i) \cdot \begin{bmatrix} x \\ y \\ z \end{bmatrix}$$

The first equation gives $y = ((1+i)/2)x$, the third gives $z = ((-1+i)/2)x$, and we find the eigenvector $\langle 1, (1+i)/2, (-1+i)/2 \rangle$.

For the eigenvalue $2-i$, the first equation gives $y = ((1-i)/2)x$ and the third gives $z = ((-1-i)/2)x$. So we see that the eigenvalue $2-i$ has eigenvector $\langle 1, (1-i)/2, (-1-i)/2 \rangle$.

13.1.2. (a) The eigenvalues of matrix B are the roots of the characteristic equation

$$0 = (1-\lambda) \cdot (1-\lambda) - 3 \cdot 2 = \lambda^2 - 2\lambda - 5, \text{ that is, } \lambda = 1 \pm \sqrt{6}$$

The eigenvalues of the matrix C are the roots of the the characteristic equation

$$0 = (2-\lambda) \cdot (1-\lambda) - 3 \cdot 1 = \lambda^2 - 3\lambda - 1, \text{ that is, } \lambda = (3 \pm \sqrt{13})/2$$

(b) The eigenvalues of the matrix D are the roots of the characteristic equation

$$\begin{aligned}
0 &= \det \begin{bmatrix} 1-\lambda & 2 & 0 & 0 \\ 3 & 1-\lambda & 0 & 0 \\ 0 & 0 & 2-\lambda & 1 \\ 0 & 0 & 3 & 1-\lambda \end{bmatrix} \\
&= (1-\lambda) \cdot \det \begin{bmatrix} 1-\lambda & 0 & 0 \\ 0 & 2-\lambda & 1 \\ 0 & 3 & 1-\lambda \end{bmatrix} - 2 \cdot \det \begin{bmatrix} 3 & 0 & 0 \\ 0 & 2-\lambda & 1 \\ 0 & 3 & 1-\lambda \end{bmatrix} \\
&= (1-\lambda)^2 \cdot \det \begin{bmatrix} 2-\lambda & 1 \\ 3 & 3-\lambda \end{bmatrix} - 2 \cdot 3 \cdot \det \begin{bmatrix} 2-\lambda & 1 \\ 3 & 1-\lambda \end{bmatrix} \\
&= ((1-\lambda)^2 - 2 \cdot 3) \cdot \det \begin{bmatrix} 2-\lambda & 1 \\ 3 & 1-\lambda \end{bmatrix} \\
&= \det \begin{bmatrix} 1-\lambda & 2 \\ 3 & 1-\lambda \end{bmatrix} \cdot \det \begin{bmatrix} 2-\lambda & 1 \\ 3 & 1-\lambda \end{bmatrix} \\
&= \left(\lambda^2 - 2\lambda - 5\right) \cdot \left(\lambda^2 - 3\lambda - 1\right)
\end{aligned}$$

The second and third equalities come from expanding along the first rows of the matrices. To interpret the last equality, for the product of two factors to be 0, at least one of the factors must be 0. Consequently, the eigenvalues of C are the eigenvalues of A together with those of B.

(c) The blocks of 0s in C show that the determinant of the C is the product of the determinants of A and B, and the same is true if we subtract λ from the diagonal entries of A, B, and C.

Exercises

13.1.1. Compute the determinant of the matrix M of Eq. (13.1) by expanding along the second and third rows of M, and by expanding along the first, second, and third columns of M.

13.1.2. Find the eigenvalues and eigenvectors of the matrices D, E, and F of Eq. (13.2). If it is not 0, take the first entry of each eigenvector to be 1.

$$D = \begin{bmatrix} 1 & 2 & -1 \\ 0 & 2 & 1 \\ 2 & 1 & 1 \end{bmatrix} \quad E = \begin{bmatrix} -1 & 2 & 0 \\ 0 & 2 & 1 \\ 1 & 0 & 2 \end{bmatrix} \quad F = \begin{bmatrix} 1 & 0 & -2 \\ 0 & 2 & 1 \\ 1 & 0 & 2 \end{bmatrix} \tag{13.2}$$

13.1.3. (a) Find the eigenvalues of the matrix G of Eq. (13.3).
(b) Find the eigenvalues of the matrix H of Eq. (13.3).
(c) Find the eigenvalues of the matrix I of Eq. (13.3).
(d) Compare these with the eigenvalues of matrices A and B of the practice problems.

$$G = \begin{bmatrix} 1 & 2 & 0 & 0 \\ 3 & 1 & 0 & 0 \\ 1 & 0 & 2 & 1 \\ 0 & 0 & 3 & 1 \end{bmatrix} \quad H = \begin{bmatrix} 1 & 2 & 0 & 0 \\ 3 & 1 & 0 & 0 \\ 0 & 0 & 2 & 1 \\ 1 & 0 & 3 & 1 \end{bmatrix} \quad I = \begin{bmatrix} 1 & 2 & 0 & 0 \\ 3 & 1 & 0 & 0 \\ 1 & 1 & 2 & 1 \\ 1 & 1 & 3 & 1 \end{bmatrix} \tag{13.3}$$

13.1.4. (a) Find the eigenvalues of the matrix J of Eq. (13.4).
(b) Compare these with the eigenvalues of matrices A and B of the practice problems.
(c) Compare this result with that of Exercise 13.1.3. Explain the source of the difference.

$$J = \begin{bmatrix} 1 & 2 & 0 & 1 \\ 3 & 1 & 0 & 0 \\ 1 & 0 & 2 & 1 \\ 0 & 0 & 3 & 1 \end{bmatrix} \quad K = \begin{bmatrix} 0 & 0 & 1 & 2 \\ 0 & 0 & 3 & 1 \\ 2 & 1 & 0 & 0 \\ 3 & 1 & 0 & 0 \end{bmatrix} \quad L = \begin{bmatrix} 1 & 2 & 3 & 1 \\ 0 & 1 & 0 & 0 \\ 0 & 0 & 1 & 0 \\ 2 & 1 & 3 & 1 \end{bmatrix} \tag{13.4}$$

13.1.5. (a) Find the eigenvalues of the matrix K of Eq. (13.4).
(b) Compare these with the eigenvalues of matrices A and B of the practice problems.

13.1.6. (a) Find the eigenvalues of the matrix L of Eq. (13.4).
(b) Compare these with the eigenvalues of matrices A and B of the practice problems.

13.1.7. With the matrices of Eq. (13.2), find the eigenvalues of D^2, E^2, and F^2. Compare these with the eigenvalues of D, E, and F.

13.1.8. With the matrices of Eq. (13.2), find the eigenvectors of D^2, E^2, and F^2. Compare these with the eigenvectors of D, E, and F.

In Exercises 13.1.9 and 13.1.10, for each matrix M of Eqs. (13.5) and (13.6), the differential equation $d\vec{x}/dt = M\vec{x}$ has a fixed point at the origin. Is the origin asymptotically stable for any of these three matrices in each exercise? Give a reason for your answer.

13.1.9.

$$(a)\; M = \begin{bmatrix} 1 & 2 & -1 \\ 0 & 1 & 1 \\ 0 & 1 & -1 \end{bmatrix} \quad (b)\; M = \begin{bmatrix} 0 & 2 & 1 \\ 1 & 0 & 1 \\ 1 & 1 & 0 \end{bmatrix} \quad (c)\; M = \begin{bmatrix} 0 & 2 & 0 \\ 1 & 1 & 0 \\ 1 & 1 & 3 \end{bmatrix} \qquad (13.5)$$

13.1.10.

$$(a)\; M = \begin{bmatrix} 1 & 2 & 1 & 2 \\ 2 & 1 & 0 & 0 \\ 0 & 0 & 3 & 2 \\ 0 & 0 & 2 & 3 \end{bmatrix} \qquad (b)\; M = \begin{bmatrix} 1 & -2 & 1 & 2 \\ -2 & 1 & 0 & 0 \\ 0 & 0 & -3 & 2 \\ 0 & 0 & 2 & -3 \end{bmatrix}$$

$$(c)\; M = \begin{bmatrix} -1 & -2 & 1 & 2 \\ 2 & -2 & 0 & 0 \\ 0 & 0 & -3 & 2 \\ 0 & 0 & 2 & -3 \end{bmatrix} \qquad (d)\; M = \begin{bmatrix} 1 & 2 & -1 & -2 \\ -2 & 1 & 0 & 0 \\ 0 & 0 & 3 & -2 \\ 0 & 0 & -2 & 3 \end{bmatrix} \qquad (13.6)$$

13.2 SIR CALCULATIONS

Now we'll return to the SIR model of Example 3.2.4, locate the fixed points, and determine their stability as a function of the model parameters.

Example 13.2.1. *Fixed points and cycles for the SIR model.* The model is Eq. (3.8), rewritten here to save flipping through the pages:

$$\frac{dS}{dt} = \beta S - \epsilon SI - \delta S \qquad \frac{dI}{dt} = \epsilon SI - \gamma I - \delta I \qquad \frac{dR}{dt} = \gamma I - \delta R \tag{13.7}$$

The fixed points are the solutions of

$$0 = \beta S - \epsilon SI - \delta S \qquad 0 = \epsilon SI - \gamma I - \delta I \qquad 0 = \gamma I - \delta R$$

We find two solutions, $(0,0,0)$ and

$$(S_*, I_*, R_*) = \left(\frac{\delta + \gamma}{\epsilon}, \frac{\beta - \delta}{\epsilon}, \frac{(\beta - \delta)\gamma}{\epsilon\delta} \right)$$

In order for each coordinate of this second fixed point to be positive (being populations, S, I, and R must be non-negative), we must have $\beta > \delta$. We'll use this when we test the stability of the second fixed point.

To test the stability of the fixed points we use the derivative matrix

$$D\vec{F}(S,I,R) = \begin{bmatrix} \beta - \epsilon I - \delta & -\epsilon S & 0 \\ \epsilon I & \epsilon S - \gamma - \delta & 0 \\ 0 & \gamma & -\delta \end{bmatrix}$$

The eigenvalues of $D\vec{F}(0,0,0)$ are $\beta - \delta$, $-\delta$, and $-\delta - \gamma$. The constants β, γ, and δ all are real and positive. Recall that the origin is an asymptotically stable fixed point if all the eigenvalues are negative (or have negative real parts, inapplicable in this example with real eigenvalues). The requirement for negative eigenvalues gives three conditions:

$$\beta < \delta,\ \delta > 0 \text{ (always true), and } -\delta < \gamma \text{ (always true)}$$

So only one condition, $\beta < \delta$, is needed to guarantee that the origin is an asymptotically stable fixed point. This condition, $\beta < \delta$, just means that the per capita birth rate is less than the per capita death rate. It is sensible that if the death rate exceeds the birth rate, the whole population will disappear.

The eigenvalues of $D\vec{F}(S_*, I_*, R_*)$ are

$$\lambda_1 = -\delta, \text{ and } \lambda_\pm = \pm\sqrt{(\delta + \gamma)(-\beta + \delta)}$$

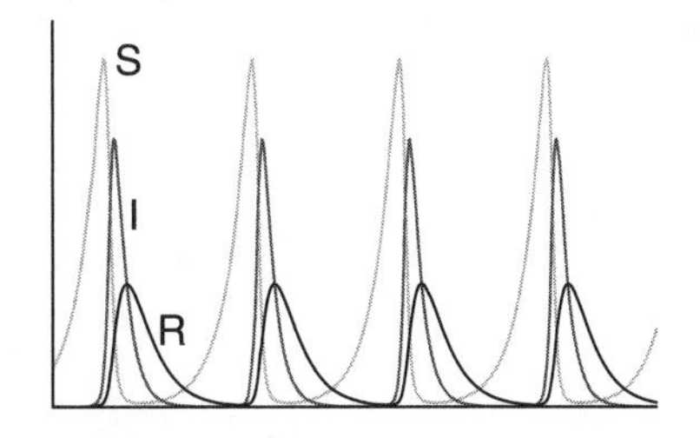

Figure 13.1. Plots of S, I, and R.

Because $\delta > 0$, the first eigenvalue is negative. So far, so good. But the condition $\beta > \delta$, mentioned above, guarantees that the second and third eigenvalues are purely imaginary. This is the one case where the eigenvalues of the derivative matrix don't tell us anything about the stability of the fixed point, but a pair of imaginary eigenvalues suggests periodic motion. In Fig. 13.1 we plot the S, I, and R curves for $\beta = 2$, $\delta = \gamma = \epsilon = 1$. Then $\lambda_{\pm} = \pm i\sqrt{2}$. Note the periodicity of the plots. The curves of this figure are similar to the Lotka-Volterra population curves of Fig. 7.13. Repeating growth patterns for S, I, and R mean periodicity of the population dynamics. □

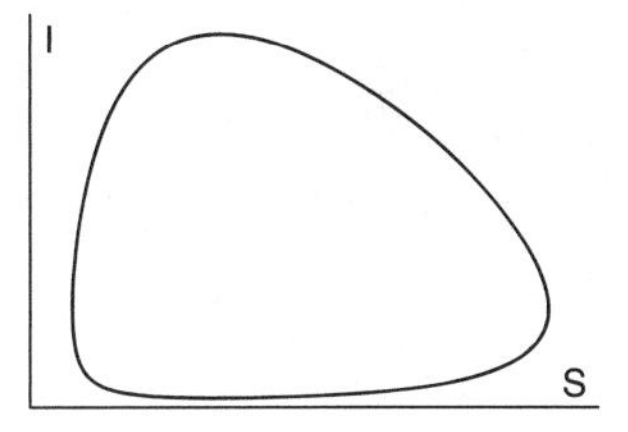

Figure 13.2. The S-I plot of Example 13.2.1.

In one sense this result is not so surprising: the R population has no effect on the S and I populations, so this is really a 2-dimensional system that drives an additional 1-dimensional system. In Fig. 13.2 we plot the trajectory $(S(t), I(t))$. This appears to be a closed trajectory, certainly suggested, but not proven, by the pair of imaginary eigenvalues. In fact, if we compute the eigenvalues for the derivative of the S', I' system,

$$\frac{dS}{dt} = \beta S - \epsilon SI - \delta S \qquad \frac{dI}{dt} = \epsilon SI - \gamma I - \delta I$$

at the fixed point (S_*, I_*), we get the eigenvalues $\lambda_{\pm}$.

Why don't we find a trapping region and apply the Poincaré-Bendixson theorem? In Fig. 13.2 the initial value is $(S, I) = (0.5, 0.5)$. If we start with an initial point closer to the S-axis, say $(0.5, 0.25)$ or $(0.5, 0.1)$, the trajectory also is periodic, extending farther in the S-direction and farther in the I-direction, without bound if our experiments are valid indicators. So no region could be a trapping region; we just have a family of ever-larger closed curves.

Now let's try a true 3-dimensional system, the *SEIS* model.

Example 13.2.2. *The SEIS model.* Remember, $E =$ exposed $=$ latent. Here are the model equations.

$$S' = \beta S - \delta S - \epsilon SI + \gamma I \quad E' = \epsilon SI - \eta E - \delta E \quad I' = \eta E - \delta I - \gamma I$$

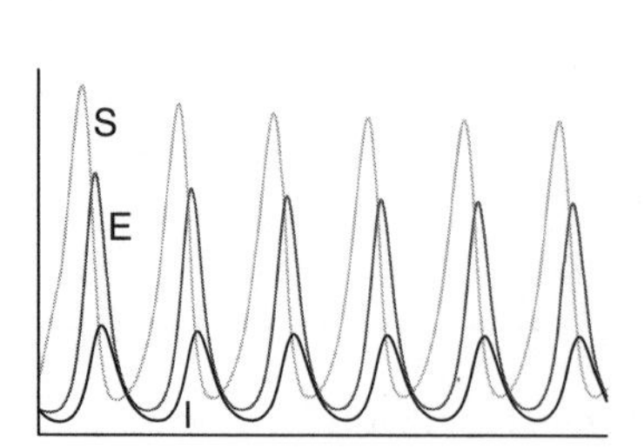

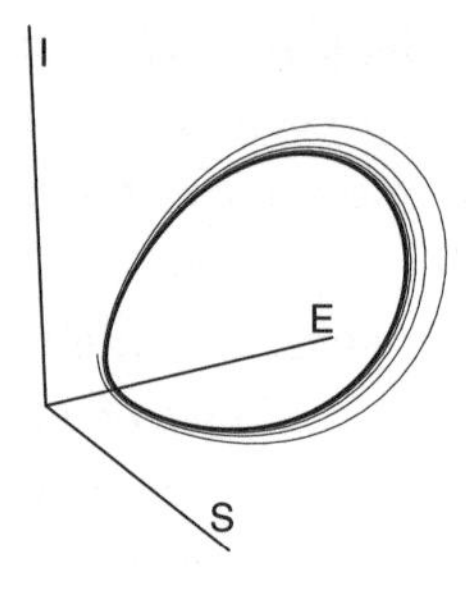

Figure 13.3. Left: plots of the S, E, and I curves. Right: the trajectory. Both are for Example 13.2.2. Mathematica code is in Appendix B.12.

Encounters between S and I move susceptibles to the latent population at a rate given by the principle of mass action, at a per capita rate η latents move to the infected population, and at a per capita rate γ infecteds move to the susceptible population (no immunity upon recovery).

The fixed points are $(0,0,0)$ and (S_*, E_*, I_*) where

$$S_* = \frac{(\delta+\gamma)(\delta+\eta)}{\epsilon\eta}, \quad E_* = \frac{(\beta-\delta)(\delta+\gamma)^2(\delta+\eta)}{\delta\epsilon\eta(\delta+\gamma+\eta)}, \text{ and}$$
$$I_* = \frac{(\beta-\delta)(\delta+\gamma)(\delta+\eta)}{\delta\epsilon(\delta+\gamma+\eta)}$$

so E_* and I_* are positive only if $\beta > \delta$. The derivative matrix is

$$D\vec{F} = \begin{bmatrix} \beta-\delta-\epsilon I & 0 & -\epsilon S+\gamma \\ \epsilon I & -\eta-\delta & \epsilon S \\ 0 & \eta & -\delta-\gamma \end{bmatrix}$$

The general form of the eigenvalues is complicated, so we'll compute them for the parameters $\beta = 2$, $\epsilon = 3$, $\delta = 1$, $\gamma = 0.5$, and $\eta = 0.75$, the parameters we used to plot the curves in Fig. 13.3. The eigenvalues at the fixed point $(0,0,0)$ are $\lambda = -1.75$, -1.5, and 1, so this fixed point is unstable. At the fixed point $(S_*, E_*, I_*) \approx (1.1667, 0.7778, 0.3889)$, the eigenvalues are $\lambda \approx -3.4779$, and $0.0306 \pm 0.8682i$, so this fixed point is unstable, too. The second plot of Fig. 13.3 shows a limit cycle. If we could find a trapping region, then because all the fixed points are unstable, we could apply the Poincaré-Bendixson theorem to deduce the existence of a limit cycle. But not so fast. The Poincaré-Bendixson theorem is a result about differential equations in 2 dimensions. In Sect. 13.7 we'll see that in fact the Poincaré-Bendixson theorem doesn't extend to 3 dimensions. We can find fixed points and test their stability with the eigenvalues of the derivative matrix, and we can plot population curves and trajectories. There are other tools, but for now we'll practice computing eigenvalues and plotting curves. □

As far as stability of fixed points is concerned, the constructs in 2 dimensions extend exactly as we expect to 3 or more dimensions. In Sects. 13.4 and 13.5 we'll see applications to the dynamics of virus populations and of immune response.

Practice Problems

13.2.1. For the SIR model where susceptibles give birth to susceptibles and infecteds to infecteds but recovereds are sterile,
(a) locate the fixed points, and
(b) determine their stability. If eigenvalues don't give enough information, plot the S, I, and R curves and interpret what you see.

13.2.2. An SIR model with a logistic limitation to the susceptible population can be modeled by the equations

$$S' = S - \epsilon SI - S^2 \qquad I' = \epsilon SI - \gamma I \qquad R' = \gamma I - R$$

where ϵ is the probability that an S and I encounter moves a susceptible to the infected population, and γ is the recovery rate of infecteds.
(a) Find the fixed points of the system. Denote by (S_*, I_*, R_*) the fixed point with all coordinates non-zero.
(b) Find the eigenvalues of the derivative matrix at each fixed point.
(c) Plot the S, I, and R curves for $\epsilon = 1$ and $\gamma = 0.15$, and for $\epsilon = 1$ and $\gamma = 0.9$. For both, take $S(0) = 0.75$, $I(0) = 0.25$, and $R(0) = 0$.
(d) Map the region of the ϵ-γ plane where the fixed point (S_*, I_*, R_*) takes on each type of dynamics.

Practice Problem Solutions

13.2.1. The equations for this SIR model are

$$S' = \beta S - \delta S - \epsilon SI \qquad I' = \beta I + \epsilon SI - \delta I - \gamma I \qquad R' = \gamma I - \delta R$$

(a) The fixed points are the solutions of

$$\beta S - \delta S - \epsilon SI = 0 \qquad \beta I + \epsilon SI - \delta I - \gamma I = 0 \qquad \gamma I - \delta R = 0$$

that is, $(0,0,0)$ and

$$(S_*, I_*, R_*) = \left(\frac{\delta + \gamma - \beta}{\epsilon}, \frac{\beta - \delta}{\epsilon}, \frac{(\beta - \delta)\gamma}{\delta\epsilon} \right)$$

(b) The fixed point stability is determined by the eigenvalues of the derivative matrix,

$$\vec{DF} = \begin{bmatrix} \beta-\delta-\epsilon I & -\epsilon S & 0 \\ \epsilon I & \beta+\epsilon S-\delta-\gamma & 0 \\ 0 & \gamma & -\delta \end{bmatrix}$$

unless two are imaginary. The eigenvalues at $(S,I,R) = (0,0,0)$ are $\lambda = \beta-\delta$, $-\delta$, and $\beta-\delta-\gamma$ (this is easy: $\vec{DF}(0,0,0)$ is a triangular matrix so the diagonal entries are the eigenvalues); the eigenvalues at (S_*, I_*, R_*) require a bit more work and are $\lambda = -\delta$ and $\lambda = \pm\sqrt{(\beta-\delta)(\beta-\delta-\gamma)}$.

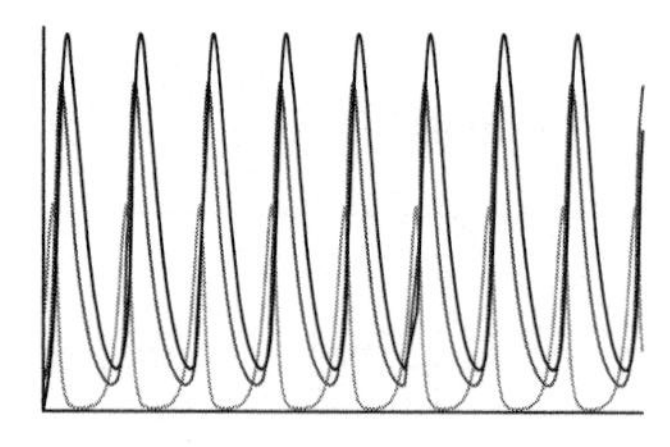

Figure 13.4. S, I, and R curves in the imaginary eigenvalue range.

The eigenvalues at $(0,0,0)$ are real and the origin is unstable if $\beta > \delta$ because then at least one of the eigenvalues is positive. If $\beta < \delta$, all three eigenvalues are negative and the origin is asymptotically stable.

Recall that for I_* and R_* to be positive we need $\beta > \delta$. Then the eigenvalues of (S_*, I_*, R_*) are real (and two negative, one positive) if $\beta < \delta$ (but as mentioned, we exclude this case) and if $\beta > \delta+\gamma$, so in that range the fixed point is unstable. For $\delta < \beta < \delta+\gamma$ the second and third eigenvalues are purely imaginary. In Fig. 13.4 we plot the S (lightest), I (middle darkness), and R (darkest) curves for $\beta = 2$, $\epsilon = 1$, $\delta = 1$, and $\gamma = 1.4$, inside the $\delta < \beta < \delta+\gamma$ range and so with two imaginary eigenvalues. The curves look periodic.

Here's a question to think over: If one of the three curves S, I, and R is periodic, must the other two also be? We'll explore this a bit in Exercises 13.2.9 and 13.2.10.

13.2.2. (a) The fixed points are the solutions of

$$0 = S(1-\epsilon I - S) \qquad 0 = I(\epsilon S - \gamma) \qquad 0 = \gamma I - R$$

The third equation gives $R = \gamma I$; the first gives $S = 0$ or $S = 1-\epsilon I$; the second gives $I = 0$ or $S = \gamma/\epsilon$. We must take one condition from the first equation and one from the second equation, so four potential combinations:

- $S = 0$ and $I = 0$, so also $R = 0$.
- $S = 0$ and $S = \gamma/\epsilon$, impossible.
- $S = 1-\gamma I$ and $I = 0$, so $S = 1$ and $R = 0$.
- $S = 1-\gamma I$ and $S = \gamma/\epsilon$, so $I = (\epsilon-\gamma)/\epsilon^2$ and $R = \gamma(\epsilon-\gamma)/\epsilon^2$.

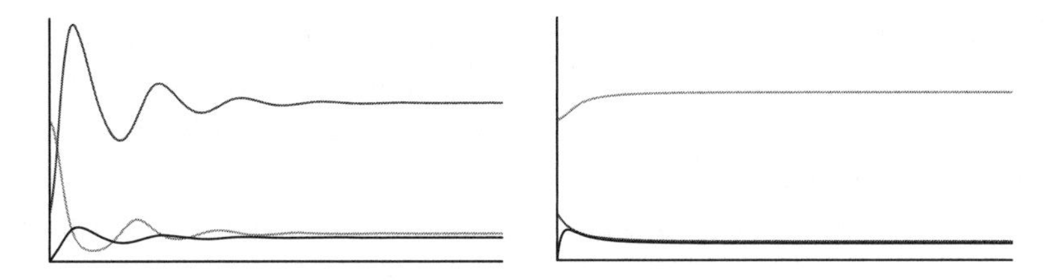

Figure 13.5. S, I, and R curves for $\epsilon = 1$ and $\gamma = 0.15$ (left) and $\epsilon = 1$, $\gamma = 0.9$ (right).

That is, the fixed points are

$$(0,0,0),\ (1,0,0),\ \text{and}\ (S_*, I_*, R_*) = \left(\frac{\gamma}{\epsilon}, \frac{\epsilon - \gamma}{\epsilon^2}, \frac{\gamma(\epsilon - \gamma)}{\epsilon^2}\right)$$

Note that in order to have $I_* > 0$ and $R_* > 0$ we must have $\epsilon > \gamma$. We'll use this point in a moment.

(b) The derivative matrix is

$$D\vec{F}(S,I,R) = \begin{bmatrix} 1 - \epsilon I - 2S & -\epsilon S & 0 \\ \epsilon I & \epsilon S - \gamma & 0 \\ 0 & \gamma & -1 \end{bmatrix}$$

The matrix $D\vec{F}(0,0,0) - \lambda I$ is lower triangular, so the eigenvalues of $D\vec{F}(0,0,0)$ are the diagonal entries of $D\vec{F}(0,0,0)$, that is, $1, -\gamma, -1$. This fixed point is unstable regardless of the value of γ.

Expanding down the third column of $D\vec{F}(1,0,0) - \lambda I$ we find that the eigenvalues are -1, -1, and $\epsilon - \gamma$. This fixed point is unstable if $\epsilon > \gamma$ and asymptotically stable if $\epsilon < \gamma$.

By expanding down the third column of $D\vec{F}(S_*, I_*, R_*) - \lambda I$ we find the eigenvalues

$$-1,\ \frac{-\gamma \pm \sqrt{-4\epsilon^2\gamma + \gamma^2 + 4\epsilon\gamma^2}}{2\epsilon}$$

We'll study the stability of this fixed point in part (d).

(c) In Fig. 13.5 we see plots of the S (lightest), I, and R (darkest) curves. For both these sets of parameter values, the fixed points $(0,0,0)$ and $(1,0,0)$ are unstable. The eigenvalues of $D\vec{F}(S_*, I_*, R_*)$ are

$$\begin{aligned} &-1, -0.075 \pm 0.3491i \quad \text{for } \epsilon = 1, \gamma = 0.15 \\ &-1, -0.1146, -0.7854 \quad \text{for } \epsilon = 1, \gamma = 0.9 \end{aligned}$$

So in both cases this fixed point is asymptotically stable. The imaginary components of the $\epsilon = 1, \gamma = 0.15$ eigenvalues are responsible for the oscillations in the left graph of Fig. 13.5, the negative real parts are responsible for the damping of these oscillations.

(d) The fixed point (S_*, I_*, R_*) is an asymptotically stable spiral if the two eigenvalues

$$\lambda_{\pm} = \frac{-\gamma \pm \sqrt{-4\epsilon^2\gamma + \gamma^2 + 4\epsilon\gamma^2}}{2\epsilon}$$

are complex with negative real parts, and an asymptotically stable node if these two eigenvalues are real and negative. First let's address the real versus complex issue. The eigenvalues are real if $-4\epsilon^2\gamma + \gamma^2 + 4\epsilon\gamma^2 > 0$ and complex if $-4\epsilon^2\gamma + \gamma^2 + 4\epsilon\gamma^2 < 0$. The boundary between these regions is the curve determined by

$$0 = -4\epsilon^2\gamma + \gamma^2 + 4\epsilon\gamma^2 = \gamma(-4\epsilon^2 + \gamma + 4\epsilon\gamma)$$

Because γ is positive, this condition requires that the second factor is 0. Solving for γ we find

$$\gamma = \frac{4\epsilon^2}{1 + 4\epsilon}$$

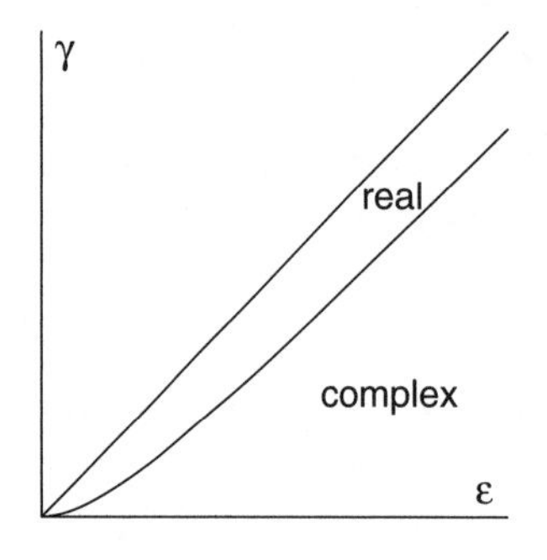

Figure 13.6. The parameter space for $\lambda_{\pm}$.

This is the curve in Fig. 13.6. For (ϵ, γ) below this curve, $\lambda_{\pm}$ are complex. For (ϵ, γ) above this curve, but still below $\gamma = \epsilon$ (remember, we said we'd use $\gamma < \epsilon$ here), the eigenvalues $\lambda_{\pm}$ are real. All that remains is to show they are negative. Obviously, λ_- is negative. To show λ_+ is negative, we must show $\sqrt{-4\epsilon^2\gamma + \gamma^2 + 4\epsilon\gamma^2} < \gamma$. A tiny bit of algebra shows that this is implied by $4\epsilon\gamma(-\epsilon + \gamma) < 0$, that is, $\gamma < \epsilon$, our second use of this condition in one paragraph.

Exercises

13.2.1. Find qualitative explanations for the main features of Fig. 13.1. Why does the S curve rise, then fall, then rise again? Same question for the I curve and the R curve. Explain the relationship between the growth of one and the decline of another.

13.2.2. The SIR model where susceptibles give birth to susceptibles, infecteds to infecteds, and recovereds to recovereds, but at half the per capita birth rate as that of the susceptibles and infecteds, so infection and immunity are passed from mother to child, has these equations:

$$S' = \beta S - \epsilon SI - \delta S \quad I' = \beta I + \epsilon SI - \gamma I - \delta I \quad R' = (\beta/2)R + \gamma I - \delta R$$

Assume $\beta \neq \delta$ and $\beta \neq \gamma + \delta$.
(a) Locate the fixed points.
(b) Find the conditions on the parameters that guarantee all the non-origin fixed points have all coordinates positive.
(c) Find the eigenvalues of the derivative matrix at the fixed points.
(d) Plot the S, I, and R curves for $\beta = 1.3$, $\delta = 0.8$, $\epsilon = 0.9$, and $\gamma = 0.4$.

13.2.3. In Practice Problem 13.2.2 how would the $S(t)$, $I(t)$, and $R(t)$ curves change if the last equation were altered from $R' = \gamma I - R$ to $R' = 0.5\gamma I - R$? Don't run a simulation. Rather, think through what changes and what doesn't.

13.2.4. For this modification of the SIR model,

$$S' = S - aSI \quad I' = aSI - bI - I^2 \quad R' = cI - R$$

where a, b, and c are positive constants,
(a) find the fixed points.
(b) Determine the stability of the non-zero fixed point as a function of the parameters a and b. Draw a map of the stabilities and types in the first quadrant of the a-b plane.

13.2.5. For this modification of the SIR model,

$$S' = S - aSI \quad I' = aSI - bI \quad R' = cI - R - R^2$$

where a, b, and c are positive constants,
(a) find the fixed points.
(b) Determine the stability of the fixed points.

13.2.6. Suppose a susceptible must encounter two infecteds in order to potentially become infected. For this SIR system,

$$S' = S - aSI^2 \quad I' = aSI^2 - bI \quad R' = bI - R$$

where a and b are positive constants,
(a) find the fixed points.
(b) Determine their stability. For the non-zero fixed point, map the regions of stabilities and types in the first quadrant of the a-b plane.

13.2.7. Suppose two infections are present and recovery from either gives immunity to both. Also, only susceptibles are reproductive. Model equations are

$$S' = (\beta - \delta)S - \epsilon_1 SI_1 - \epsilon_2 SI_2 \qquad I_1' = \epsilon_1 SI_1 - (\delta + \gamma)I_1$$
$$I_2' = \epsilon_2 SI_2 - (\delta + \gamma)I_2 \qquad R' = \gamma(I_1 + I_2) - \delta R$$

(a) Find the fixed points.
(b) Show the origin is asymptotically stable if $\beta < \delta$.
(c) If $\beta > \delta$ show that exactly one of the non-origin fixed points has a positive eigenvalue, and which fixed point has a positive eigenvalue depends on the relative values of ϵ_1 and ϵ_2.

13.2.8. For the model of Exercise 13.2.7, plot the I_1 and I_2 curves for
(a) $\beta = 1$, $\delta = 0.9$, $\gamma = 0.5$, $\epsilon_1 = 0.5$, and $\epsilon_2 = 0.9$, and
(b) $\beta = 1$, $\delta = 0.9$, $\gamma = 0.5$, $\epsilon_1 = 0.9$, and $\epsilon_2 = 0.5$.
For both, take $S(0) = 0.6$, $I_1(0) = 0.2$, and $I_2(0) = 0.2$.

The next two problems involve this system.

$$x_1' = -a_1x_1y_1 + c_1x_1 \qquad x_2' = -a_2x_2y_2 + c_2x_2$$
$$y_1' = a_1x_1y_1 - b_1y_1 + e(y_2 - y_1) \qquad y_2' = a_2x_2y_2 - b_2y_2 + e(y_1 - y_2)$$
$$z_1' = b_1y_1 - d_1z_1 \qquad z_2' = b_2y_2 - d_2z_2$$

13.2.9. In this system set $a_1 = 0.6$, $b_1 = 0.5$, $c_1 = 0.2$, $d_1 = 0.1$, $a_2 = 1.8$, $b_2 = 0.4$, $c_2 = 0.1$, and $d_2 = 0.2$. For the simulations set $x_1(0) = x_2(0) = 0.8$, $y_1(0) = y_2(0) = 0.2$, and $z_1(0) = z_2(0) = 0$.
(a) Set $e = 0$ and plot $y_1(t)$ and $y_2(t)$ for $0 \le t \le 100$. Notice that both curves are periodic, but they have different periods.
(b) This may not be such a surprise, because with $e = 0$ we have two uncoupled systems with different parameters. So set $e = 0.3$ and plot the $y_1(t)$ and $y_2(t)$ curves for $0 \le t \le 100$. What do you observe?

13.2.10. (a) Now set $c_2 = 0.5$ and leave all the other parameters unchanged. Plot the $y_1(t)$ and $y_2(t)$ curves for $0 \le t \le 100$ for $e = 0$ and for $e = 0.3$. Do these curves appear to have the same periods?
(b) Now return to an SIR model with constant population size N. Suppose $S(t) = \sin(t)$. Ignore the differential equations, so all we know is that for all t, $S(t) + I(t) + R(t) = N$ and that $S(t) = \sin(t)$. Must $I(t)$ and $R(t)$ also be periodic with period 2π? Give a proof or a counterexample.

(c) Continuing with the hypotheses of (b), if also $R(t)$ is periodic with period 2π, must $I(t)$ be periodic with period 2π?

13.3 MICHAELIS-MENTEN KINETICS

One of the most familiar models of enzyme kinetics is that of Leonor Michaelis and Maud Menten. Michaelis-Menten kinetics models [234] the action of an enzyme E on a substrate S to form a complex ES that can split back into E and S or can form a product P and return the enzyme E. (What's the substrate? Nothing geological; it's the substance on which the enzyme acts. The sugar amylose is the substrate for the enzyme amylase.) The reaction is written as

$$S + E \underset{b}{\overset{a}{\rightleftarrows}} ES \overset{c}{\rightarrow} P + E \tag{13.8}$$

where a, b, and c are the rate constants for these reactions. Typically, the back-reaction $ES \leftarrow P + E$ is ignored, at least on time scales relevant to the problem at hand.

Guided by Eq. (13.8) and the principle of mass action, we'll write a set of four coupled nonlinear differential equations, plot some solution curves, show how to reduce the dynamics to a 2-dimensional system, and derive the familiar Michaelis-Menten rate law

$$\frac{d[P]}{dt} = \frac{V_{max}[S]}{K_m + [S]} \tag{13.9}$$

Here $[E]$, $[S]$, $[ES]$, and $[P]$ denote the concentrations of E, S, ES, and P, V_{max} is the maximum rate of production of the product P, and $K_m = (b+c)/a$ is the *Michaelis constant*. We'll see that K_m is the substrate concentration that gives $d[P]/dt = (1/2)V_{max}$.

To derive the Michaelis-Menten differential equations, we first note that Eq. (13.8) carries the same sort of information as the graphs of Example 3.2.4 on the SIR model. Then we see

$$\frac{d[S]}{dt} = -a[E][S] + b[ES] \tag{13.10}$$

$$\frac{d[E]}{dt} = -a[E][S] + (b+c)[ES] \tag{13.11}$$

$$\frac{d[ES]}{dt} = a[E][S] - (b+c)[ES] \tag{13.12}$$

$$\frac{d[P]}{dt} = c[ES] \tag{13.13}$$

For example, substrate S disappears at a rate governed by substrate-enzyme encounters, and the number of these encounters is proportional to $[E][S]$ by the principle of mass action. Substrate is added when the complex ES breaks down into substrate and enzyme.

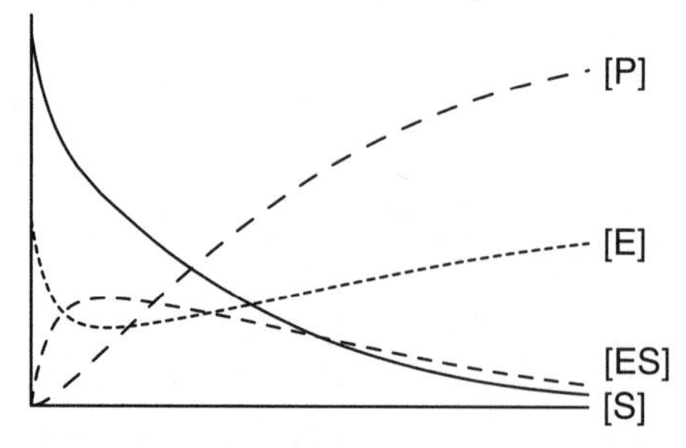

Figure 13.7. Solution curves for the system (13.10)–(13.13).

We know how to generate numerical solutions to these equations. Mathematica code is given in Sect. B.13. In Fig. 13.7 we see solution curves for the system (13.10)–(13.13). The $[S]$ curve is solid, the $[E]$ curve has the narrowest dashes, the $[ES]$ curve has wider dashes, and the $[P]$ curve has the widest dashes.

What do we see? The substrate $[S]$ curve (the solid curve) decreases steadily and is concave up. Does it approach an asymptote? The product $[P]$ curve (widest dashes) grows steadily and eventually is concave down. Does it approach an asymptote? We'll see.

The $[E]$ (narrow dashes) and $[ES]$ curves (wider dashes) seem to exhibit the opposite behavior to one another. In fact, this is easily seen by adding Eqs. (13.11) and (13.12):

$$\frac{d[E]}{dt} + \frac{d[ES]}{dt} = \big(-a[E][S] + (b+c)[ES]\big) + \big(a[E][S] - (b+c)[ES]\big) = 0$$

so we see that

$$\frac{d}{dt}\big([E] + [ES]\big) = \frac{d[E]}{dt} + \frac{d[ES]}{dt} = 0$$

That is, $[E] + [ES]$ is constant, so for example,

$$[E(t)] = [E(0)] + [ES(0)] - [ES(t)] = [E(0)] - [ES(t)] \tag{13.14}$$

because before the enzyme is introduced the ES complex is absent, so $[ES(0)] = 0$. Then we can write the right-hand side of Eqs. (13.10) and (13.12) in terms of $[S]$ and $[ES]$ alone

$$\frac{d[S]}{dt} = -a\big([E(0)] - [ES]\big)[S] + b[ES] \tag{13.15}$$

$$\frac{d[ES]}{dt} = a\big([E(0)] - [ES]\big)[S] - (b+c)[ES] \tag{13.16}$$

What about the product equation (13.13)? The concentration $[P]$ doesn't occur on the right side of Eqs. (13.10), (13.11), and (13.12), so we can solve these without a solution for $[P]$. Once we have a solution for $[ES]$, we can solve Eq. (13.13) directly.

That is, the dynamics of the Michaelis-Menten model are contained in the two equations (13.15) and (13.16). To study the long-term behavior of systems in the plane, we know how to start: find the nullclines, then the fixed points, and then test the fixed point stability. The $[S]$-nullcline and $[ES]$-nullcline are

$$[ES] = \frac{a[E(0)][S]}{a[S]+b} \quad \text{and } [ES] = \frac{a[E(0)][S]}{a[S]+b+c}$$

Because the constants a, b, and c are positive, the only fixed point is the origin. The derivative matrix is

$$D\vec{F}([S],[ES]) = \begin{bmatrix} -a[E(0)]+a[ES] & a[S]+b \\ a[E(0)]-a[ES] & -a[S]-b-c \end{bmatrix}$$

and at the origin the derivative matrix is

$$D\vec{F}(0,0) = \begin{bmatrix} -a[E(0)] & b \\ a[E(0)] & -b-c \end{bmatrix}$$

The eigenvalues are complicated, but with a little work we can see that both are negative and the origin is an asymptotically stable node. So $[S]$ and $[ES]$ go to 0, and by Eq. (13.14), $[E] \to [E(0)]$.

The analysis of Michaelis and Menten continued with the assumption that the substrate exceeded the enzyme supply, so the ES complex would form rapidly and reach an approximate equilibrium, that is, $d[ES]/dt \approx 0$. This is called the *quasi-steady state hypothesis*. Set the right side of Eq. (13.16) equal to 0 and solve for $[ES]$,

$$[ES] = \frac{a[E(0)][S]}{b+c+a[S]}$$

Substitute this expression for $[ES]$ into Eq. (13.13) to obtain the *Michaelis-Menten rate law*

$$\frac{d[P]}{dt} = c\frac{[E(0)][S]}{((b+c)/a)+[S]} = \frac{V_{max}[S]}{K_m+[S]} \tag{13.17}$$

where $V_{max} = c[E(0)]$.

In essence, their analysis separated the variables into two time scales: $[E]$ and $[ES]$ are fast variables, while $[S]$ and $[P]$ are slow variables. The fast variables achieve their equilibrium, and then the slow variables evolve against the background of the fixed fast variables. This approach has been adapted to many other systems. See [150] for an instructive review.

To see that V_{max} is the maximum growth rate of $[P]$, take the large $[S]$ limit of the right-hand side of Eq. (13.17). This gives

$$\lim_{[S]\to\infty} \frac{d[P]}{dt} = \lim_{[S]\to\infty} \frac{V_{max}[S]}{K_m+[S]} = V_{max}$$

Finally, suppose $[S] = K_m$. Then

$$\frac{d[P]}{dt} = \frac{V_{max}[S]}{K_m + [S]} = \frac{V_{max}K_m}{K_m + K_m} = \frac{V_{max}}{2}$$

Though some care must be used in its application, the Michaelis-Menten rate law (13.17) gives a quick estimate of the rate of product production from the substrate concentration. This kinetic model can be applied in settings other than enzyme catalysis of reactions. One example is the transport of nutrients across cell membranes as described in Sect. 7.1 of [104]. Other treatments of Michaelis-Menten kinetics are in Chapter 7 of [104], Chapter 1 of [107}, and Chapter 6 of [246]. This simple model has a wide reach. How amazing that evolution has expressed this underlying principle in such a wide range of systems.

Practice Problems

13.3.1. Find $d[P]/dt$ when the substrate concentration is $[S] = K_m/2$. Assume the Michaelis-Menten rate law (13.17) can be applied.

13.3.2. Fig. 13.8 was generated by $a = 0.2$, $b = 0.03$, $c = 0.05$, $[S(0)] = 1$, $[E(0)] = 0.5$, $[ES(0)] = [P(0)] = 0$. Double the value of a and leave all other parameters unchanged. How do the solution curves change?

Practice Problem Solutions

13.3.1. Apply the Michaelis-Menten rate law (13.17). Then

$$\frac{d[P]}{dt} = \frac{V_{max}[S]}{K_m + [S]} = \frac{V_{max}K_m/2}{K_m + K_m/2} = \frac{V_{max}}{3}$$

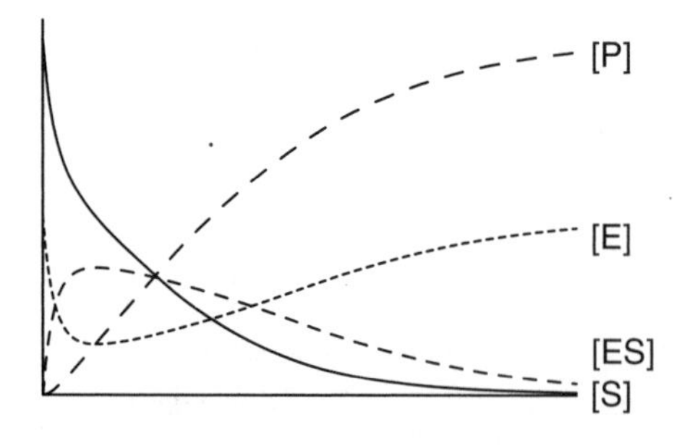

Figure 13.8. Solution curves for Practice Problem 13.3.2.

13.3.2. When we compare these plots with those of Fig. 13.7, we see that the $[S]$-curve goes to 0 more quickly. Also, the vertical space between the local minimum on the $[E]$-curve and the local maximum on the $[ES]$-curve has increased. Finally, the $[E]$-curve crosses the $[S]$-curve between the two times the $[E]$-curve crosses the $[SE]$-curve. In Fig. 13.7 the $[E]$-curve crosses the $[S]$-curve after both times the $[E]$-curve crosses the $[ES]$-curve.

Exercises

13.3.1. As an alternative to the computation of the derivative matrix eigenvalues, plot the nullclines of Eqs. (13.15) and (13.16). For your graphs take $a = 1$, $b = 1$, $c = 1$, and $[E(0)] = 1$. In each region of the first quadrant defined by the nullclines, determine whether the vector field points NE, NW, SW, or SE. From this can you deduce that the origin is an asymptotically stable fixed point?

13.3.2. Assume (13.17) can be applied. Find $d[P]/dt$ when
(a) $[S] = K_m/V_{max}$
(b) $[S] = V_{max}/K_m$.
Verify that the calculated values of $d[P]/dt$ satisfy $d[P]/dt < V_{max}$.

13.3.3. Assume that $d[P]/dt$ is a constant A and that (13.17) can be applied. Find an expression for the substrate concentration $[S]$ as a function of A.

13.3.4. For reactions that satisfy (13.17), find the value of $[S]$ that gives the maximum value of $d[P]/dt$, or show there is no maximum.

13.3.5. For reactions that satisfy (13.17), if $[S_2] = 2[S_1]$, does the following relation hold?

$$\left.\frac{d[P]}{dt}\right|_{[S_2]} = 2\left.\frac{d[P]}{dt}\right|_{[S_1]}$$

Give a reason to support your answer.

13.3.6. (a) Plot the graphs for $a = 0.2$, $b = 0.06$, $c = 0.05$, $[S(0)] = 1$, $[E(0)] = 0.5$, $[ES(0)] = [P(0)] = 0$.
(b) Plot the graphs for $a = 0.2$, $b = 0.03$, $c = 0.10$, $[S(0)] = 1$, $[E(0)] = 0.5$, $[ES(0)] = [P(0)] = 0$.
(c) Compare these plots to Figs. 13.7 and 13.8. What changes? What doesn't?

13.3.7. (a) Plot the graphs for $a = 0.4$, $b = 0.06$, $c = 0.05$, $[S(0)] = 1$, $[E(0)] = 0.5$, $[ES(0)] = [P(0)] = 0$. (We've doubled two of a, b, and c.)
(b) Plot the graphs for $a = 0.2$, $b = 0.06$, $c = 0.10$, $[S(0)] = 1$, $[E(0)] = 0.5$, $[ES(0)] = [P(0)] = 0$.
(c) Plot the graphs for $a = 0.4$, $b = 0.03$, $c = 0.10$, $[S(0)] = 1$, $[E(0)] = 0.5$, $[ES(0)] = [P(0)] = 0$.
(d) Explain what changes and what doesn't.

13.3.8. Suppose one enzyme molecule must interact with two substrate molecules to produce a complex $E2S$ that gives rise to one enzyme molecule and two product molecules.
(a) Write the differential equations for $d[S]/dt$, $d[E]/dt$, $d[E2S]/dt$, and $d[P]/dt$.
(b) Write the $d[S]/dt$ and $d[E2S]/dt$ equations in terms of only $[S]$ and $[E2S]$.
(c) Sketch the nullclines for the $d[S]/dt$, $d[E2S]/dt$ system. Show the origin is the only fixed point. From the directions of the vector field, deduce that the origin is an asymptotically stable fixed point.

13.3.9. (a) Plot the $[P]$, $[E]$, $[E2S]$, and $[S]$ curves for the system of of Exercise 13.3.8. Take $a = 0.2$, $b = 0.03$, $c = 0.05$, $[S(0)] = 1$, $[E(0)] = 0.5$, $[E2S(0)] = [P(0)] = 0$.
(b) Compare these plots with those of Fig. 13.7, which has the same parameters. What is the main differences?

13.3.10. Find an example of the Michaelis-Menten model in a system other than enzyme kinetics and the transport of nutrients across a cell membrane. This will give you another glimpse of the point mentioned at the end of this section.

13.4 VIRUS DYNAMICS BASICS

Population dynamics–predator-prey, competing species, and the like—are nice biological applications of our techniques of differential equation modeling, but you may think this has little to do with medicine. However, suppose we treat the host's body as the ecosystem, viruses as prey, and immune system agents as predators. Then general results about population dynamics can be applied to tell us something about how the virus population grows. This approach is called *virus dynamics*.

Robert May and Martin Nowak have written a wonderful book, *Virus Dynamics: Mathematical Principles of Immunology and Virology* [261], on this topic. Among other diseases, HIV [168, 224, 261, 273, 274, 275, 278, 279, 308, 342], hepatitis B [64, 259, 261], and hepatitis C [55, 256, 273, 276] have been studied using virus dynamics. We'll sketch a few of the results of this approach to HIV.

In order to do this, we'll need to understand a bit about viruses and about how the immune system works. Just the minimal amount, because in its details the immune system is very, very complicated. You've never seen *The Idiot's Guide to the Immune System* or *The Immune System for Dummies*. There's a reason you haven't.

Sect. A.15 has details about the immune system, Sect. A.22 about molecular genetics. For now, we'll give a quick sketch of what we need to know in order to understand a simple model of the progression of an HIV infection, and how the model changes under two therapies.

A virus contains a core, the home of its genetic material, either DNA or RNA. If RNA, the virus is called an *RNA virus*. A *retrovirus* is a type of RNA virus that, when it invades a cell, converts its RNA into viral DNA and inserts that into the cell's DNA. The virus core is covered with a protein coat called the *capsid*; some viruses have an *envelope*, a lipid membrane that covers the capsid. Alone, a virus cannot reproduce; to copy itself a virus must hijack the reproductive mechanism of another cell. To recognize the kind of cell it has evolved to attack, a virus uses the *cell surface receptors*, proteins that span the cell membrane, allowing communication between the cell and its surroundings. When a virus encounters an appropriate receptor it binds to the receptor. Some RNA viruses make the cell construct viral RNA and proteins, assemble them into a new virus particle, and release it from the cell. Other viruses insert their DNA (or their RNA, after it is converted to DNA) into the cell's DNA, so the cell begins to make copies of the virus.

To fight against viruses, bacteria, and other invaders, evolution has produced a complex, and mostly effective, immune system. The immune system has three parts: physical barriers (the skin and the digestive, respiratory, and reproductive mucosa), the innate immune system (this includes the complement system, macrophages, and neutrophils), and the adaptive immune system (this includes the lymphocytes: B cells, which mature in the bone marrow, and T cells, which mature in the thymus). The innate immune system is hard-wired to recognize and rapidly attack pathogens that have become familiar over our evolutionary history. The adaptive immune system can recognize and attack unfamiliar pathogens. The number of possible pathogens is about 10^8. How the adaptive immune system can be ready to fight almost all of them with only a few billion lymphocytes is an evolutionary triumph that we'll sketch in Sect. A.15.

Here we'll sketch a few aspects of the adaptive immune system. B cells and T cells recognize something as foreign and dangerous when their receptors bind to invaders' peptides (protein fragments) presented in one of several ways. Alone this is not sufficient: there are fail-safes that guard against unwarranted responses. When these conditions are met, the cell is *activated*. Then B cells make *antibodies*, proteins that mark the invaders for attack by other cells. Two additional T cell receptors are *CD*4 and *CD*8 receptors. Mature T cells generally express one or the other of these receptor types. *Helper T cells* express *CD*4 and secrete *cytokines*, chemical signals that direct other immune system agents in the attack. *Killer T* *cells* express *CD*8 and kill the invaders. HIV attacks helper T cells, weakening the immune system to the point that common infections can be deadly.

Now we'll build a mathematical model of an HIV infection. (Technically, we'll model HIV-1, more prevalent and pathogenic than HIV-2.) First, let's look at a tiny bit of history.

In the early 1980s, researchers at the CDC noticed an increase in deaths from infectious diseases that normally are controlled handily by the immune system. The ailment was named acquired immune deficiency syndrome (AIDS) and an intense search for its cause began. In 1983 a group led by Françoise Barré-Sinoussi and Luc Montagnier at the Institut Pasteur in Paris found a retrovirus in tissue from the lymph nodes of an AIDS patient. At that time, the only known human retroviruses were human T cell leukemia virus (HTLV) 1 and 2. Initially, both the group in Paris and a group led by Robert Gallo at the NIH in Bethesda, Maryland, thought the retrovirus was related to HTLV-1, but soon it was found to be a new type of retrovirus, in 1986 named human immunodeficiency virus (HIV). Robin Weiss [81] and his group at University College London showed that HIV attacks the immune system through helper T cells. The continuing story of the quest for treatments for HIV is filled with heartbreak, frantic work, wrong turns, brilliant insights, and strong personalities. It is the subject of many books, but not this one. The Wikipedia article on the history of HIV/AIDS is a good place to start, if you're interested in that story.

As we mentioned, HIV targets helper T cells. When a *virion*—an infective virus particle outside a host cell, moving freely and able to infect other cells—encounters an activated helper T cell, the viral envelope binds with the *CD*4 receptor on the T cell membrane. The viral core enters the T cell, an enzyme called *reverse transcriptase* copies the viral RNA into DNA, and another enzyme called *integrase* inserts this DNA into the host cell's DNA. Then the T cell makes copies of the viral RNA. New virions form within the cell, then bud off, carrying some of the T cell membrane as the viral envelope. Usually, the host cell dies after a few thousand virions have budded from its surface.

Without treatment an HIV infection progresses through three stages:

- The *initial acute phase* shows a high viral load and presents flu-like symptoms.
- The *second phase* sees a lower, nearly constant viral load, called the *set point*, while the helper T cell population declines slowly. This phase can last for up to a decade. During most of this phase, the patient is asymptomatic. Blood counts are the only indicator that there is a problem.
- The *third phase* is AIDS, immune system failure and death from opportunistic infections or uncommon cancers.

With all the complex details of HIV infections, why would we think that a simple mathematical model could tell us anything useful? Certainly, the fine details can't be captured. But the hope of modeling is that some general aspects of virus dynamics result from simple relations between major elements of the populations. Let's see what we can find this way.

First, there's a straightforward application of modeling, with some subtle experiments to assess parameter values. Early in the study of HIV infection, the asymptomatic second phase was a puzzle. What was going on then? Is the virus quiescent? After all, the chickenpox virus can stay quietly in nerve cells for many years, perhaps eventually giving rise to shingles, which my sister says can fill a couple of very unpleasant weeks. If the HIV virus were inactive in the second stage, therapy would be more challenging. Early treatments most often slowed the progression of the disease for only a limited time, so a common approach was to delay treatment until the third phase. Results were not encouraging.

In the mid 1990s, David Ho, Alan Perelson, and their coworkers [168, 273, 275, 278] began to analyze the dynamics of HIV during the second phase of the infection. The simplest model is $dV/dt = P - cV$, where $V(t)$ is the virus concentration at time t, P is the uninhibited virus production rate, and c is the virus clearance rate. In the second phase, the viral load is approximately a constant, V_0, so $dV/dt = 0$ and $P = cV_0$. Twenty HIV-positive patients in the second phase of the infection were given ABT-538 (Ritonavir), a protease inhibitor. These work by causing newly infected helper T cells to produce only noninfectious viruses (soon we'll say a bit more about this therapy), so after the pre-treatment virions have been cleared and the remaining infectious T cells have died, the viral load should drop dramatically. That's exactly what was observed. After a short delay, the viral load plummeted and the helper T cell population rose substantially. Some subtlety is involved, but the exponential drop in the viral load gives an estimate of the clearance rate c. Then with the set point V_0, Perelson and Ho found the virus production rate $P = cV_0$. While there is some variation among patients, the average virus production rate throughout the entire (untreated) second phase is about 10^{10} virions per day. Every day for up to ten years the body produces ten billion new HIV virus particles. And because the set point is about constant, every day the immune system destroys ten billion HIV virus particles. Far from quiescent, the second phase of an HIV infection is a pitched battle. The immune system needs all the help it can get, so Ho and Perelson recommended that treatment begin as soon as possible in the second phase.

A similar analysis was reported in [45, 376]. The early days of virus dynamics models were complex, in no small part because they involved many scientists and many theories. A clear description of a path to this application of mathematical modeling is in Chapter 8 of Steven Strogatz's enjoyable book *Infinite Powers: How Calculus Reveals the Secrets of the Universe* [352], also presented in [353]. Some more technical details are provided in [273]. Really, though, you should read all of [352]: every page contains delights.

To build differential equations models, we'll start with an extension of the SIS model of Exercise 3.2.5, and apply this to infection spread in the helper T cell population through contact between susceptible helper T cells and virions. Use these populations:

$$x = \text{uninfected helper T cells,}$$
$$y = \text{productively infected helper T cells, and}$$
$$z = \text{virions}$$

In the SIS model, the transition between the susceptible population and the infected population is due to interactions between members of the susceptible and infected populations. Unlike the SIS model, in this HIV model the transition between x and y is driven by encounters of x and z. Uninfected helper T cells are produced at a constant rate β. The x, y, and z populations have per capita death rates δ_x, δ_y, and δ_z, also called the rate at which these populations are *cleared*. The constant γ is the probability that an encounter between a virion and an uninfected helper T cell makes the helper T cell productively infected. Here we'll again use the principle of mass action. Finally, κ is the per capita rate at which infected cells produce virions. Then the model is

$$\frac{dx}{dt} = \beta - \delta_x x - \gamma xz \qquad \frac{dy}{dt} = \gamma xz - \delta_y y \qquad \frac{dz}{dt} = \kappa y - \delta_z z \tag{13.18}$$

The constant κ can be expressed as $N\delta_y$, where N is the *burst size*, the total number of virions produced by an infected cell over its lifetime. Because the mean lifetime of an infected cell is $1/\delta_y$ (recall Prop. 3.2.1), $N\delta_y$ is the virion production rate. That is, $\kappa = N\delta_y$. Also, this gives an expression for the burst size:

$$N = \frac{\kappa}{\delta_y} \tag{13.19}$$

Can this simple model capture any observed features of a real HIV infection? To start, we'll compute the basic reproductive ratio R_0, the average number of infected cells that are produced by a single infected cell near the start of the infection. We first encountered R_0 in Example 3.2.5, where uninfected cells became infected through direct contact with an infected cell. As we've seen, in this virus dynamics model, infection does not occur by direct interaction of infected and uninfected cells, but through the intermediary of virions. Here are the steps of the calculation.

1. One infected helper T cell produces virions at the rate κ. From Prop. 3.2.1 we see that the average lifetime of an infected helper T cell is $1/\delta_y$, so one infected helper T cell produces κ/δ_y virions.
2. One virion produces infected cells at the rate γx (that is, γxz with $z = 1$). Applying Prop. 3.2.1 again, we see that the average lifetime of a virion is $1/\delta_z$, so from the time it buds off of an infected cell until the time it is cleared, one virion produces $\gamma x/\delta_z$ infected cells.
3. Combining steps 1 and 2, we see that on average one infected helper T cell will give rise to

$$\text{(virions per infected cell)} \cdot \text{(cells infected per virion)} = \frac{\kappa}{\delta_y} \cdot \frac{\gamma x}{\delta_z} \tag{13.20}$$

   infected helper T cells.
4. Near the start of the infection, both y and z are nearly 0, so x is nearly constant. Then $x' = 0$ and from the first equation of Eq. (13.18), the x population is about β/δ_x.

Substituting in the value of x from step 4, we find

$$R_0 = \frac{\kappa}{\delta_y} \cdot \frac{\gamma x}{\delta_z} = \frac{\kappa\gamma\beta}{\delta_y\delta_z\delta_x} \tag{13.21}$$

As we mentioned in Example 3.2.5, the infection will die out if $R_0 < 1$. Starting from a population of N infected cells, if $R_0 < 1$ how long until the infection disappears? In units of the time required for one infected cell to infect another cell, we see that with each successive time unit,

$$N \to N \cdot R_0 \to (N \cdot R_0) \cdot R_0 = N \cdot R_0^2 \to N \cdot R_0^3 \to \cdots \to N \cdot R_0^T$$

When the population of infected cells drops to 1, the infection ends. Solving $N \cdot R_0^T = 1$ for T gives $T = \ln(N)/\ln(1/R_0)$.

If $R_0 > 1$, the infection spreads. Because the viral load (the size of the z population) has great clinical significance, we'd like to check if this model matches, at least qualitatively, the first two phases of observed HIV infections. In Fig. 13.9 we plot the $z(t)$ curve for parameters with $R_0 = 0.8$ (solid curve), $R_0 = 4.0$ (short dashes), and

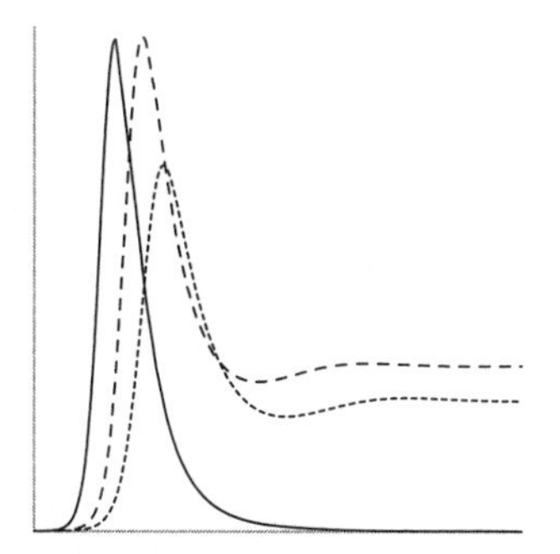

Figure 13.9. Plots of $z(t)$.

$R_0 = 4.8$ (long dashes). The equilibrium z in the $R_0 = 4.0$ graph is about 1.5×10^6 virions; the equilibrium z-value in the $R_0 = 4.8$ graph is about 1.9×10^6.

Let's see what the fixed points can tell us. Recall that the fixed points are the intersections of the nullclines, three in the case of Eq. (13.18). We find there are two fixed points:

$$(x_0, y_0, z_0) = \left(\frac{\beta}{\delta_x}, 0, 0\right) \quad \text{and} \tag{13.22}$$
$$(x_1, y_1, z_1) = \left(\frac{\delta_y\delta_z}{\gamma\kappa}, \frac{\gamma\kappa\beta - \delta_y\delta_x\delta_z}{\delta_y\gamma\kappa}, \frac{\gamma\kappa\beta - \delta_y\delta_x\delta_z}{\delta_y\gamma\delta_z}\right)$$

The eigenvalues of the derivative matrix

$$D\vec{F}(x,y,z) = \begin{bmatrix} -\delta_x - \gamma z & 0 & -\gamma x \\ \gamma z & -\delta_y & \gamma x \\ 0 & \kappa & -\delta_z \end{bmatrix}$$

at both fixed points are a real mess. Maybe expressing some of the derivative matrix entries in terms of R_0 will simplify the eigenvalue calculation. At the fixed point $(\beta/\delta_x, 0, 0)$ the derivative matrix is

$$\begin{bmatrix} -\delta_x & 0 & -\gamma\beta/\delta_x \\ 0 & -\delta_y & \gamma\beta/\delta_x \\ 0 & \kappa & -\delta_x \end{bmatrix} = \begin{bmatrix} -\delta_x & 0 & -\delta_y\delta_z R_0/\kappa \\ 0 & -\delta_y & \delta_y\delta_z R_0/\kappa \\ 0 & \kappa & -\delta_z \end{bmatrix}$$

The R_0 version might look worse, but the eigenvalues,

$$-\delta_x, \text{ and } \lambda_\pm = \frac{-(\delta_y + \delta_z) \pm \sqrt{(\delta_y - \delta_z)^2 + 4\delta_y\delta_z R_0}}{2}, \tag{13.23}$$

are easier to interpret:

- If $R_0 < 1$, then

  $$\sqrt{(\delta_y - \delta_z)^2 + 4\delta_y\delta_z R_0} < \sqrt{(\delta_y - \delta_z)^2 + 4\delta_y\delta_z} = \sqrt{(\delta_y + \delta_z)^2} = \delta_y + \delta_z$$

  where the last equality holds because both δ_y and δ_z are positive. From this we see that both λ_+ and λ_- are negative. That is, if $R_0 < 1$ all three eigenvalues of $D\vec{F}(x_0, y_0, z_0)$ are negative and this fixed point is asymptotically stable.
- If $R_0 > 1$, then a similar argument shows

  $$\sqrt{(\delta_y - \delta_z)^2 + 4\delta_y\delta_z R_0} > \delta_y + \delta_z$$

  and so λ_+ is positive. Because it has at least one positive eigenvalue, the fixed point is unstable. It's a bit like a saddle point: trajectories move toward the fixed point along the plane spanned by the eigenvectors of the eigenvalues $-\delta_x$ and λ_-, then run away from the fixed point along the line determined by the eigenvector of λ_+.

For (x_1, y_1, z_1), we know $x_1 > 0$. In terms of the original expression of the coordinates, $y_1 > 0$ and $z_1 > 0$ gives the complicated condition $\gamma\kappa\beta - \delta_y\delta_x\delta_z \geq 0$. This, too, may be easier to understand if we express the coordinates of this fixed point in terms of R_0 and $x_0 = \beta/\delta_x$, the equilibrium value of uninfected helper T cells in the absence of infected cells and HIV virions. Then with a little algebra we see

$$(x_1, y_1, z_1) = \left(\frac{x_0}{R_0}, (R_0 - 1)\frac{\delta_x\delta_z}{\gamma\kappa}, (R_0 - 1)\frac{\delta_x}{\gamma}\right) \tag{13.24}$$

and so y_1 and z_1 are positive if and only if $R_0 > 1$.

For the R_0 values for the systems pictured in Fig. 13.9, Eq. (13.24) shows that both y_1 and z_1, the equilibrium value of the the infected cell population and of the viral load, are positive in both dashed graphs, but negative in the solid graph. What are we to take from that? Obviously a body cannot contain a negative number of infected cells or of viruses. When $y = z = 0$, from the second and third equations of (13.18), we see that $y' = z' = 0$ so y and z will decrease to 0 and stay there; the system has reached the equilibrium (x_0, y_0, z_0). (See Exercise 13.4.1.) So, sometimes we need to think a bit about the implications of a numerical solution. Don't take every number at its face value.

What values of R_0 are found in HIV-1 patients? In 1999, Susan Little and coworkers [212] estimated R_0 lies between 7 and 34. In 2010, Perelson and coworkers [308] studied 47 patients infected with HIV-1 and obtained a median value of $R_0 = 8.0$, with the first and third quartiles 4.9 and 11. This is a bit lower than Little's estimate, but still large enough to be very bad news.

Back to math. Can we test the stability of the fixed point (x_1, y_1, z_1) by the eigenvalues of the derivative matrix, reformulated in terms of R_0? Sadly, no, an analytic expression for the eigenvalues still is a mess. We'll have to be content with numerics.

We'll use the parameters that generated the short-dash curve in Fig. 13.9. These are $\beta = 10^5$, $\delta_x = 0.1$, $\gamma = 2 \times 10^{-7}$, $\delta_y = 0.5$, $\kappa = 100$, and $\delta_z = 5$, the parameters that generate Fig. 3.3 of Nowak and May's book [261]. If you compare their picture to ours and are puzzled by the differences, note the scale on the vertical axis of their graphs. Their scale is logarithmic; ours is linear. With these parameters, the fixed point is $(x_1, y_1, z_1) = (1.25 \times 10^5, 1.75 \times 10^5, 3.5 \times 10^6)$, the derivative matrix is

$$\begin{bmatrix} -0.8 & 0 & -0.025 \\ 0.7 & -0.5 & 0.025 \\ 0 & 100 & -5 \end{bmatrix}$$

and the eigenvalues are

$$\lambda_1 = -5.56497, \lambda_{\pm} = -0.367015 \pm 0.423923i$$

Because one eigenvalue is negative and the complex eigenvalues have negative real parts, this fixed point is asymptotically stable, with a spiral aspect being responsible for the wobbles in the dashed trajectories of Fig. 13.9. By this and similar calculations we can show that the limiting viral load in this figure and those obtained by varying some parameters reflect the observed dynamics of this disease.

In the two dashed graphs of Fig. 13.9 we see that the viral load settles down to its non-zero equilibrium value, and that value increases with R_0. Here's one reason that this is interesting. The length of time a patient spends in the approximately symptom-free second phase varies from a few months to many years, with the average time about ten years. The transition from the second phase to the third phase occurs when the helper T cell count has fallen below 200 per microliter. Death follows, usually quickly. There is a good correlation between the viral load in the first year of infection and the time spent in the second phase: the higher the viral load, the shorter the time in the second phase. So our simple model can reproduce much of this behavior, but the model can't explain all observations, and no one expects that it reflects all the complex biochemistry of HIV infection. The model does suggest that some of the dynamics of this disease can result from fairly general considerations. Also, a modification of the model can be used to project the effectiveness of proposed treatments. Let's see how.

An early HIV treatment was the drug AZT, a reverse transcriptase inhibitor. Recall that reverse transcriptase copies the viral RNA into viral DNA, which integrase inserts in the helper T cell DNA. Without converting viral RNA to DNA, an infected helper T cell will not produce virions and the infection will not spread. Good news, right? For a little while: on average, AZT treatment added about six months to the lives of AIDS patients. The problem is that a point mutation in the HIV reverse transcriptase gene can produce a new enzyme that does not bind to AZT. This mutated HIV is immune to AZT.

A different approach is to use protease inhibitors. These don't stop HIV particles from infecting helper T cells, but they cause infected helper T cells to shed virions that are not infectious. Treatment with protease inhibitors produced similar results: the viral load drops significantly, then recovers after a few months. The reason is the same as we saw with reverse transcriptase inhibitor treatments: a point mutation of the HIV genome can make it immune to the protease inhibitor used in the treatment.

We'll show how to alter the model (13.18) to account for reverse transcriptase inhibitors and for protease inhibitors. Then we'll see how these treatments produce the observed dip in the viral load. Finally, we'll see that combinations of three drugs have produced some much-needed reason for hope among those infected with HIV.

Reverse transcriptase inhibitor treatment

Treating HIV with reverse transcriptase inhibitors prevents viral RNA from being converted to DNA, so it cannot be inserted into the host cell's DNA. Then when a virion encounters a helper T cell, the cell does not become infected in the sense that it no longer produces virions. Approved for use in 1987, AZT was the first reverse transcriptase inhibitor used to treat HIV. Since then, others have been developed.

Nowak and May modeled reverse transcriptase inhibitor treatment by setting $\gamma = 0$ in Eq. (13.18). This gives

$$\frac{dx}{dt} = \beta - \delta_x x \qquad \frac{dy}{dt} = -\delta_y y \qquad \frac{dz}{dt} = \kappa y - \delta_z z \tag{13.25}$$

We see that the x population decouples from the y and z populations, and all the equations are linear. We can solve the x equation by separation of variables and substitution:

$$\begin{aligned} \frac{dx}{\beta - \delta_x x} &= dt \\ -\frac{1}{\delta_x}\frac{du}{u} &= dt && \text{substitute } u = \beta - \delta_x x \\ \beta - \delta_x x = u &= e^{-\delta_x t} e^C && \text{integrate and exponentiate} \\ \beta - \delta_x x(0) &= e^C && \text{take } t = 0 \end{aligned}$$

Then we can solve the penultimate equation for x, with e^C given by the last equation.

$$x(t) = \frac{\beta}{\delta_x} - \frac{1}{\delta_x} e^{-\delta_x t}\big(\beta - \delta_x x(0)\big)$$

Note that we also could apply the methods of Sect. 4.5 and deduce that $x = \beta/\delta_x$ is a stable fixed point.

The y equation is simpler still. Separation of variables gives

$$y(t) = y(0)e^{-\delta_y t}$$

With this solution for y we can rewrite the z equation as

$$\frac{dz}{dt} + \delta_z z = \kappa y(0) e^{-\delta_y t}$$

This is a non-autonomous differential equation. We can solve it by the series method of Sect. 10.9 or by the method of integrating factors of Sect. A.1. We'll derive this solution in Example A.1.4. It is

$$z(t) = \frac{\kappa y(0)}{\delta_z - \delta_y}\left(e^{-\delta_y t} - e^{-\delta_z t}\right) + z(0)e^{-\delta_z t} \qquad (13.26)$$

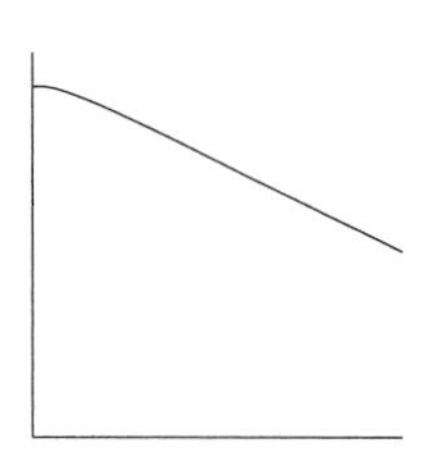

Figure 13.10. Log($z(t)$).

In Fig. 13.10 we plot Log($z(t)$) as a function of t. Why the logarithm? From Eq. (13.26) we see that for large t the graph of $z(t)$ will fall off exponentially. We expect the average lifetime of a virion is less than the average lifetime of an infected helper T cell, that is, $1/\delta_z < 1/\delta_y$. It follows that $e^{-\delta_z t}$ decreases more rapidly than $e^{-\delta_y t}$. Then eventually the graph of Log($z(t)$) is close to a straight line with slope $-\delta_y$.

We see this in the figure, but also notice the fairly flat start of the graph. Nowak and May call this the *shoulder phase* of the virion population decay. If the reverse transcriptase inhibitor is not completely effective, the decay of the virion population is slower.

But this works only for a while. Eventually, HIV will mutate so the therapy is ineffective, these mutated HIV viruses will infect cells, and the infection will take off again.

Protease inhibitor treatment

Another approach is to treat HIV with protease inhibitors, early treatments reported by George Shaw and coworkers [376] and David Ho and coworkers [168]. Protease inhibitors cause infected cells to produce noninfectious viruses. To model this treatment, we refine the model by subdividing the z population into two subpopulations:

$$z_i = \text{infectious virions, and } z_n = \text{noninfectious viruses}$$

We'll assume both virus populations z_i and z_n are cleared at the same rate, δ_z. Then the model is

$$\frac{dx}{dt} = \beta - \delta_x x - \gamma x z_i \qquad \frac{dy}{dt} = \gamma x z_i - \delta_y y \qquad (13.27)$$

$$\frac{dz_i}{dt} = -\delta_z z_i \qquad \frac{dz_n}{dt} = \kappa y - \delta_z z_n$$

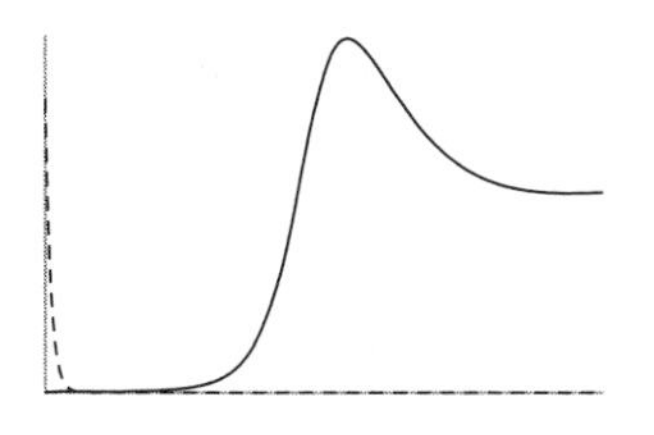

Figure 13.11. z_i and z_n.

Also, we'll assume the treatment is completely effective, so after treatment all infected helper T cells produce only noninfectious viruses. For a while, there still will be an infectious virion population, produced by helper T cells infected before treatment. After treatment, no new infectious virions are produced, so as expressed in the third equation of Eq. (13.27), the z_i population undergoes exponential decay.

In Fig. 13.11 plot the z_i (dashed) and z_n (solid) curves by solving numerically Eq. (13.27). A glance at the equations shows that the z_i population should decay to 0. In Exercise 13.4.2 you'll explore the long-term behavior of the z_n population. Mathematica code to generate this figure is in Sect. B.14.

In the relatively short time after treatment until the z_i population clears, let's assume the x population is approximately constant, $x = x_0$. This isn't exactly right, but if the z_i population clears quickly enough, it's close. We can apply the method of integrating factors (Sect. A.1) to solve the y equation, obtaining a sum of the exponentials $e^{-\delta_y t}$ and $e^{-\delta_z t}$. Then we can apply the method of integrating factors again to solve the z_n equation. In addition to sums of exponentials, we obtain a product of t and an exponential. In this short time window, the dynamics of the z_n population are more complicated than a sum of exponentials.

As with reverse transcriptase inhibitors, protease inhibitors work for a while, until HIV mutates to a version that does not respond to this treatment. This is an important chapter in the story, but not at all the last word.

We'll end this section by discussing how some of the model parameters, the clearance rates, can be measured. Then we'll say a bit about mutation rates and the efficacy of combination drug therapies.

Measuring the clearance rate δ_w of a species w can be difficult, because usually new w cells are created as old w cells die. For populations governed by exponential decay—for example, infectious virions and infected helper T cells after the z_i population has dropped to 0—we don't have the complication of population growth masking clearance, but still, direct measurement of the clearance rate can be subtle. Easier is to measure the half-life, the time $t_{1/2}$ for which $w(t_{1/2}) = (1/2)w(0)$. Recall that exponential decay implies $w(t) = w(0)e^{-\delta_w t}$, so we see that

$$(1/2)w(0) = w(t_{1/2}) = w(0)e^{-\delta_w t_{1/2}}$$

Then the half-life is

$$t_{1/2} = \frac{\ln(1/2)}{-\delta_w} = \frac{\ln(2)}{\delta_w}$$

Recall from Prop. 3.2.1 that the clearance rate δ_w is the reciprocal of the mean lifetime, so the mean lifetime and the half-life are related by

$$t_{1/2} = \ln(2) \times \text{mean lifetime}$$

Half-lives are not too difficult to approximate. Measure the population w and start the clock running. Call this value $w(0)$. Measure the concentration at equal time intervals T_1, T_2, T_3, and so on. Then the time T_n where $w(T_n)$ is closest to $w(0)/2$ is an approximation of the half-life. In this way we can approximate the clearance rates of virions and infected helper T cells.

The HIV genome is about 10^4 bases long. In 1995 Louis Mansky and Nobel laureate Howard Temin [222] measured the HIV mutation rate at about 3×10^{-5}. In 2015 José Cuevas and coworkers [80] measured a much higher rate, but only a few percent of the mutations appear in virions in the plasma, suggesting that many of these mutations are defective. HIV-1 exhibits the highest mutation rate observed in any organism. This allows HIV to evolve drug resistance rapidly, and makes it very difficult to treat. For example, a single point mutation of the reverse transcriptase gene can make that enzyme immune to the effects of a particular treatment. From the genome length and the mutation rate, the duration of the effectiveness of a treatment can be estimated. As mentioned, early single-drug trials prolonged lives by about six months.

Starting in 1995, great success was achieved with therapies consisting of a protease inhibitor and two reverse transcriptase inhibitors. Resistance to these combinations would require three simultaneous point mutations, a very unlikely occurrence. Pinar Iyidogan and Karen Anderson [181] give a survey of recent developments in antiretroviral therapies and the evolution of resistance.

The simple mathematical models presented in this section can reproduce some of the observed features of HIV infection, for example, the initial exponential increase of viral load, the slow down to a maximum, and then the decay to a steady state, as illustrated in Fig. 13.9. But these simple equations cannot predict when the treatment they model ceases to be effective. Math is good for some things, but it cannot do everything. Our goal is to find some things math can do, and to recognize some things math cannot do, at least not yet.

Practice Problems

13.4.1. Take the untreated HIV model (13.18) and derive the initial exponential growth rate of the virion population z. Assume $R_0 > 1$. Also assume x is constant and takes the value β/δ_x. In this regime, find the ratio of the growth rates of the y and z populations.

13.4.2. For the untreated HIV model (13.18) at the equilibrium (x_1, y_1, z_1) of Eq. (13.22), show that the probability p that a virion infects new helper T cells is the reciprocal of the burst size N. Show that increasing the per capita death rate δ_y of the infected population increases p.

Practice Problem Solutions

13.4.1. If $x = \beta/\delta_x$ is constant and y and z are small, then (13.18) reduces to

$$y' = \gamma xz - \delta_y y = \frac{\gamma\beta}{\delta_x} z - \delta_y y \qquad z' = \kappa y - \delta_z z \tag{13.28}$$

Next, compute the derivative matrix,

$$D\vec{F} = \begin{bmatrix} -\delta_y & \dfrac{\gamma\beta}{\delta_x} \\ \kappa & -\delta_z \end{bmatrix}$$

After some simplification we see that the eigenvalues are $\lambda_\pm$ of Eq. (13.23)

$$\lambda_\pm = \frac{-(\delta_y + \delta_z) \pm \sqrt{(\delta_y - \delta_z)^2 + 4R_0\delta_y\delta_z}}{2}$$

The eigenvectors of λ_+ are

$$\begin{bmatrix} y \\ z \end{bmatrix} = \begin{bmatrix} y \\ y\epsilon_+ \end{bmatrix} \text{ where } \epsilon_+ = \frac{\delta_x}{\gamma\beta} \cdot \frac{-(\delta_y + \delta_z) + \sqrt{(\delta_y - \delta_z)^2 + 4R_0\delta_y\delta_z}}{2}$$

Because Eq. (13.28) is a linear 2-dimensional equation, we can apply the methods of Sect. 8.4 to see that the solutions starting on the line through the origin and in the direction of this eigenvector are

$$\begin{bmatrix} y(t) \\ z(t) \end{bmatrix} = \begin{bmatrix} y(0) \\ y(0)\epsilon_+ \end{bmatrix} e^{\lambda_+ t}$$

Both the infected helper T cell population y and the virion population z grow at the rate $e^{\lambda_+ t}$. The ratio of these populations is $z(t)/y(t) = \epsilon_+$. In the yz plane the only fixed point is the origin. Here the trajectories run away to infinity along this line. Of course, before long the x population will change by a substantial amount and this 2-dimensional model no longer is valid.

13.4.2. At equilibrium, on average each infected cell gives rise to one new infected cell. This is how the infected cell population remains constant. An infected cell produces a burst of $N = \kappa/\delta_y$ virions. To get only one new infected cell from this infected cell, only one of these N virions can infect a new cell. So the probability p that one of these virions infects a new cell is $p = 1/N$.

Recall Eq. (13.19), $N=\kappa/\delta_y$. From this we see that $p=\delta_y/\kappa$, and consequently, increasing δ_y increases p.

Exercises

13.4.1. Show that as $R_0 \to 1$ the fixed point (x_1, y_1, z_1) of Eq. (13.24) approaches the fixed point (x_0, y_0, z_0) of Eq. (13.22).

13.4.2. In the model of protease inhibitor treatment of HIV given by Eq. (13.27) suppose the infectious virion population z_i has decayed to 0. Describe the long-term behavior of the noninfectious virus population z_n. Give a reason to support your answer.

13.4.3. Suppose $w' = -\delta w$ and $\delta = 2$. If $w(0) = 1000000$, find the time required for w to drop to 500000. Find the time required for w to drop from 500000 to 250000. Find the time required for w to drop from 250000 to 12500. Explain why you didn't need to do the last two calculations.

13.4.4. For the model (13.25) of HIV treated with a reverse transcriptase inhibitor,
(a) find the fixed point,
(b) calculate the derivative matrix of (13.25), and
(c) find the eigenvalues of the derivative matrix and determine the stability of the fixed point.

13.4.5. For the model (13.27) of HIV treated with a protease inhibitor,
(a) find the fixed point,
(b) calculate the derivative matrix of (13.25), and
(c) find the eigenvalues of the derivative matrix and determine the stability of the fixed point.

13.4.6. For the model (13.27) of HIV treated with a protease inhibitor, in the early stage of the infection suppose x is constant, with the value β/δ_x. Show the resulting 3-dimensional system for y', z_i' and z_n' is linear. Find the eigenvalues of the derivative matrix and show the origin is asymptotically stable. How is this consistent with the plot of Fig. 13.11?

13.4.7. For the untreated HIV model (13.18), suppose $R_0 \gg 1$. Show that for the fixed point (x_1, y_1, z_1) of (13.22), the equilibrium populations of infected cells and of virions, y_1 and z_1 are approximately independent of γ, the probability that a virion infects an uninfected cell.

13.4.8. In the untreated HIV model (13.18), suppose $\delta_y \gg \delta_x$. That is, infected cells die more rapidly than do uninfected cells. (We say the virus is highly *cytopathic*.)
(a) Show that the equilibrium population of infected cells is much smaller than the equilibrium population of uninfected cells before the start of the infection.
(b) Show that the equilibrium population of virions decreases with increasing δ_y. Assume all other parameters are held constant.

13.4.9. In the untreated HIV model (13.18), take $\beta = 100000$, $\delta_x = 0.1$, $\gamma = 0.0000002$, $\kappa = 50$, and $\delta_z = 5$. Plot the y curve for $0 \leq t \leq 100$, for $\delta_y = 0.2, 0.3, \ldots, 0.9$. Comment on the relation between the clearance rate of infected cells and the amplitude of the oscillatory convergence of the y population to its equilibrium value.

13.4.10. In the untreated HIV model (13.18), take $\beta = 100000$, $\delta_x = 0.1$, $\delta_y = 0.5$, $\kappa = 50$, and $\delta_z = 5$. Plot the y curve for $0 \leq t \leq 100$, for $\gamma = 0.0000002$ to 0.0000022 in steps of 0.0000004. Comment on the effect of increasing γ on the maximum peak of infected population, the value of R_0, and the equilibrium value of the infected population.

13.5 IMMUNE RESPONSE DYNAMICS

In this section we'll study how the presence of virions and infected cells initiates an immune response. Specifically, we'll model the dynamics of CTLs, cytotoxic (cell-killing) T lymphocytes. Generally, these are the killer T cells. We know that the presence of infected cells amplifies the production of killer T cells sensitive to the infectious agent, but at what rate? We'll study one model of *CTL response*, the number of virus-specific CTL present at any time, and then mention a few others.

Sketches of some background in the immune system and molecular genetics are in Sects. A.15 and A.22.

The virus population replicates according to the basic virus dynamics model of Sect. 13.4. As before, x denotes the population of uninfected cells, y the population of infected cells, and z the population of virions. In addition to these we add w, the population of CTLs. One model is this modification of (13.18):

$$\frac{dx}{dt} = \beta - \delta_x x - \gamma xz \qquad \frac{dy}{dt} = \gamma xz - \delta_y y - \epsilon yw \tag{13.29}$$
$$\frac{dz}{dt} = \kappa y - \delta_z z \qquad \frac{dw}{dt} = \eta yw - \delta_w w$$

The first three equations are the model (13.18), with the addition of the term $-\epsilon yw$ to the dy/dt equation. This term represents the rate at which CTLs clear infected cells, using the principle of mass action to measure the likelihood of contacts between infected cells and CTLs, and ϵ is the probability that a contact will result in clearance of the infected cell. The per capita CTL clearance rate is δ_w, and the CTL production rate is ηyw, proportional to the number of encounters between infected cells and CTLs (and the principle of mass action, yet again). Note that CTLs are produced only in the presence of infected cells.

We'll start with the fixed points, the intersections of the four nullclines. There are three. First,

$$(x_0, y_0, z_0, w_0) = \left(\frac{\beta}{\delta_x}, 0, 0, 0\right)$$

the equilibrium population of uninfected cells in the absence of any infection and consequently the absence of a CTL response; second,

$$(x_1, y_1, z_1, w_1) = \left(\frac{\delta_y \delta_z}{\gamma\kappa}, \frac{\beta}{\delta_y} - \frac{\delta_x \delta_z}{\gamma\kappa}, \frac{\beta\kappa}{\delta_y \delta_z} - \frac{\delta_x}{\gamma}, 0\right)$$

the equilibrium population of uninfected cells, infected cells, and virions, when a CTL response has not been activated; and third, when the CTL response is activated

$$(x_2, y_2, z_2, w_2) = \left(\frac{\beta\eta\delta_z}{\delta_x \delta_z \eta + \delta_w \gamma\kappa}, \frac{\delta_w}{\eta}, \frac{\delta_w \kappa}{\delta_z \eta}, \frac{\beta\gamma\eta\kappa}{\epsilon(\delta_x \delta_z \eta + \delta_w \gamma\kappa)} - \frac{\delta_y}{\epsilon}\right)$$

Note that (x_0, y_0, z_0, w_0) and (x_1, y_1, z_1, w_1) are the fixed points of (13.22), our untreated HIV model of Sect. 13.4, with no CTL response, $w = 0$, coordinate added.

Why do we need the second fixed point? It's known that the infected cell population must exceed a threshold in order to initiate a CTL response. The model (13.29) captures this threshold behavior in the dw/dt equation:

$$\frac{dw}{dt} = \eta yw - \delta_w w = w(\eta y - \delta_w)$$

So $dw/dt > 0$ only if $y > \delta_w/\eta = y_2$.

Of course, before the CTL response can be activated, the infection must take off, and this requires $R_0 > 1$, where $R_0 = \kappa\gamma\beta/\delta_x\delta_y\delta_z$ is the basic reproductive ratio (13.21) for the untreated infection. When the CTL response is activated, the basic reproductive ratio is

$$R_0' = \frac{\kappa\gamma\beta}{\delta_x(\delta_y + \epsilon w_2)\delta_z}$$

The only difference is that the δ_y factor in the denominator of R_0 is replaced by $\delta_y + \epsilon w_2$. To see why, note that $dy/dt = \gamma xz - \delta_y y$ in (13.18) is replaced by $dy/dt = \gamma xz - (\delta_y + \epsilon w)y$ in (13.29). In the derivation of R_0 (Eq. (13.20)), the factor virions per infected cell is

(virion production rate per infected cell) · (lifetime of an infected cell)

The virion production rate per infected cell still is κ. The average lifetime of an infected cell is the reciprocal of its clearance rate (Prop. 3.2.1), and in the dy/dt equation of (13.29) we see the clearance rate of an infected cell is $\delta_y + \epsilon w_2$: to the natural per capita mortality rate δ_y we add the per capita rate ϵw_2 at which CTLs clear infected cells, when the CTLs are at their equilibrium value. Substituting in the value of w_2, with a bit of algebra we find

$$R_0' = \frac{\kappa\gamma\beta}{(\delta_y + \epsilon w_2)\delta_x\delta_z} = 1 + \frac{\gamma\kappa\delta_w}{\eta\delta_x\delta_z}$$

Because $R_0' > 1$, we see that the infection will not die out.

Before we continue our analysis, let's look at some graphs. On the left of Fig. 13.12 we plot the number of infected cells $y(t)$, along with a horizontal line at height $y_2 = \delta_w/\eta$. We plot this horizontal line for two reasons. One is to show that the $y(t)$ curve converges to y_2 through damped oscillations. We'll find complex eigenvalues of the derivative matrix in a moment. The second reason is to show that whenever $y(t) > y_2$ the CTL response, shown in the right graph of Fig. 13.12, is activated. Note this graph also oscillates, and settles down to w_2, the horizontal line in that graph. How do the oscillations of $y(t)$ compare with those of $w(t)$?

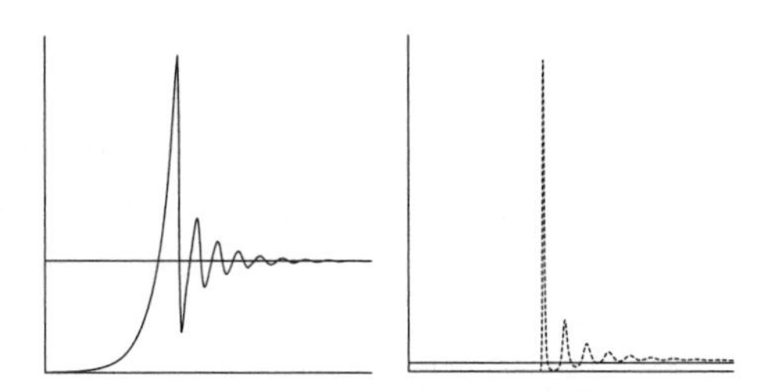

Figure 13.12. Left: $y(t)$. Right: $w(t)$.

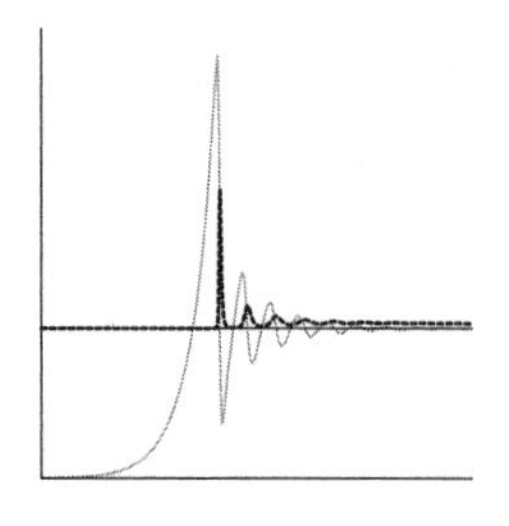

Figure 13.13. Both $y(t)$ and $w(t)$.

To see this, it's easier to plot the graphs of $y(t)$ and $w(t)$ together. We do this in Fig. 13.13. (Mathematica code is in Sect. B.15.) Unlike Fig. 13.12, here both graphs have the same vertical scale, but to make the comparison easier, the $w(t)$ graph has been translated vertically. Note that each rise of of the infected cell population (solid gray curve) above $y = y_2$ causes a rise in the CTL population (dark dashed curve), but after a time lag. These dynamics of the y and w populations are reminiscent of the Lotka-Volterra predator-prey system, but rather than closed orbits, in this CTL model we see convergence to an asymptotically stable fixed point.

The parameters for the graphs of Figs. 13.12 and 13.13 are $\beta = 100000$, $\delta_x = 0.1$, $\gamma = 0.0000002$, $\delta_y = 0.5$, $\epsilon = 0.02$, $\kappa = 50$, $\delta_z = 5$, $\eta = 0.002$, and $\delta_w = 50$. With these values, the fixed points are

$$\begin{array}{llll} x_1 = 156250 & y_1 = 168750 & z_1 = 2.7 \times 10^6 & w_1 = 0 \\ x_2 = 555556 & y_2 = 25000 & z_2 = 399750 & w_2 = 638.9 \end{array}$$

The eigenvalues of the derivative matrix evaluated at the fixed point (x_2, y_2, z_2, w_2) are

$$-5.0477, \; -0.1799, \; -6.6152 \pm 24.2733i$$

Two eigenvalues are negative; the other two are complex with negative real parts. So at these parameters, this fixed point is asymptotically stable with some spiral behavior, hence the wobbles in the graphs of Figs. 13.12 and 13.13.

Another comment about plotting the graphs of Figs. 13.12 and 13.13. We used $x(0) = 1000000$, $y(0) = 0$, $z(0) = 2000$, and $w(0) = 0.0002$. The presence of virions will generate infected cells, regardless of the initial populations. But look at the last equation of (13.29): if $w(0) = 0$, then $w(t) = 0$ for all $t \geq 0$. And we know there must be some killer T cells with appropriate receptors in order to initiate a CTL response.

Also note that the coordinates of the fixed point (x_1, y_1, z_1, w_1) are those of the fixed point (x_2, y_2, z_2, w_2) with w_2 set to 0. For the parameters given, notice that

$$x_2 > x_1, \; y_2 < y_1, \; z_2 < z_1$$

That is, this CTL response

- increases the population of uninfected cells, and
- decreases the population of infected cells and of virions.

This is what we expect evolution would produce.

How general is this? Can we find a simple expression for

$$\frac{\text{equilibrium infected population without CTL response}}{\text{equilibrium infected population with CTL response}} = \frac{y_1}{y_2}$$

in terms of R_0 and R_0'? Yes we can. First, observe that

$$\frac{y_1}{y_2} = \frac{(\beta/\delta_y) - (\delta_x\delta_y/\gamma\kappa)}{\delta_w/\eta} = \frac{\eta(\beta\gamma\kappa - \delta_x\delta_y\delta_z)}{\delta_w\delta_y\gamma\kappa} \tag{13.30}$$

Now if $R_0 < 1$ the infection dies out on its own, and if $R_0' < 1$ the CTL response will stop the infection. These cases are straightforward to understand. What's interesting is the case $R_0 > 1$ and $R_0' > 1$. Then

$$\frac{R_0 - 1}{R_0' - 1} = \frac{\dfrac{\kappa\gamma\beta}{\delta_x\delta_y\delta_z} - 1}{\left(1 + \dfrac{\gamma\kappa\delta_w}{\eta\delta_x\delta_z}\right) - 1} = \frac{\eta(\beta\gamma\kappa - \delta_x\delta_y\delta_z)}{\delta_w\delta_y\gamma\kappa}$$

and consequently

$$R_0' = 1 + \frac{y_2}{y_1}(R_0 - 1) \tag{13.31}$$

For example, in order to achieve a 50% reduction in the equilibrium population of infected cells (that is, $y_2 = y_1/2$), the CTL response reproductive ratio R_0' must equal $(1/2)(R_0 + 1)$. In Exercise 13.5.5 we'll find that Eq. (13.31) implies a relation between η and δ_w, the parameters of the w equation, and the parameters of R_0. This suggests adjustments to the CTL parameters in order to reduce the infected cell population, but the model does not hint at how to achieve these parameter adjustments. Math can take us only so far; biology and chemistry do have many necessary things to say.

Practice Problems

13.5.1. The CTL response can take different forms, for example, $dw/dt = \eta(y) - \delta_w w$, where

$$\eta(y) = \begin{cases} 0 & \text{if } y = 0 \\ \eta_0 & \text{if } y > 0 \end{cases}$$

Here η_0 is a positive constant, the rate of production of CTLs when a CTL response is initiated.

(a) Write the equations for the system with this CTL response.
(b) Find the fixed points (with $y > 0$) of this system when the CTL response is 0 and when the CTL response is η_0.
(c) By comparing the coordinates of these fixed points, assess the effectiveness of the CTL response.

13.5.2. Another form of CTL response is

$$\frac{dw}{dt} = \eta y - \delta_w w$$

With these parameters

$$\beta = 50000,\ \delta_x = 0.05,\ \gamma = 0.0000002,\ \delta_y = 0.5,\ \epsilon = 0.01,$$
$$\kappa = 50,\ \delta_z = 5,\ \eta = 0.001,\ \delta_w = 30$$

(a) Plot the x-, y-, z-, and w- curves for $0 \le t \le 100$.
(b) Find the fixed points of this system. There is an abstract solution to the fixed point equation, but it's a real mess. It is much easier to do this numerically, and considerably simpler to interpret the results.
(c) Find the eigenvalues of the derivative matrix at these fixed points. Use the eigenvalues to interpret the graphs of part (a).

Practice Problem Solutions

13.5.1. (a) The equations for the system with this CTL response are

$$\frac{dx}{dt} = \beta - \delta_x x - \gamma xz \qquad \frac{dy}{dt} = \gamma xz - \delta_y y - \epsilon yw$$
$$\frac{dz}{dt} = \kappa y - \delta_z z \qquad \frac{dw}{dt} = \eta(y) - \delta_w w$$

(b) When the CTL response is 0, the fixed point with $y > 0$ is

$$(x_1, y_1, z_1, w_1) = \left(\frac{\delta_y\delta_z}{\gamma\kappa}, \frac{\beta}{\delta_y} - \frac{\delta_x\delta_z}{\gamma\kappa}, \frac{\beta\kappa}{\delta_y\delta_z} - \frac{\delta_x}{\gamma}, 0\right)$$

When the CTL response is η_0, the fixed point is $(x_2, y_2, z_2, w_2) =$

$$\left(\frac{\delta_y\delta_z}{\gamma\kappa} + \frac{\delta_z\epsilon\eta}{\delta_w\gamma\kappa}, \frac{\beta\delta_w}{\delta_w\delta_y + \epsilon\eta} - \frac{\delta_x\delta_z}{\gamma\kappa}, \frac{\beta\kappa\delta_w}{\delta_z(\delta_y\delta_w + \epsilon\eta)} - \frac{\delta_x}{\gamma}, \frac{\eta}{\delta_w}\right)$$

(c) Remember that all the parameters are positive. Then certainly $x_2 > x_1$, so CTL response increases the equilibrium value of uninfected cells. Next, $y_1 > y_2$ is equivalent to

$$\frac{\beta}{\delta_y} > \frac{\beta\delta_w}{\delta_w\delta_y + \epsilon\eta}$$

which follows by cross-multiplying. Finally, $z_1 > z_2$ is equivalent to

$$\frac{\beta\kappa}{\delta_y\delta_z} > \frac{\beta\kappa\delta_w}{\delta_z(\delta_y\delta_w + \epsilon\eta)}$$

Again, this follows by cross-multiplying. So we see that this CTL response increases the population of uninfected cells, and decreases the populations of infected cells and of virions, but does not eliminate these populations.

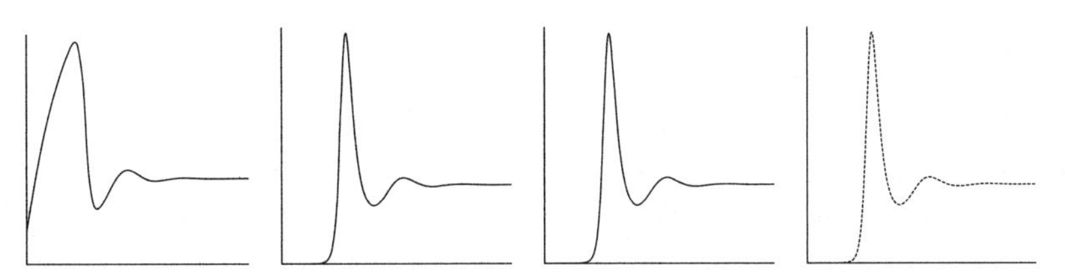

Figure 13.14. Left to right: plots of $x(t)$, $y(t)$, $z(t)$, and $w(t)$.

13.5.2. (a) With these initial values, $x(0) = 50000$, $y(0) = 0$, $z(0) = 2000$, $w(0) = 0.1$, the plots are shown in Fig. 13.14.
(b) We find the fixed points by solving the four nullcline equations simultaneously. One solution has all four coordinates negative, so we'll discard that one. The other two are

$$x_1 = 1000000,\ y_1 = 0,\ z_1 = 0,\ w_1 = 0$$
$$x_2 = 261752,\ y_2 = 70510,\ z_2 = 705104,\ w_2 = 2.35035$$

The first fixed point is the equilibrium value $x = \beta/\delta_x$ of uninfected cells in the absence of infection. The second fixed point is more interesting.
(c) For this CTL response, the derivative matrix is

$$D\vec{F} = \begin{bmatrix} -\delta_x - \gamma z & 0 & -\gamma x & 0 \\ \gamma z & -\delta_y - \epsilon w & \gamma x & -\epsilon y \\ 0 & \kappa & -\delta_z & 0 \\ 0 & \eta & 0 & -\delta_w \end{bmatrix}$$

The eigenvalues of the derivative matrix at the fixed point (x_1, y_1, z_1, w_1) are

$$-30, -6.63104, 1.13104, -0.05$$

and the eigenvalues at the fixed point (x_2, y_2, z_2, w_2) are

$$-29.976, -5.53877, -0.0998887 \pm 0.246526i$$

The one positive eigenvalue for (x_1, y_1, z_1, w_1) is enough to show that fixed point is unstable. Even a tiny bit of virus sends the system sailing away from this fixed point. On the other hand, two of the eigenvalues for (x_2, y_2, z_2, w_2) are negative and the other two are complex with negative real parts, so this fixed point is asymptotically stable. The pair of complex eigenvalues guarantee the wobbles in the graphs of Fig. 13.14.

Exercises

In these problems we'll say CTL3 is the system defined by Eq. (13.29), CTL2 is the system with the dw/dt equation replaced by $dw/dt = \eta y - \delta_w w$ (Practice Problem 13.5.2), and CTL1 is the system with the dw/dt equation replaced by $dw/dt = \eta(y) - \delta_w w$ (Practice Problem 13.5.1).

13.5.1. Take the parameters $\beta = 100000$, $\delta_x = 0.1$, $\gamma = 0.0000002$, $\delta_y = 0.5$, $\epsilon = 0.002$, $\kappa = 80$, $\delta_z = 0.5$, $\eta = 0.002$, $\delta_w = 0.5$. Plot the $y(t)$ curve for CTL1, CTL2, and CTL3 for $0 \le t \le 30$. Describe the asymptotic behavior of $y(t)$. Which model has the lowest equilibrium population of infected cells? Which exhibits the highest maximum before converging to the equilibrium value?

13.5.2. Repeat Exercise 13.5.1 taking $\gamma = 0.0000001$ for $0 \le t \le 30$.

13.5.3. Repeat Exercise 13.5.1 taking $\kappa = 60$ for $0 \le t \le 30$.

13.5.4. For the CTL3 model,
(a) show that the equilibrium population of uninfected cells, x_2, is an increasing function of η.
(b) Show that $\lim_{\eta\to\infty} x_2 = \beta/\delta_x$.
(c) Show that $\lim_{\eta\to\infty} y_2 = \lim_{\eta\to\infty} z_2 = 0$.

13.5.5. Recall that y_1 is the equilibrium infected population without a CTL response and y_2 is the equilibrium infected population with a CTL response. Suppose that $y_2 = y_1/2$. How might δ_w or η be changed to guarantee that $y_2 = y_1/4$?

13.5.6. For the CTL3 model, is it necessarily the case that the activation of the CTL response lowers the equilibrium infected cell population? That is, must $y_2 < y_1$? Give a reason if you think the answer is "Yes"; give an example if you think the answer is "no".

13.5.7. In the CTL1 model with a CTL response active,
(a) show that the equilibrium population of infected cells, y_2, is an increasing function of δ_w.
(b) Show $\lim_{\delta_w\to\infty} y_2 = y_1$.
(c) Interpret large δ_w in terms of the mean lifetime of a CTL cell.
(d) In this model, must y_2 be less than y_1?

13.5.8. In the CTL1 model, suppose y_1 is the equilibrium infected population without a CTL response and y_2 is the equilibrium infected population with a CTL response active.
(a) Find expressions for y_2 and y_1.
(b) Compute $\lim_{\eta \to 0} y_1/y_2$. Explain why this result makes sense.

13.5.9. For CTL1, CTL2, and CTL3 take the parameters $\beta = 100000$, $\delta_x = 0.1$, $\gamma = 0.0000002$, $\delta_y = 0.5$, $\epsilon = 0.002$, $\kappa = 50$, $\delta_z = 5$, $\eta = 0.002$, $\delta_w = 50$.
(a) Find the fixed points with all coordinates positive .
(b) Compute the eigenvalues of the derivative matrix of CTL1, CTL2, and CTL3 at these fixed points. Determine the stability of these fixed points. Comment on the similarity of the eigenvalues for CTL1 and CTL2.

13.5.10. For the CTL3 model, find a relation between z_1/z_2, the ratio of equilibrium virion populations without and with a CTL response, and the basic reproductive ratios R_0 and R_0'. Compare this relationship with that of Eq. (13.31).

13.6 THE HODGKIN-HUXLEY EQUATIONS

The transmission of electrical impulses (action potentials) along axons was studied by Alan Hodgkin and Andrew Huxley in the giant axon of the squid *Loligo*. (A few authors transpose this to the "axon of the giant squid." Typically, *Loligo* are under 2 feet long, hardly giant. On the other hand, *Loligo*'s giant axon is giant, with a diameter of about 0.5 mm, a thousand times the diameter of the axons in our brains.) Their results were published in [172], the last in a series of five papers [173, 169, 170, 171, 172]. Hodgkin and Huxley were awarded the 1963 Nobel Prize in Physiology or Medicine for their brilliant work.

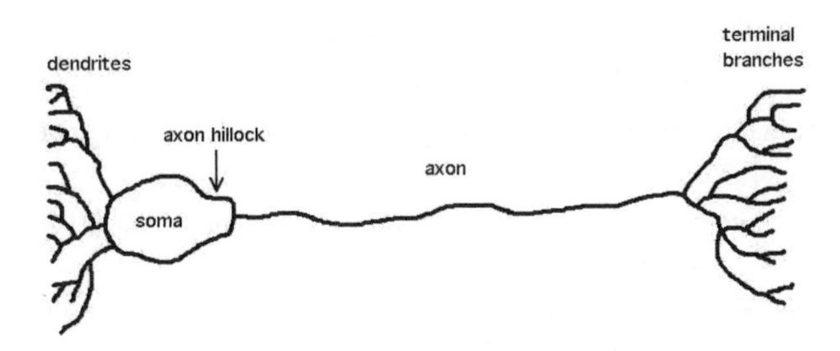

Figure 13.15. A schematic representation of a neuron.

Fig. 13.15 shows a sketch of some of the main components of a neuron. Signals from other neurons are collected in the dendrites and integrated in the cell body, the *soma*. If the integrated signal satisfies a threshold condition, a signal is initiated at the *axon hillock* and sent down the axon to the terminal branches. There the signal is chemically transmitted across the synapses to dendrites of other neurons, where it contributes to integration in those neurons. If the threshold conditions are met, more signals travel down more axons.

Take about 86 billion neurons. (How do we know this number? Make a soup [164] of the brain, take a sample, and count the number of neuron nuclei in the sample [163].) Put them together with just the right connections, a few thousand per neuron [225], and there you are. Really, you. Your memories, your skills, loves, fears, dreams, worries—all are made of patterns of electrochemical impulses traveling through the network of these connected neurons.

Ignoring this gigantic network issue, Hodgkin and Huxley focused on a simpler problem: how do neurons work? What's the mechanism of this magic? Building on earlier work including [74], Hodgkin and Huxley used an experimental technique called a *voltage clamp*. They inserted a tiny electrode down the length of an axon (hence the choice of the *Loligo* axon with its large diameter), held the electrical potential constant, and measured the current flow for different values of the electrical potential. We'll say a bit more about voltage clamps in Sect. A.14.

Before we describe the Hodgkin-Huxley model, we'll list some physiological observations of the functioning of a neuron.

- Relative to its surroundings, the inside of the cell has a lower concentration of Na^+ ions and a higher concentration of K^+ ions. Along the axon, active ion pumps (ADP-powered protein complexes that transport ions across a cell membrane in the direction opposite the ion concentration gradient) maintain a *resting potential* of about -70mV.
- The electrical potential is related to ionic concentrations through the *Nernst equation*

$$V = \frac{kT}{q}\log\left(\frac{\text{outside concentration}}{\text{inside concentration}}\right)$$

  where k is the Boltzmann constant, T is the temperature, and q is the charge of an ion.
- The ions flow through channels in the cell membrane, with separate channels for K^+ and Na^+.
- If the integrated impulses exceed a threshold (about -20mV), then the axon Na^+ channels open quickly. The flow of Na^+ into the cell further increases the potential, until it becomes positive.
- More slowly, the K^+ channels open, allowing the K^+ ions to flow out of the cell, lowering the potential.
- Also slowly, the Na^+ blockers begin to act, reducing the influx of Na^+ ions.

- The potential undershoots the resting potential, then slowly recovers. See Fig. 13.16.
- At the site adjacent to where these events occurred, the potential exceeds the threshold so the process occurs at this location.
- This produces a *traveling wave* spike down the axon.

Based on their observations and measurements, Hodgkin and Huxley designed a circuit analogue of the electrical activity of the axon. Applying standard circuit theory laws, they then derived a family of equations to model the change in electrical potential. By careful experiments and numerical simulations they determined the factors in the functional dependence of the ionic currents. This completed their derivation of the Hodgkin-Huxley equations. Then they tested the predictions of their equations against their experiments.

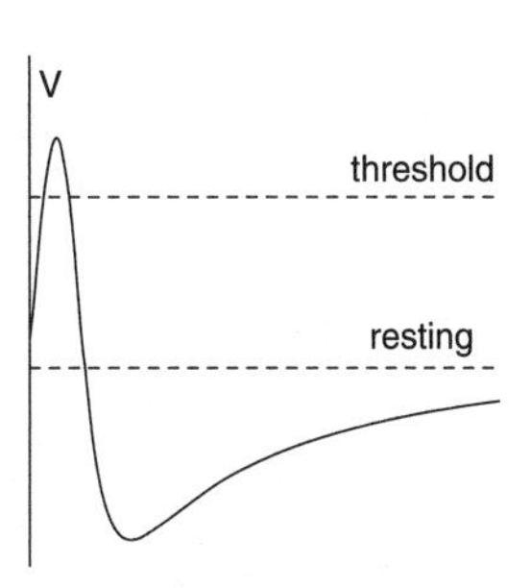

Figure 13.16. Schematic representation of a firing potential.

At the simplest level, the Hodgkin-Huxley equations are the relation of the current flowing through transmembrane channels to the rate of change of the electrical potential difference across the membrane,

$$I = C\frac{dV}{dt} + g_{Na}(V)(V - V_{Na}) + g_K(V)(V - V_K) + g_L(V - V_L) \qquad (13.32)$$

where C is the capacitance of the membrane; $g_{Na}(V)$, $g_K(V)$, and $g_L(V)$ are the sodium, potassium, and "leakage" ("leftover" makes sense, too) ionic conductances, all functions of V; and V_{Na}, V_K, and V_L are the sodium, potassium, and leakage resting potentials determined by the Nernst equation.

By extensive careful experimentation, Hodgkin and Huxley found the ionic conductances take these forms

$$g_{Na}(V) = \bar{g}_{Na}m^3h \qquad \text{and} \qquad g_K(V) = \bar{g}_K n^4$$

where $\bar{g}_{Na}$, $\bar{g}_K$, and g_L are constants, n is the activation of K^+ channels, m is the activation of Na^+ channels, and h is the inactivation of Na^+ channels. Moreover, the variables m, n, and h satisfy

$$\frac{dn}{dt} = \alpha_n(V)(1-n) - \beta_n(V)n \tag{13.33}$$

$$\frac{dm}{dt} = \alpha_m(V)(1-m) - \beta_m(V)m \tag{13.34}$$

$$\frac{dh}{dt} = \alpha_h(V)(1-h) - \beta_h(V)h \tag{13.35}$$

with

$$\alpha_n(V) = \frac{0.01(V+10)}{e^{(V+10)/10}-1} \qquad \beta_n(V) = 0.125e^{V/80}$$

$$\alpha_m(V) = \frac{0.1(V+25)}{e^{(V+25)/10}-1} \qquad \beta_m(V) = 4e^{V/18}$$

$$\alpha_h(V) = 0.07e^{V/20} \qquad \beta_h(V) = \frac{1}{e^{(V+30)/10}+1}$$

In the absence of an applied current, Eq. (13.32) becomes

$$\frac{dV}{dt} = -\frac{1}{C}\Big(g_{Na}(V)(V-V_{Na}) + g_K(V)(V-V_K) + g_L(V-V_L)\Big) \tag{13.36}$$

The Hodgkin-Huxley equations are Eq. (13.33) through Eq. (13.36). If an external current I is applied, then Eq. (13.36) is modified by subtracting I inside the large parentheses on the right-hand side.

Simulations reveal that for brief time periods, n and h vary more slowly than V and m. Fitzhugh [117] applied geometric techniques to the (V,m) plane, holding n and h constant at their equilibrium values. We saw his results in Sect. 7.6. This approximation is instructive not just for what it reveals about the Hodgkin-Huxley equations, but also because it draws attention to the fact that some systems consist of subsystems that operate on different time scales, and focusing on a single time scale can reveal a lot.

How did Fitzhugh do this? We'll use the parameters that Fitzhugh (page 870 of [117]) and Hodgkin and Huxley (Table 3 of [172]) used:

$$\bar{g}_{Na} = 120, \bar{g}_K = 36, g_L = 0.3, V_{Na} = -115, V_K = 12, V_L = -10.6, C = 1$$

Then solving the Hodgkin-Huxley equations numerically we find that n and h approach limiting values $n_\infty \approx 0.318$ and $h_\infty \approx 0.596$. Now we can find the m-nullcline equation by setting $dm/dt = 0$ in Eq. (13.34) and solving for m,

$$m_1(V) = \frac{\alpha_m(V)}{\alpha_m(V) + \beta_m(V)} \tag{13.37}$$

For the V-nullcline, we can set $dV/dt = 0$ in Eq. (13.36) and solving for m,

$$m_2(V) = \left(\frac{-\bar{g}_K n_\infty^4 (V-V_K) - g_L(V-V_L)}{\bar{g}_{Na} h_\infty (V-V_{Na})}\right)^{1/3} \tag{13.38}$$

On the left of Fig. 13.17 we see a plot of the m-nullcline (dashed) and V-nullcline (solid). One fixed point is on the upper left, two are on the lower right. To make these two clearer, the right image of the figure is a magnification of the lower right of the left graph. We see three fixed points, and because $0 \leq m \leq 1$ in the Hodgkin-Huxley equations, these three are the only fixed points. We'll be content with numerical expressions for these fixed points. Code to find these is in Appendix B.16.

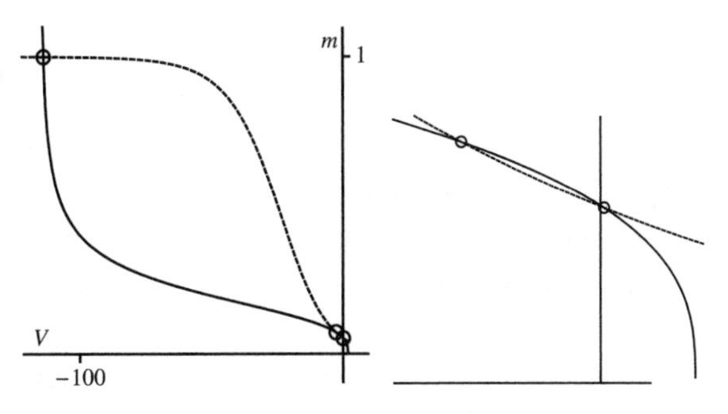

Figure 13.17. The m and V nullclines.

$$p_1 = (-113.9, 0.999),\ p_2 = (-2.685, 0.072),\ p_3 = (0.071, 0.051)$$

In Practice Problem 13.6.1 and Exercise 13.6.1 we'll determine the type and stability of these fixed points.

Time for some pictures. In Fig. 13.18 we see three graphs for the Hodgkin-Huxley equations with the parameters listed above (the list starts with $\bar{g}_{Na} = 120$). Also, all have initial values $V(0) = 10$, $m(0) = 0.015$, $n(0) = 0.577$, and $h(0) = 0.874$. The applied currents (in gray) are

$$I_1(t) = 0 \text{ (left)},\ I_2(t) = \begin{cases} 0 & \text{for } t < 6 \\ 3 & \text{for } 6 \leq t \leq 8 \\ 0 & \text{for } t > 8 \end{cases} \text{ (center)},\ I_3(t) = \sin(t) \text{ (right)}$$

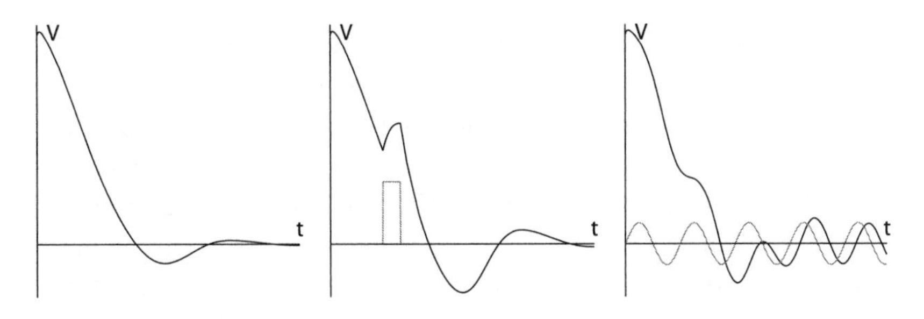

Figure 13.18. Three V vs. t graphs with applied current I_1 (left), I_2 (center), and I_3 (right).

In order to plot these graphs, we have solved the full set of Hodgkin-Huxley equations, not just the reduced 2-dimensional (V, m) system. In the left we see that V undergoes an oscillatory decay to 0, or close to 0. In the center we see that the current pulse, plotted in gray, temporarily reverses the initial decline in V, but as soon as the pulse stops, we see a return to the oscillatory decay of the left graph. In the right, we see the V graph initially decays but then begins to oscillate sinusoidally, though as with the predator-prey equations, here V and I are out of phase. Will these graphs eventually become in phase? Will the amplitude of the V graph eventually match that of the I graph? We—that is, you—may explore these questions in the exercises.

Figure 13.19. Left: a V versus t graph. Center: an m versus V graph. Right: two m versus V plots with slightly different starting points. The scales of the center and right graphs are different: the elliptical blob in the lower right of the right image is approximately the location and extent of the center figure.

In Fig. 13.19 we use a more complex driving current,

$$I_4(t) = \sin(\pi t) + \sin(\pi t/\sqrt{2})$$

two sine functions whose frequencies are not rational multiples of one another. (We included the factor of π to compress the time scale of the plots.) Here the parameters are the same, except now $C = 1.9$. The left graph is a plot of V versus t. This appears to be much more complicated than our earlier plots of V versus t. The center graph is a plot of $(V(t), m(t))$. This approach is common in dynamical systems. If we plot $(V(t), m(t))$ for the Hodgkin-Huxley equations with driving current I_3 (See Exercise 13.6.4), the graph is much simpler. Sometimes complex plots of this type are taken to suggest chaotic dynamics. But recall from Sect. 2.1 that sensitivity to initial conditions is a feature of chaos. The right graph of Fig. 13.19 shows short time plots from two initial points, identical except that $m(0) = 0.045$ for the dashed curve, and

$m(0) = 0.046$ for the solid (gray) curve. Initially, the curves stay nearby and make a large excursion to the upper left of the graph, then both converge to this blob and there appear to diverge. To show some of the details of this divergence, we plot the curves only after they arrive at the blob. This apparent divergence suggests, but does not prove, chaotic behavior.

To illustrate the range of behaviors of the Hodgkin-Huxley equations, Kazuyuki Aihara and Gen Matsumoto [2] show that in certain parameter ranges chaotic behavior is observed for periodically stimulated Hodgkin-Huxley equations, and the physical neurons (from the squid *Doryteuthis bleekeri*) they model. Given the complexity and nonlinearity of the Hodgkin-Huxley equations, it is no surprise that they can exhibit chaotic dynamics. But still it does take careful work to establish the existence of chaos. Examples include [24, 127, 148, 210, 321, 330].

The complexity of these equations gives a small hint of the work involved in deriving them. Because this is a 4-dimensional system, we've seen that visualizing the solution can take some effort. We must plot four separate graphs and seek relations between the curves. This is easy enough now, but Hodgkin and Huxley did their numerical work on a desk calculator, a mechanical device with an electric motor to drive the gears and a front panel covered with buttons. Now, almost 70 years later, writing the code to generate the figures of this section took me about 15 minutes; generating the figures took seconds. How far we've come. I wonder what's next.

Practice Problems

13.6.1. (a) Write Eqs. (13.34) and (13.36) for $n = n_\infty$ and $h = h_\infty$.
(b) Compute the derivative matrix of the dm/dt and dV/dt system.
(c) Determine the stability and type of the fixed point $p_3 = (0.073, 0.051)$.

13.6.2. For the full Hodgkin-Huxley system, Eqs. (13.33)–(13.36), with the usual parameters

$$\bar{g}_{Na} = 120, \bar{g}_K = 36, g_L = 0.3, V_{Na} = -115, V_K = 12, V_L = -10.6$$

set $C = 0.1$. Take the driving current to be a periodic square pulse that begins with $I_5(t) = 0.1$ for $0 \leq t \leq 10$ and $I_5(t) = 0$ for $10 < t < 20$.
(a) Plot the $(V(t), m(t))$ curve for $50 \leq t \leq 250$.
(b) Plot the $V(t)$ and $m(t)$ curves for the same t-range.
(c) Can you explain some features of the graph of (a) from those of the graphs of (b)? "No," is not the answer I'm expecting here.

Practice Problem Solutions

13.6.1. (a) We see that Eq. (13.34) is unchanged, but Eq. (13.36) is modified by $n = n_\infty$ and $h = h_\infty$, so the system becomes

$$\frac{dm}{dt} = \alpha_m(V)(1-m) - \beta_m(V)m$$
$$\frac{dV}{dt} = \frac{-1}{C}\left(\bar{g}_{Na}m^3h_\infty(V-V_{Na}) + \bar{g}_K n_\infty^4(V-V_K) + g_L(V-V_L)\right)$$

(b) The derivative matrix is

$$D\vec{F} = \begin{bmatrix} -\alpha_m - \beta_m & \alpha_m'(1-m) - \beta_m' m \\ \frac{-1}{C}\bar{g}_{Na}3m^2h_\infty(V-V_{Na}) & \frac{-1}{C}\left(\bar{g}_{Na}m^3h_\infty + \bar{g}_K n_\infty^4 + g_L\right) \end{bmatrix}$$

where

$$\alpha_m' = \frac{-0.01Ve^{(V+25)/10} - 0.15e^{(V+25)/10} - 0.1}{\left(e^{(V+25)/10} - 1\right)^2} \quad \text{and} \quad \beta_m' = \frac{2}{9}e^{V/18}$$

(c) At $V = 0.073$, we find $\alpha_m' = -0.015$ and $\beta_m' = 0.223$. Then at $m = 0.051$ and $V = 0.073$ the derivative matrix is

$$\begin{bmatrix} -4.239 & -0.026 \\ -64.219 & 0.659 \end{bmatrix}$$

The eigenvalues are $\lambda = -4.559$ and $\lambda = 0.978$, so this fixed point is a saddle point.

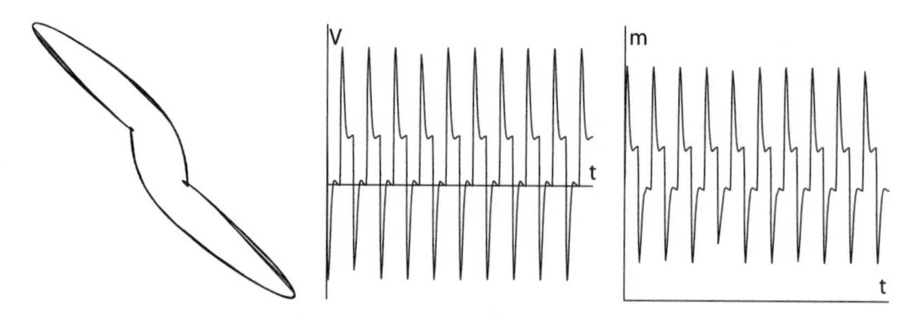

Figure 13.20. Left to right: plots of the $(V(t), m(t))$, $V(t)$, and $m(t)$ curves for Practice Problem 13.6.2. The scales of V and m are significantly different.

13.6.2. In Sect. B.16 we'll code the periodic current I_5.
(a) A plot of the $(V(t), m(t))$ curve is shown on the left of Fig. 13.20.
(b) Plots of the $V(t)$ and $m(t)$ curves are shown in the center and the right of Fig. 13.20.

(c) The $V(t)$ and $m(t)$ plots have identical t-axes, so we can superimpose the graphs to see in what way the graphs interact. The small wiggles that look like cubic curves in both the V and m graphs line up in time and account for the two small wiggles of the (V,m) graph. At other locations, one of V and m decreases while the other increases, accounting for the two large loops in the (V,m) graph.

Exercises

13.6.1. Determine the stability and type of fixed point of the system of Practice Problem 13.6.1 at the fixed points $p_1 = (-113.9, 0.999)$ and $p - 2 = (-2.685, 0.072)$.

13.6.2. (a) How does the V-nullcline equation (13.38) change if a constant current $I = 1$ is added to Eq. (13.36) of the Hodgkin-Huxley equations?
(b) Find the fixed points.

13.6.3. Determine the stability and type of the fixed points found in Exercise 13.6.2.

13.6.4. Again assume $n = n_\infty$ and $h = h_\infty$ are constant. Plot trajectories in the $V - m$ plane when the current I_3 is added to the dV/dt equation. Do your graphs support the existence of a limit cycle for this system?

13.6.5. Repeat the steps of Practice Problem 13.6.2 for $100 \leq t \leq 300$, with
(a) $C = 0.01$, and with
(b) $C = 0.001$.

13.6.6. In the Hodgkin-Huxley equation (13.36), change $g_{Na}(V) = \bar{g}_{Na} m^3 h$ to $g_{Na}(V) = \bar{g}_{Na} m^4 h$. Use the parameters $\bar{g}_{Na} = 120$, $\bar{g}_K = 36$, $g_L = 0.3$, $V_{Na} = -115$, $V_K = 12$, $V_L = -10.6$, and $C = 1$. Suppose $h = h_\infty \approx 0.596$ and $n = n_\infty \approx 0.318$.
(a) Plot the V-nullcline and the m-nullcline.
(b) Find numerical values of the coordinates of the fixed points.
(c) Compute the derivative matrix.
(d) Determine the stability and type of each fixed point.

13.6.7. Continue with the system of Exercise 13.6.6. To the dV/dt equation add the current I_5.
(a) Plot the $(V(t), m(t))$ graph for $100 \leq t \leq 300$.
(b) Plot the curves $V(t)$ and $m(t)$ for $100 \leq t \leq 300$.

13.6.8. Repeat Exercise 13.6.7 using the current I_3 instead of I_5.

13.6.9. Compare the $(V(t), m(t))$, $V(t)$, and $m(t)$ plots for the Hodgkin-Huxley equations (13.33)–(13.36) to those of the system with $g_{Na}(V) = \bar{g}_{Na} m^4 h$. Use the parameters $\bar{g}_{Na} = 120$, $\bar{g}_K = 36$, $g_L = 0.3$, $V_{Na} = -115$, $V_K = 12$, $V_L = -10.6$, and $C = 1$ for both systems. Use I_5 of Practice Problem 13.6.2. Do you see significant differences between these plots?

13.6.10. To the Hodgkin-Huxley equations (13.33)–(13.36) with parameters $\bar{g}_{Na} = 120$, $\bar{g}_K = 36$, $g_L = 0.3$, $V_{Na} = -115$, $V_K = 12$, $V_L = -10.6$, and $C = 1$, add the current $I_6(t) = 0$ for $0 \leq t < 10$ and for $20 < t \leq 60$, and $I_6(t) = 1$ for $10 \leq t \leq 20$.
(a) On the same graph plot $n(t)$ and $I_6(t)$. On the same graph plot $m(t)$ and $I_6(t)$. On the same graph plot $h(t)$ and $I_6(t)$. On the same graph plot $V(t)$ and $I_6(t)$. For all graphs take $0 \leq t \leq 60$.
(b) For which plots, n, m, h, and V, does the start and stop of the current pulse appear to generate abrupt changes in the direction of the plot?

13.7 CHAOS IN PREDATOR-PREY SYSTEMS

The Poincaré-Bendixson theorem guarantees autonomous differential equations in the plane cannot exhibit chaotic solutions. Familiar now is the presence of chaos in some systems of nonlinear differential equations of 3 or more dimensions. Although many think that modeling chaotic motion is contemporary mathematics, as we mentioned in Sect. 2.4, many of these ideas are present in Poincaré's work on celestial mechanics [286] from the 1890s, and these ideas were rediscovered several times in the 20th century [152, 53, 217, 227].

The bad news of chaos is that it signals the impossibility of long-term prediction: inevitable tiny uncertainties in the initial state eventually will give rise to a wide range of future states at a given time. This is bad enough, but in Sect. 13.8 we'll explore chaotic behavior in the Duffing equation, a system that exhibits a different kind of unpredictabilty: final state sensitivity. That is, trajectories eventually converge to one of several stable fixed points, but to which fixed point depends very delicately on the starting point.

In Sect. 13.9 we'll give an example in which chaotic behavior is clinically beneficial. Despite the confusion some authors have about this issue, chaos is distinct from randomness. Robert Devaney [83] characterizes chaos by three properties. We mentioned these in Sect. 2.4 and recall them here.

Sensitivity to initial conditions: tiny changes in initial values can result in trajectories that eventually diverge.
Periodic trajectories are dense: arbitrarily close to every trajectory is a periodic trajectory.
Mixing: given any two regions in phase space, a trajectory in one region eventually enters the other region.

That the $r = 4$ logistic map $L(x) = rx(1 - x)$ exhibits these three features is an instructive, and fairly involved, result given in Chapter 1 of [83], for example. We sketched a simpler example in Sect. 2.4.

The good news of chaos is a consequence of the second and third properties: chaos is filled with order, and regardless of the current state, a small perturbation can redirect the trajectory to approach a periodic trajectory. You may think this idea for temporarily directing a chaotic system to a periodic trajectory is like those silly "economics" problems in calculus 1—if we suppose the cost of producing x units is the function $C(x)$ and the revenue from selling x units is the function $R(x)$, at what level should production be set in order to maximize profit?—that left you wondering, "Wait, how in the world do they know those functions?" If that's what you think, don't worry. Remarkably, Edward Ott, Celso Grebogi, and James Yorke found a way [267] to implement this process without any underlying model of the dynamics. Their method doesn't need equations that describe the system. A sequence of measurements of the system are enough. We'll see how to do this in Sect. 13.9.

In this section, we'll see a visual way to recognize chaos in differential equations, specifically in a system with one predator and two prey species. In Sects. 7.3 and A.4 we see that the trajectories of one predator–one prey systems often consist of cycles. This makes sense: the predator population should lag the prey population. As the prey population increases, predators capture more prey, giving more energy for the production of new predators. As the predator population rises, the prey are over hunted and the prey population drops. This decrease in the predator's food supply causes the predator population to drop. Fewer predators mean fewer prey are captured, so the prey population increases, and we've returned to the start of this description.

What isn't clear from this qualitative sketch is whether the predator and prey populations will cycle, their trajectories forever chasing one another, or will spiral in to a stable equilibrium. The clever trick presented in Sect. A.4 showed the trajectories are cycles. While these are not the same as limit cycles detected by the Poincaré-Bendixson theorem of Sect. 9.7, they are another example of periodic behavior in a 2-dimensional system. Converge to a fixed point, diverge to infinity, cycle, or converge to a cycle pretty much exhausts the list of available dynamics for 2-dimensional differential equations. ("Pretty much" because there can be trajectories that start near a saddle point in the unstable direction and end at a saddle point in the stable direction. See Exercise 13.7.1.) With more than two species can we see more complex behaviors? Yes, indeed.

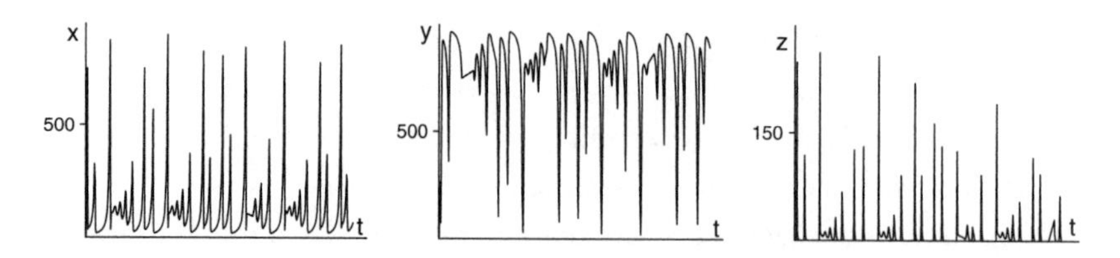

Figure 13.21. Population curves for Gilpin's model (13.39).

Michael Gilpin [137] modified the predator-prey system to include a second prey species, and logistic interactions between species competing for the same resources. Denoting prey populations by x and y, and the predator population by z, Gilpin's model is

$$\begin{aligned} x' &= r_1 x - x(a_{11}x + a_{12}y + a_{13}z) \\ y' &= r_2 y - y(a_{21}x + a_{22}y + a_{23}z) \\ z' &= r_3 z - z(a_{31}x + a_{32}y + a_{33}z) \end{aligned} \tag{13.39}$$

In Fig. 13.21 we plot the population curves $x(t)$, $y(t)$, and $z(t)$ for $0 \leq t \leq 2000$ using the parameters of (13.40). Note the range of x- and y-values is about between 0 and 1000, while the range of z-values is about between 0 and 300. Also note all three populations rise and fall abruptly.

$$\begin{array}{llll} r_1 = 1 & a_{11} = 0.001 & a_{12} = 0.001 & a_{13} = 0.01 \\ r_2 = 1 & a_{21} = 0.0016 & a_{22} = 0.001 & a_{23} = 0.001 \\ r_3 = -1 & a_{31} = -0.005 & a_{32} = -0.0005 & a_{33} = 0 \end{array} \tag{13.40}$$

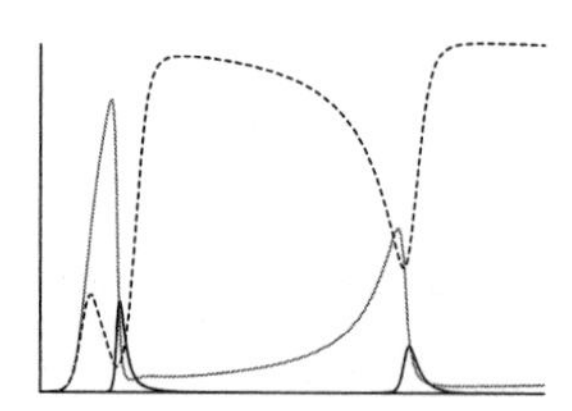

Figure 13.22. Short population curves for (13.39).

Certainly the graphs of Fig. 13.21 are too cramped horizontally for us to see relations between the curves. To help understand how these populations dance together, in Fig. 13.22 we plot these curves for a shorter time range, $0 \leq t \leq 100$, the x-curve in gray, the y-curve dashed, and the z-curve in solid black. We see that as long as the predator population (z) is low, the prey populations rise. When the predator population rises, the prey populations fall, x more rapidly than y (see Exercise 13.7.2). Falling prey populations lead to a falling predator population. The y population recovers more quickly than the x population, and the predator population rises more in response to an increase in the x population than to an increase in the y population because $|a_{31}| > |a_{32}|$. More can be read from this figure and the form (13.39) of Gilpin's model, but this is enough for now.

In Fig. 13.23 we plot a trajectory $(x(t), y(t), z(t))$ for $0 \leq t \leq 2000$. This appears to be much more complicated than a simple limit cycle. The trajectories spiral toward the small x and z, large y corner, but then are thrown back down to small y, only to spiral back toward large y. Closer scrutiny reveals that the trajectories do not lie on a surface in 3-dimensional space, even though a quick glance suggests they lie on a cone. In fact, far from what appears to be the apex of the cone, nearby trajectories are stretched apart and then folded over near the apex. Which sheet of the fold the trajectories follow depends delicately on where they cross the fold. This "stretching and folding" was a commonly invoked description of the mechanism of chaos in the early 1980s. Mathematica code to generate Fig. 13.23 is in Sect. B.17.

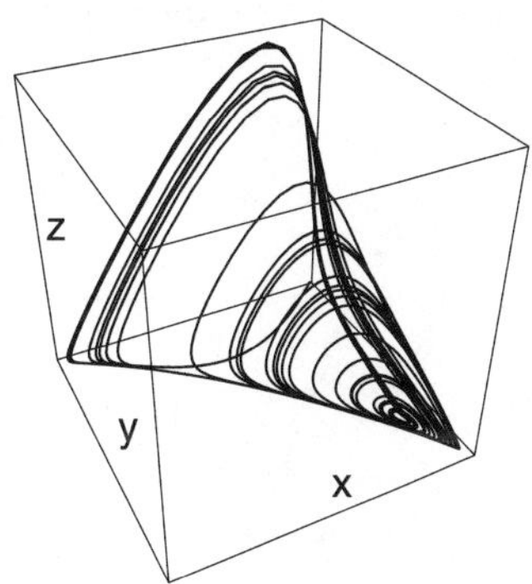

Figure 13.23. A trajectory for (13.39).

One more comment. Recall that one of the features of chaos is that periodic trajectories are dense. Periodic trajectories are simple closed curves. Recall the term "closed" means that the trajectory ends where it begins, so you can run round and round a closed trajectory forever, unless a fixed point lies in the curve. Also recall the term "simple" means that the trajectory does not cross itself. These can be circles or ellipses or warped and bumpy circles, or knots. Look up "knot (mathematics)" on Wikipedia to see some examples. The trajectory we see in Fig. 13.23 suggests there may be knotted periodic trajectories for Gilpin's model. Joan Birman and Robert Williams [42] proved that the chaotic Lorenz weather model has knotted periodic trajectories, so Gilpin's model may as well. Complications within complications, but we won't pursue this any further.

The lesson to take so far is that while a two-species predator-prey system is completely predictable, this cannot necessarily be said for three-species systems. To be sure, if we know the parameters and the starting values exactly, and make no mistakes in our calculations, we (or more likely, our computers) can draw the trajectories until we get bored. So what's the problem?

The problem is that we don't know the model parameters or the starting values exactly. Every physical measurement must of necessity involve some uncertainty. What about integer variables, the number of rabbits in the population, for example. Surely if you count 215 rabbits, there can't really be 215.73 rabbits. True, but how do you know you counted all the rabbits? Did you quietly look under every bush and rock? Did you go over the whole countryside with ground-penetrating radar, looking for rabbit burrows? And if you did, I want to know who funded this strange project, a zero-error rabbit census. I've got a few proposals they might consider.

Every measurement has uncertainties. For some systems these uncertainties don't matter, don't change the prediction by much more than the magnitude of the initial uncertainty. For example, in the one-predator-species, one-prey-species model, small errors have the effect that instead of going along one cycle, the system follows a nearby cycle that stays about the same distance from the point on the original trajectory.

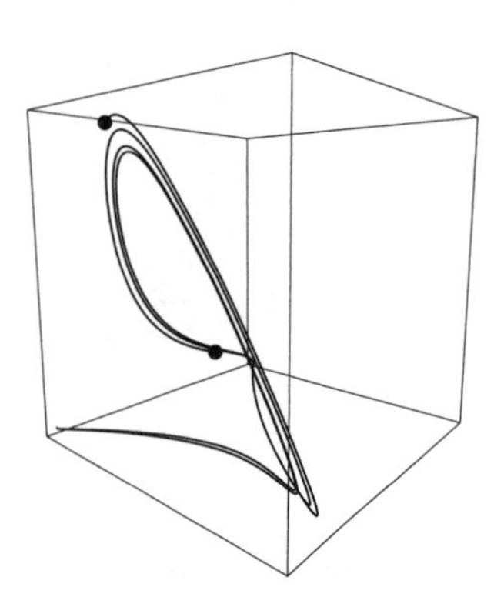

Figure 13.24. Sensitivity to initial conditions for (13.39).

In contrast, Fig. 13.24 shows two trajectories, starting from nearby initial points, $(x(0), y(0), z(0)) = (1,1,1)$ and $(x(0), y(0), z(0)) = (1.1,1,1)$, for $0 \leq t \leq 100$. The trajectories stay nearby for a while, but then diverge. The dots mark the $t = 100$ points for the two trajectories. We can predict for some time, but not forever. A rough measure of the "amount of chaos" might be the prediction horizon, the duration over which small uncertainties have small effects.

Adding one more prey species can open up dynamics much more complex than the neat periodic trajectories we see in the Lotka-Volterra system. While 2 dimensions are well understood, in so many ways 3 dimensions remain a mystery.

Practice Problems

13.7.1. (a) Find the fixed points of Gilpin's model (13.39) with parameters (13.40).
(b) Compute the derivative matrix of (13.39).
(c) Determine the stability of the fixed point $(0,0,0)$.

13.7.2. In Gilpin's model (13.39) and (13.40),
(a) change a_{21} from 0.0016 to 0.0017. Does this system look chaotic? Plot the population curves to make your assessment.
(b) Change a_{23} from 0.001 to 0.003. Does this system look chaotic?

13.7.3. (a) Find the fixed points of both systems of Practice Problem 13.7.2.
(b) If any fixed point has all three coordinates positive, find the eigenvalues of the derivative matrix evaluated at that fixed point.
(c) Comment on the eigenvalues of the fixed point and the perceived presence or absence of chaos for the system with this fixed point.

Practice Problem Solutions

13.7.1. (a) The fixed points of Gilpin's model are the solutions of

$$\begin{aligned}
0 &= r_1 x - x(a_{11}x + a_{12}y + a_{13}z) = x - x(0.001x + 0.001y + 0.01z) \\
0 &= r_2 y - y(a_{21}x + a_{22}y + a_{23}z) = y - y(0.0016x + 0.001y + 0.001z) \\
0 &= r_3 z - z(a_{31}x + a_{32}y + a_{33}z) = -z - z(-0.005x - 0.0005y)
\end{aligned}$$

Numerical solutions with all coordinates non-negative are $(x(0), y(0), z(0)) = (0,0,0)$, $(1000,0,0)$, $(0,1000,0)$, $(200,0,80)$, and $(120,800,8)$.
(b) The derivative matrix is

$$D\vec{F}(x,y,z) = \begin{bmatrix} m_{11} & -a_{12}x & -a_{13}x \\ -a_{21}y & m_{22} & -a_{23}y \\ -a_{31}z & -a_{32}z & m_{33} \end{bmatrix}$$

where

$$\begin{aligned}
m_{11} &= r_1 - 2a_{11}x - a_{12}y - a_{13}z \\
m_{22} &= r_2 - a_{21}x - 2a_{22}y - a_{23}z \\
m_{33} &= r_3 - a_{31}x - a_{32}y - 2a_{33}z
\end{aligned}$$

(c) At the fixed point $(0,0,0)$ the derivative matrix is

$$D\vec{F}(0,0,0) = \begin{bmatrix} r_1 & 0 & 0 \\ 0 & r_2 & 0 \\ 0 & 0 & r_3 \end{bmatrix} = \begin{bmatrix} 1 & 0 & 0 \\ 0 & 1 & 0 \\ 0 & 0 & -1 \end{bmatrix}$$

The eigenvalues are $\lambda = 1, 1, -1$ so the origin is an unstable fixed point.

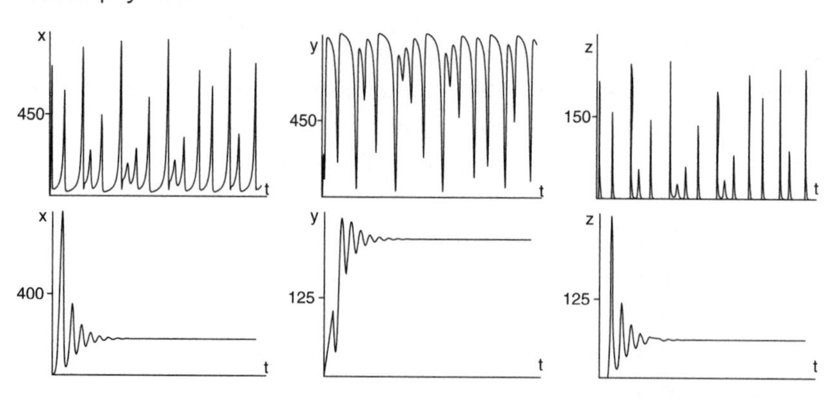

Figure 13.25. Population curves for Practice Problem 13.7.2. Top row plots are for the parameters of (a); bottom row for (b).

13.7.2. The first row of Fig. 13.25 shows the x, y, and z population curves for the parameters of (a), with $0 \leq t \leq 1000$ and with initial values $x(0) = y(0) = z(0) = 1$. We see low values for x and z punctuated by excursions to high values, and high values of y punctuated by excursions to low values. The times of the excursions don't appear to follow any obvious pattern, so it's plausible this system exhibits chaotic dynamics. There are quantitative ways to test this. Perhaps the most familiar is called the Liapunov exponent, which measures the average divergence of nearby trajectories. We sketched this for the logistic map in Sect. 2.4 and describe it for differential equations in sect. A.19. But for now, we'll be content to see that the population curves do not converge to a fixed point or exhibit a periodic pattern.

The second row of Fig. 13.25 shows the x, y, and z population curves for the parameters of (b), with $0 \leq t \leq 200$ and with initial values $x(0) = y(0) = z(0) = 1$. We used the shorter time span to make more apparent the oscillatory convergence to the fixed point. And yes, here the graphs are all we need: they certainly show convergence to a fixed point.

13.7.3. (a) The fixed points of Gilpin's model with parameters (a) are

$$(0,0,0),(0,1000,0),(1000,0,0),(200,0,80),(121.622,783.784,9.45946),$$
$$(0,2000,-1000);$$

the fixed points of Gilpin's model with parameters (b) are

$$(0,0,0),(1000,0,0),(0,333.33,0),(200,0,80),(178.182,218.182,60.3636),$$
$$(0,2000,-5000),(1428.57,-428.571),(0,2000,-5000).$$

For both sets of parameters, only one fixed point has all coordinates positive. For (a) it's $P_a = (121.622, 783.784, 9.45946)$, for (b) it's $P_b = (178.182, 218.182, 60.3636)$.

(b) With the parameters of (a), the eigenvalues of the derivative matrix evaluated at P_a are $\lambda = -0.8384, 0.1624 \pm 0.1199i$. With the parameters of (b), the eigenvalues of the derivative matrix evaluated at P_b are $\lambda = -0.6454, -0.03914 \pm 0.7116i$.

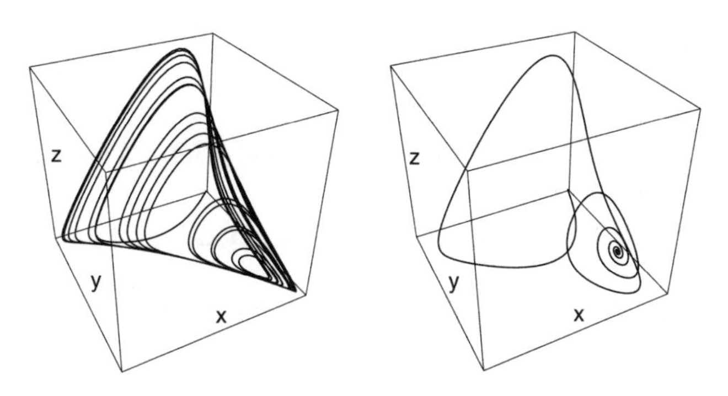

Figure 13.26. Trajectories for Practice Problem 13.7.2. Parameters of (a) on the left, parameters of (b) on the right.

(c) For (a) the complex eigenvalues have positive real parts, and the real eigenvalue is negative. This means that trajectories move in toward the fixed point along the eigenvector of the negative eigenvalue, then spiral away, only to return again and again. The left image of Fig. 13.26 exhibits these features, consistent with chaotic dynamics.

For (b) the complex eigenvalues have negative real parts, and the real eigenvalue is negative. This guarantees the fixed point is asymptotically stable; The complex pair shows the trajectories spiral in to the fixed point. No chaos here.

Exercises

13.7.1. For the system $x' = x^2 - 1, y' = -xy$

(a) Show the fixed points are $(\pm 1, 0)$.

(b) Show both fixed points are saddle points.

(c) Show these saddle points are connected by a straight trajectory.

13.7.2. Referring to Fig. 13.22, Eq. (13.39), and parameters (13.40), explain why the x population falls more than the y population when the z population rises.

13.7.3. In Gilpin's model (13.39) and (13.40),
(a) change a_{11} from 0.001 to 0.002. Does this system look chaotic? Plot the population curves to make your assessment.
(b) Change a_{11} from 0.001 to 0.0005. Does this system look chaotic?

13.7.4. In Gilpin's model (13.39) and (13.40),
(a) change a_{12} from 0.001 to 0.002. Does this system look chaotic? Plot the population curves to make your assessment.
(b) Change a_{12} from 0.001 to 0.0005. Does this system look chaotic?

13.7.5. In Gilpin's model (13.39) and (13.40),
(a) change a_{13} from 0.01 to 0.02. Does this system look chaotic? Plot the population curves to make your assessment.
(b) Change a_{13} from 0.01 to 0.005. Does this system look chaotic?

13.7.6. Determine the stability of the fixed point $(200, 0, 80)$ of Gilpin's model (13.39) and (13.40).

13.7.7. For the systems of Exercise 13.7.3 (a) and (b),
(a) find the fixed points of each system.
(b) If any fixed point has all three coordinates positive, find the eigenvalues of the derivative matrix evaluated at that fixed point.
(c) Comment on the eigenvalues of the fixed point and the perceived presence or absence of chaos for the system with this fixed point.

13.7.8. For the systems of Exercise 13.7.4 (a) and (b),
(a) find the fixed points of both systems.
(b) If any fixed point has all three coordinates positive, find the eigenvalues of the derivative matrix evaluated at that fixed point.
(c) Comment on the eigenvalues of the fixed point and the perceived presence or absence of chaos for the system with this fixed point.

13.7.9. For the systems of Exercise 13.7.5 (a) and (b),
(a) find the fixed points of both systems.
(b) If any fixed point has all three coordinates positive, find the eigenvalues of the derivative matrix evaluated at that fixed point.
(c) Comment on the eigenvalues of the fixed point and the perceived presence or absence of chaos for the system with this fixed point.

13.7.10. (a) Modify Gilpin's model to represent two predator (y and z) and one prey (x) species. To make the signs clear, take all the coefficients (r_1 through a_{33}) positive. Assume all the logistic terms have negative effects on growth. That is, the model includes the terms $-a_{11}x^2$, $-a_{22}y^2$, and $-a_{33}z^2$. Assume the two predator species do not interact.
(b) Alter the model of (a) so z is a superpredator, that is, z preys on both x and y, but y preys only on x.

13.8 THE DUFFING EQUATION AND FINAL STATE SENSITIVITY

A system exhibits *stochastic resonance* if the addition of a small amount of noise (the "stochastic" part) can enhance the detection of weak signals (the "resonance" part). This may seem counterintuitive, but you might have experienced it yourself. If you are listening to a podcast with the volume too low, the presence of quiet background noise can make the podcast voices audible. We'll sketch a few biological applications and give a mathematical illustration of stochastic resonance in Sect. A.20. That example will be based on the driven Duffing equation, which we'll study in this section.

The Duffing equation is

$$x'' + ax' + bx + cx^3 = 0 \tag{13.41}$$

In mechanical systems modeled by this equation, the ax' term represents damping by friction, so $a \geq 0$. Doesn't friction have a dampening effect? Sure, but note that the x' term is on the same side of the equation as the x'' term. The bx and cx^3 terms represent the restoring force, nonlinear so long as $c \neq 0$.

For the moment, assume $a = 0$, that is, no friction. Then we can write the Duffing equation as

$$x'' = -bx - cx^3 = -\frac{d}{dx}\left(b\frac{x^2}{2} + c\frac{x^4}{4}\right) = -\frac{d}{dx}V(x)$$

That is, the frictionless Duffing equation states that acceleration is the negative of the derivative of the potential function $V(x) = bx^2/2 + cx^4/4$.

In Fig. 13.27 we plot the potential V as a function of x with $a = 0$, $b = -2$, and $c = 4$. With $b < 0$ and $c > 0$, the potential is a quartic curve with two local minima. This is called a *two-well potential.*

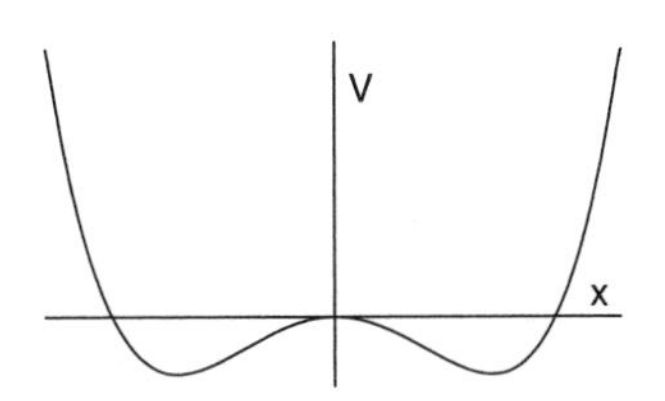

Figure 13.27. The potential for (13.41).

We'll begin with the dynamics of the Duffing equation. First, we'll apply the method of Eqs. (9.25) and (9.26) to turn the Duffing equation (with friction) into a pair of first-order equations:

$$x' = y \qquad y' = -ay - bx - cx^3 \tag{13.42}$$

In Fig. 13.28 we plot the nullclines for $a = 0$, $b < 0 < c$ (left) and $a > 0$, $b < 0 < c$ (right).

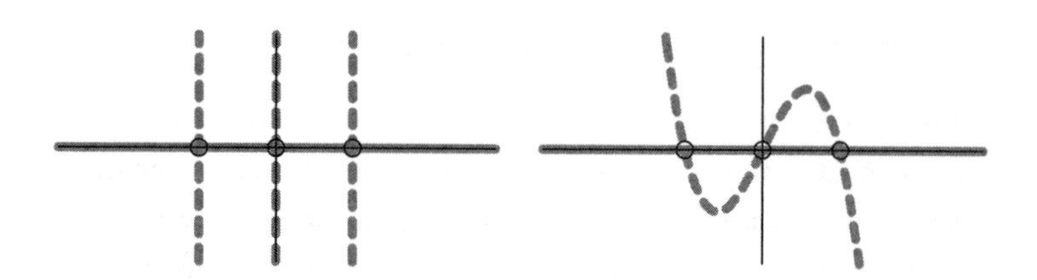

Figure 13.28. Nullclines for the Duffing system (13.42). On the left, $a = 0$; on the right, $a = 1$.

Regardless of the value of a, the $b < 0 < c$ Duffing system (13.42) has three fixed points: $(0,0)$, $(\pm\sqrt{-b/c},0)$. The derivative matrix is

$$D\vec{F}(x,y) = \begin{bmatrix} 0 & 1 \\ -b - 3cx^2 & -a \end{bmatrix}$$

The eigenvalues of $D\vec{F}(0,0)$ are $\lambda_\pm = \left(-a \pm \sqrt{a^2 - 4b}\right)/2$. The origin is a saddle point because $a > 0$ and $b < 0$, so $\sqrt{a^2 - 4b} > a$. The eigenvalues of $D\vec{F}(\pm\sqrt{-b/c},0)$ are $\lambda_\pm = \left(-a \pm \sqrt{a^2 + 8b}\right)/2$. These eigenvalues are complex with negative real parts if $a^2/8 < -b$ (remember that $b < 0$), so in this range these fixed points are asymptotically stable spirals. If $a^2/8 > -b$ the eigenvalues are real and both negative (because $a^2 + 8b < a^2$), so both fixed points are asymptotically stable nodes.

We could have done all of this in Sect. 8.5, so why study the Duffing equation now? Isn't this chapter about higher-dimensional systems? It is; what we've done about the Duffing equation is background for the system that really interests us, the forced Duffing equation

$$x'' + ax' + bx + cx^3 = d\cos(\omega t + \varphi) \tag{13.43}$$

with the constants b negative, and a and c positive. The right-hand side is the forcing term. In a physical system, this alters the acceleration x'', and so represents a force applied to the system. Think of the forcing term as laterally shaking the potential graph of Fig. 13.27.

A simple system modeled by the forced Duffing equation is an elastic metal ribbon suspended over two magnets and shaken horizontally with amplitude d, frequency ω, and phase φ. (The phase represents where in the shaking process we start the clock for the trajectory we follow.) See Fig. 13.29. The cx^3 term reflects the departure from Hooke's law for this apparatus. What story does physical intuition tell about this system?

Without forcing ($d = 0$), the metal ribbon will come to rest over one of the magnets. Even with some forcing, the ribbon still will come to slightly jiggly "rest" over one of the magnets, but to find over which might be tricky. This system may exhibit *final state sensitivity*, more pernicious than the sensitivity to initial conditions of chaos because with the forced Duffing equation the eventual behavior is (approximately) stable, but which equilibrium is reached can depend very delicately on the position and speed of the ribbon at the start of its journey.

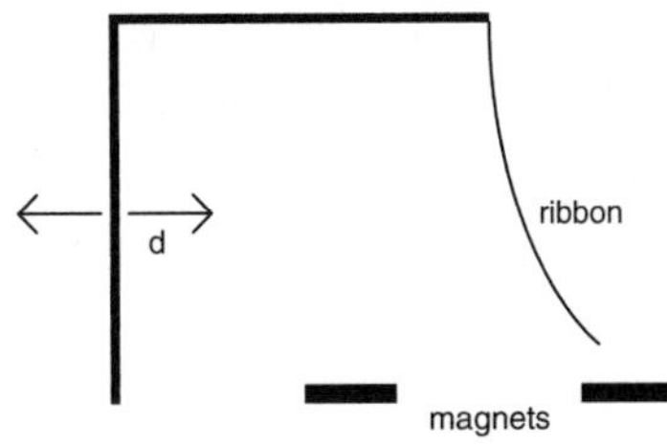

Figure 13.29. A mechanical system modeled by Eq. (13.43).

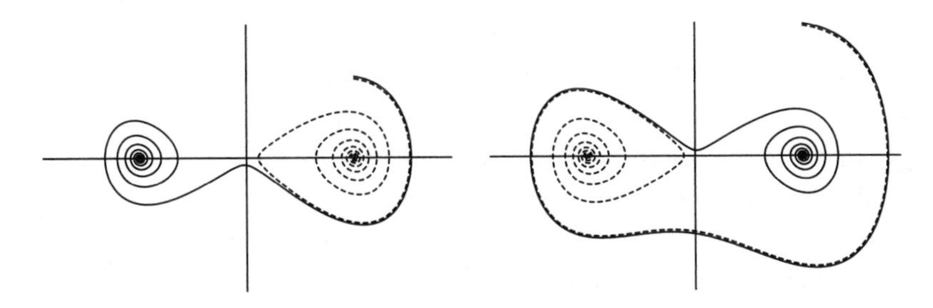

Figure 13.30. Pairs of trajectories for the unforced Duffing system (13.42).

For example, in Fig. 13.30 we see pairs of trajectories for the unforced Duffing system (13.42), with $a = 0.015$, $b = -0.5$, and $c = 0.5$, so the fixed points are $(0,0)$ and $(\pm 1,0)$. The horizontal axis is x, the vertical x'. On the left the dashed curve is the trajectory that starts at $(x,x') = (1,0.71)$; the solid curve starts at $(x,x') = (1,0.73)$. The dashed curve converges to $(1,0)$, the solid curve to $(-1,0)$. With this tiny difference, the trajectories converge to different fixed points. On the right the trajectories start at $(x,x') = (1,1.17)$ (dashed, converge to $(-1,0)$) and $(1,1.19)$ (solid, converge to $(1,0)$). With a lot more work (or some clever programming and a reasonably fast computer), we could make a map of which initial conditions lead to which fixed points. However, often a bit of thinking can save any amount of brute force, and this is one of those times.

For these values of a, b, and c, the origin is a saddle point and the other two fixed points are asymptotically stable spirals. We know that two trajectories converge to the saddle point, and we know these are unstable, in the sense that any path even the tiniest distance away from these trajectories will flow in toward the saddle point, then move away from the saddle point.

We take a different approach: move a small distance in the direction of an eigenvector $\langle p,q\rangle$, of the saddle point negative eigenvalue and follow the trajectory from that point backward in time. In Sect. B.18 we'll see how to code Mathematica to generate time-reversed solutions. The idea is simple: to go from $t=0$ back to $t=-10$, use the time limits $\{t,-10,0\}$. Because we specify $x(0)$ and $y(0)$, the numerical solution of the differential equation must go backwards in time.

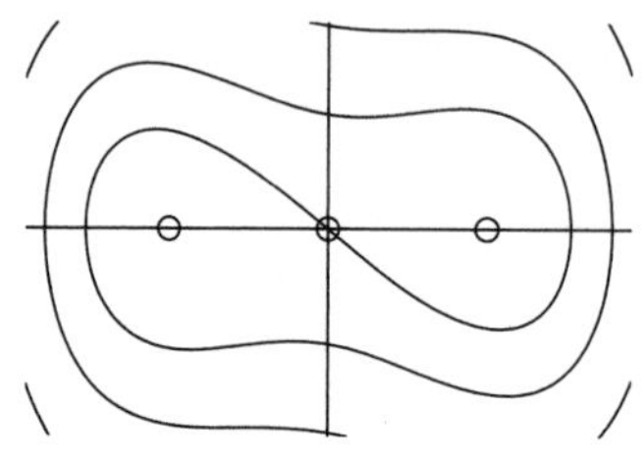

Figure 13.31. Separatrices for Eq. (13.41).

Repeat this with the initial point a small distance in the direction $\langle -p,-q\rangle$. These curves are called the *separatrices* of the system. In Fig. 13.31 we see the separatrices for $a=0.15$, $b=-0.5$, and $c=0.5$. Because trajectories cannot cross (Sect. A.2.), the separatrices divide the plane into two regions, those initial values from which the trajectories flow to $(-\sqrt{-b/c},0)$, and those from which the trajectories flow to $(\sqrt{-b/c},0)$. These sets of initial values are called the *basins of attraction* of the fixed points $(\pm\sqrt{-b/c},0)$.

To understand the behavior shown in Fig. 13.30, on the left of Fig. 13.32 we plot the separatrices and the two trajectories from the right image of Fig. 13.30. The small box in the left image is magnified on the right. Because the separatrices are trajectories and trajectories cannot cross, we see why the separatrices split trajectories into those that converge to one fixed point and those that converge to the other.

These basins of attraction are not so complicated, and in fact seem to be pretty well-behaved. For any given set of parameters, finding the separatrices to any accuracy desired is not so difficult, and the separatrices bound the basins of attraction. To see more interesting behavior, we must turn on the forcing term in the Duffing equation.

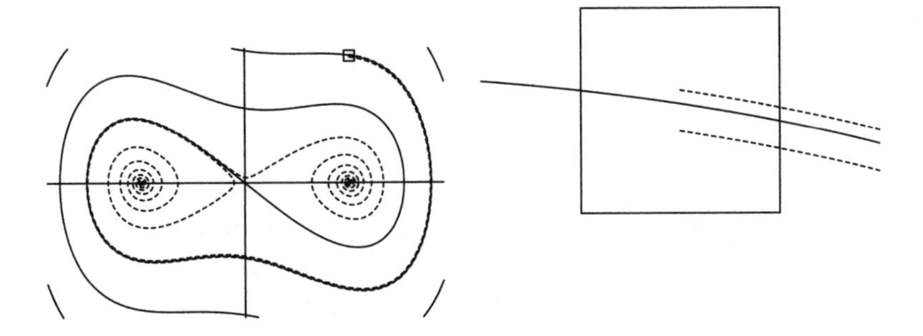

Figure 13.32. Separatrices and trajectories.

First, how can the Duffing equation exhibit chaos, when the Poincaré-Bendixson theorem precludes chaos for differential equations in the plane? (We're close to answering the question of why this section is in the chapter on high-dimensional systems.) A bit more care is needed: the Poincaré-Bendixson theorem applies to *autonomous* differential equations in the plane. To turn the forced Duffing equation into an autonomous system, we need to treat time as a third variable, $z = t$. Then Eq. (13.43) becomes

$$x' = y \qquad y' = -ay - bx - cx^3 + d\cos(\omega z + \varphi) \qquad z' = 1 \tag{13.44}$$

In Fig. 13.33 we see 3-dimensional plots of the forced Duffing equation for $a = 0.15$, $b = -0.5$, $c = 0.5$, $d = 3.3$, $\omega = 1$, and $\varphi = 0$, with initial value $(x, x', t) = (1, 0.73, 0)$ (left) and $(1, 0.74, 0)$ (right).

These trajectories appear quite different, but superimposing the graphs is not a promising approach. Figuring out these 3-dimensional pictures individually is hard enough; combining them would be a real headache. However, if

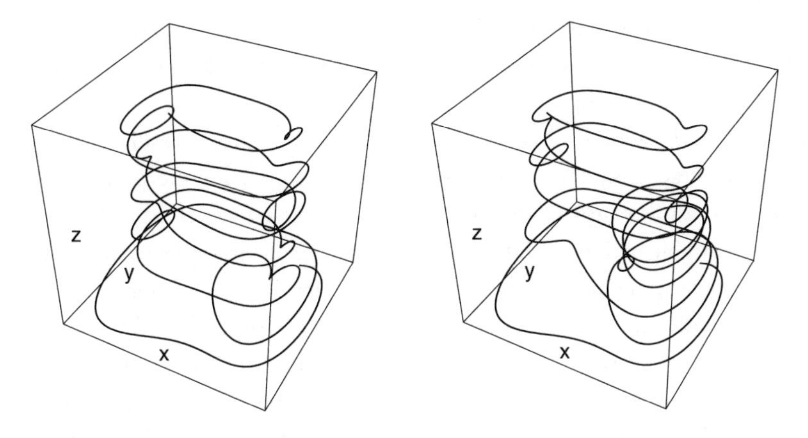

Figure 13.33. Trajectories of the forced Duffing equation, plotted in 3 dimensions.

the projection to the xt-plane or to the yt-plane shows sensitivity to initial conditions, we'll be satisfied. The right image of Fig. 13.34 is the projection of the two trajectories of Fig. 13.33 into the xt-plane; the dashed curve is the projection of the trajectory from $(1, 0.74, 0)$. The projections diverge fairly quickly, then return to nearby paths for a while, only to diverge again. This rough pattern would continue for as long as we care to run the differential equations solver.

The left image of Fig. 13.34 is a projection of the left image of Fig. 13.33 to the xy-plane. It is another way to see that a non-autonomous planar equation properly is a 3-dimensional system: in 2 dimensions we have trajectories appearing to cross one another, something we know is impossible. Mathematica code for the right graph is in Sect. B.18.

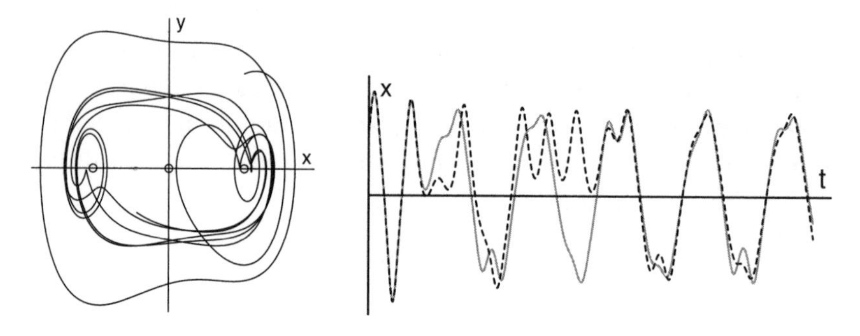

Figure 13.34. Left: a projection of the left image of Fig. 13.33 to the xy-plane. Right: projections of both images of Fig. 13.33 to the xt-plane.

Our visual intuition can be supported by Liapunov exponents, described for differential equations in Sect. A.19. For now we'll say that these exponents measure the divergence in different directions of nearby trajectories. Trajectories separated in time only, for example, $\vec{r}(t) = (x(t), x'(t), t)$ and $\vec{s}(t) = \vec{r}(t+1)$, keep the same separation in the time direction, so in that direction the Liapunov exponent is 0. If another Liapunov exponent is positive, initially nearby trajectories diverge, a signature of chaos. Finally, in order for the trajectories to converge to an attractor of the type we saw in Sect. 13.7, necessary for the presence of unstable periodic solutions characteristic of chaotic systems, the sum of the Liapunov exponents must be negative. Together, these conditions require at least 3 dimensions, another way to understand that in order for a differential equation to exhibit chaotic trajectories, it must have at least three variables.

In Fig. 13.35 we see the basins of attraction for the two asymptotically stable fixed points. As the forcing amplitude increases, the basins become more intricately mixed. For initial points near the boundary of the basins, tiny

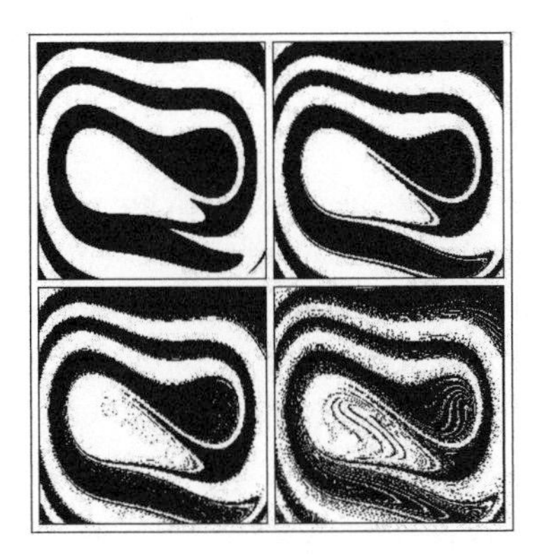

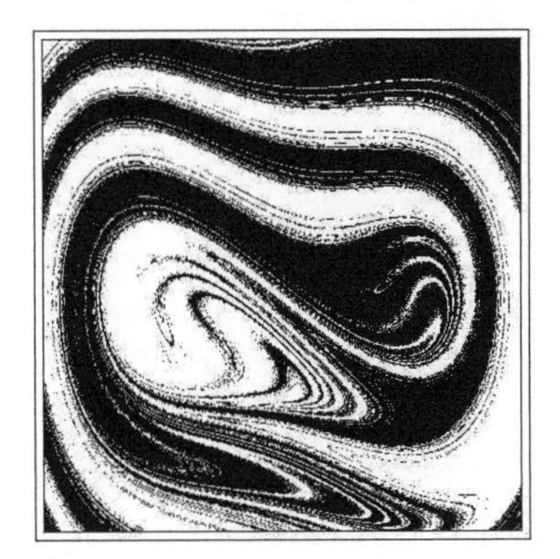

Figure 13.35. Left: basins of attraction as the forcing amplitude increases. Right: a magnification of the lower right picture on the left.

uncertainties in the initial conditions can swamp our ability to predict to which fixed point the trajectories converge.

Another visualization scheme is available for differential systems with periodic forcing, (13.44), for example. We can plot the points $(x(t), x'(t))$ for $\omega t = 0, 2\pi, 4\pi, 6\pi, \ldots$. This is called a *Poincaré section*. Other time choices are possible, for example $\omega t = 2n\pi + \varphi$, for $n = 0, 1, 2, \ldots$, and for a constant

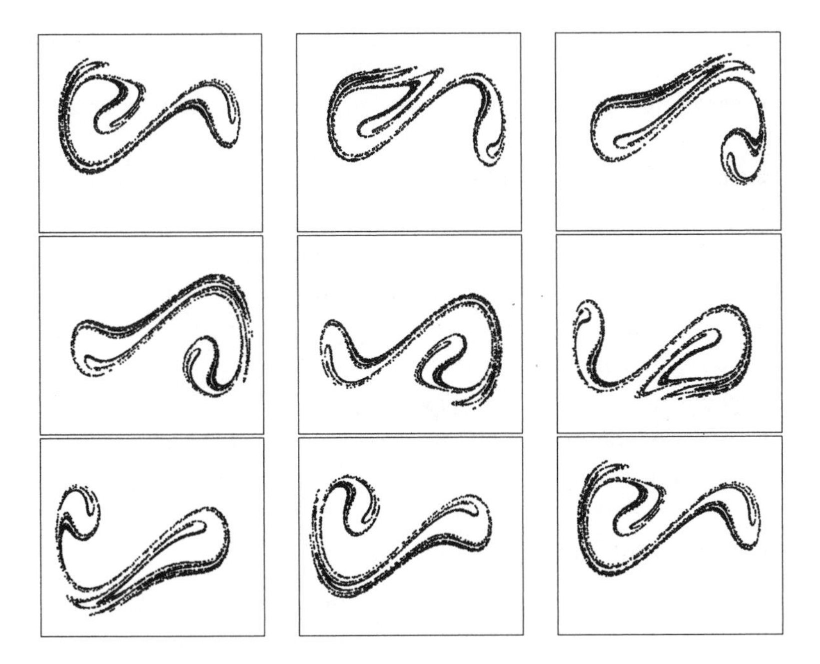

Figure 13.36. Poincaré sections of the forced Duffing equation.

$\varphi, 0 \leq \varphi < 2\pi$. For example, Fig. 13.36 shows Poincaré sections for the forced Duffing system with $a = 0.15$, $b = -1$, $c = 1$, $d = 0.3$, $\omega = 1$, and $\varphi = k\pi/4$ for $k = 0, 1, \ldots 8$.

These pictures are best seen in an animation, making clear the stretching and folding present in so many chaotic systems. In the static medium of a book, this is the best I can do. Well, not really. I could print these images, one at a time, on the upper right corner of successive pages. Then flipping through the pages would give a crude manual animation. I tried this, but the result did not match my expectations, so I'll leave the animation to your imagination, or your programming.

Practice Problems

13.8.1. In the forced Duffing system (13.44) with $\omega = 1$ and $\varphi = 0$,
(a) if $(x(0), y(0), z(0)) = (x(2\pi), y(2\pi), z(2\pi))$, show that for all $t \geq 2\pi$, $(x(t), y(t), z(t)) = (x(t - 2\pi), y(t - 2\pi), z(t - 2\pi))$. Even though when unpacked in 3 dimensions the forced Duffing system has no closed trajectories, the repeating path of this trajectory makes it sensible to call the trajectory periodic.
(b) Now take also $b = -c < 0$. Show that with these parameters, the unforced Duffing system (13.42) has fixed points $(-1, 0)$, $(0, 0)$, and $(1, 0)$.
(c) Make a schematic sketch of a periodic trajectory that loops twice around $(x, x') = (-1, 0)$, once around $(1, 0)$, and once around both together.
(d) In how many combinatorially distinct ways can such a trajectory be arranged?

13.8.2. In the forced Duffing equation (13.43), set $a = 0.15$, $b = -0.5$, $c = 0.5$, $d = 0.5$, $\omega = 1$, and $\varphi = 0$.
(a) Show that the 3-dimensional system (13.44) has no fixed points.
(b) For each value of t, $0 \leq t \leq 2\pi$, find the numerical value $x(t)$ of the real fixed point of Eq. (13.44), that is, the intersections of the x- and y-nullclines, treating the value of t as a constant.
(c) Plot the population curve $(x(t), t)$ for $0 \leq t \leq 2\pi$.
(d) Compare the plot of the curve $(x(t), 0, t)$ with a trajectory of Eq. (13.43) and interpret the resulting image.

Practice Problem Solutions

13.8.1. (a) This follows from the uniqueness of solutions (Sect. A.2), because $\cos(t) = \cos(t - 2\pi)$ and the system (13.44) is unaltered by the change of variables $s = t - 2\pi$. For example, by the chain rule $dx/ds = (dx/dt)(dt/ds) = dx/dt$.

(b) For the unforced Duffing system (13.42) the x-nullcline is $y = 0$ and the y-nullcline is $0 = -ay - bx - cx^3 = -x(b + cx^2)$. Then the fixed points occur at $x = 0$ and $x = \pm\sqrt{-b/c} = \pm 1$ because $b = -c$. Consequently, the fixed points are $(-1,0)$, $(0,0)$, and $(1,0)$.

(c) A schematic plot is shown in Fig. 13.37. Note that the initial point and the terminal point of the trajectory have the same x- and y-coordinates. The dashed lines are at $x = -1, y = 0$ and $x = 1, y = 0$.

(d) To shorten the notation, let l, r, and b denote paths that loop around $(-1,0)$, around $(1,0)$, and around both $(-1,0)$ and $(1,0)$. Then the combinatorial possibilities are *llrb*, *llbr*, *lrlb*, *lblr*, *lrbl*, *lbrl*, *rlbl*, *blrl*, *rbll*, and *brll*. There are 10. For simple problems, sometimes counting is the best approach.

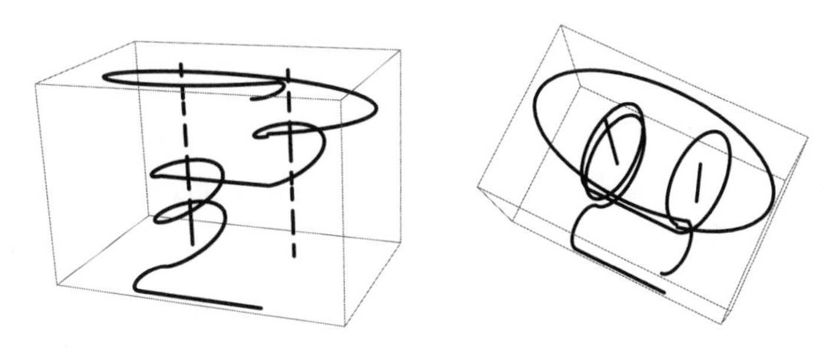

Figure 13.37. Two views of the schematic plot of Practice Problem 13.8.1.

13.8.2. (a) This is easy once we recall that the fixed points are the intersections of the nullclines, because there is no z-nullcline, because $z' = 1$ and so $z' \neq 0$.

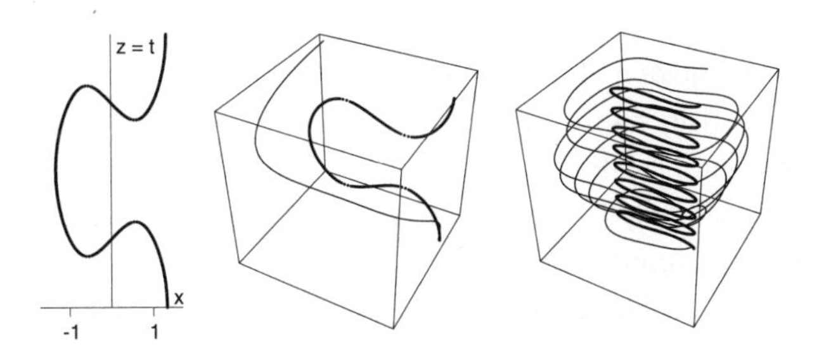

Figure 13.38. The $(x(t), 0, t)$ curve and trajectories for Practice Problem 13.8.1.

(b) From Eq. (13.44) we see that the x-nullcline equation is $y = 0$, so the y-nullcline equation becomes

$$bx + cx^3 = d\cos(\omega t + \varphi) \quad \text{that is,} \quad -0.5x + 0.5x^3 = 0.5\cos(t)$$

For each value of t, we solve this equation numerically. (Mathematica can give the abstract solution of the equivalent $-x + x^3 = \cos(t)$, but it's a mess.) For $t = 0$ we obtain $1.32472, -0.662359 \pm 0.56228i$. For $t = 0.1$ we obtain 1.32355, $-0.661773 \pm 0.560204i$, and so on. In Sect. B.18 we'll see how to automate the process of finding the real solutions $x(t)$.
(c) The first graph of Fig. 13.38 is a plot of $(x(t), t)$ for $0 \le t \le 2\pi$.
(d) The second and third graphs of Fig. 13.38 are plots of $(x(t), 0, t)$ and a trajectory of Eq. (13.44) for $0 \le t \le 2\pi$ (second graph) and for $0 \le t \le 50$. We see that the trajectory encompasses the fixed point graph, and eventually the trajectory achieves its smallest and largest values of x at about the same t-values when the graph $(x(t), 0, t)$ achieves its smallest and largest x-values. That is, although the trajectory amplitude is larger than that of the $(x(t), 0, t)$ curve, their frequencies and phases appear to be synchronized.

Exercises

13.8.1. Plot the region in the (a, b) plane where the non-zero fixed points of Eq. (13.42) are asymptotically stable spirals. (Hint: tr-det plane.)

13.8.2. In Eq. (13.44) with $\omega = 1$, $\varphi = 0$, and $c = -b$,
(a) make a schematic sketch of a periodic trajectory that loops three times around $(x, x') = (-1, 0)$ and twice around $(1, 0)$.
(b) Make a schematic sketch of a periodic trajectory that loops five times around $(1, 0)$ and once around both $(-1, 0)$ and $(1, 0)$.
(c) In how many combinatorially distinct ways can the trajectories of (a) and of (b) be arranged?

13.8.3. Continue with the situation of Exercise 13.8.2.
(a) In how many combinatorially distinct ways can a trajectory loop once around $(-1, 0)$ and three times around $(1, 0)$?
(b) In how many combinatorially distinct ways can a trajectory loop twice around $(-1, 0)$ and six times around $(1, 0)$?
(c) Explain why the number you found in (b) is not twice the number you found in (a).

13.8.4. In the forced Duffing system (13.44), change the cx^3 term to cx^2.
(a) For the unforced system ($d = 0$), find the fixed points.
(b) Determine the stability and type of the fixed point at the origin as a function of a and b.

(c) For $a = 0.05$, $b = -0.5$, $c = 0.5$, $d = 0.1$, $\omega = 1$, and $\varphi = 0$, plot a trajectory for $0 \le t \le 200$. Does this appear to be chaotic or periodic? Support your answer.

13.8.5. Continuing Exercise 13.8.4, test for sensitivity to initial conditions by superimposing the $x(t)$ graph with $(x(0), y(0), z(0)) = (1.0, 0, 0)$ for $0 \le t \le 200$ and the $x(t)$ graph with $(x(0), y(0), z(0)) = (1.1, 0, 0)$. In Sect. B.18 we'll give code to plot the difference of the x-coordinates of these plots.

13.8.6. In the forced Duffing system (13.44), change the bx term to bx^2.
(a) For the unforced system ($d = 0$), find the fixed points.
(b) For $a = 0.05$, $b = -0.5$, $c = 0.5$, $d = 0.1$, $\omega = 1$, and $\varphi = 0$, plot a trajectory for $0 \le t \le 200$. Does this appear to be chaotic or periodic?
(c) Test for sensitivity to initial conditions by superimposing the $x(t)$ graph with $(x(0), y(0), z(0)) = (1.0, 0, 0)$ for $0 \le t \le 200$ and the $x(t)$ graph with $(x(0), y(0), z(0)) = (1.1, 0, 0)$.

13.8.7. Repeat parts (b) and (c) of Exercise 13.8.6 with $d = 0.1$ changed to $d = 0.2$.

13.8.8. In the forced Duffing equation (13.43) does doubling the driving frequency ω halve the time between successive x spikes and y spikes? Devise some simulations and report your results.

13.8.9. In the forced Duffing system (13.44), set $a = 0.05$, $b = -0.5$, $c = 0.5$, $d = 3.3$, $\omega = 1$, and $\varphi = 0$.
(a) For a sample of t-values, $0 \le t \le 2\pi$, find the numerical value $x(t)$ of the real fixed point of Eq. (13.43), that is, the intersections of the x- and y-nullclines, treating the value of t as a constant.
(b) Plot the curve $(t, x(t))$ for $0 \le t \le 2\pi$.
(c) Compare the plot of the curve $(x(t), 0, t)$ with a trajectory of Eq. (13.43) and interpret the resulting image.

13.8.10. Repeat Exercise 13.8.9 for $a = 0.1$, $b = -2.5$, $c = 0.1$, $d = 2.3$, $\omega = 1$, and $\varphi = 0$.

13.9 CONTROLLING CHAOS

In Sects. 2.4, 13.7, and 13.8 we saw some examples of chaos in discrete and continuous systems. In particular, we saw that chaotic systems exhibit sensitivity to initial conditions. When chaos entered popular culture in the mid-1980s, well-described in James Gleick's book *Chaos: Making a New Science* [140], one of the clearest messages was the lack of long-term predictability in chaotic systems. Every effort to measure initial conditions must necessarily include some uncertainty: a measurement is good only to some number of decimal digits. Beyond that, measurement can tell us nothing. Say we can measure x to three decimal digits. So if we measure $x(0) = 0.234$, the actual value might be $x(0) = 0.2343$ or $x(0) = 0.2344$. Then by sensitivity to initial conditions, eventually the trajectories starting from these two values of $x(0)$ will diverge, and because we don't know from which value the system actually started, the possibility of close prediction vanishes with this divergence. If we can't make long-term predictions of a chaotic system's behavior, how can we possibly find an intervention to force the system into a desirable state. Seems hopeless, doesn't it? Not at all. Not even a little bit.

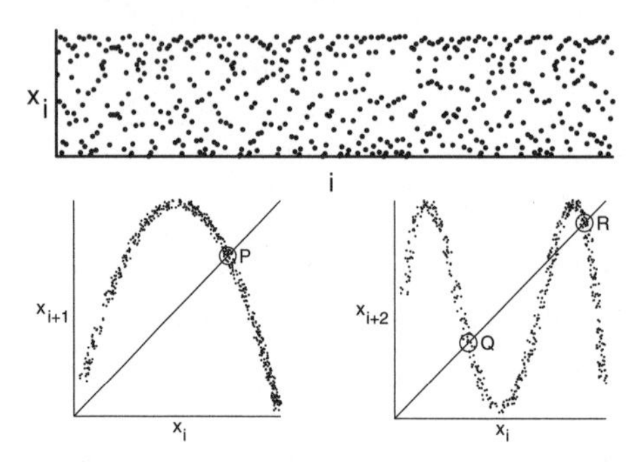

Figure 13.39. Top: time series. Bottom: first and second return maps.

In Sect. 2.4 we saw that chaotic systems exhibit properties in addition to sensitivity to initial conditions. We'll focus on the denseness of (necessarily unstable) periodic trajectories. In Sect. 13.7 we mentioned that Edward Ott, Celso Grebogi, and James Yorke developed the *OGY method* [267] for temporarily stabilizing any of the unstable periodic trajectories, *without having a model of the underlying dynamics*. Here's how.

Suppose we have a time series, $x_1, x_2, \ldots, x_N$, for instance, intervals between successive heartbeats. The top of Fig. 13.39 shows a cartoon example. This is the data we have; we have no model of its source. (Well, of course, I do: I wrote the code to generate the figures. But you don't know the code, so we'll forget it for the rest of this section.) What can we do? First, we can find the approximate locations of fixed points and 2-cycles by plotting points of the *first return map* (x_i, x_{i+1}) and of the *second return map* (x_i, x_{i+2}). Recalling the role of the diagonal line for locating fixed points and cycle points (Sects. 2.2, 2.3), we approximate the fixed point at P, the intersection of the first return map with the diagonal, and the 2-cycle points at Q and R, the intersections of the second return map with the diagonal. (The other intersection of the second return map with the diagonal is the fixed point P.) See the lower images of Fig. 13.39.

We'll describe the OGY method to stabilize the fixed point $P = (x_*, x_*)$. To stabilize cycles is similar: just replace the first return map with the second, third, ..., depending on the length of the cycle we want to stabilize. First, we find a linear function that approximates how the point (x_i, x_{i+1}) changes to the point (x_{i+1}, x_{i+2}). That is, we use linear regression to find the best values for a, b, c, and d that satisfy

$$\begin{bmatrix} x_{i+2} \\ x_{i+1} \end{bmatrix} = \begin{bmatrix} a & b \\ c & d \end{bmatrix} \begin{bmatrix} x_{i+1} \\ x_i \end{bmatrix} \tag{13.45}$$

Call this 2×2 matrix M. Typically, the fixed point P is a saddle point, so M has an eigenvalue $\lambda_u > 0$ and an eigenvalue $\lambda_s < 0$. Call an eigenvector $\vec{e}_u$ of λ_u an *unstable eigenvector*, and an eigenvector $\vec{e}_s$ of λ_s a *stable eigenvector*.

Write

$$\vec{v}_n = \begin{bmatrix} x_{n+1} \\ x_n \end{bmatrix} \quad \vec{v}_{n+1} = \begin{bmatrix} x_{n+2} \\ x_{n+1} \end{bmatrix} \quad \text{and} \quad \vec{v}_* = \begin{bmatrix} x_* \\ x_* \end{bmatrix}$$

Then Eq. (13.45) can be written as $\vec{v}_{n+1} = M\vec{v}_n$. Because $\vec{v}_* = M\vec{v}_*$ we have

$$\vec{v}_{n+1} - \vec{v}_* = M(\vec{v}_n - \vec{v}_*) \tag{13.46}$$

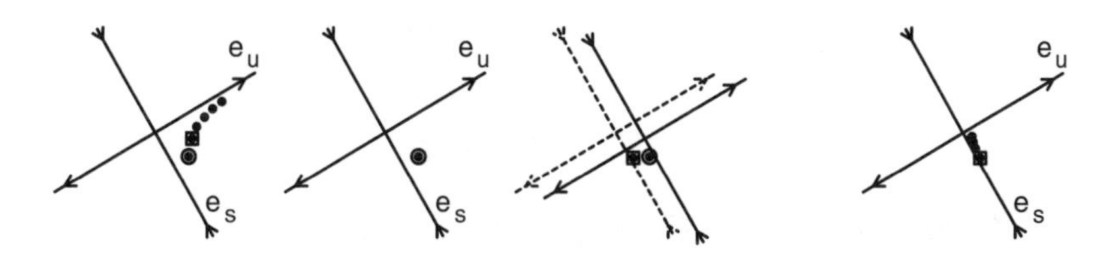

Figure 13.40. A sketch of the OGY method. The dot enclosed in a circle is $\vec{v}_n$; the dot enclosed in a square is $\vec{v}_{n+1}$.

Suppose $\vec{v}_n$ is near the stable eigenvector $\vec{e}_s$. Then, as seen in the first image of Fig. 13.40, successive iterates approach the fixed point but eventually get pushed away in the direction of the unstable eigenvectors. We assume the system has some accessible parameter, s, and that we can approximate how the fixed point moves as s changes: that is, we can generate a new return map and compare the new fixed point to the old fixed point. Suppose the system is in the state shown in the second image of Fig 13.40. Change s so $\vec{v}_n$ iterates to $\vec{v}_{n+1}$, which lies almost on the stable eigenvector of the original system. See the third image of Fig. 13.40, where the stable and unstable directions of the original system are shown dashed and those of the new system are solid lines. Now return s to its original value and successive iterates will move along the stable eigenvector toward the fixed point. See the fourth image of Fig. 13.40.

Of course, these iterates are not exactly on the stable eigenvector, so eventually the iterates move away in an unstable direction. When the iterates have moved beyond a preset tolerance, adjust s again to move the current state to the stable direction of the original system. And so on. Every time the system moves too far from the fixed point, adjust s for one time step to move it back to the stable direction of the original system.

How do we find the system parameter adjustment? First, translate the coordinates so $(x_*, x_*) = (0,0)$ when $s = 0$ and let $\vec{z}(s)$ denote the fixed point for $s \neq 0$. As the fixed point moves, Eq. (13.46), now written as

$$\vec{v}_{n+1} - \vec{z}(s) = M(\vec{v}_n - \vec{z}(s)) \tag{13.47}$$

remains valid. For small s, $\vec{z}(s)$ should change approximately linearly with s. That is, let $\vec{g} = (d\vec{z}/ds)|_{s=0}$. Then for small s we have

$$\frac{\vec{z}(s)}{s} = \frac{\vec{z}(s) - \vec{z}(0)}{s - 0} \approx \left.\frac{d\vec{z}}{ds}\right|_{s=0} = \vec{g}$$

and

$$\vec{z}(s) \approx s\vec{g} \quad \text{for small } s \tag{13.48}$$

Recall that our eigenvectors $\vec{e}_s$ and $\vec{e}_u$ are column vectors and are multiplied on the right side of the matrix:

$$M\vec{e}_s = \lambda_s \vec{e}_s \quad \text{and} \quad M\vec{e}_u = \lambda_u \vec{e}_u$$

Take these to be unit vectors, so $\|\vec{e}_u\| = \sqrt{e_{u,1}^2 + e_{u,2}^2} = 1$ and $\|\vec{e}_s\| = 1$.

Now we'll make a matrix W whose columns are these eigenvectors

$$W = \begin{bmatrix} \vec{e}_u & \vec{e}_s \end{bmatrix} = \begin{bmatrix} e_{u,1} & e_{s,1} \\ e_{u,2} & e_{s,2} \end{bmatrix}$$

In Exercise 13.9.1 we'll show that $\det W \neq 0$ if and only if the eigenvectors $\vec{e}_u$ and $\vec{e}_s$ are not parallel. Recall from Eq. (8.9) that W is invertible if and only if $\det(W) \neq 0$.

Next, take $\vec{f}_u$ and $\vec{f}_s$ to be the rows of the matrix W^{-1}. In terms of the row and column vectors, the condition $W^{-1}W = I$ can be written as

$$\vec{f}_u \vec{e}_u = 1, \quad \vec{f}_u \vec{e}_s = 0, \quad \vec{f}_s \vec{e}_u = 0, \quad \text{and} \quad \vec{f}_s \vec{e}_s = 1 \tag{13.49}$$

If you're wondering what sort of multiplication $\vec{f}_u \vec{e}_u$ is, it's just familiar matrix multiplication

$$\begin{bmatrix} f_{u,1} & f_{u,2} \end{bmatrix} \begin{bmatrix} e_{u,1} \\ e_{u,2} \end{bmatrix} = f_{u,1}e_{u,1} + f_{u,2}e_{u,2}$$

What happens if we multiply $\vec{f_u}$ and $\vec{f_u}$ in the other order, column on the left, row on the right? We get a matrix:

$$\begin{bmatrix} e_{u,1} \\ e_{u,2} \end{bmatrix} \begin{bmatrix} f_{u,1} & f_{u,2} \end{bmatrix} = \begin{bmatrix} e_{u,1}f_{u,1} & e_{u,1}f_{u,2} \\ e_{u,2}f_{u,1} & e_{u,2}f_{u,2} \end{bmatrix}$$

Also, we need to write M this way

$$M = \lambda_u \vec{e}_u \vec{f}_u + \lambda_s \vec{e}_s \vec{f}_s \tag{13.50}$$

The proof isn't hard, but it requires a bit of linear algebra, so we'll postpone it till Sect. A.12.3, where we'll see why we made the matrix W from the eigenvectors $\vec{e}_u$ and $\vec{e}_s$.

Recall that the *dot product* of vectors $\vec{a} = \langle a_1, a_2 \rangle$ and $\vec{b} = \langle b_1, b_2 \rangle$ is

$$\vec{a} \cdot \vec{b} = a_1 b_1 + a_2 b_2 \tag{13.51}$$

An equivalent, perhaps more familiar, formulation is

$$\vec{a} \cdot \vec{b} = \|\vec{a}\| \, \|\vec{b}\| \cos(\theta) \tag{13.52}$$

where θ is the (smaller) angle between $\vec{a}$ and $\vec{b}$. (We'll derive this in Sect. 17.1.) Note that $\|\vec{b}\| \cos(\theta)$ is the length of the component of $\vec{b}$ in the direction of $\vec{a}$ (draw a picture with $\vec{a}$ and $\vec{b}$ having the same starting point), so if $\vec{a} \cdot \vec{b} = 0$, the component of $\vec{b}$ in the direction of $\vec{a}$ has length 0. We'll mention that multiplying a row vector (on the right) by a column vector (on the left) had the same result as taking the dot product of the vectors when both are row vectors.

$$\begin{bmatrix} a_1, a_2 \end{bmatrix} \begin{bmatrix} b_1 \\ b_2 \end{bmatrix} = a_1 b_1 + a_2 b_2$$

For the condition that $\vec{x}_{n+1}$ lies on the stable eigenvector, Ott, Grebogi, and Yorke find s to make $\vec{f}_u \vec{v}_{n+1} = 0$. (Alternately, one could take $\vec{e}_u \cdot \vec{x}_{n+1} = 0$. See Sects. A.12.4 and A.12.5.) Next they multiply both sides of Eq. (13.47) on the left by the row vector $\vec{f}_u$

$$\begin{aligned}
\vec{f}_u(\vec{v}_{n+1} - \vec{z}(s)) &= \vec{f}_u M(\vec{v}_n - \vec{z}(s)) \\
\vec{f}_u(\vec{v}_{n+1} - s\vec{g}) &= \vec{f}_u M(\vec{v}_n - s\vec{g}) && \text{by Eq. (13.48)} \\
\vec{f}_u \vec{v}_{n+1} - s\vec{f}_u \vec{g} &= \vec{f}_u(\lambda_u \vec{e}_u \vec{f}_u + \lambda_s \vec{e}_s \vec{f}_s)(\vec{v}_n - s\vec{g}) && \text{by Eq. (13.50)} \\
\vec{f}_u \vec{v}_{n+1} - s\vec{f}_u \vec{g} &= \lambda_u \vec{f}_u(\vec{v}_n - s\vec{g}) && \text{by Eq. (13.49)}
\end{aligned}$$

Then they set $\vec{f}_u \vec{v}_{n+1} = 0$ and solve the last equation for s to obtain the *OGY formula*

$$s = \frac{\lambda_u}{\lambda_u - 1} \frac{\vec{f}_u \vec{v}_{n+1}}{\vec{f}_u \vec{g}} \tag{13.53}$$

In Exercise 13.9.10 we'll derive an alternate formula for parameter adjustment from the condition $\vec{e}_u \cdot \vec{v}_{n+1} = 0$. We'll see that the OGY formula is simpler and easier to implement.

If the matrix M is *symmetric*, that is, $m_{ij} = m_{ji}$, then $\vec{e}_u$ and $\vec{f}_u$ are parallel vectors in the plane (the calculation is straightforward but messy, so we'll sketch it in Sect. A.12.4), so $\vec{e}_u \cdot \vec{v}_{n+1} = 0$ if and only if $\vec{f}_u \vec{v}_{n+1} = 0$. If M is close to symmetric, then $\vec{e}_u$ and $\vec{f}_u$ are close to parallel. See Practice Problem 13.9.1 and Exercises 13.9.2 and 13.9.3 for some examples, and Appendix A.12.4 for a derivation.

Though the idea sketched in Fig. 13.40 is easy to understand, we can visualize control of chaos more directly by temporarily stabilizing cycles in the chaotic logistic map. The adjustable system parameter is the s of $L(x) = sx(1-x)$, but what is the equivalent of the line determined by the stable eigenvector? For a saddle point, every point on this line iterates to the fixed point. For the logistic map, the corresponding construction is the collection of points that iterate toward the fixed point.

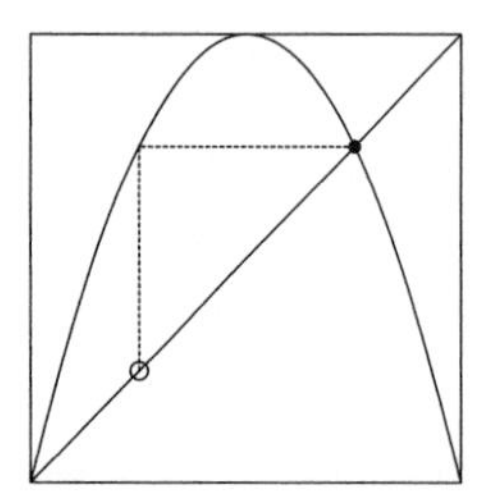
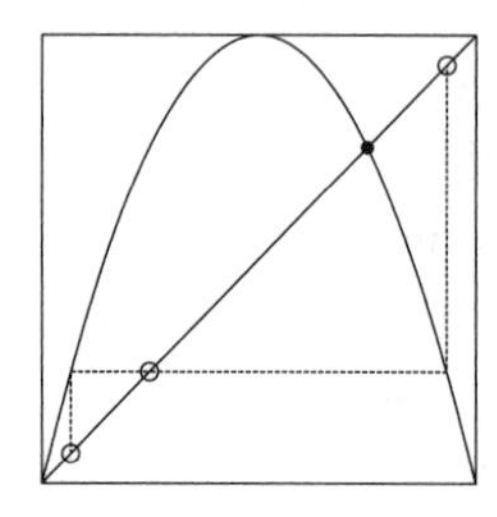
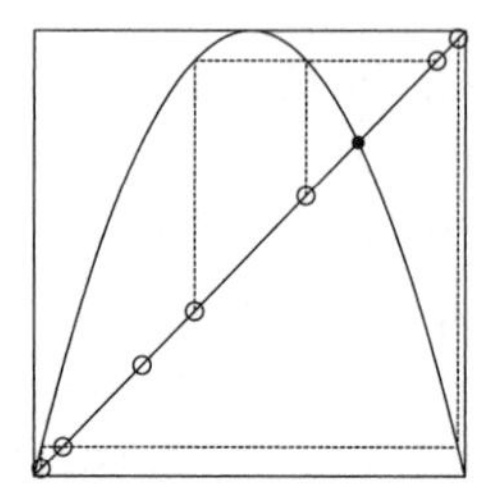

Figure 13.41. Left to right: one iterate to the fixed point, two iterates, three iterates.

Fig. 13.41 shows the first few steps in the construction of this collection for the $s = 4$ logistic map. The fixed points are $x = 0$ and $x = 3/4$. To find points that iterate to the fixed point $x = 3/4$, we can run the graphical iteration in reverse: horizontally to the graph, then vertically to the diagonal. In the first image, we find the point that iterates to $x = 3/4$. Algebraically, we solve $L(x) = 3/4$ and obtain $x = 1/4$ and $x = 3/4$. In the second image we find the two points that iterate to $x = 1/4$. Algebraically, we solve $L(x) = 1/4$ and obtain $x = (2 \pm \sqrt{3})/4$. In the third image we find two points that iterate to each of the points found in the second image. We can continue as long as we like, to generate a large collection of points that will iterate to the fixed point. (If $s < 4$, points very close to $x = 1$ aren't in the image of L, so graphical iteration cannot be reversed from these points; algebraically, the quadratic equation has complex roots.)

Here's how to implement control for the logistic map. The first image of Fig. 13.42 shows the $s = 4$ logistic map, the (unstable) fixed point the black dot, the current iterate ($x = 0.1$ in this example) the gray dot, and a point ($x = 0.25$) that iterates to the fixed point shown by the circle. We want to adjust the logistic map so $x = 0.1$ iterates to $x = 0.25$. Because we know the dynamics, rather than use the OGY formula, we just solve

$$0.25 = L(0.1) = s \cdot 0.1 \cdot (1 - 0.1) \quad \text{so} \quad s = 2.7778$$

The second image of Fig. 13.42 shows one iterate of the $s = 2.7778$ logistic map, which sends $x = 0.1$ to $x = 0.25$. In the third image, the logistic map is adjusted back to $s = 4$ and $x = 0.25$ iterates to the fixed point.

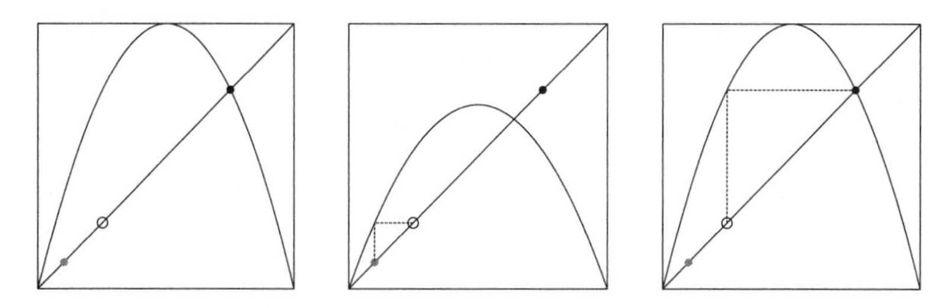

Figure 13.42. Controlling the logistic map to an unstable fixed point.

In Fig. 13.43 we see two implementations of control of the logistic map to stabilize the unstable fixed point at $x = 0.75$. We'll make an improvement of the algorithm in a moment, but first we'll interpret these figures. In the left image, when an iterate x gets within $\delta = 0.02$ of $x = 0.25$, solve $0.25 = L(x)$ for s and adjust s. Then after one iteration, return s to $s = 4$. Because this iterate is not exactly 0.75, later iterates eventually will diverge from this fixed point. Control is reapplied only when x_n comes within δ of 0.25. On the left we see some long periods of wandering before the iterates get close enough to 0.25 to activate another instance of control. Can we reduce the duration of the wandering intervals?

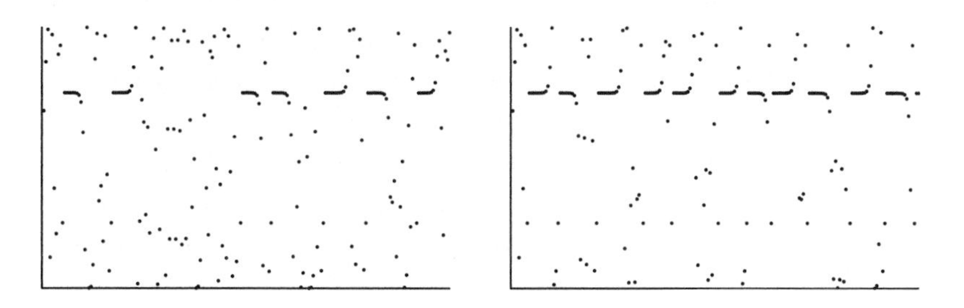

Figure 13.43. Implementations of control to an unstable fixed point.

One approach is illustrated in the right image of Fig. 13.43. Instead of waiting until an iterate falls within δ of 0.25, we activate control when an iterate falls within δ of 0.25, $(2+\sqrt{3})/4$, or $(2-\sqrt{3})/4$. This reduces the wait between the times when control can be applied. Using even more points that iterate to the fixed point, we can further reduce the wait between controls.

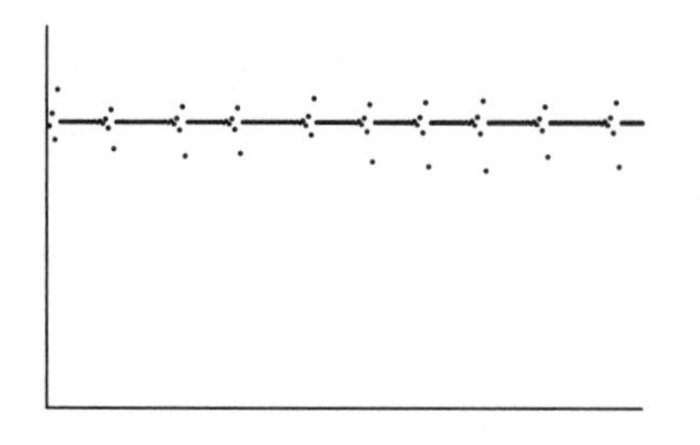

Figure 13.44. A better control strategy.

Better yet is this: set another threshold, ϵ, and once the iterates are controlled to near the fixed point, reapply control as soon as an iterate gets farther than ϵ away from the fixed point. In Fig. 13.44 we see an example of this approach, with $\delta = 0.02$ again and with $\epsilon = 0.06$. The time away from the fixed point is small. Smaller values of ϵ will give smaller excursions from the fixed point before control is reapplied. This is the control strategy applied in the OGY method.

Mathematica code for Figs. 13.43 and 13.44 is in Sect. B.19.

Control can be applied to stabilize cycles as well as fixed points. Recall that in Fig. 13.39 we saw that we can locate 2-cycles because they are fixed points of the second return map (x_n, x_{n+2}). Similarly, points on 3-cycles are fixed points of the third return map (x_n, x_{n+3}), and so on. We'll do some examples in Practice Problem 13.9.2 and in Exercises 13.9.4–13.9.7.

In experimental situations, locating (approximate) fixed points and cycles are straightforward applications of the return maps. But to estimate the change of fixed point or cycle with the system parameter, the vector $\vec{g}$ of the OGY formula, a learning phase is needed. Make a few small parameter changes and note the variations in the fixed point location.

The first experimental implementation of the OGY method was reported in 1990 by William Ditto, S. N. Rauseo, and Mark Spano [88], who stabilized chaotic oscillations of a flexible metal ribbon in an oscillating magnetic field. They had no detailed model of the system dynamics, but generated the time series return map from strain gauges on the base of the ribbon, then applied the OGY method to stabilize the chaotically oscillating ribbon to a fixed point and to a 2-cycle. Other examples followed rapidly: controlling chaos in chemical reactions [284], in lasers [136], in electronic circuits [179], and in wildlife populations [270].

Although these control strategies were developed to control chaos, they also can be used to suppress undesirable nonchaotic dynamics. An application to a cardiac model is given in [60].

Many physiological systems lack an obvious accessible control parameter. The OGY method of control is to move the fixed point so the next iterate lands very near the stable direction of the original fixed point, then return the fixed point to its original position. Without an accessible system parameter, how to move the fixed point is not so clear. To deal with this circumstance, in [130] Alan Garfinkel and coworkers report another approach, *proportional perturbation feedback* (PPF), based on the OGY idea. Rather than move the fixed point, they move the next iterate so it lies on the stable direction of the fixed point. In fact, the PPF formulas are similar to the OGY formulas. Their cardiac dynamics study provides a clear example. Garfinkel et al. use a portion of the interventricular septum of a rabbit heart, forced to arrhythmia by dosing the fluid surrounding the heart tissue with ouabain, a plant-derived poison, sometimes used on East African hunting arrows, that inhibits sodium-potassium transmembrane ion pumps thereby altering membrane voltage and causing cardiac arrhythmias. The data points are pairs of successive interbeat intervals, (I_{n-1}, I_n); regular heartbeats correspond to fixed points $I_n = I_{n-1}$. First comes a learning phase (lasting from 5 to 60 seconds in experiments) during which the fixed point and its stable and unstable directions are found. When a data point falls close to a fixed point, the next interbeat interval is shortened by applying a single electrical stimulus before the next beat would have occurred, with the timing selected so the next data point lies on the stable direction of the fixed point. The system approaches the fixed point, but eventually moves away. Then control is reapplied. In eight of eleven experiments this method replaced the arrhythmia with periodicity. In some instances, PPF control also eliminated the shortest interbeat intervals, suggesting that control may be an effective way to combat tachycardia. Another description of these experiments is in [379].

In [322] this method is applied to *anticontrol*, reducing the periodic behavior of a system, in this case spontaneously bursting in vitro networks of neurons. These findings suggest application to in vivo epileptic foci.

In [89] Ditto and coworkers reported on experiments to control atrial fibrillation (AFib) in twenty-five patients at Emory University. AFib was induced in each patient and an unstable fixed point determined from the first return map of the interbeat intervals. Again lacking an accessible system parameter, the PPF method was used to stabilize AFib. This worked in three-quarters of the patients, though in about half of these, the control was short-lived.

In [63] the authors point out that the hypothesis that AFib is well modeled by low-dimensional chaos, and consequently subject to OGY or PPF control, is not so well supported by recent data. Rather, AFib is better described as a high-dimensional combination of stochastic and nonlinear deterministic processes. Current control techniques are unlikely to be effective for AFib, but more sophisticated methods may work, and the current methods may work on other problems. For example, David Christini and coworkers [63] report that in 52 of 54 trials with 5 patients, *adaptive control* techniques (no learning phase is required) successfully suppressed cardiac alternans, showing that in at least some instances, control is effective in treating arrhythmias.

Adaptive control methods were developed in [61] and reported in [62]. This approach is important for clinical applications. Time spent in a learning phase before controlling cardiac arrhythmia would be unwelcome. Another variation is *dynamic control*, which allows for drifting system parameters. In [155] this method is applied to control cardiac alternans in a section of rabbit heart.

A *continuous control* strategy similar to OGY was applied by Bianca Ferreira and coworkers [115] to control irregular dynamics of surrogate ECG signals synthesized by three nonlinear oscillators coupled with a time delay.

Control of chaos is a very active area of research. The most successful applications have been in physical sciences, probably because almost every biological system has some high-dimensional elements. Whether or not control can be applied depends on how central the high-dimensional aspect is to the part of the process we wish to control. Do the high-dimensional dynamics contribute a tiny jiggle, wild swings swamping the low-dimensional behavior, or something in between? The OGY method will work in the first case, won't in the second, and might in the third. In this case, "try it and see" is the best advice.

When radically new techniques are developed, science grows by trying all sorts of applications. Some will work, some won't. Studying this distinction refines our ability to determine conditions necessary for the new technique. The ability to treat AFib by chaos control was a long shot, but the potential value of a positive outcome justifies the effort.

Science has an old saying, source unknown to me: If everything you try works, you aren't trying things that are crazy enough. A little bit of crazy can spark creativity.

Practice Problems

13.9.1. Here we'll use the matrices (i) and (ii)

$$\text{(i)} \quad \begin{bmatrix} 1 & 2 \\ 2 & 1 \end{bmatrix} \qquad \text{(ii)} \quad \begin{bmatrix} 1 & 2+\epsilon \\ 2 & 1 \end{bmatrix}$$

where ϵ is a positive constant.
(a) For matrix (i) show that $\vec{e}_u$ and $\vec{f}_u$ are parallel.
(b) For matrix (ii) compute the angle between $\vec{e}_u$ and $\vec{f}_u$.

13.9.2. For this problem and for Exercises 13.9.4–13.9.7 we'll use the $r = 2$ tent map (2.4)

$$T(x) = \begin{cases} 2 \cdot x & \text{for } x \leq 1/2 \\ 2 - 2 \cdot x & \text{for } x \geq 1/2 \end{cases}$$

because the arithmetic is easier than with the logistic map.
(a) Find the 2-cycle points x_a and x_b of this tent map.
(b) If the iterate x_n is near x_a, find the formula to adjust r so $x_{n+1} = x_a$.

Practice Problem Solutions

13.9.1. (a) The eigenvalues of matrix (i) are $\lambda_u = 3$ and $\lambda_s = -1$. The vector $\vec{e}_u$ is the eigenvector corresponding to λ_u, so

$$\vec{e}_u = \begin{bmatrix} 1 \\ 1 \end{bmatrix} \quad \text{The eigenvector corresponding to } \lambda_s \text{ is } \vec{e}_s = \begin{bmatrix} 1 \\ -1 \end{bmatrix}$$

In the derivation of the OGY formula, we used unit eigenvectors. We needn't bother with this here because to establish that two vectors are parallel, vector length is unimportant. To find $\vec{f}_u$, form the matrix $U = \langle \vec{e}_u \vec{e}_s \rangle$ and compute U^{-1}. Then $\vec{f}_u$ is the top row of the inverse.

$$U = \begin{bmatrix} 1 & 1 \\ 1 & -1 \end{bmatrix} \quad \text{so} \quad U^{-1} = \begin{bmatrix} \frac{1}{2} & \frac{1}{2} \\ \frac{1}{2} & -\frac{1}{2} \end{bmatrix} \quad \text{and} \quad \vec{f}_u = \begin{bmatrix} \frac{1}{2} & \frac{1}{2} \end{bmatrix}$$

We see that viewed as vectors in the plane, $\vec{e}_u$ and $\vec{f}_u$ are parallel.
(b) For matrix (ii) the eigenvalues are $\lambda_u = 1 + \sqrt{4+2\epsilon}$ and $\lambda_s = 1 - \sqrt{4+2\epsilon}$. The eigenvectors are

$$\vec{e}_u = \begin{bmatrix} 1 \\ \sqrt{\dfrac{2}{2+\epsilon}} \end{bmatrix} \quad \text{and} \quad \vec{e}_s = \begin{bmatrix} 1 \\ -\sqrt{\dfrac{2}{2+\epsilon}} \end{bmatrix}$$

Next, U and U^{-1} are

$$U = \begin{bmatrix} 1 & 1 \\ \sqrt{\frac{2}{2+\epsilon}} & -\sqrt{\frac{2}{2+\epsilon}} \end{bmatrix} \quad \text{and} \quad U^{-1} = \begin{bmatrix} \frac{1}{2} & \frac{\sqrt{2+\epsilon}}{2\sqrt{2}} \\ \frac{1}{2} & -\frac{\sqrt{2+\epsilon}}{2\sqrt{2}} \end{bmatrix}$$

Then the vector $\vec{f}_u = \begin{bmatrix} \frac{1}{2} & \frac{\sqrt{2+\epsilon}}{2\sqrt{2}} \end{bmatrix}$. To find the angle between $\vec{e}_u$ and $\vec{f}_u$, solve Eq. (13.52) for $\cos(\theta)$,

$$\cos(\theta) = \frac{\vec{f}_u \vec{e}_u}{\|\vec{f}_u\|\,\|\vec{e}_u\|} = \frac{2\sqrt{4+2\epsilon}}{4+\epsilon}$$

The left image of Fig. 13.45 is a plot of $\cos(\theta)$ as a function of ϵ for $0 \le \theta \le 20$; the right image is a graph of θ (in radians). For a modest range of ϵ values, we see that $\vec{f}_u$ and $\vec{e}_u$ are nearly parallel.

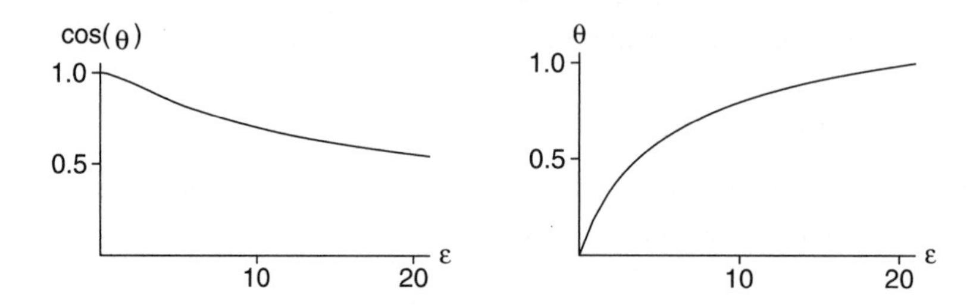

Figure 13.45. Left: $\cos(\theta)$ as a function of ϵ. Right: θ as a function of ϵ.

13.9.2. (a) The 2-cycle points are the fixed points of $T^2(x) = T(T(x))$. We'll start with a formula for T^2:

$$T(T(x)) = \begin{cases} 2 \cdot T(x) & \text{for } T(x) \le 1/2 \\ 2 - 2 \cdot T(x) & \text{for } T(x) \ge 1/2 \end{cases}$$

$$= \begin{cases} 2 \cdot 2x & \text{for } 0 \le 2x \le 1/2 \text{ and } 0 \le x \le 1/2 \\ 2 \cdot (2-2x) & \text{for } 0 \le 2-2x \le 1/2 \text{ and } 1/2 \le x \le 1 \\ 2 - 2 \cdot 2x & \text{for } 2x \ge 1/2 \text{ and } 0 \le x \le 1/2 \\ 2 - 2 \cdot (2-2x) & \text{for } 2-2x \ge 1/2 \text{ and } 1/2 \le x \le 1 \end{cases}$$

$$= \begin{cases} 4x & \text{for } 0 \le x \le 1/4 \\ 4 - 4x & \text{for } 3/4 \le x \le 1 \\ 2 - 4x & \text{for } 1/4 \le x \le 1/2 \\ -2 + 4x & \text{for } 1/2 \le x \le 3/4 \end{cases}$$

Here, for example, the conditions $0 \leq 2x \leq 1/2$ and $0 \leq x \leq 1/2$ arise because we want $0 \leq T(x) \leq 1/2$ and also $T(x) = 2x$.

The graph of T^2 intersects the diagonal line once in each interval $[0,1/4]$, $[1/4,1/2]$, $[1/2,3/4]$, and $[3/4,1]$. In the first and third intervals, these intersections are the fixed points of T; the other two intersections are the 2-cycle for T. So

$$x_a = 2 - 4x_a, \quad \text{so } x_a = 2/5$$

$$x_b = 4 - 4x_b, \quad \text{so } x_b = 4/5$$

(b) Suppose x_n is near x_a. Then $x_n < 1/2$ and so $T(x_n) = r \cdot x_n$. The value of r to implement control is $r \cdot x_n = 2/5$ and so $r = 2/(5x_n)$.

Exercises

13.9.1. Show that the vectors $\vec{e}_u$ and $\vec{e}_s$ are parallel if and only if $\det W = 0$. Hint: recall Eqs. 13.51 and 13.52.

Use these matrices for Exercises 13.9.2 and 13.9.3.

$$\text{(a)} \begin{bmatrix} 1 & 3+\epsilon \\ 3 & 1 \end{bmatrix} \quad \text{(b)} \begin{bmatrix} 1 & \epsilon \\ 1 & 1 \end{bmatrix} \quad \text{(c)} \begin{bmatrix} 1 & 2+\epsilon \\ 3 & 1 \end{bmatrix} \quad \text{(d)} \begin{bmatrix} -1 & 1+\epsilon \\ 0 & 1 \end{bmatrix}$$

13.9.2. For matrices (a) and (b), compute the angle between $\vec{f}_u$ and $\vec{e}_u$ as a function of ϵ.

13.9.3. For matrices (c) and (d), compute the angle between $\vec{f}_u$ and $\vec{e}_u$ as a function of ϵ.

13.9.4. Continuing with Practice Problem 13.9.2,
(a) suppose x_n is near x_a. Find the adjustment of r to make $x_{n+1} = x_b$.
(b) Suppose x_n is near x_b. Find the adjustment of r to make $x_{n+1} = x_b$.
(c) Suppose x_n is near x_b. Find the adjustment of r to make $x_{n+1} = x_a$.

13.9.5. Find the coordinates of the two 3-cycles $\{x_d, x_e, x_f\}$ and $\{x_g, x_h, x_i\}$ of $T(x)$ for $r = 2$. Suppose $x_d < x_g$.

13.9.6. Using the coordinates of the 3-cycles found in Exercise 13.9.5,
(a) suppose x_n is near x_d. Find the adjustment of r to make $x_{n+1} = x_e$.
(b) Suppose x_n is near x_f. Find the adjustment of r to make $x_{n+1} = x_e$.
(c) Suppose x_n is near x_d. Find the adjustment of r to make $x_{n+1} = x_h$.

13.9.7. For the $r = 2$ tent map, adjust r producing $x_{n+1} = T(x_n)$, then return to $r = 2$.
(a) Suppose $x_n < 1/4$. Find the adjustment of r so that $x_{n+2} = 2/3$.
(b) Suppose $x_n < 1/4$. Find the adjustment of r so that $x_{n+3} = 2/3$.

13.9.8. Suppose

$$S(x) = \begin{cases} r \cdot x & \text{for } 0 \leq x < 1/3 \\ r \cdot x - r/3 & \text{for } 1/3 \leq x < 2/3 \\ r \cdot x - 2r/3 & \text{for } 2/3 \leq x \leq 1 \end{cases}$$

(a) Show that $x = 1/2$ is a fixed point for the $r = 3$ function.
(b) Suppose $x_n < 1/3$. Find the adjustment of r so that $x_{n+1} = 1/2$.
(c) Suppose x_n is close to 0. Find the adjustment of r so that $x_{n+1} < 1/3$ and, returning r to 3, $x_{n+2} = 1/2$.

13.9.9. Find a recent reference to an application of controlling chaos to cardiac arrhythmias. Write a summary of the method and results.

13.9.10. Take the dot product of both sides of

$$\vec{v}_{n+1} - s\vec{g} = (\lambda_u \vec{e}_u \vec{f}_u + \lambda_s \vec{e}_s \vec{f}_s)(\vec{v}_n - s\vec{g})$$

with $\vec{e}_u$, set $\vec{e}_u \cdot \vec{v}_{n+1} = 0$, and derive another formula for parameter adjustment to control chaos. Compare this formula with Eq. (13.53).

Chapter 14 Stochastic models

The reductionist approach to science was the dominant voice during the era of Newton and Descartes. Resolve a system into its simplest pieces, figure out how the pieces work, then glue them back together. Reductionism is a bit more complicated, but that's the core idea. It worked well for simple problems in mechanics, but not so much with more complex systems, especially living systems. Mary Shelley's fantasy is just a fantasy. Sew together body parts and jump start with electricity does not produce a living, breathing, thinking person. And adequately complex nonliving systems can exhibit surprising behavior. The rowhammer problem in dynamic RAM is a good example. (If rowhammer is unfamiliar, Google or https://xkcd.com/1938 are good starts.)

Even in systems where reductionism could work in principle, the calculations can be too daunting. Too many trees obscure the forest. The air in a room provides an extreme example. Assuming air molecules are tiny billiard balls interacting through collisions governed by Newtonian mechanics, if we knew the position and momentum of every molecule in the room, we could predict the future movement of every molecule in the room. But who cares? Are you interested in the movement of a particular nitrogen molecule? I'm not. Averages of densities and movements of all the molecules give the pressure and temperature in the room. These are features important to us. Probability is a tool we use to investigate aggregate properties.

In this chapter, we study systems that can be in several, maybe many, states, observe or deduce the probability of moving from one state to another, and sometimes use this to predict the long-term behavior of the system, at least as far as the partition into these states is concerned. For this, probabilities suffice: we need not know the underlying mechanics.

14.1 MARKOV CHAINS

A stochastic process is any sequence of measurements, ordered by time, and related in a way that involves only probabilistic prediction of the next measurement from the current. The apparent lack of determinism can come from inherent uncertainties or from our incomplete understanding of process details. Think of a sequence of coin tosses, or the intervals between successive beats of your heart, or the daily number of new cases of flu in Manhattan.

Suppose we partition the set of possible values of each measurement into N subsets. (Recall a partition means every value lies in exactly one of these subsets.) Call these subsets the *states* of the system, and call the collection of states the *state space*. Also, suppose we measure time in discrete steps: $t = 0, t = 1, t = 2, \ldots$. This last assumption is only for convenience; there are continuous-time stochastic processes. At time t let X_t denote the state in which the measurement lies. A stochastic process configured in this way is called a *Markov chain* if the state X_{t+1} depends on the state X_t and not on X_s for any $s < t$, and this dependence is the same for all t. In other words, neither the value of t nor how the system got to state X_t has any bearing on what happens next, that is, on X_{t+1}. There are more general models in which the state transition probabilities change with time. We won't pursue these, but Google can take you there.

Markov chains are good models of many biological processes. For example, whether or not a transmembrane channel of a cell opens at a given time depends on only the energy configuration of gating molecules, not on how the molecules got to that configuration. We'll see that we can figure out some things about the long-term behavior of a Markov chain from the mathematics of the chain. Eigenvalues and eigenvectors will be useful here; we didn't learn about them just to solve some differential equations.

The dynamics of an N-state Markov chain are specified by N^2 *transition probabilities*. These are conditional probabilities of the form

$$P(X_{t+1} = i | X_t = j) \text{ which we'll denote as } P(j \to i) \text{ or } p_{ij}$$

Some authors reverse the subscript order. We use this order because it agrees with the order of states written in the conditional probability. In addition, this is the subscript order we'll use for Leslie matrices in Sect. 14.3. Reversing the roles of rows and columns of a matrix half-way through the chapter did not appear to be a good idea. So we'll stick with reading p_{ij} as the probability of going from state j to state i.

We'll look at examples after we've established some basic properties.

Given a list of all states, the transition graph of a Markov chain has one node for each state and one arrow for each transition. In Fig. 14.1 each node is represented by a circle labeled with the state number. The arrow $j \to i$ would be labeled with the transition probability $P(j \to i) = p_{ij}$.

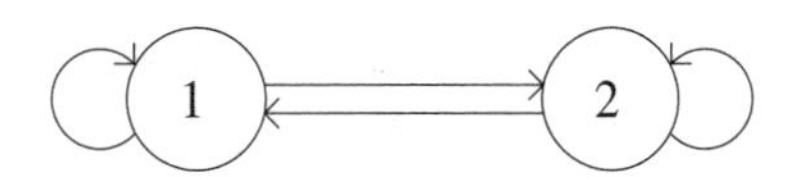

Figure 14.1. A transition graph.

To begin to understand the effects of these probabilities, we consider three examples. In each, state 1 corresponds to the instruction "move up one step" and state 2 to the instruction "move down one step." Here are the transition probabilities for these examples.

Example 1 $P(1 \to 1) = P(2 \to 2) = P(1 \to 2) = P(2 \to 1) = 0.5$

Example 2 $P(1 \to 1) = P(2 \to 2) = 0.7,\ P(1 \to 2) = P(2 \to 1) = 0.3$

Example 3 $P(1 \to 1) = P(2 \to 2) = 0.3,\ P(1 \to 2) = P(2 \to 1) = 0.7$

The choice of transitions, quantified by these probabilities, produces a sequence of moves up and down. We can represent each sequence graphically in the obvious way: time along the horizontal axis, the result of up and down moves along the vertical axis. Plots of 5000 points of these examples are shown in Fig. 14.2. Example 1 is an approximation of 1-dimensional Brownian motion. In Example 2, the motion tends to stay in the same direction, giving rise to longer excursions up and down. In Example 3, the motion tends to reverse direction, giving much shorter runs of consecutive upward or downward motion.

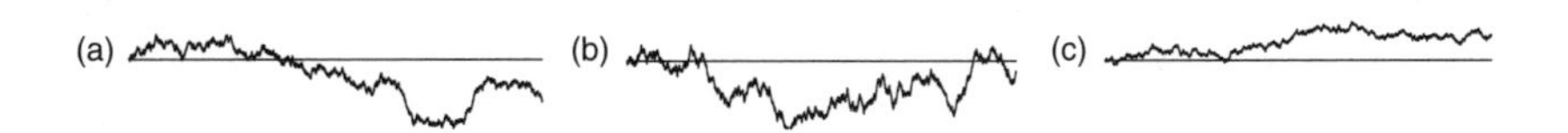

Figure 14.2. Time series plots for Examples 1, 2, and 3.

For now we'll stay with a system with two states. At time step t we'll say the fraction of the population in state 1 is A_t and the fraction in state 2 is B_t. Because every member of the population must be in state 1 or 2, $A_t + B_t = 1$. Then at the next time step, $t+1$, we expect the corresponding fractions A_{t+1} and B_{t+1} to be given by

$$A_{t+1} = P(1 \to 1) \cdot A_t + P(2 \to 1) \cdot B_t = p_{11} \cdot A_t + p_{12} \cdot B_t$$
$$B_{t+1} = P(1 \to 2) \cdot A_t + P(2 \to 2) \cdot B_t = p_{21} \cdot A_t + p_{22} \cdot B_t \qquad (14.1)$$

This should make sense. For example, in order to be in state 1 at time $t+1$, we must be in state 1 at t and stay in state 1, or we must be in state 2 at t and make the transition from 2 to 1.

Then $[p_{ij}]$ is the *transition matrix*, and Eq. (14.1) can be written as

$$\begin{bmatrix} A_{t+1} \\ B_{t+1} \end{bmatrix} = \begin{bmatrix} p_{11} & p_{12} \\ p_{21} & p_{22} \end{bmatrix} \begin{bmatrix} A_t \\ B_t \end{bmatrix} = P \begin{bmatrix} A_t \\ B_t \end{bmatrix}$$

We use the column index to indicate the current state (the "from" state) and the row index to indicate the next state (the "to" state). Note that each column of the matrix must sum to 1, because, for example, from state 1 the system must go to either state 1 or state 2 and so $p_{11} + p_{21} = 1$.

Starting from A_0 and B_0 and iterating, we obtain

$$\begin{bmatrix} A_{t+1} \\ B_{t+1} \end{bmatrix} = P \begin{bmatrix} A_t \\ B_t \end{bmatrix} = \cdots = P^{t+1} \begin{bmatrix} A_0 \\ B_0 \end{bmatrix} \qquad (14.2)$$

This seems plausible, but there's a subtle point hidden in this expression: the probability of going from state j to state i in $t+1$ steps is

$$\left[P^{t+1}\right]_{ij} = p_{ij}^{(t+1)} \text{ not } \left(p_{ij}\right)^{t+1} \qquad (14.3)$$

That is, the probability is the (i,j) element of the matrix P^{t+1}, not the $(t+1)$st power of the (i,j) element of the matrix P. We'll work through this for $t = 1$; higher values of t use no additional ideas, just more ink. To go from j to i in $t+1 = 2$ steps we must follow one of two paths:

$$j \to 1 \to i \text{ or } j \to 2 \to i \qquad (14.4)$$

That is, the event $j \to i$ in 2 steps is the union of the two events of Eq.(14.4), mutually exclusive because if you go from j to 1 to i, you do not also go from j to 2 to 1. Then by Eq. (11.5),

$$p_{ij}^{(2)} = P(\{j \to 1 \to i\} \cup \{j \to 2 \to i\})$$
$$= P(j \to 1 \to i) + P(j \to 2 \to i) \qquad (14.5)$$

Now because they occur in successive time steps, the event $j \to 1$ is independent of the event $1 \to i$. Similarly, $j \to 2$ is independent of $2 \to i$, so by Eq. (11.6),

$$P(j \to 1 \to i) = P(j \to 1) \cdot P(1 \to i) = p_{1j} \cdot p_{i1}$$
$$P(j \to 2 \to i) = P(j \to 2) \cdot P(2 \to i) = p_{2j} \cdot p_{i2} \qquad (14.6)$$

Combining Eqs. (14.5) and (14.6) we find

$$p_{ij}^{(2)} = p_{1j} \cdot p_{i1} + p_{2j} \cdot p_{i2} = p_{i1} \cdot p_{1j} + p_{i2} \cdot p_{2j} = \sum_{k=1}^{2} p_{ik} \cdot p_{kj} \tag{14.7}$$

But this is just the formula for multiplying two matrices, specifically, multiplying P by itself, so

$$[p_{ij}^{(2)}] = P^2$$

Repeating this argument many times (or arguing by induction), we find that for all $t \geq 1$,

$$[p_{ij}^{(t)}] = P^t \tag{14.8}$$

and we see that Eq. (14.2) is valid. For a Markov chain with n states, writing out the n^2 equations of Eq. (14.8) and expressing P^t as $P^{t-s} \cdot P^s$, we have the *Chapman-Kolmogorov equations*,

$$p_{ij}^{(t)} = \sum_{k=1}^{n} p_{ik}^{(t-s)} p_{kj}^{(s)} \tag{14.9}$$

central to much of what we'll do about Markov chains.

For some Markov chains P and some initial distributions $\vec{v}$, there is an *equilibrium distribution* $\vec{w}$ that satisfies

$$\lim_{m\to\infty} P^m \vec{v} = \vec{w}$$

Does this limit exist, and if it does, do we get the same $\vec{w}$ for all $\vec{v}$? We'll see.

A matrix of non-negative entries for which each column sums to 1 is called a *stochastic matrix*. As mentioned above, some authors reverse the roles of rows and columns in their matrix representation of Markov chains.

Example 14.1.1. *Equilibrium distribution for a two-state system.* Find the equilibrium distribution of states for the transition matrix

$$P = \begin{bmatrix} 0.7 & 0.4 \\ 0.3 & 0.6 \end{bmatrix}$$

To compute the distribution of states far into the future, we need a high power of P. (Okay, almost always we'll be able to avoid computing high powers of P, but this one time we'll approach the problem this way.) As long as we take the power to be 2^n for some n, this is fairly easy (for a computer) to do.

$$P^2 = P \cdot P, \quad P^4 = P^2 \cdot P^2, \quad \ldots, \quad P^{1024} = P^{512} \cdot P^{512}, \text{ and so on.}$$

In this way, we obtain

$$P^4 = \begin{bmatrix} 0.5749 & 0.5668 \\ 0.4251 & 0.4332 \end{bmatrix}, \quad P^8 = P^{16} = \cdots = \begin{bmatrix} 0.571429 & 0.571429 \\ 0.428571 & 0.428571 \end{bmatrix}$$

So already we have a surprise. The powers of P, at least these powers, seem to have converged to a fixed matrix. Call this matrix P^∞. By observing that, for example, the sequence $P^3, P^5, P^9, \ldots, P^{2^n+1}$ converges to the same limit, we see that all high enough powers of P, not just P^{2^n}, converge to P^∞.

Now take some initial distribution of states 1 and 2 and find the equilibrium distribution of these states. Let's start with $A_0 = 0.5$ and $B_0 = 0.5$. Then the equilibrium distribution is a column of P^∞:

$$\begin{bmatrix} A_\infty \\ B_\infty \end{bmatrix} = P^\infty \begin{bmatrix} 0.5 \\ 0.5 \end{bmatrix} = \begin{bmatrix} 0.571429 \\ 0.428571 \end{bmatrix}$$

Here's another surprise. If we take $A_0 = 0.3$ and $B_0 = 0.7$ we get

$$\begin{bmatrix} A_\infty \\ B_\infty \end{bmatrix} = P^\infty \begin{bmatrix} 0.3 \\ 0.7 \end{bmatrix} = \begin{bmatrix} 0.571429 \\ 0.428571 \end{bmatrix}$$

What about $A_0 = 0$ and $B_0 = 1$?

$$\begin{bmatrix} A_\infty \\ B_\infty \end{bmatrix} = P^\infty \begin{bmatrix} 0 \\ 1 \end{bmatrix} = \begin{bmatrix} 0.571429 \\ 0.428571 \end{bmatrix}$$

In fact, so long as $B_0 = 1 - A_0$ (which must be true for 2-state chains), we get

$$\begin{bmatrix} A_\infty \\ B_\infty \end{bmatrix} = P^\infty \begin{bmatrix} A_0 \\ B_0 \end{bmatrix} = \begin{bmatrix} 0.571429 \\ 0.428571 \end{bmatrix} \tag{14.10}$$

No matter where we start, we wind up with the same distribution. □

From this example we are led to several questions.

Question 1. Why do we call this limiting distribution the equilibrium distribution?

Question 2. Does every Markov chain have an equilibrium distribution?

Question 3. If a Markov chain has an equilibrium distribution, is that distribution unique?

Question 4. If a Markov chain has an equilibrium distribution, can we find the equilibrium distribution without computing a high power of P?

If we assume there is a limiting matrix $P^\infty = \lim_{n\to\infty} P^n$, the first question is easy to answer. Because

$$\begin{bmatrix} A_\infty \\ B_\infty \end{bmatrix} = P^\infty \begin{bmatrix} A_0 \\ B_0 \end{bmatrix} \quad \text{for any initial distribution} \quad \begin{bmatrix} A_0 \\ B_0 \end{bmatrix}$$

we can take the initial distribution to be the equilibrium distribution

$$\begin{bmatrix} A_\infty \\ B_\infty \end{bmatrix} = P^\infty \begin{bmatrix} A_\infty \\ B_\infty \end{bmatrix} \text{ so } P\begin{bmatrix} A_\infty \\ B_\infty \end{bmatrix} = P(P^\infty)\begin{bmatrix} A_\infty \\ B_\infty \end{bmatrix} = P^\infty \begin{bmatrix} A_\infty \\ B_\infty \end{bmatrix} = \begin{bmatrix} A_\infty \\ B_\infty \end{bmatrix}$$

because $P(P^\infty) = P^\infty$. This shows the equilibrium distribution is a fixed point of P. We can apply Eq. (14.10) to show that this fixed point is stable: for any initial distribution, the iterated application of P generates a sequence of vectors that converges to this fixed point.

$$\begin{bmatrix} A_\infty \\ B_\infty \end{bmatrix} = P^\infty \begin{bmatrix} A_0 \\ B_0 \end{bmatrix} = \left(\lim_{m\to\infty} P^m\right)\begin{bmatrix} A_0 \\ B_0 \end{bmatrix} = \lim_{m\to\infty}\left(P^m \begin{bmatrix} A_0 \\ B_0 \end{bmatrix}\right)$$

The second question also is easy, though the answer isn't what we might want. The answer is "No." Here's an example.

Example 14.1.2. *Distributions that cycle.* We'll look at the behavior of any initial distribution under repeated application of this matrix $P = \begin{bmatrix} 0 & 1 \\ 1 & 0 \end{bmatrix}$. The first few iterates are

$$P\begin{bmatrix} A_0 \\ B_0 \end{bmatrix} = \begin{bmatrix} B_0 \\ A_0 \end{bmatrix}, P^2\begin{bmatrix} A_0 \\ B_0 \end{bmatrix} = \begin{bmatrix} A_0 \\ B_0 \end{bmatrix}, P^3\begin{bmatrix} A_0 \\ B_0 \end{bmatrix} = \begin{bmatrix} B_0 \\ A_0 \end{bmatrix}, P^4\begin{bmatrix} A_0 \\ B_0 \end{bmatrix} = \begin{bmatrix} A_0 \\ B_0 \end{bmatrix}$$

and so on. There's no equilibrium distribution: the iterates just keep oscillating from one to the other. Fortunately, it turns out that this is the worst problem that can happen. If a Markov chain does not have a cycle, then it has an equilibrium distribution. □

The answer to the third question is "No." Different initial distributions can lead to different equilibrium distributions. Here's an example.

Example 14.1.3. *Multiple equilibrium distributions.* For all non-negative A_0 and B_0 with $A_0 + B_0 = 1$,

$$\begin{bmatrix} 0.5 & 0.5 & 0 & 0 \\ 0.5 & 0.5 & 0 & 0 \\ 0 & 0 & 0.5 & 0.5 \\ 0 & 0 & 0.5 & 0.5 \end{bmatrix}\begin{bmatrix} A_0 \\ B_0 \\ 0 \\ 0 \end{bmatrix} = \begin{bmatrix} 0.5 \\ 0.5 \\ 0 \\ 0 \end{bmatrix} = \vec{v}_1$$

$$\begin{bmatrix} 0.5 & 0.5 & 0 & 0 \\ 0.5 & 0.5 & 0 & 0 \\ 0 & 0 & 0.5 & 0.5 \\ 0 & 0 & 0.5 & 0.5 \end{bmatrix}\begin{bmatrix} 0 \\ 0 \\ A_0 \\ B_0 \end{bmatrix} = \begin{bmatrix} 0 \\ 0 \\ 0.5 \\ 0.5 \end{bmatrix} = \vec{v}_2$$

Consequently, any initial distribution with the last two entries 0 converges (in one step) to the equilibrium distribution $\vec{v}_1$, and any initial distribution with the first two entries 0 converges (in one step) to the equilibrium distribution $\vec{v}_2$. So we see that this Markov chain has at least two equilibrium distributions. Try an experiment to see what happens if all four entries of the initial distribution are non-zero.

But really, this isn't such a surprise: the Markov chain consists of two pieces—states 1 and 2, and states 3 and 4—that don't talk with one another. And again, something like this is the worst that can happen. □

The answer to the fourth question is "Yes." You must have noticed in Example 14.1.1 that each column of P^{∞} is equal to the vector of the eventual distribution. This is a clue.

Example 14.1.4. *Eigenvectors and equilibrium distributions.* The matrix P of Example 14.1.1 has eigenvalues 1 and 0.3. An eigenvector for 1 is $\begin{bmatrix} 1 & 3/4 \end{bmatrix}^{\text{tr}}$. Here the superscript "tr" denotes the *transpose* defined by $[a_{ij}]^{\text{tr}} = [a_{ji}]$, so the transpose of a row vector is a column vector. This cannot be a distribution of states of the system because the entries of this eigenvector do not sum to 1. Because any non-zero multiple of an eigenvector still is an eigenvector, we can form the *unit eigenvector* by dividing each entry by the sum of the eigenvector entries:

$$\begin{bmatrix} 1/(1+3/4) \\ 3/4/(1+3/4) \end{bmatrix} = \begin{bmatrix} 4/7 \\ 3/7 \end{bmatrix} = \begin{bmatrix} A_{\infty} \\ B_{\infty} \end{bmatrix} \approx \begin{bmatrix} 0.571429 \\ 0.428571 \end{bmatrix}$$

In some settings, a unit vector is a vector $\begin{bmatrix} a & b \end{bmatrix}^{\text{tr}}$ of Euclidean length 1, that is, $\sqrt{a^2+b^2} = 1$. When the vector represents the distribution of the population among the system states, the vector entries sum to 1.

Is it just a coincidence that the entries of the eigenvector of $\lambda = 1$ give the eventual distribution? Of course not. Here's why. An eigenvector of the eigenvalue 0.3 is $\begin{bmatrix} 1 & -1 \end{bmatrix}^{\text{tr}}$. In a moment we'll see that given any initial distribution A_0 and B_0, there are constants c and d giving

$$\begin{bmatrix} A_0 \\ B_0 \end{bmatrix} = c \begin{bmatrix} 4/7 \\ 3/7 \end{bmatrix} + d \begin{bmatrix} 1 \\ -1 \end{bmatrix} = \begin{bmatrix} 4/7 & 1 \\ 3/7 & -1 \end{bmatrix} \begin{bmatrix} c \\ d \end{bmatrix} = N \begin{bmatrix} c \\ d \end{bmatrix}$$

where N is the matrix whose columns are eigenvectors of P. So long as the matrix N is invertible, that is, $\det(N) \neq 0$, the coefficients c and d can be found. For systems in the plane, $\det(N) \neq 0$ if the eigenvectors of P are not multiples of one another, certainly the case here. Then

$$\begin{bmatrix} A_m \\ B_m \end{bmatrix} = P^m \begin{bmatrix} A_0 \\ B_0 \end{bmatrix} = P^m \left(c \begin{bmatrix} 4/7 \\ 3/7 \end{bmatrix} + d \begin{bmatrix} 1 \\ -1 \end{bmatrix} \right)$$
$$= cP^n \begin{bmatrix} 4/7 \\ 3/7 \end{bmatrix} + dP^m \begin{bmatrix} 1 \\ -1 \end{bmatrix}$$
$$= c \cdot 1^m \begin{bmatrix} 4/7 \\ 3/7 \end{bmatrix} + d \cdot 0.3^m \begin{bmatrix} 1 \\ -1 \end{bmatrix} \to c \cdot \begin{bmatrix} 4/7 \\ 3/7 \end{bmatrix} \quad \text{as } m \to \infty$$

So we see that every initial population of states evolves to a multiple of the eigenvector of the eigenvalue 1. □

You might wonder about the constant c and the fact that both $A_0 + B_0 = 1$ and $A_m + B_m = 1$. In this situation we always can take $c = 1$. Here's why. Because $B_0 = 1 - A_0$,

$$\begin{bmatrix} A_0 \\ B_0 \end{bmatrix} = \begin{bmatrix} A_0 \\ 1 - A_0 \end{bmatrix} = 1 \begin{bmatrix} 4/7 \\ 3/7 \end{bmatrix} + d \begin{bmatrix} 1 \\ -1 \end{bmatrix}$$

Then the first component equation is $A_0 = 4/7 + d$, and the second component equation is $1 - A_0 = 3/7 - d$, equivalent to the first. So A_0 determines d and the population converges monotonically to the unit dominant eigenvector, A_m increasing and B_m decreasing if $d > 0$ (if $A_0 < 4/7$), A_m decreasing and B_m increasing if $d < 0$.

One way the result of Example 14.1.4 can be generalized is Prop. 14.1.1. The setting is that P has one eigenvalue, which we might as well call λ_1, strictly larger than the magnitudes of all the others. In Sect. 14.2 we'll find simple conditions on P that guarantee this.

Proposition 14.1.1. Suppose $\lambda_1, \dots, \lambda_n$ are the eigenvalues of P with eigenvectors $\vec{v}_1, \dots, \vec{v}_n$, suppose $\lambda_1 > |\lambda_i|$ for $2 \leq i \leq n$, and suppose the matrix A whose columns are $\vec{v}_1, \dots, \vec{v}_n$ is invertible. Then for any vector $\vec{u}$, there is a constant c_1 for which

$$\lim_{m \to \infty} \left(\frac{1}{\lambda_1^m} P^m \vec{u} - c_1 \vec{v}_1 \right) = 0 \tag{14.11}$$

Before we see the proof, let's figure out why Eq. (14.11) is a generalization of Example 14.1.4. The role of the eigenvalue $\lambda = 1$ was not so obvious in that example because a high power of 1 still is 1. If $\lambda_1 > 1$ we must be a bit more careful. We can interpret Eq. (14.11) as showing that for any initial distribution $\vec{u}$, $P^m \vec{u}$ converges to $c_1 \lambda_1^m \vec{v}_1$. That is, the population grows at the rate λ_1 and the distribution converges to $\vec{v}_1$.

Proof. The proof is just an elaboration of Example 14.1.4, modified to allow for the possibility that $\lambda_1 \neq 1$.

Given any vector $\vec{u}$ representing the initial distribution of the system into its various states, because A is invertible there are constants $c_1, c_2, \ldots, c_n$ for which

$$\vec{u} = c_1\vec{v}_1 + c_2\vec{v}_2 + \cdots + c_n\vec{v}_n = \left[\vec{v}_1\vec{v}_2\ldots\vec{v}_n\right]\begin{bmatrix} c_1 \\ c_2 \\ \cdots \\ c_n \end{bmatrix} = A\begin{bmatrix} c_1 \\ c_2 \\ \cdots \\ c_n \end{bmatrix}$$

Here's why: multiply A^{-1} on the left of both sides of the equation. Then the coefficients $c_1, c_2, \ldots, c_n$ are the entries of $A^{-1}\vec{u}$. When any vector $\vec{u}$ can be represented as a unique linear combination of $\vec{v}_1, \ldots, \vec{v}_n$, we say $\vec{v}_1, \ldots, \vec{v}_n$ form a *basis* for the states of the system.

Now we're ready for the final calculation

$$\begin{aligned}
&\lim_{m\to\infty}\left(\frac{1}{\lambda_1^m}P^m\vec{u} - c_1\vec{v}_1\right) \\
&= \lim_{m\to\infty}\left(\frac{1}{\lambda_1^m}P^m\left(c_1\vec{v}_1 + c_2\vec{v}_2 + \cdots + c_n\vec{v}_n\right) - c_1\vec{v}_1\right) \\
&= \lim_{m\to\infty}\left(\frac{1}{\lambda_1^m}\left(c_1\lambda_1^m\vec{v}_1 + c_2\lambda_2^m\vec{v}_2 + \cdots + c_n\lambda_n^m\vec{v}_n\right) - c_1\vec{v}_1\right) \\
&= \lim_{m\to\infty}\left(c_1\vec{v}_1 + c_2\left(\frac{\lambda_2}{\lambda_1}\right)^m\vec{v}_2 + \cdots + c_n\left(\frac{\lambda_n}{\lambda_1}\right)^m\vec{v}_n - c_1\vec{v}_1\right) = 0
\end{aligned}$$

For $2 \leq i \leq n$, $\dfrac{|\lambda_i|}{\lambda_1} < 1$ and so $\lim_{m\to\infty}\left(\dfrac{\lambda_i}{\lambda_1}\right)^m = 0$. This gives the last equality. □

A (necessarily positive) eigenvalue strictly larger than the magnitudes of all the other eigenvalues is called the *dominant eigenvalue*. Prop. 14.1.1 shows that if there is a dominant eigenvalue, the corresponding unit eigenvector describes the long-term distribution of the states of the system. In population models, the dominant eigenvalue λ_1 can be greater than 1, and Prop. 14.1.1 then shows that λ_1 is the eventual growth rate of the population. In Sect. 14.2 we'll discuss conditions guaranteeing the existence of a dominant eigenvalue.

But before we get to the theorem of Sect. 14.2, let's see a few other things we can figure out about Markov chains. We'll need some notation. The states of a Markov chain can be partitioned into two classes, *transient* and *ergodic*. When a Markov chain leaves the transient class, it cannot return; when a Markov chain enters the ergodic class, it cannot exit. An ergodic class can contain absorbing states and periodic states. A state i is *absorbing* if $p_{ii} = 1$, that is, if state i can be followed only by itself. States 1, 2, $\ldots, q$ are *periodic* if $p_{21} = 1$, $p_{32} = 1$, $\ldots$,

$p_{q\,q-1} = 1$, and $p_{1q} = 1$. States that can be visited again and again are called *recurrent*. Every Markov chain with a finite number of states must have at least one recurrent state: under repeated iteration, eventually we run out of new states to enter and must revisit at least one.

The ergodic class is called the absorbing class by some authors, but this can lead to confusion because the ergodic class need not consist only of absorbing states; it also can include cycles.

In the remainder of this section we'll outline conditions that guarantee the existence of an equilibrium distribution. Sect. A.13 provides other results and their proofs, including the expected number of steps before a Markov chain enters its ergodic class and the expected number of steps until a chain revisits a state in its ergodic class. For now we'll include one simple example, formulated more elegantly in Sect. A.13.

Proposition 14.1.2. Suppose a system starts in state i, that is, $X_0 = i$. In the first m iterates, the number T of times that $X_k = j$ has $\mathbb{E}(T) = \sum_{k=1}^{m} \left(P^k\right)_{ij}$.

Proof. This proposition is about expected values, so let's define $Y_k = 1$ if $X_k = j$ and $Y_k = 0$ if $X_k \neq j$. Then

$$\mathbb{E}(T) = \mathbb{E}(Y_1|X_0 = i) + \cdots + \mathbb{E}(Y_m|X_0 = i)$$

Now for each k, by Eq. (11.14) we have

$$\begin{aligned}\mathbb{E}(Y_k|X_0 = i) &= 0 \cdot P(Y_k = 0|X_0 = i) + 1 \cdot P(Y_k = 1|X_0 = i)\\ &= P(Y_k = 1|X_0 = i) = P(X_k = j|X_0 = i) = p_{ji}^{(k)} = \left(P^k\right)_{ji}\end{aligned}$$

where the last equality follows from Eq. (14.8). □

To study the existence of equilibrium distributions, first we'll analyze the exceptions that occurred in our answers to Questions 2 and 3. For this, we need to generalize the notion of periodic states. A Markov *chain* is *periodic* if there are $q \geq 2$ disjoint sets of states $A_1, A_2, \ldots, A_q$ with

$$A_1 \to A_2 \to \cdots \to A_q \to A_1$$

where, for example, $A_1 \to A_2$ means that for each state $i \in A_1$, $p_{ji} = 0$ unless $j \in A_2$. That is, every state of A_1 has a positive transition probability only to states of A_2. Time for an example.

Example 14.1.5. *Periodic and non-periodic Markov chains.* Here are transition matrices for Markov chains, one periodic, the other aperiodic.

$$P_1 = \begin{bmatrix} 0 & 0 & 0 & 0 & \frac{1}{2} & \frac{1}{2} \\ 0 & 0 & 0 & 0 & \frac{1}{2} & \frac{1}{2} \\ \frac{1}{2} & \frac{1}{2} & 0 & 0 & 0 & 0 \\ \frac{1}{2} & \frac{1}{2} & 0 & 0 & 0 & 0 \\ 0 & 0 & \frac{1}{2} & \frac{1}{2} & 0 & 0 \\ 0 & 0 & \frac{1}{2} & \frac{1}{2} & 0 & 0 \end{bmatrix} \qquad P_2 = \begin{bmatrix} \frac{1}{2} & \frac{1}{3} & 0 & 0 & 0 & \frac{1}{3} \\ \frac{1}{2} & \frac{1}{3} & 0 & 0 & 0 & 0 \\ 0 & \frac{1}{3} & \frac{1}{2} & \frac{1}{3} & 0 & 0 \\ 0 & 0 & \frac{1}{2} & \frac{1}{3} & 0 & 0 \\ 0 & 0 & 0 & \frac{1}{3} & \frac{1}{2} & \frac{1}{3} \\ 0 & 0 & 0 & 0 & \frac{1}{2} & \frac{1}{3} \end{bmatrix} \tag{14.12}$$

For the Markov chain with transition matrix P_1, the disjoint sets of states are $A_1 = \{1,2\}$, $A_2 = \{3,4\}$, and $A_3 = \{5,6\}$. The eigenvalues of P_1 are 1, $-1/2 \pm i\sqrt{3}/2, 0, 0, 0$, and the unit eigenvector of $\lambda = 1$ is $\vec{v}_1 = [1/6, 1/6, 1/6, 1/6, 1/6, 1/6]^{tr}$. If we aren't careful, we'd say that $\vec{v}_1$ is the equilibrium distribution of P_1 and with time the population gets evenly spread among the six states.

The first hint that something is off is the eigenvalues $\lambda_2 = -1/2 + i\sqrt{3}/2$ and $\lambda_3 = -1/2 - i\sqrt{3}/2$. Note that $|\lambda_2| = |\lambda_3| = 1$, so we do not have $\lambda_1 > |\lambda_i|$ for all other eigenvalues λ_i. Then we cannot apply the reasoning of Prop. 14.1.1 and of Example 14.1.4. If we start from any distribution with nonzero values only in A_1 or only in A_2 or only in A_3, successive applications of P_1 will cycle through those sets of states. While it is true that the distribution with all values $1/6$ is a fixed point (in distributions; individual elements cycle through the states), this fixed point is unstable. Nearby distributions iterate to cycling distributions. Subtract 0.01 from the first element of $\vec{v}_1$ and add 0.01 to the last element. Apply P_1 seven times to this vector. What do you find? The Markov chain with transition matrix P_1 has no equilibrium distribution.

The matrix P_2 warns against a common mistake. With the same sets A_1, A_2, A_3, elements of A_1 stay in A_1 or go to A_2, elements of A_2 stay in A_2 or go to A_3, and elements of A_3 stay in A_3 or go to A_1. Looks cyclic, doen't it? The unit eigenvector for the eigenvalue $\lambda_1 = 1$ is $\begin{bmatrix} 4/21 & 3/21 & 4/21 & 3/21 & 4/21 & 3/21 \end{bmatrix}^{tr}$. Picking a few random initial distributions, under repeated application of P_2 all converge to this eigenvector. This Markov chain does have an equilibrium distribution, not really a surprise because the eigenvalues of P_2 are 1, $0.7776 \pm 0.2000i$, $0.05575 \pm 0.2000i$, and -0.1667, so $\lambda = 1$ is the dominant eigenvalue. To understand the difference between the long-term dynamics of P_1 and P_2, look closely at the definition of periodic Markov chain. Because P_2 allows $A_1 \to A_1$, $A_2 \to A_2$, and $A_3 \to A_3$, the corresponding Markov chain is not periodic. The point is to read definitions carefully. □

A Markov chain that is not periodic is aperiodic. The easiest example of an aperiodic Markov chain is one with all the transition probabilities positive. Then for any partition $A_1, A_2, \dots, A_q$ of states, we don't have $A_1 \to A_2$ because every element of A_1 has a nonzero probability of going to every state of the whole Markov chain, not just to those states in A_2.

A set C of states is *closed* if for all $i \in C$ and for all $j \notin C$, $p_{ji} = 0$. That is, the probability of moving from a state in C to a state outside of C is 0. The ergodic class is closed, as is an absorbing state. For another example, in the Markov chain with the transition matrix in Example 14.1.3, we see the sets $C_1 = \{1,2\}$ and $C_2 = \{3,4\}$ are closed.

The usual approach for dealing with Markov chains having at least two closed sets (do you see why closed sets must be disjoint?) is to treat each closed set as its own Markov chain.

We'll mention one more result. If i is a recurrent state and $\vec{v}$ is the right unit eigenvector of $\lambda = 1$, then

$$\mathbb{E}(\text{number of time steps needed to go from } i \text{ to } i) = 1/v_i$$

The proof is sketched in Prop. A.13.6, but a rough idea is straightforward. Wait long enough that all the transient states no longer are visited, and suppose there are no absorbing states so we don't get stuck in a single state. All that remains are recurrent states, and the average time between returns to each state is about the reciprocal of the time spent in that state.

Now we're ready to state the main result about Markov chains with a finite number of states. Suppose P is the transition matrix. Then

Theorem 14.1.1. If the state space of this Markov chain does not have two or more disjoint closed subsets, then the chain has a unique equilibrium distribution $\vec{v}$ determined by $P\vec{v} = \vec{v}$ where $\vec{v}$ is the unit eigenvector. If the Markov chain is aperiodic, then for all i and j, $v_i = \lim_{n\to\infty} p_{ij}^{(n)}$.

We've already seen these results demonstrated in examples. The point of stating the theorem is to spell out the conditions that guarantee a Markov chain has an equilibrium distribution. We won't attempt the proof, even a sketch. Consult Feller [114], Kemeny and Snell [195], and Tijms [357] for all sorts of fascinating aspects of Markov chains.

Although Andrey Markov applied his technique only to linguistics, specifically to the study of the transition probabilities between vowels and consonants, Markov chains are an effective way to model some of the apparently random behavior of biological systems. We'll discuss instances in Chapter 16.

Practice Problems

We'll use these transition matrices.

$$P_1 = \begin{bmatrix} 0.5 & 0.1 \\ 0.5 & 0.9 \end{bmatrix} \quad P_2 = \begin{bmatrix} 0.5 & 0.9 \\ 0.5 & 0.1 \end{bmatrix} \quad P_3 = \begin{bmatrix} 0.5 & 0.1 & 0 \\ 0.4 & 0.2 & 0.2 \\ 0.1 & 0.7 & 0.8 \end{bmatrix}$$

$$P_4 = \begin{bmatrix} 0.7 & 0 & 0 & 0 & 0 \\ 0.1 & 0.5 & 0.3 & 0 & 0 \\ 0 & 0.5 & 0.7 & 0 & 0 \\ 0.2 & 0 & 0 & 0.8 & 0.2 \\ 0 & 0 & 0 & 0.2 & 0.8 \end{bmatrix}$$

14.1.1. Find the equilibrium distribution for P_1 and P_2.

14.1.2. Find the equilibrium distribution for P_3.

14.1.3. Find the expected value of the number of times in $n = 9$ iterates that state 2 is reached, starting from $X_0 = 1$, for the Markov chain with transition matrix P_1.

14.1.4. For the process defined by this transition matrix P_4, find the equilibrium distribution if the initial distribution is entirely in state 1.

Practice Problem Solutions

14.1.1. The eigenvalues of P_1 are 1 and 0.4. An eigenvector of 1 is $\begin{bmatrix} 1 & 5 \end{bmatrix}^{tr}$. The equilibrium distribution is the unit eigenvector in this direction, $\begin{bmatrix} 1/6 & 5/6 \end{bmatrix}^{tr}$, and eventually the system has a 1/6 probability of being in state 1 and a 5/6 probability of being in state 2.
(b) The eigenvalues of P_2 are 1 and -0.4. An eigenvector of 1 is $\begin{bmatrix} 1 & 5/9 \end{bmatrix}^{tr}$. The equilibrium distribution is the unit eigenvector in that direction: $[1/(1+5/9), (5/9)/(1+5/9)]^{tr} = [9/14, 5/14]^{tr}$. That is, eventually the system has a 9/14 probability of being in state 1 and a 5/14 probability of being in state 2.

14.1.2. Finding the eigenvalues of P_3 involves solving a cubic equation, so we let Mathematica do the work. Rounding to three digits to the right of the decimal we obtain $\lambda = 1$, 0.537, and -0.037. We could let Mathematica give us an (approximate) eigenvector for $\lambda = 1$, but it's easy enough to find the exact value by solving the eigenvector equation

$$\begin{bmatrix} 0.5 & 0.1 & 0 \\ 0.4 & 0.2 & 0.2 \\ 0.1 & 0.7 & 0.8 \end{bmatrix} \begin{bmatrix} x \\ y \\ z \end{bmatrix} = 1 \cdot \begin{bmatrix} x \\ y \\ z \end{bmatrix}$$

The first two equations are

$$0.5x + 0.1y = x \qquad \text{and} \qquad 0.4x + 0.2y + 0.2z = y$$

The third equation is a consequence of the first two. The first equation gives $y = 5x$. Substituting this into the second equation and solving for z gives $z = 18x$. Then the unit eigenvector condition, $x + y + z = 1$ gives $\begin{bmatrix} 1/24 & 5/24 & 18/24 \end{bmatrix}^{tr}$. These are the probabilities of the system's eventually being in states 1, 2, and 3.

14.1.3. We'll apply Prop. 14.1.2, so we need the sum $p_{21}^{(1)} + p_{21}^{(2)} + \cdots + p_{21}^{(9)}$. These are the $(2,1)$ entries of the matrices $P_1, P_1^2, \dots, P_1^9$. Rounded to three digits to the right of the decimal, these are

$$P_1 = \begin{bmatrix} 0.5 & 0.1 \\ 0.5 & 0.9 \end{bmatrix} \qquad P_1^2 = \begin{bmatrix} 0.3 & 0.14 \\ 0.7 & 0.86 \end{bmatrix} \qquad P_1^3 = \begin{bmatrix} 0.22 & 0.156 \\ 0.78 & 0.844 \end{bmatrix}$$

$$P_1^4 = \begin{bmatrix} 0.188 & 0.162 \\ 0.812 & 0.838 \end{bmatrix} \qquad P_1^5 = \begin{bmatrix} 0.175 & 0.165 \\ 0.825 & 0.835 \end{bmatrix} \qquad P_1^6 = \begin{bmatrix} 0.170 & 0.166 \\ 0.830 & 0.834 \end{bmatrix}$$

$$P_1^7 = \begin{bmatrix} 0.168 & 0.166 \\ 0.832 & 0.834 \end{bmatrix} \qquad P_1^8 = \begin{bmatrix} 0.167 & 0.167 \\ 0.833 & 0.833 \end{bmatrix} \qquad P_1^9 = \begin{bmatrix} 0.167 & 0.167 \\ 0.833 & 0.833 \end{bmatrix}$$

Then the expected number of visits is the sum of the $(2,1)$ entries of these matrices, including P,

$$0.5 + 0.7 + 0.78 + 0.812 + 0.825 + 0.83 + 0.832 + 0.833 + 0.833 \approx 6.945$$

14.1.4. First note that with each iteration 0.3 of the population of state 1 leaves state 1, and nothing from any other state enters state 1, so eventually state 1 will empty out. Where does it go? From the transition matrix we see that 0.1 goes into state 2, and 0.2 goes into state 4. Next, we see that states 2 and 3 make transitions among only themselves, and also states 4 and 5 make transitions among only themselves. So, the entire population winds up in states $\{2,3\} = A$ and in states $\{4,5\} = B$. Twice as much goes into B as goes into A. From this, we can deduce that $2/3$ of the population goes into B and $1/3$ into A. If this isn't clear, we'll give another derivation near the end of this solution.

The transition matrices for $\{2,3\} = A$ and for $\{4,5\} = B$ are

$$P_A = \begin{bmatrix} 0.5 & 0.3 \\ 0.5 & 0.7 \end{bmatrix} \qquad P_B = \begin{bmatrix} 0.8 & 0.2 \\ 0.2 & 0.8 \end{bmatrix}$$

These matrices have unit eigenvectors $\begin{bmatrix} 3/8 & 5/8 \end{bmatrix}^{tr}$ and $\begin{bmatrix} 1/2 & 1/2 \end{bmatrix}^{tr}$ corresponding to the dominant eigenvalue $\lambda = 1$. Because 1/3 of the initial population goes into A and 2/3 of the initial population goes into B, the equilibrium distribution is

$$\begin{aligned} &\begin{bmatrix} 0 & (1/3)\cdot(3/8) & (1/3)\cdot(5/8) & (2/3)\cdot(1/2) & (2/3)\cdot(1/2) \end{bmatrix}^{tr} \\ &= \begin{bmatrix} 0 & 1/8 & 5/24 & 1/3 & 1/3 \end{bmatrix}^{tr} \end{aligned}$$

Finally, let's find another way to see that 1/3 of the initial population winds up in A. In the first iteration, 1/10 of the initial population goes into A. In the second iteration, 1/10 of the 7/10 that remain in state 1 goes into A. In the third iteration, 1/10 of the $(7/10)^2$ that remain in state 1 goes into A. By now the pattern should be clear. This much of state 1 goes into A:

$$\frac{1}{10} + \frac{1}{10}\cdot\frac{7}{10} + \frac{1}{10}\cdot\left(\frac{7}{10}\right)^2 + \cdots = \frac{1}{10}\cdot\frac{1/10}{1-7/10} = \frac{1}{3}$$

Exercises

14.1.1. Find the equilibrium distributions for the Markov chains with transition matrices (a), (b), and (c).

$$\text{(a)} \begin{bmatrix} 0.1 & 0.9 \\ 0.9 & 0.1 \end{bmatrix} \quad \text{(b)} \begin{bmatrix} 0.2 & 0.8 \\ 0.8 & 0.2 \end{bmatrix} \quad \text{(c)} \begin{bmatrix} p & 1-p \\ 1-p & p \end{bmatrix} \quad \text{for } 0 < p < 1$$

14.1.2. Find the equilibrium distributions for the Markov chains with transition matrices (a) and (b).

$$\text{(a)} \begin{bmatrix} 0.7 & 0.6 \\ 0.3 & 0.4 \end{bmatrix} \qquad \text{(b)} \begin{bmatrix} 0.5 & 0 & 0 \\ 0.5 & 0.7 & 0.6 \\ 0 & 0.3 & 0.4 \end{bmatrix}$$

(c) Explain the equilibrium distribution of (b) in terms of that of (a).

14.1.3. Find the equilibrium distributions for the Markov chains with these transition matrices. Drawing the transition graph before calculating eigenvalues may help. Express your answers as fractions, not decimals.

$$\text{(a)} \begin{bmatrix} 0.5 & 0.1 & 0 \\ 0.4 & 0.2 & 1 \\ 0.1 & 0.7 & 0 \end{bmatrix} \quad \text{(b)} \begin{bmatrix} 0.3 & 0.5 & 1 \\ 0.7 & 0.5 & 0 \\ 0 & 0 & 0 \end{bmatrix}$$

$$\text{(c)} \begin{bmatrix} 0.3 & 0.4 & 0.2 & 0 \\ 0.7 & 0.6 & 0 & 1 \\ 0 & 0 & 0 & 0 \\ 0 & 0 & 0.8 & 0 \end{bmatrix} \quad \text{(d)} \begin{bmatrix} 0.3 & 0.4 & 0 \\ 0.7 & 0.5 & 0.2 \\ 0 & 0.1 & 0.8 \end{bmatrix}$$

14.1.4. Find the equilibrium distributions for the Markov chains with these transition matrices. Again, use the transition graph as a guide for the calculation.

$$\text{(a)} \begin{bmatrix} 0.4 & 0.2 & 0.3 \\ 0.4 & 0.8 & 0.3 \\ 0.2 & 0 & 0.4 \end{bmatrix} \quad \text{(b)} \begin{bmatrix} 0.5 & 0.6 & 0.5 \\ 0.3 & 0.1 & 0.2 \\ 0.2 & 0.3 & 0.3 \end{bmatrix}$$

$$\text{(c)} \begin{bmatrix} 0.3 & 0.4 & 0.5 & 0 \\ 0.3 & 0.2 & 0.2 & 0.5 \\ 0.4 & 0.4 & 0.3 & 0.5 \\ 0 & 0 & 0 & 0 \end{bmatrix} \quad \text{(d)} \begin{bmatrix} 0 & 0 & 0 & 0 \\ 0.2 & 0.5 & 0.5 & 0.5 \\ 0.4 & 0.5 & 0.5 & 0 \\ 0.4 & 0 & 0 & 0.5 \end{bmatrix}$$

14.1.5. For the Markov chain determined by transition matrices (a), (b), and (c), from which state, $X_0 = 1$, $X_0 = 2$, or $X_0 = 3$, should we start to to have the highest number of expected visits to state 2 in $n = 5$ iterates? Can you find a rough explanation for your answer? Something other than, “I did the calculation and this is what I got.”

$$\text{(a)} \begin{bmatrix} 0.5 & 0.2 & 0.2 \\ 0.5 & 0 & 0.3 \\ 0 & 0.8 & 0.5 \end{bmatrix} \quad \text{(b)} \begin{bmatrix} 0.5 & 0.2 & 0.2 \\ 0 & 0.8 & 0.5 \\ 0.5 & 0 & 0.3 \end{bmatrix} \quad \text{(c)} \begin{bmatrix} 0.5 & 0 & 0.5 \\ 0.5 & 0.5 & 0 \\ 0 & 0.5 & 0.5 \end{bmatrix}$$

14.1.6. For the Markov chain determined by these transition matrices, which state has the longest expected recurrence time, which the smallest?

$$\text{(a)} \begin{bmatrix} 0.2 & 0 & 0.3 & 0.1 \\ 0.4 & 0.4 & 0.1 & 0.1 \\ 0.3 & 0.4 & 0.3 & 0 \\ 0.1 & 0.2 & 0.3 & 0.8 \end{bmatrix} \quad \text{(b)} \begin{bmatrix} 0.4 & 0.4 & 0.1 & 0.1 \\ 0.2 & 0 & 0.3 & 0.1 \\ 0.3 & 0.4 & 0.3 & 0 \\ 0.1 & 0.2 & 0.3 & 0.8 \end{bmatrix}$$

14.1.7. For the Markov chains determined by these transition matrices, find the equilibrium distribution if initially the population is entirely in state 1. Sketching the transition graphs may help.

$$(a)\begin{bmatrix} 0.3 & 0 & 0 & 0 & 0 \\ 0.4 & 0.4 & 0.1 & 0 & 0 \\ 0 & 0.6 & 0.9 & 0 & 0 \\ 0 & 0 & 0 & 0.8 & 0.5 \\ 0.3 & 0 & 0 & 0.2 & 0.5 \end{bmatrix} \quad (b)\begin{bmatrix} 0.2 & 0 & 0 & 0 & 0 \\ 0.4 & 0.4 & 0 & 0.1 & 0 \\ 0.4 & 0 & 0.6 & 0 & 0.2 \\ 0 & 0.6 & 0 & 0.9 & 0 \\ 0 & 0 & 0.4 & 0 & 0.8 \end{bmatrix}$$

14.1.8. For the Markov chains determined by these transition matrices, find the equilibrium distribution if initially the population is entirely in state 1. Sketching the transition graphs may help.

$$(a)\begin{bmatrix} 0.3 & 0 & 0 & 0 & 0 & 0 \\ 0.4 & 0.3 & 0.5 & 0 & 0 & 0 \\ 0 & 0.7 & 0.5 & 0 & 0 & 0 \\ 0.3 & 0 & 0 & 0.5 & 0.5 & 0 \\ 0 & 0 & 0 & 0.2 & 0.4 & 0.1 \\ 0 & 0 & 0 & 0.3 & 0.1 & 0.9 \end{bmatrix} \quad (b)\begin{bmatrix} 0.2 & 0 & 0 & 0 & 0 & 0 \\ 0.4 & 0.4 & 0 & 0 & 0 & 0 \\ 0 & 0.6 & 0.6 & 0 & 0.2 & 0.5 \\ 0.4 & 0 & 0 & 0 & 0 & 0 \\ 0 & 0 & 0.3 & 1 & 0.6 & 0 \\ 0 & 0 & 0.1 & 0 & 0.2 & 0.5 \end{bmatrix}$$

14.1.9. For the Markov chains determined by these transition matrices, find the equilibrium distribution if
(a) initially the population is entirely in state 1,
(b) initially the population is entirely in state 3,
(c) initially the population is entirely in state 4, and
(d) initially 0.2 of the population is in state 1, 0.4 of the population is in state 3, and 0.4 of the population is is state 4.

$$(i)\begin{bmatrix} 0.4 & 0.6 & 0 & 0 & 0 & 0 \\ 0.6 & 0.4 & 0 & 0 & 0 & 0 \\ 0 & 0 & 0.4 & 0 & 0 & 0.4 \\ 0 & 0 & 0 & 0.5 & 1 & 0 \\ 0 & 0 & 0 & 0.5 & 0 & 0 \\ 0 & 0 & 0.6 & 0 & 0 & 0.6 \end{bmatrix} \quad (ii)\begin{bmatrix} 0 & 0.5 & 0 & 0.4 & 1 & 0 \\ 0 & 0.5 & 0 & 0 & 0 & 0 \\ 1 & 0 & 0 & 0 & 0 & 0.7 \\ 0 & 0 & 0 & 0.2 & 0 & 0 \\ 0 & 0 & 1 & 0.4 & 0 & 0 \\ 0 & 0 & 0 & 0 & 0 & 0.3 \end{bmatrix}$$

14.1.10. For the Markov chains determined by this transition matrix,

$$\begin{bmatrix} 1/2 & \alpha \\ 1/2 & 1-\alpha \end{bmatrix}$$

find a numerical value for the constant α so that in $n = 4$ iterates starting from state 2 the expected number of times state 2 is reached is
(a) 2, (b) 3, and (c) 4.
(d) Can you explain your answer to (c) without using Prop. 14.1.2?

14.2 THE PERRON-FROBENIUS THEOREM

Here we'll give conditions that guarantee the largest positive eigenvalue of a matrix is strictly larger than the magnitude of all other eigenvalues.

First some background notions. An $n \times n$ matrix M is called *irreducible* if for each i and j, there is some k with $(M^k)_{ij} > 0$. We must mention a subtle point: different i and j might need different k.

We can give an alternate characterization of irreducibility by the transition graph G of M. Recall there is an edge from vertex j to vertex i (remember the order of i and j) if and only if $M_{ij} > 0$. Now recall Eq. (14.3) tells us that $(M^k)_{ij} > 0$ means there is a positive probability of going from j to i in k steps, so in the graph G there is a path of length k from vertex j to vertex i. Irreducible means that every pair of vertices in G is connected by a path of edges, but the paths may have different lengths.

A graph with a path from every vertex to every vertex is called *strongly connected*. But then what's a connected graph? It's the same idea as strongly connected, if we ignore the directions of the arrows in the graph. For example, $1 \to 2$ is not strongly connected because there's no path (with arrows) from 2 to 1. But if we ignore the direction of the arrow, the graph becomes $1-2$. Without the direction of the edge, it connects 1 to 2, 2 to 1, 1 to 2 to 1, and 2 to 1 to 2, via paths between each pair of vertices.

An $n \times n$ matrix M is *primitive* if for some k, all $(M^k)_{ij} > 0$. Here the same power k works for all i and j. Some mathematical names make sense—composite numbers are products of other numbers and eigenvectors are vectors a matrix transforms into a multiple of itself. Okay, this makes sense only if you read German. "Eigen" is the German word for "own," in the sense of belonging to the self. But "primitive"? Does this suggest the matrix is unevolved? Biologists often use the more descriptive term *power positive* for primitive matrices.

In a moment we'll see an additional condition that guarantees an irreducible matrix is primitive. First, though, the theorem.

Theorem 14.2.1. Suppose M is a primitive matrix. Then
(1) M has a dominant eigenvalue λ_1, that is, $\lambda_1 > |\lambda|$ for all other eigenvalues λ of M,
(2) λ_1 has a unit eigenvector $\vec{v}_0$ with all elements positive, and
(3) all eigenvectors of λ_1 are multiples of $\vec{v}_0$.

In 1907 Oskar Perron proved this theorem for matrices with all entries positive (that is, primitive with $k=1$). In 1912 Ferdinand Frobenius extended the theorem to irreducible matrices with all entries non-negative. The conclusion of Frobenius's extension is a bit weaker: instead of $\lambda_1 > |\lambda|$ for all other eigenvalues we have $\lambda_1 \geq |\lambda|$ for all other eigenvalues. Mostly, we'll stick with Perron's original version, extended to primitive matrices.

So, how can we tell if a matrix M is primitive? We could compute high powers of M, but how high? Right off, it might seem that there is no upper limit, that we'd need to keep multiplying more and more factors of M looking for an M^k with all entries positive. But the situation isn't that gloomy. Corollary 8.5.9 of [177] states that an $n \times n$ matrix with all entries non-negative is primitive if and only if M^{n^2-2n+2} has all positive entries. The exponent doesn't grow all that quickly—it's only 82 when $n = 10$—but we'll take another more geometrical approach, using the transition graph of M.

By examining the graph G we can tell if the matrix M is irreducible by checking if the graph is strongly connected. All we need to do is this: between every pair of vertices p and q, find a path in the graph from p to q. Unless the graph has a lot of vertices, this is easy. But "easy" rarely wins the day in math. Here's an illustration of why irreducibility won't suffice to guarantee the matrix has a dominant eigenvalue.

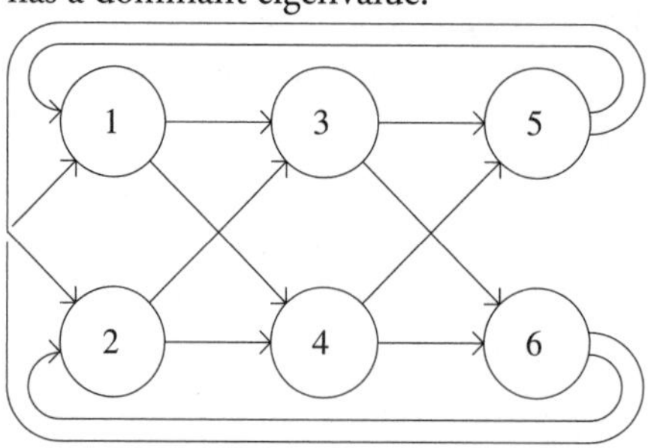

Figure 14.3. The transition graph for matrix P_1 of Example 14.1.5.

Recall transition matrix P_1 of Example 14.1.5. In Fig. 14.3 we see the transition graph for this matrix. Note that this graph is strongly connected: from any vertex you can get to any other vertex. However, this matrix is not primitive, not power positive: every power of P_1 has twenty-four 0s and twelve 1/2s. As we mentioned in Example 14.1.5, the iterates of many initial distributions do not converge to an equilibrium distribution, but cycle forever. But notice that every cycle in the transition graph has length a multiple of 3. For example, $1 \to 3 \to 5 \to 1$ and $1 \to 4 \to 6 \to 3 \to 5 \to 1$. We'll see that this is the problem.

For this, we need a concept and a lemma. The concept is this: the *greatest common divisor* of two positive integers a and b is

$$(a,b) = \text{the largest number that divides both } a \text{ and } b$$

So for example,

$$(15,12) = 3, \quad (524,126) = 2, \quad (51,47) = 1$$

In particular, two integers a and b are *relatively prime* if $(a, b) = 1$, that is, if they have no factor (other than 1) in common.

Now for the lemma, called Bézout's identity, a result from number theory. This is based on *Euclid's algorithm*, first described by Euclid though probably not discovered by him, for finding the greatest common divisor. It's nothing more than iterated division and keeping track of the remainders. What to divide by what is the clever point of the algorithm. We'll describe Euclid's algorithm and then illustrate it by showing $(524, 126) = 2$.

Suppose $a > b$. Then dividing a by b gives a quotient q_1 and a remainder r_1 with $0 \leq r_1 < b$. (The remainder always is smaller than the divisor.) Now continue, dividing b by r_1, obtaining a quotient of q_2 and remainder $r_2 < r_1$. Because the remainders keep decreasing, eventually some $r_n = 0$. Then the previous remainder r_{n-1} is the greatest common divisor. Let's write down the steps, look at an example, and then see why the algorithm works.

$$\begin{aligned}
a &= b \cdot q_1 + r_1 && \text{with } r_1 < b \\
b &= r_1 \cdot q_2 + r_2 && \text{with } r_2 < r_1 \\
r_1 &= r_2 \cdot q_3 + r_3 && \text{with } r_3 < r_2 \\
&\dots \\
r_{n-4} &= r_{n-3} \cdot q_{n-2} + r_{n-2} && \text{with } r_{n-2} < r_{n-3} \\
r_{n-3} &= r_{n-2} \cdot q_{n-1} + r_{n-1} && \text{with } r_{n-1} < r_{n-2} \\
r_{n-2} &= r_{n-1} \cdot q_n + 0
\end{aligned} \tag{14.13}$$

Here are the steps of Euclid's algorithm applied to find the greatest common divisor of 524 and 126.

$$\begin{aligned}
524 &= 126 \cdot 4 + 20 \\
126 &= 20 \cdot 6 + 6 \\
20 &= 6 \cdot 3 + 2 \\
6 &= 2 \cdot 3 + 0
\end{aligned}$$

The smallest non-zero remainder is 2, so by the algorithm, $(524, 126) = 2$.

To show the algorithm works, we must see the smallest non-zero remainder, r_{n-1}, divides both a and b, and that every other number that divides a and b also divides r_{n-1}.

From the last line of Eq. (14.13) we see r_{n-1} divides r_{n-2}. Then r_{n-1} divides both terms on the right side of the penultimate line of Eq. (14.13), so r_{n-1} divides the left side, r_{n-3}. Working our way up, at each line r_{n-1} divides both terms on the right side and so divides the left side. Eventually we arrive at the top two lines and deduce r_{n-1} divides both b and a.

Now suppose some positive $m > 1$ divides both a and b, so there are integers s and t with $a = m \cdot s$ and $b = m \cdot t$. We'll solve each line of Eq. (14.13) for the remainder and work our way down. First,

$$r_1 = a - b \cdot q_1 = m \cdot s - m \cdot t \cdot q_1 = m \cdot (s - t \cdot q_1)$$

so m divides r_1. Next,

$$r_2 = b - r_1 \cdot q_2 = m \cdot t - m \cdot (s - t \cdot q_1) \cdot q_2$$

so m divides r_2. Continuing in this way, we see that m divides r_{n-1}, so r_{n-1} is the greatest common divisor of a and b.

Now we'll prove Bézout's identity.

Lemma 14.2.1. If $(a,b) = d$, then there are unique integers x and y for which $ax + by = d$.

Proof. In our formulation of Euclid's algorithm, $(a,b) = r_{n-1}$. We'll climb back up Eq. (14.13), writing r_{n-1} as a sum of integer multiples of the left sides of two consecutive lines.

As we have seen, the penultimate line of Eq. (14.13) gives

$$(a,b) = r_{n-1} = r_{n-3} - r_{n-2} \cdot q_{n-1} \tag{14.14}$$

That is, (a,b) is a sum of integer multiples of r_{n-2} and r_{n-3}.

Solve the line of Eq. (14.13) with remainder r_{n-2} for r_{n-2} and we can rewrite Eq. (14.14) as

$$\begin{aligned}(a,b) &= r_{n-3} - (r_{n-4} - r_{n-3} \cdot q_{n-2}) \cdot q_{n-1} \\ &= r_{n-3} \cdot (1 + q_{n-2} \cdot q_{n-1}) - r_{n-4} \cdot q_{n-1} \end{aligned} \tag{14.15}$$

So (a,b) is a sum of integer multiples of r_{n-3} and r_{n-4}. Solve the line of Eq. (14.13) with remainder r_{n-3} for r_{n-3} and substitute that in Eq. (14.15). We find that (a,b) is a sum of integer multiples of r_{n-4} and r_{n-5}.

Continue upward. When we arrive at the top of Eq. (14.13), we find that (a,b) is a sum of integer multiples of a and b. □

It's instructive to follow the steps of the proof of Bézout's identity applied to $(524, 126) = 2$. You'll find $2 = 524 \cdot 19 - 126 \cdot 79$.

Now we won't use Bézout's identity directly, but rather this corollary.

Corollary 14.2.1. If $(a,b) = 1$, then for any positive integer n,
(i) there are integers u and v with $n = au + bv$, and
(ii) if $n \geq ab$, the integers u and v in $n = au + bv$ both are positive.

Proof. (i) Apply Bézout's identity to $(a,b)=1$: there are integers x and y for which $ax+by=1$. Now multiply both sides of the equation by n

$$a(xn)+b(yn)=n$$

So $u=xn$ and $v=yn$ are the desired integers.

(ii) Because $n = au + bv$, and a and b are positive, at least one of the coefficients u and v is positive. Suppose u is negative. Note that $0<v<a$ because

- if $v=a$ then $au+bv=au+ba=a(u+b)$, and
- if $v=a+1$ then $au+bv=au+b(a+1)=a(u+b)+b$, and so on.

That is, if $v \geq a$, then part of v can be absorbed into the u coefficient. Then $u<0$ implies $n=au+bv \leq bv < ba$.

Arguing similarly, if $v<0$ then again $n<ab$. So if $n \geq ab$, we see that both coefficients u and v must be positive. □

Finally we're ready for the proposition.

Proposition 14.2.1. If the transition graph G of a matrix M is strongly connected and has (at least) two cycles of relatively prime length, then M is primitive.

Proof. Suppose one cycle C_1 in G has length a, another cycle C_2 has length b, and $(a,b)=1$. Also, let n denote the number of vertices in the graph G. We'll show that any pair of vertices p and q in G can be connected by a path of length $3(n-1)+ab$. Because every pair of vertices can be connected by paths of length $3(n-1)+ab$, every element of $M^{3(n-1)+ab}$ is positive, that is, M is primitive.

Now start with the vertex p. Because the graph G is strongly connected, there is a path P_1 from p to a point x in the cycle C_1, another path P_2 from x to a point y in the cycle C_2, and a third path P_3 from y to q. Because G has n vertices, any two distinct vertices can be connected by a path of length $n-1$. If, for example, p and x are not distinct but rather are the same, then they can be connected by a path of length 0. We're not saying G has a loop at p, but rather that p already is a vertex in the cycle C_1. So in either case each of the three paths has length

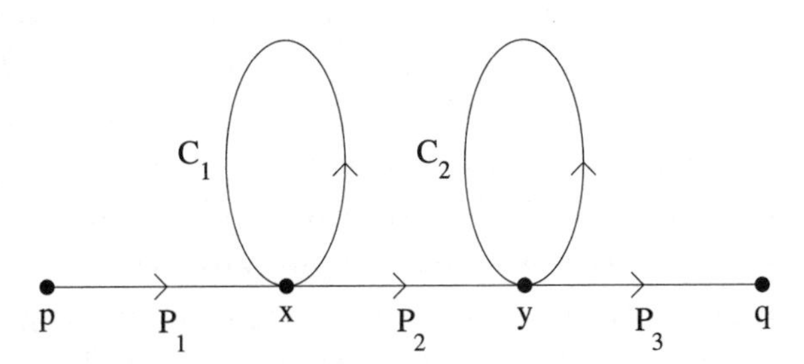

Figure 14.4. From p to x by P_1, k_1 times around C_1, from x to y by P_2, k_2 times around C_2, and from y to q by P_3.

$\leq (n-1)$. We want to construct a path of length $3(n-1)+ab$ from p to q. So far, the length of the path we have from p to q is

$$\text{length}(P_1)+\text{length}(P_2)+\text{length}(P_3) \leq 3(n-1)$$

To construct a path from p to q of length $3(n-1)+ab$, we must add length

$$3(n-1)+ab-(\text{length}(P_1)+\text{length}(P_2)+\text{length}(P_3)) \geq ab$$

This is where we apply Cor. 14.2.1. Any length $\geq ab$ can be achieved by some number $k_1 \geq 0$ of runs around C_1 and some number $k_2 \geq 0$ of runs around C_2. Fig. 14.4 shows a path of length $3(n-1)+ab$. □

The Perron-Frobenius Theorem 14.2.1 and Prop. 14.2.1 are the tools we'll use to show that some Leslie matrices of Sect. 14.3 have dominant eigenvalues. For those Leslie matrices that do not satisfy the hypotheses of the Perron-Frobenius theorem, we will have an alternative, Prop. 14.3.1, with a weaker conclusion, but still sufficient for our purposes.

The Perron-Frobenius theorem also is useful in our study of Markov chains. In these applications we'll make use of an additional result.

Proposition 14.2.2. The eigenvalues λ of any square stochastic matrix M satisfy $|\lambda| \leq 1$.

Proof. Recall the entries of a stochastic matrix are non-negative, and the entries of each column sum to 1. Also, in Sect. 14.4 we'll see that left eigenvectors and right eigenvectors have the same eigenvalues. For this proof, left eigenvectors are more useful.

Suppose M has an eigenvalue λ with $|\lambda| > 1$, and left eigenvector $\vec{w} = \langle w_1 \dots w_n \rangle$ that satisfies $\vec{w}M = \lambda\vec{w}$. Each entry of $\vec{w}M$ has the form $w_1 m_{1j} + w_2 m_{2j} + \cdots + w_n m_{nj}$ where $m_{1j} + m_{2j} + \cdots + m_{nj} = 1$. That is, each entry of the left-hand side is a convex combination of $w_1, w_2, \dots, w_n$. Consequently, each of these is $\leq \max\{w_i\}$. To see why this is true, draw the $n=2$ and $n=3$ cases. Put w_1 on the x-axis and w_2 on the y-axis. Then $w_1 m_{1j} + w_2 m_{2j}$ lies on the line segment between these points, so no coordinate is larger than the maximum of w_1 and w_2. In 3 dimensions the convex combinations form a triangle.

So, each entry of $\vec{w}M$ is $\leq \max\{w_i\}$. But if $\lambda > 1$, at least one entry of $\lambda\vec{w}$ is $> \max\{w_i\}$. Impossible. This shows $\lambda \leq 1$. We'll leave it to you (or Google) to show that $|\lambda| \leq 1$. □

Practice Problems

14.2.1. If a matrix M is primitive and invertible, must M^{-1} be primitive? Give a reason if you think the answer is "Yes"; give an example if you think the answer is "No."

14.2.2. For each of these transition matrices M, find two cycles with relatively prime lengths, or show there are no such cycles. For those that have two cycles with relatively prime lengths, find the smallest k for which M^k has all entries positive. To simplify arithmetic, take all non-zero entries to be 1.

$$\text{(a)} \begin{bmatrix} 0 & 1 & 1 & 0 \\ 1 & 0 & 0 & 1 \\ 1 & 0 & 0 & 1 \\ 0 & 1 & 1 & 0 \end{bmatrix} \quad \text{(b)} \begin{bmatrix} 0 & 1 & 1 & 0 \\ 1 & 0 & 0 & 1 \\ 0 & 1 & 0 & 0 \\ 0 & 0 & 1 & 1 \end{bmatrix} \quad \text{(c)} \begin{bmatrix} 0 & 1 & 1 & 0 \\ 0 & 0 & 1 & 1 \\ 1 & 0 & 0 & 0 \\ 0 & 0 & 0 & 1 \end{bmatrix}$$

Practice Problem Solutions

14.2.1. Recalling the formula (8.9) for the inverse of a 2×2 matrix, Eq. (8.9), we see that if all the matrix entries are positive, half of the inverse entries are negative. This suggests the answer is "No." Let's try a simple example. Take

$$M = \begin{bmatrix} \frac{1}{3} & \frac{2}{3} \\ \frac{2}{3} & \frac{1}{3} \end{bmatrix},$$

Then

$$M^{-1} = \begin{bmatrix} -1 & 2 \\ 2 & -1 \end{bmatrix} \quad M^{-2} = \begin{bmatrix} 5 & -4 \\ -4 & 5 \end{bmatrix} \quad M^{-3} = \begin{bmatrix} -13 & 14 \\ 14 & -13 \end{bmatrix}$$

It's easy to see the pattern and conclude that all powers of M^{-1} have two negative and two positive terms. That is, M^{-1} is not primitive, while M is primitive because all its entries are positive.

14.2.2. Fig. 14.5 shows the transition graphs for these matrices.

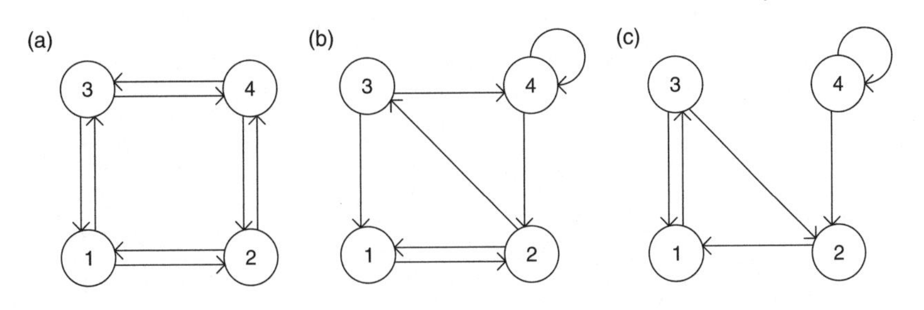

Figure 14.5. The transition graphs for the matrices of Practice Problem 14.2.2.

In (a) we see 2-cycles, $1 \to 2 \to 1$, for example, 4-cycles, $1 \to 2 \to 4 \to 2 \to 1$, and 6-cycles, $1 \to 2 \to 4 \to 3 \to 4 \to 2 \to 1$, but no odd cycles are obvious. It seems clear from the graph that we can't get from any vertex back to itself in an odd number of steps, but we can tell this from the matrix, too. Recall there is a path of length k from i to j if $(M^k)_{ji} \neq 0$, so there is a k-cycle if some element of the diagonal of M^k is non-zero. Here are some powers of M:

$$M^2 = \begin{bmatrix} 2 & 0 & 0 & 2 \\ 0 & 2 & 2 & 0 \\ 0 & 2 & 2 & 0 \\ 2 & 0 & 0 & 2 \end{bmatrix} \quad M^3 = \begin{bmatrix} 0 & 4 & 4 & 0 \\ 4 & 0 & 0 & 4 \\ 4 & 0 & 0 & 4 \\ 0 & 4 & 4 & 0 \end{bmatrix} \quad M^4 = \begin{bmatrix} 8 & 0 & 0 & 8 \\ 0 & 8 & 8 & 0 \\ 0 & 8 & 8 & 0 \\ 8 & 0 & 0 & 8 \end{bmatrix}$$

The non-zero diagonal entries in M^2 and M^4 give another way to see there are 2- and 4-cycles; the 0s on the diagonal of M^3 show there are no 3-cycles. The pattern is clear, and it's easy to see that every odd power of M has only 0s along its diagonal. All the cycles have even length, so there are no cycles of relatively prime lengths.

In (b) we see a 2-cycle, $1 \to 2 \to 1$, and a 3-cycle, $1 \to 2 \to 3 \to 1$. It's easy to find a path from every vertex to every vertex, so by Prop. 14.2.1 we know this matrix is primitive. Here are the first few powers of M:

$$M^2 = \begin{bmatrix} 1 & 1 & 0 & 1 \\ 0 & 1 & 2 & 1 \\ 1 & 0 & 0 & 1 \\ 0 & 1 & 1 & 1 \end{bmatrix} \quad M^3 = \begin{bmatrix} 1 & 1 & 2 & 2 \\ 1 & 2 & 1 & 2 \\ 0 & 1 & 2 & 1 \\ 1 & 1 & 1 & 2 \end{bmatrix} \quad M^4 = \begin{bmatrix} 1 & 3 & 3 & 3 \\ 2 & 2 & 3 & 4 \\ 1 & 2 & 1 & 2 \\ 1 & 2 & 3 & 3 \end{bmatrix}$$

We see that $n = 4$ is the smallest n for which M^n has all terms positive. This is quite a bit better than $3(4-1) + 2 \cdot 3 = 15$, the bound provided by Prop. 14.2.1. The bound of Corollary 8.5.9 of [177], $n^2 - 2n + 2 = 10$ for $n = 4$, is better than this 15, but not so good as the 4 that we've just found.

In (c) we see a 2-cycle, $1 \to 3 \to 1$, and a 3-cycle, $3 \to 2 \to 1 \to 3$, so this matrix has two cycles whose lengths are relatively prime. Here are the first few powers of M:

$$M^2 = \begin{bmatrix} 1 & 0 & 1 & 1 \\ 1 & 0 & 0 & 1 \\ 0 & 1 & 1 & 0 \\ 0 & 0 & 0 & 1 \end{bmatrix} \quad M^3 = \begin{bmatrix} 1 & 1 & 1 & 1 \\ 0 & 1 & 1 & 1 \\ 1 & 0 & 1 & 1 \\ 0 & 0 & 0 & 1 \end{bmatrix} \quad M^4 = \begin{bmatrix} 1 & 1 & 2 & 2 \\ 1 & 0 & 1 & 2 \\ 1 & 1 & 1 & 1 \\ 0 & 0 & 0 & 1 \end{bmatrix}$$

None of these have all positive entries, and in fact all have the last row $0, 0, 0, 1$. It's easy to see that if A and B are 4×4 matrices both with last row $0, 0, 0, 1$, then the last row of AB also is $0, 0, 0, 1$. That is, for the M of part (c), for all n,

M^n has at least three entries that are 0. What went wrong? Doesn't Prop. 14.2.1 apply here? The conclusion of a proposition holds whenever the hypotheses are satisfied. The proposition has two hypotheses: that the graph has two cycles with relatively prime lengths, and that the graph is strongly connected. As we can see from transition graph (c), the second hypothesis is not satisfied: from vertex 4 we can go only to vertex 4 and to no other vertex. It is no surprise, then, that this matrix is not primitive.

Exercises

14.2.1. (a) Find the greatest common divisor (a, b) for

$$a = 105,\ b = 45;\ a = 321,\ b = 123;\ a = 471,\ b = 121;\ a = 654,\ b = 136$$

(b) For each greatest common divisor of (a), write (a, b) as the sum of integer multiples of a and b.

14.2.2. Which of these matrices are primitive?

$$\text{(a)} \begin{bmatrix} 0 & 1 & 0 & 1 \\ 1 & 0 & 1 & 0 \\ 0 & 1 & 0 & 1 \\ 1 & 0 & 1 & 0 \end{bmatrix} \quad \text{(b)} \begin{bmatrix} 1 & 1 & 0 & 1 \\ 1 & 0 & 1 & 0 \\ 0 & 1 & 0 & 1 \\ 1 & 0 & 1 & 0 \end{bmatrix} \quad \text{(c)} \begin{bmatrix} 1 & 1 & 1 & 0 \\ 1 & 1 & 1 & 1 \\ 1 & 1 & 1 & 0 \\ 0 & 0 & 1 & 0 \end{bmatrix} \quad \text{(d)} \begin{bmatrix} 1 & 1 & 1 & 0 \\ 1 & 1 & 1 & 0 \\ 1 & 1 & 1 & 0 \\ 0 & 0 & 1 & 1 \end{bmatrix}$$

14.2.3. Which of these matrices are primitive?

$$\text{(a)} \begin{bmatrix} 1 & 1 & 1 & 1 \\ 1 & 0 & 0 & 0 \\ 0 & 1 & 0 & 0 \\ 0 & 0 & 1 & 0 \end{bmatrix} \quad \text{(b)} \begin{bmatrix} 1 & 1 & 0 & 1 \\ 1 & 0 & 0 & 0 \\ 0 & 1 & 0 & 0 \\ 0 & 0 & 1 & 0 \end{bmatrix} \quad \text{(c)} \begin{bmatrix} 1 & 0 & 0 & 1 \\ 1 & 0 & 0 & 0 \\ 0 & 1 & 0 & 0 \\ 0 & 0 & 1 & 0 \end{bmatrix} \quad \text{(d)} \begin{bmatrix} 1 & 1 & 1 & 0 \\ 1 & 0 & 0 & 0 \\ 0 & 1 & 0 & 0 \\ 0 & 0 & 1 & 0 \end{bmatrix}$$

14.2.4. Show matrix (a) is not primitive. Show matrix (b) is primitive, not by computing powers of the transition matrix, but by applying Prop. 14.2.1.

$$\text{(a)} \begin{bmatrix} 0 & 0 & 1 & 0 \\ 1 & 0 & 0 & 0 \\ 0 & 0 & 0 & 1 \\ 0 & 1 & 0 & 0 \end{bmatrix} \qquad \text{(b)} \begin{bmatrix} 0 & 0 & 1 & 0 \\ 1 & 0 & 0 & 0 \\ 0 & 0 & 0 & 1 \\ 0 & 1 & 0 & 1 \end{bmatrix}$$

14.2.5. Show the matrices (a) and (b) are not primitive. For each, show that by changing an appropriate pair of 0s to 1s, the resulting matrix is primitive. Think of Prop. 14.2.1 to guide your choices. Can changing a single entry of (a) or (b) from 0 to 1 make the resulting matrix primitive?

$$\text{(a)}\begin{bmatrix}0&1&0&0&0\\1&0&0&0&0\\0&0&0&0&1\\0&0&1&0&0\\0&0&0&1&0\end{bmatrix}\qquad\text{(b)}\begin{bmatrix}1&0&0&0&0\\0&0&0&0&1\\0&1&0&0&0\\0&0&1&0&0\\0&0&0&1&0\end{bmatrix}$$

14.2.6. (a) By an example, show that the product of primitive matrices need not be primitive. Think of Exercise 14.2.4 to guide your choices.
(b) Show that if A and B are $n \times n$ primitive matrices that commute (i.e., $AB = BA$), then the product AB is primitive.
(c) Must primitive $n \times n$ matrices commute if their product is primitive?

14.3 STRUCTURED POPULATIONS

A more nuanced view of population growth can be achieved when we account for the fact that reproduction does not occur for the very young or the very old. Ecologist Patrick Leslie [206, 207] devised a model to capture the dynamics of a population's age strata. While we'll focus mostly on this interpretation, the techniques can be applied to other population structures. For example, mathematicians Odo Diekmann and Hans Heesterbeek (also a professor of farm animal health) adapted Leslie's approach to study the spread of diseases [85]. Applying the sensitivity analysis of Sect. 14.4 to disease populations can allow epidemiologists to determine how to deploy limited resources to best reduce the rate of the spread of an epidemic. Perhaps these techniques can be applied through virus dynamics to suggest effective ways to direct the immune system for controlling the growth of a virus population.

Because we're interested in the population growth rate, usually only females, mothers and daughters, are counted. For these calculations, males are mostly irrelevant. The total population will be about twice the value produced by the model.

Suppose the population is divided into $A+1$ age strata (classes), numbered 0 through A. For $i = 0, 1, \ldots, A$,

$$N_i(t) = \text{the population in age class } i \text{ at time } t$$

The units of time are the durations of each age class, for simplicity all taken to be the same. For example, each age class can be indexed by the age year, $i = 0$ for newborns, $i = 1$ for one-year-olds, and so on. Set

p_i = probability that a member of class i survives to enter class $i+1$

f_i = average number of (female) newborns (in class $i=0$) produced by females in class i

Then all the $N_i(t)$ contribute to the newborns $N_0(t+1)$ at time $t+1$ by

$$N_0(t+1) = f_0 N_0(t) + f_1 N_1(t) + \cdots + f_A N_A(t) \tag{14.16}$$

For all other age classes $i = 1, \dots, A$, $N_i(t+1)$ is the fraction of the previous age class population $N_{i-1}(t)$ that survives to enter age class i at time $t+1$. Note that the members of $N_i(t)$ have either died or have moved on to contribute to $N_{i+1}(t+1)$. Then for $i > 0$,

$$N_i(t+1) = p_{i-1} N_{i-1}(t) \tag{14.17}$$

The information of Eqs. (14.16) and (14.17) can be encoded as a matrix equation

$$\begin{bmatrix} N_0(t+1) \\ N_1(t+1) \\ \dots \\ \dots \\ N_A(t+1) \end{bmatrix} = \begin{bmatrix} f_0 & f_1 & \cdots & f_{A-1} & f_A \\ p_0 & 0 & \dots & 0 & 0 \\ 0 & p_1 & \dots & 0 & 0 \\ & & \dots & & \\ 0 & 0 & \dots & p_{A-1} & 0 \end{bmatrix} \begin{bmatrix} N_0(t) \\ N_1(t) \\ \dots \\ N_{A-1}(t) \\ N_A(t) \end{bmatrix} \tag{14.18}$$

In more compact form this is $\vec{N}(t+1) = L\vec{N}(t)$. In this model, the transition matrix L is called the *Leslie matrix* and the corresponding transition graph is called the *life-cycle graph*.

The eigenvalues of the Leslie matrix are the roots of a simple characteristic equation

$$\begin{aligned} \lambda^{A+1} &= f_0\lambda^A + f_1 p_0 \lambda^{A-1} + f_2 p_0 p_1 \lambda^{A-2} + f_3 p_0 p_1 p_2 \lambda^{A-3} + \cdots \\ &\quad + f_{A-1} p_0 p_1 p_2 \cdots p_{A-2}\lambda + f_A p_0 p_1 p_2 \cdots p_{A-2} p_{A-1} \end{aligned} \tag{14.19}$$

In Sect. A.18 we'll calculate the characteristic polynomial for $A = 5$, where the Leslie matrix is large enough to show all the complications of the general case, but small and specific enough to understand easily.

Except for the $A = 1$ case, the Leslie matrix eigenvalues do not admit a simple analytic form. We might think that the number of 0s in the Leslie matrix would make tractable the calculation of its eigenvalues, but sadly this is not the case. Unless enough of the p_i or f_i are 0 to simplify the calculation (for example, if $f_0 = f_2 = f_3 = \cdots = f_A = 0$, then the eigenvalues are $\pm\sqrt{p_0 f_1}$, and $A-1$ copies of 0), we must find the eigenvalues numerically.

Many Leslie matrices, for example, if $f_A = 0$ in the form displayed in Eq. (14.18), are not power positive. (Draw the transition graph. You'll find no path from state A to any other state.) In these cases, the Perron-Frobenius theorem cannot be applied. However, we do have this.

Proposition 14.3.1. Every Leslie matrix of the form in Eq. (14.18) has a unique positive eigenvalue λ_1, and for every complex eigenvalue $a+ib$, $\lambda_1 > a$.

Proof. Divide both sides of the characteristic equation (14.19) by λ^{A+1}:

$$1 = \frac{f_0}{\lambda} + \frac{f_1 p_0}{\lambda^2} + \frac{f_2 p_0 p_1}{\lambda^3} + \cdots + \frac{f_A p_0 p_1 \cdots p_{A-1}}{\lambda^{A+1}} \tag{14.20}$$

This is called the *Lotka-Euler equation.* Now let $g(\lambda)$ denote the right-hand side of this equation. That is,

$$g(\lambda) = \frac{f_0}{\lambda} + \frac{f_1 p_0}{\lambda^2} + \frac{f_2 p_0 p_1}{\lambda^3} + \cdots + \frac{f_A p_0 p_1 \cdots p_{A-1}}{\lambda^{A+1}} \tag{14.21}$$

Then we want to show that the equation $g(\lambda) = 1$ has only one positive solution.

Recall that in the eigenvector equation $L\vec{N} = \lambda\vec{N}$, the entries of $\vec{N}$ are the age class populations, necessarily non-negative, so only non-negative λ are of interest.

Because each term of $g(\lambda)$ has non-negative numerator, we see that

$$\lim_{\lambda\to 0} g(\lambda) = \infty \quad \text{and} \quad \lim_{\lambda\to\infty} g(\lambda) = 0$$

Next, compute the first and second derivatives of g

$$\frac{dg}{d\lambda} = -\frac{f_0}{\lambda^2} - 2\frac{f_1 p_0}{\lambda^3} - 3\frac{f_2 p_0 p_1}{\lambda^4} - \cdots - (A+1)\frac{f_A p_0 p_1 \cdots p_{A-1}}{\lambda^{A+2}}$$

$$\frac{d^2 g}{d\lambda^2} = 2\frac{f_0}{\lambda^3} + 6\frac{f_1 p_0}{\lambda^4} + 12\frac{f_2 p_0 p_1}{\lambda^5} + \cdots + (A+1)(A+2)\frac{f_A p_0 p_1 \cdots p_{A-1}}{\lambda^{A+3}}$$

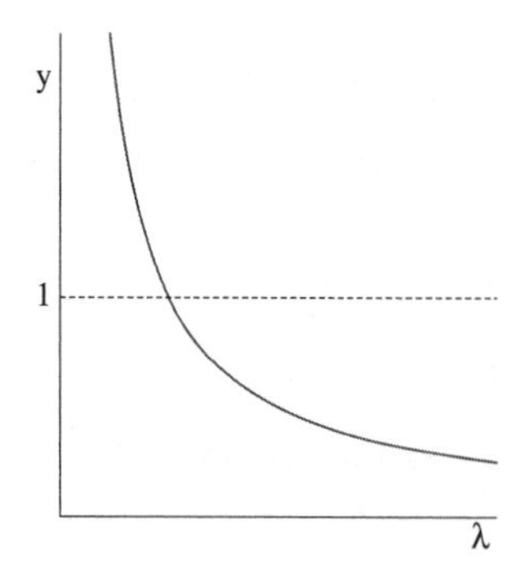

Figure 14.6. The graph of $y = g(\lambda)$.

We see that $dg/d\lambda < 0$ and $d^2g/d\lambda^2 > 0$. Consequently, the graph of $y = g(\lambda)$ is decreasing and concave up, as pictured in Fig. 14.6. Together with the limiting behavior as $\lambda \to 0$ and $\lambda \to \infty$ mentioned above, we see the graph of $y = g(\lambda)$ crosses the line $y = 1$ exactly once, so the characteristic equation for the Leslie matrix has exactly one positive solution.

We must mention one more point. The Lotka-Euler equation (14.20) is obtained by dividing both sides of the Leslie matrix characteristic equation (14.19) by λ^{A+1}, which we can do as long as $\lambda \neq 0$. Any proof based on the Lotka-Euler equation cannot rule out $\lambda = 0$ as an eigenvalue.

Now let's look at the relative magnitudes of the largest positive eigenvalue and the real part of any complex eigenvalue $\lambda = a + ib$. We'll use the polar representation and Euler's formula

$$\lambda = e^{\alpha + i\beta} = e^{\alpha} e^{i\beta} = e^{\alpha}(\cos(\beta) + i\sin(\beta))$$

where $\beta = \tan^{-1}(b/a)$ because

$$\tan(\beta) = \frac{\sin(\beta)}{\cos(\beta)} = \frac{e^{\alpha}\sin(\beta)}{e^{\alpha}\cos(\beta)} = \frac{b}{a}$$

and $\alpha = \ln(\sqrt{a^2 + b^2})$ because $\tan(\beta) = b/a$ implies $\cos(\beta) = a/\sqrt{a^2 + b^2}$ (draw the right triangle with base a and altitude b), and so

$$a = e^{\alpha}\cos(\beta) = e^{\alpha}\frac{a}{\sqrt{a^2 + b^2}} \text{ implies } e^{\alpha} = \sqrt{a^2 + b^2}$$

The expression for $g(\lambda)$ in Eq. (14.21) involves λ^{-n}, so we need to understand these for complex λ. First, use the polar representation

$$\lambda^{-n} = (e^{\alpha} e^{i\beta})^{-n} = (e^{\alpha})^{-n}(e^{i\beta})^{-n}$$

and focus on the second factor:

$$(e^{i\beta})^{-n} = (e^{i(-\beta)})^{n} = (\cos(-\beta) + i\sin(-\beta))^{n}$$

Every point $\cos(-\beta) + i\sin(-\beta)$ lies on the circle $|z| = 1$, so $(\cos(-\beta) + i\sin(-\beta))^n$ lies on this circle. Then $-1 \leq \text{Re}((e^{i\beta})^{-n}) \leq 1$ and

$$\text{Re}(\lambda^{-n}) = (\text{Re}(e^{\alpha})^{-n})\text{Re}((e^{i\beta})^{-n}) = (e^{\alpha})^{-n}\text{Re}((e^{i\beta})^{-n})$$

Because $\text{Re}((e^{i\beta})^{-n}) \leq 1$ this gives

$$\text{Re}(\lambda^{-n}) \leq (e^{\alpha})^{-n} \tag{14.22}$$

Now if λ is complex, so is $g(\lambda)$, and in this situation $g(\lambda) = 1$ really is two equations:

$$\text{Re}(g(\lambda)) = 1 \quad \text{and} \quad \text{Im}(g(\lambda)) = 0 \tag{14.23}$$

We'll use the first of these and Eq. (14.21),

$$\begin{aligned}\text{Re}(g(\lambda)) &= \text{Re}\big(f_0\lambda^{-1} + f_1 p_0 \lambda^{-2} + \cdots + f_A p_0 p_1 \cdots p_{A-1}\lambda^{-(A+1)}\big)\\ &= f_0\text{Re}(\lambda^{-1}) + f_1 p_0 \text{Re}(\lambda^{-2}) + \cdots + f_A p_0 p_1 \cdots p_{A-1}\text{Re}(\lambda^{-(A+1)})\end{aligned}$$

Because all the coefficients are non-negative, by Eq. (14.22) we have

$$\begin{aligned}\text{Re}(g(\lambda)) &\leq f_0(e^{\alpha})^{-1} + f_1 p_0 (e^{\alpha})^{-2} + \cdots + f_A p_0 p_1 \cdots p_{A-1}(e^{\alpha})^{-(A+1)}\\ &= g(e^{\alpha})\end{aligned} \tag{14.24}$$

Now because $\lambda = e^\alpha \cos(\beta) + ie^\alpha \sin(\beta)$ is complex and not real, the imaginary part is not 0, so $\sin(\beta) \neq 0$. Then β is not an integer multiple of π, so $\cos(\beta) < 1$. This is the point we'll use to establish $a < \lambda_1$.

For complex eigenvalues λ we know

$$g(\lambda_1) = 1 = \mathrm{Re}(g(\lambda)) \leq g(e^\alpha)$$

where the second equality follows from Eq. (14.23) and the inequality from Eq. (14.24). Because g is a decreasing function, $g(\lambda_1) \leq g(e^\alpha)$ implies that $e^\alpha \leq \lambda_1$. Finally, $a = e^\alpha \cos(\beta) < e^\alpha$ because $\cos(\beta) < 1$ and so

$$a < e^\alpha \leq \lambda_1$$

as promised. □

As in Sect. 14.1, this largest eigenvalue λ_1 of the Leslie matrix is the dominant eigenvalue, and as in Sect. 14.1, the entries of the unit eigenvector of the dominant eigenvalue are the *stable age distribution* of the population. Remember, in this setting the term "unit eigenvector" means the eigenvector whose entries, necessarily non-negative, sum to 1, not the eigenvector with Euclidean length 1.

A difference between the conclusions of the Perron-Frobenius theorem and those of Prop. 14.3.1 is that we no longer are guaranteed that $\lambda_1 > |\lambda|$ for all other eigenvalues λ. Here are two Leslie matrices that illustrate this.

$$M_1 = \begin{bmatrix} 0 & 5/4 & 0 & 0 \\ 3/4 & 0 & 0 & 0 \\ 0 & 1/2 & 0 & 0 \\ 0 & 0 & 1/4 & 0 \end{bmatrix} \qquad M_2 = \begin{bmatrix} 0 & 0 & 5/4 & 0 \\ 7/8 & 0 & 0 & 0 \\ 0 & 3/4 & 0 & 0 \\ 0 & 0 & 1/8 & 0 \end{bmatrix}$$

The eigenvalues of M_1 are $\pm\sqrt{15}/4, -\sqrt{15}/4, 0, 0$. The dominant eigenvalue is $\lambda_1 = \sqrt{15}/4$, and certainly $\lambda_1 = |-\sqrt{15}/4|$.

The eigenvalues of M_2 are complicated; numerical approximations are $0.936109, -0.468055 \pm i0.810694, 0$. The dominant eigenvalue λ_1 is 0.936109, and again we find $\lambda_1 = |-0.468055 \pm i0.810694|$. However, as we expect from the second part of Prop. 14.3.1, $\lambda_1 > -0.468055$, the real part of the complex eigenvalues. As far as the relative magnitudes of Leslie matrix eigenvalues are concerned, Prop. 14.3.1 is good enough for the applications we have in mind.

All the 0s in a Leslie matrix do not simplify the task of finding eigenvalues, but they are a help in finding the eigenvectors. We'll illustrate this with an $A = 3$ Leslie matrix. How to extend to the more general case will be clear.

Using the form of the Leslie matrix in Eq. (14.18), the $A = 3$ eigenvector equation becomes

$$\begin{bmatrix} f_0 & f_1 & f_2 & f_3 \\ p_0 & 0 & 0 & 0 \\ 0 & p_1 & 0 & 0 \\ 0 & 0 & p_2 & 0 \end{bmatrix} \begin{bmatrix} N_0 \\ N_1 \\ N_2 \\ N_3 \end{bmatrix} = \lambda \begin{bmatrix} N_0 \\ N_1 \\ N_2 \\ N_3 \end{bmatrix} \tag{14.25}$$

where to reduce some of the visual complication we have not written the (t) in $N_i(t)$.

One row at a time, this gives

$$f_0N_0 + f_1N_1 + f_2N_2 + f_3N_3 = \lambda N_0$$
$$p_0N_0 = \lambda N_1, \quad p_1N_1 = \lambda N_2, \text{ and } \quad p_2N_2 = \lambda N_3$$

Solving the last three equations for N_1, N_2, and N_3 we find

$$N_1 = \frac{p_0N_0}{\lambda}, \ N_2 = \frac{p_1N_1}{\lambda} = \frac{p_1p_0N_0}{\lambda^2}, \ N_3 = \frac{p_2N_2}{\lambda} = \frac{p_2p_1p_0N_0}{\lambda^3}$$

By the way, substituting these expressions for N_1, N_2, and N_3 into Eq. (14.25), and assuming $N_0 \neq 0$, the first row becomes the characteristic equation, Eq. (14.19), for this Leslie matrix.

Substituting in these expressions for N_1, N_2, and N_3, the unit eigenvector equation $N_0 + N_1 + N_2 + N_3 = 1$ becomes

$$N_0 + \frac{p_0N_0}{\lambda} + \frac{p_1p_0N_0}{\lambda^2} + \frac{p_2p_1p_0N_0}{\lambda^3} = 1$$

Solving for N_0 and substituting into the expressions for N_1, N_2, and N_3, then multiplying numerator and denominator of each entry by λ^3, the unit eigenvector for every eigenvalue λ of this Leslie matrix is $[v_1, v_2, v_3, v_4]$, where

$$v_1 = \frac{\lambda^3}{D}, \quad v_2 = \frac{p_0\lambda^2}{D}, \quad v_3 = \frac{p_0p_1\lambda}{D}, \quad v_4 = \frac{p_0p_1p_2}{D}, \quad \text{and} \tag{14.26}$$
$$D = \lambda^3 + p_0\lambda^2 + p_0p_1\lambda + p_0p_1p_2$$

How to generalize this for larger values of A is clear.

We've been too long without an example. Let's see a simple way to use this formula, and some interpretations of the calculations.

Example 14.3.1. *The stable distribution of a simple age-structured population.* Suppose a population has four age classes (so $A = 3$), with
$f_0 = 0$: members of the first age class are not reproductive.
$f_1 = 1.5$: an average member of age class 2 contributes 1.5 newborns.
$f_2 = 1.2$: an average member of age class 3 contributes 1.2 newborns.
$f_3 = 0$: members of the age class 4 are not reproductive.
$p_0 = 0.9$: 90% of the members of age class 1 survive to age class 2.

$p_1 = 0.7$: 70% of the members of age class 2 survive to age class 3.
$p_2 = 0.3$: 30% of the members of age class 3 survive to age class 4.

For this system, the Leslie matrix is

$$L = \begin{bmatrix} 0 & 1.5 & 1.2 & 0 \\ 0.9 & 0 & 0 & 0 \\ 0 & 0.7 & 0 & 0 \\ 0 & 0 & 0.3 & 0 \end{bmatrix}$$

Numerically we find the eigenvalues are approximately

$$1.37792, -0.688958 \pm 0.272012i, 0$$

So the dominant eigenvalue is $\lambda = 1.37792$, and from Eq. (14.26) we see the stable age distribution is

$$N_0 \approx 0.486,\ N_1 \approx 0.317,\ N_2 \approx 0.161,\ N_4 \approx 0.035$$

The λ value tells us that with each time step the population size is multiplied by a factor of about 1.378, that is, this population experiences about a 38% increase. Pretty fast-growing. Moreover, the bulk of the population is in the first two age classes, with only about 3.5% in the top age class. With so few in the last age class, we could ask if $A = 3$ is an appropriate model. If we drop the last age class and modify Eq. (14.26) so the denominators of the eigenvector entries are $\lambda^2 + p_0\lambda + p_0p_1$, along with the corresponding changes to the numerators, we find first that the list of eigenvalues is unchanged, except for losing the 0 eigenvalue, so the dominant eigenvalue is unchanged. Consequently, the population continues to grow at the same (high) rate. Next, the stable age distribution is

$$N_0 \approx 0.504,\ N_1 \approx 0.329,\ N_2 \approx 0.167$$

which is pretty close to those numbers for the $A = 3$ model. More telling is to calculate the ratios N_0/N_1 and N_1/N_2 for both models:

$$A = 3: \quad N_0/N_1 \approx 1.533,\ N_1/N_2 \approx 1.969$$
$$A = 2: \quad N_0/N_1 \approx 1.532,\ N_1/N_2 \approx 1.970$$

So for this model, the top age class is unimportant, an observation that is personally discouraging because that's the age class to which I belong in the human population.

The eigenvector of the dominant eigenvalue suggests something a bit less gloomy. Around half of the population is in the first age class. In human and many other populations, these are the children, an age class that needs parental care. So while a population represented by this model doesn't care much for its geezers, it does protect its youth. □

Some variations: the stratification of a population need not be by age. Size classes or developmental stages (for example, the egg, larva, pupa, and adult stages of insects) can be used, and as mentioned, there are epidemiological applications. Susceptibility can vary with age class, as can the probability of transmission because of the number and type of contacts, and also the mean duration of the time in the infected state. A class-specific refinement of the SIR model can give more accurate results.

With classifiers other than age, the Leslie matrix need not have the form we saw in Eq. (14.18). For example, with size classes we find at least two departures from the class transitions encoded in the Leslie matrix of Eq. (14.18): with growth spurts an individual can jump more than one size class in a time step, and an individual can move down in size class because we shrink as we age. With this stratification by size, the characteristic equation can be more complicated than Eq. (14.19), and in general we must repeat the derivation of properties for every variant of a Leslie matrix. Still, stratification of a population can give a more detailed view of the dynamics. For some cases it can be an enlightening technique.

Practice Problems

14.3.1. For an $A = 3$ population model with the parameters

$$f_0 = 0,\ f_1 = 1.1,\ f_2 = 1.2,\ f_3 = 0,\ p_0 = 0.8,\ p_1 = 0.8,\ p_2 = 0.3$$

find the eventual growth factor, the percentage growth rate, and the stable age distribution.

14.3.2. In each part of this problem, begin with the parameters from Practice Problem 14.3.1.
(a) What value must f_2 take to eventually give a constant population? Estimate f_2 to two digits to the right of the decimal.
(b) If the values of f_0, f_1, f_2, and f_3 are doubled, does the eventual population percentage growth rate double?
(c) If the values of p_0, p_1, and p_2, are halved, does the eventual population percentage growth rate halve? Why couldn't we double the p_i values and ask if the eventual population percentage growth rate doubled?

Practice Problem Solutions

14.3.1. With these parameters, the Leslie matrix is

$$L = \begin{bmatrix} 0 & 1.1 & 1.2 & 0 \\ 0.8 & 0 & 0 & 0 \\ 0 & 0.8 & 0 & 0 \\ 0 & 0 & 0.3 & 0 \end{bmatrix}$$

Numerically we find the eigenvalues are approximately $1.227, -0.614\pm 0.499i, 0$, so eventually the population grows by a factor of about 1.227 per time step; that is, the population percentage growth rate is about 22.7% per time step. That's still very rapid growth. Using Eq. (14.26) we find the stable age distribution is about

$$N_0 \approx 0.459,\ N_1 \approx 0.299,\ N_2 \approx 0.195,\ N_3 \approx 0.048$$

14.3.2. (a) In terms of the Leslie matrix, a constant population means that the dominant eigenvalue is $\lambda = 1$. Numerically we find

$$f_2 = 0.187 \text{ gives } \lambda = 0.999849 \text{ and } f_2 = 0.188 \text{ gives } \lambda = 1.00015$$

Estimated two places to the right of the decimal, the dominant eigenvalue is $\lambda = 1$ when $f_2 \approx 0.19$.
(b) Doubling the values of the f_i changes the dominant eigenvalue from 1.227 to 1.642, so the population percentage growth rate increases from 22.7% to 64.7%, more than doubled. Doubled growth rates in the earlier age strata increase later strata populations, and so have compound effects on the growth rates of these strata.
(c) Halving all the p_i reduces the dominant eigenvalue from 1.227 to 0.821, so the population percentage growth rate drops from +22.7% to −17.9%. Lower class-to-class survival rates have compound effects on the total population percentage growth rates.

We can't double all the values of the p_i from Practice Problem 14.3.1 because being the proportion of population class i that survives to enter class $i+1$, p_i cannot be greater than 1. In Practice Problem 14.3.1, $p_0 = p_1 = 0.8$ and $p_2 = 0.3$, so while p_2 could be doubled and not be greater than 1, the same cannot be said for p_0 and p_1.

Exercises

14.3.1. For populations with these parameters, find the eventual growth factor, percentage growth rate, and the stable age distribution.
(a) $A = 2, f_0 = 0.2, f_1 = 1.1, f_2 = 0.5; p_0 = 0.7, p_1 = 0.7$.
(b) $A = 2, f_0 = 0.2, f_1 = 1.1, f_2 = 0.5; p_0 = 0.7, p_1 = 0.2$.
(c) $A = 3, f_0 = 0, f_1 = 1.5, f_2 = 0.2, f_3 = 0; p_0 = 0.8, p_1 = 0.7, p_2 = 0.6$.
(d) $A = 3, f_0 = 0, f_1 = 1.5, f_2 = 0.4, f_3 = 0; p_0 = 0.8, p_1 = 0.7, p_2 = 0.6$.
(e) $A = 3, f_0 = 0.2, f_1 = 1.5, f_2 = 0.2, f_3 = 0; p_0 = 0.8, p_1 = 0.7, p_2 = 0.6$.

14.3.2. Suppose $f_1 \neq 0$ and $f_2 \neq 0$ in an $A = 3$ population model. For every positive r, is the eventual population percentage growth rate unchanged if f_1 is replaced by f_1/r and f_2 is replaced by $f_2 \cdot r$? Give a counterexample if you think the answer is "No"; give a reason if you think the answer is "Yes."

14.3.3. Consider an $A = 3$ population with these parameters:

$$f_0 = 0, f_1 = 2, f_2 = 0.25, f_3 = 0; p_0 = 0.8, p_1 = 0.5, p_2 = 0.2$$

Find the dominant eigenvalue and the stable age distribution. Notice that N_3 is not equal to $p_0 \cdot p_1 \cdot p_2$. Every member of the population has started in age class 0, so why isn't the fraction of the population in age class 3 the product of the probabilities of making the transition from class 0 to 1, from class 1 to 2, and from class 2 to 3? Find an explanation.

14.3.4. Does dividing all the p_i and f_i by 2 divide the eventual population growth rate by 2? Give a counterexample if you think the answer is "No"; give a reason if you think the answer is "Yes."

14.3.5. Can a Leslie matrix model an oscillating population? Give a reason to support your answer.

14.3.6. If a population has an even number of states, does Prop. 14.3.1 imply that at least one eigenvalue must be 0? Recall that complex eigenvalues occur in pairs and that the dominant eigenvalue is positive. Give a counterexample if you think the answer is "No"; give a reason if you think the answer is "Yes."

14.4 REPRODUCTIVE VALUE AND SENSITIVITY ANALYSIS

Most of the eigenvectors we've studied are called *right eigenvectors* because in the eigenvector equation $M\vec{v} = \lambda\vec{v}$ the eigenvector is multiplied on the right side of the matrix, so $\vec{v}$ is written as a column vector. Recall that we can have row vectors $\vec{w}$ that are left eigenvectors if they satisfy

$$\vec{w}M = \lambda\vec{w}$$

We'll see that left and right eigenvectors of a matrix can be used in productive ways, but first we should ask if there are left and right eigenvalues. Happily, left and right eigenvectors have the same eigenvalues, the roots of the familiar characteristic equation. The reason is almost identical to our derivation of the characteristic equation for right eigenvectors. For a non-zero row vector $\vec{w}$ to be a left eigenvector for M, we must have

$$\vec{w}M = \lambda\vec{w}, \text{ so } \vec{w}M - \lambda\vec{w} = \vec{0}, \ \vec{w}M - \lambda\vec{w}I = \vec{0}, \text{ and } \vec{w}(M - \lambda I) = \vec{0}$$

In order for this last equation to have a non-zero solution $\vec{w}$, the matrix $M - \lambda I$ must not be invertible. That is, $\det(M - \lambda I) = 0$. But this is just the characteristic equation for right eigenvectors.

To illustrate how different left and right eigenvectors can be, we'll find them for a 3×3 primitive matrix.

Example 14.4.1. *Unit left and right eigenvectors.* Take

$$M = \begin{bmatrix} 3/4 & 0 & 1/4 \\ 1/4 & 1/2 & 0 \\ 0 & 1/2 & 3/4 \end{bmatrix}$$

The matrix entries are non-negative, the entries of each column sum to 1, and an easy matrix multiplication shows every entry of M^2 is positive. Then the Perron-Frobenius theorem shows $\lambda = 1$ has the largest magnitude among all the eigenvalues. In fact, the eigenvalues of M are $1, 1/2 \pm i/4$. We have plenty of practice finding right eigenvectors. The unit right eigenvector for $\lambda = 1$ has $x = z = 2/5$ and $y = 1/5$.

Now we'll find the unit left eigenvector for the eigenvalue $\lambda = 1$ of M.

$$\begin{bmatrix} a & b & c \end{bmatrix} \begin{bmatrix} 3/4 & 0 & 1/4 \\ 1/4 & 1/2 & 0 \\ 0 & 1/2 & 3/4 \end{bmatrix} = 1 \begin{bmatrix} a & b & c \end{bmatrix}$$

Multiplying the row vector to the left of M by the first column of M gives

$$a(3/4) + b(1/4) = a, \quad \text{that is,} \quad a = b$$

Multiplying the row vector by the second and third columns of M gives $c = b$ and $a = c$. Any of these three conditions is a consequence of the other two. The unit eigenvector condition, $a + b + c = 1$, then gives $\vec{w} = [1/3, 1/3, 1/3]$, different from the right eigenvector. □

Computer algebra systems such as Mathematica usually find right eigenvectors. In Sect. B.20 we'll show a way to find left eigenvectors with Mathematica.

Why do we need left eigenvectors as well as right eigenvectors? In Sect. 14.1 we saw that the entries of the unit right eigenvector give the proportion of the stable population in each class. Do the entries of the unit left eigenvector have a similar interpretation? That the answer is "yes" is not much of a surprise.

Suppose we have a structured population with class occupancies $\vec{N}(t) = \langle N_1(t), \dots, N_k(t) \rangle$ at time t, that the occupancy vector evolves according to the transition matrix M by

$$\vec{N}(t) = M\vec{N}(t-1) = \cdots = M^t \vec{N}(0), \tag{14.27}$$

that λ_1 is the dominant eigenvalue with left eigenvector $\vec{w}_1$ and right eigenvector $\vec{v}_1$, and that the matrix A whose columns are the right eigenvectors $\vec{v}_1, \dots, \vec{v}_k$ for the eigenvalues $\lambda_1, \dots, \lambda_k$ of M satisfies $\det(A) \neq 0$. We've encountered this condition in Prop. 14.1.1, where we saw it guaranteed that the vectors $\vec{v}_i$ form a basis. Then,

Proposition 14.4.1. The entry $(\vec{w}_1)_i$ of the unit left eigenvector $\vec{w}_1$ of the dominant eigenvalue λ_1 gives the proportion of the long-term population descending from a single individual in state i.

Proof. We want to interpret the entries of the left unit eigenvector $\vec{w}_1$, so we'll take the dot product of $\vec{w}_1$ with the column vector $\vec{N}(t)$ that gives the population in each class at time t. Then for any initial populations $\vec{N}(0)$,

$$\begin{aligned}
\vec{w}_1\vec{N}(t) &= \vec{w}_1(M^t\vec{N}(0)) && \text{by Eq. (14.27)}\\
&= \vec{w}_1(M^t(c_1\vec{v}_1 + c_2\vec{v}_2 + \cdots + c_k\vec{v}_k)) && \text{eigenvectors form a basis}\\
&= \vec{w}_1(c_1M^t\vec{v}_1 + c_2M^t\vec{v}_2 + \cdots + c_kM^t\vec{v}_k)\\
&= \vec{w}_1(c_1\lambda_1^t\vec{v}_1 + c_2\lambda_2^t\vec{v}_2 + \cdots + c_k\lambda_k^t\vec{v}_k)\\
&= \lambda_1^t\vec{w}_1(c_1\vec{v}_1 + c_2(\lambda_2/\lambda_1)^t\vec{v}_2 + \cdots + c_k(\lambda_k/\lambda_1)^t\vec{v}_k)
\end{aligned}$$

Because $\lambda_1 > |\lambda_i|$ for $2 \le i \le k$, we see $\lim_{t\to\infty}(\lambda_i/\lambda_1)^t = 0$ and so

$$\lim_{t\to\infty}\frac{\vec{w}_1\vec{N}(t)}{\lambda_1^t} = c_1\vec{w}_1\vec{v}_1$$

If this argument seems familiar, look again at Prop. 14.1.1.

Now suppose we change the initial population by adding 1 individual to state i. We can represent this by adding to $\vec{N}(0)$ the column vector $\vec{e}_i$ which consists of 0s and a 1 in position i. Repeating the argument above, we arrive at

$$\begin{aligned}
\vec{w}_1\vec{N}(t) &= \vec{w}_1(M^t(\vec{N}(0) + \vec{e}_i))\\
&= \vec{w}_1(M^t\vec{N}(0) + M^t\vec{e}_i)\\
&= \vec{w}_1((c_1\lambda_1^t\vec{v}_1 + c_2\lambda_2^t\vec{v}_2 + \cdots + c_k\lambda_k^t\vec{v}_k) + M^t\vec{e}_i)\\
&= \lambda_1^t\vec{w}_1(c_1\vec{v}_1 + c_2(\lambda_2/\lambda_1)^t\vec{v}_2 + \cdots + c_k(\lambda_k/\lambda_1)^t\vec{v}_k) + \vec{w}_1M^t\vec{e}_i\\
&= \lambda_1^t\vec{w}_1(c_1\vec{v}_1 + c_2(\lambda_2/\lambda_1)^t\vec{v}_2 + \cdots + c_k(\lambda_k/\lambda_1)^t\vec{v}_k) + \lambda_1^t\vec{w}_1\vec{e}_i\\
&= \lambda_1^t\vec{w}_1(c_1\vec{v}_1 + c_2(\lambda_2/\lambda_1)^t\vec{v}_2 + \cdots + c_k(\lambda_k/\lambda_1)^t\vec{v}_k) + \lambda_1^t(\vec{w}_1)_i
\end{aligned}$$

where we have used $\vec{w}_1M^t = \lambda_1^t\vec{w}_1$ because $\vec{w}_1$ is a left eigenvector of λ_1, and also $\vec{w}_1\vec{e}_i = (\vec{w}_1)_i$, the ith entry of $\vec{w}_1$. Then we have

$$\lim_{t\to\infty}\frac{\vec{w}_1(M^t(\vec{N}(0) + \vec{e}_i))}{\lambda_1^t} = c_1\vec{w}_1\vec{v}_1 + (\vec{w}_1)_i$$

That is, increasing the initial population by 1 individual in state i increases the population growth rate by $(\vec{w}_1)_i$, supporting the name "reproductive value" for the right eigenvector of the dominant eigenvalue. □

But wait a moment. Which right eigenvector? Remember, any non-zero multiple of an eigenvector still is an eigenvector. We'll look at the ratio of the coefficient of λ_1^t with and without the additional individual in state i. First, we need to recall two facts about multiplying vectors by constants:

$$\langle ka, kb, kc\rangle = k\langle a, b, c\rangle$$

and

$$\langle ka, kb, kc\rangle \cdot \langle d, e, f\rangle = kad + kbe + ecf = k(ad + be + cf) = k(\langle a, b, c\rangle \cdot \langle d, e, f\rangle)$$

Then if we replace the right eigenvector $\vec{w}_1$ by $k\vec{w}_1$, the ratio of coefficients is

$$\frac{c_1(k\vec{w}_1)\vec{v}_1 + (k\vec{w}_1)_i}{c_1(k\vec{w}_1)\vec{v}_1} = \frac{k(c_1\vec{w}_1\vec{v}_1 + (\vec{w}_1)_i)}{k(c_1\vec{w}_1\vec{v}_1)} = \frac{c_1\vec{w}_1\vec{v}_1 + (\vec{w}_1)_i}{c_1\vec{w}_1\vec{v}_1}$$

So we've seen that the entries of the unit right eigenvector of the dominant eigenvalue give the proportions of the long-term population in each age class, and that the entries of the unit left eigenvector of the dominant eigenvalue give the proportion of the long-term population descending from a single individual in the corresponding age class. Taken separately, the left and right eigenvectors inform us about the long-term population.

Taken together, left and right eigenvectors have another application: sensitivity analysis of eigenvalues. Suppose some of the entries of the matrix M depend on a parameter x. Then the eigenvalues and corresponding eigenvectors also can depend on x. If (the column vector) $\vec{v}$ is a right eigenvector of M and (the row vector) $\vec{w}$ is a left eigenvector of M, both with eigenvalue λ, then

$$\vec{w}M = \vec{w}\lambda \quad \text{and} \quad M\vec{v} = \lambda\vec{v}$$

Then multiplying the $\vec{v}$ eigenvector equation on the left by $\vec{w}$, we have

$$\vec{w}M\vec{v} = \vec{w}(M\vec{v}) = \vec{w}\lambda\vec{v}$$

Now we'll use the product rule to differentiate $\vec{w}\lambda\vec{v} = \vec{w}M\vec{v}$ with respect to x, denoting d/dx by $'$:

$$\begin{aligned}\vec{w}'\lambda\vec{v} + \vec{w}\lambda'\vec{v} + \vec{w}\lambda\vec{v}\,' &= \vec{w}'M\vec{v} + \vec{w}M'\vec{v} + \vec{w}M\vec{v}\,' \\ &= \vec{w}'\lambda\vec{v} + \vec{w}M'\vec{v} + \vec{w}\lambda\vec{v}\,'\end{aligned}$$

Subtracting $\vec{w}'\lambda\vec{v}$ and $\vec{w}\lambda\vec{v}\,'$ from both sides and solving for λ' gives

$$\lambda' = \frac{\vec{w}M'\vec{v}}{\vec{w}\vec{v}} \tag{14.28}$$

A special case of Eq. (14.28) is particularly useful: suppose the variable x is m_{ij}, the (i,j) entry of the matrix M. Then

$$\frac{\partial M}{\partial m_{ij}} = [\delta_{ij}]$$

where the matrix $[\delta_{ij}]$ has all entries 0, except for a single 1 in row i and column j.

In this case, Eq. (14.28) becomes

$$\frac{\partial \lambda}{\partial m_{ij}} = \frac{\vec{w}[\delta_{ij}]\vec{v}}{\vec{w}\vec{v}}$$

Now it's easy to see that $\vec{w}[\delta_{ij}]\vec{v} = w_i v_j$, for example,

$$\begin{bmatrix} w_1 & w_2 & w_3 \end{bmatrix} \begin{bmatrix} 0 & 1 & 0 \\ 0 & 0 & 0 \\ 0 & 0 & 0 \end{bmatrix} \begin{bmatrix} v_1 \\ v_2 \\ v_3 \end{bmatrix} = \begin{bmatrix} w_1 & w_2 & w_3 \end{bmatrix} \begin{bmatrix} v_2 \\ 0 \\ 0 \end{bmatrix} = w_1 v_2$$

This gives the equation for the dependence of eigenvalues on matrix entries:

$$\frac{\partial \lambda}{\partial m_{ij}} = \frac{w_i v_j}{\vec{w}\vec{v}} \tag{14.29}$$

How is this equation useful? Unlike our simple mathematical examples where the state transition probabilities (matrix entries) are precise, in every biological or physical example the matrix entries are estimates based on measurements. Every matrix entry has some uncertainties, so because the dominant eigenvalue of a Leslie matrix is the population long-term growth rate, we'd like to know how uncertainties in the matrix entries alter the growth rate. Also, we can ask which matrix entries have the largest influence on the dominant eigenvalue.

When we model the spread of a disease, this analysis tells us which matrix entry has the largest effect on the dominant eigenvalue, the rate of disease spread, and so tells us where to focus limited resources to best slow the disease spread.

Example 14.4.2. *Dominant eigenvalue sensitivity to matrix entries.* Take

$$M = \begin{bmatrix} 0.1 & 1.5 & 0 \\ 0.7 & 0 & 0 \\ 0 & 0.4 & 0 \end{bmatrix}$$

The eigenvalues of the Leslie matrix M are about $1.076, -0.976, 0$, so the dominant eigenvalue is 1.076. The unit left eigenvector $\vec{w}$ and the unit right eigenvector $\vec{v}$ of the dominant eigenvalue are

$$\vec{w} = \begin{bmatrix} 0.418 & 0.582 & 0 \end{bmatrix} \qquad \vec{v} = \begin{bmatrix} 0.528 \\ 0.344 \\ 0.128 \end{bmatrix}$$

Then

$$\vec{w}\vec{v} = w_1 v_1 + w_2 v_2 + w_3 v_3 \approx 0.421$$

and the sensitivity of the dominant eigenvalue λ to each matrix entry is

$$\left[\frac{\partial \lambda}{\partial m_{ij}}\right] = \left[\frac{w_i v_j}{\vec{w}\vec{v}}\right] = \begin{bmatrix} 0.524 & 0.342 & 0.127 \\ 0.730 & 0.476 & 0.177 \\ 0 & 0 & 0 \end{bmatrix} \tag{14.30}$$

This suggests that the dominant eigenvalue depends most strongly on m_{21}.

What about $m_{32} = 0.4$? Can this really have no effect on the dominant eigenvalue? Easy enough to test: change m_{32} to 1 (remember, m_{32} is the probability of a member of age class 2 surviving to age class 3, so we must have $0 \leq m_{32} \leq 1$) and compute the dominant eigenvalue. We get the same value, 1.076, that we got for the dominant eigenvalue of the original matrix M. Think a bit and you'll see why m_{32} doesn't alter the dominant eigenvalue: the dominant eigenvalue is the long-term growth rate of the population, and age class 3 produces no offspring, so contributes nothing to the population growth rate. It is for this reason that some authors drop late age classes that contribute no offspring.

Changing m_{32} from 0.4 to 1 does have some effect on the right unit eigenvectors of the dominant eigenvalue.

$$m_{32} = 0.4: \quad \vec{w} = \begin{bmatrix} 0.418 & 0.582 & 0 \end{bmatrix}, \quad \vec{v} = \begin{bmatrix} 0.528 \\ 0.344 \\ 0.128 \end{bmatrix}$$

$$m_{32} = 1.0: \quad \vec{w} = \begin{bmatrix} 0.418 & 0.582 & 0 \end{bmatrix}, \quad \vec{v} = \begin{bmatrix} 0.443 \\ 0.288 \\ 0.268 \end{bmatrix}$$

The unit left eigenvectors are unchanged, as we expect from Prop. 14.4.1: w_1 and w_2 are the proportions of the long-term population descending from a single individual in the first two age classes. Because $f_2 = 0$, no one descends from individuals in the third age class and so increasing the population of the third class by changing m_{23} has no effect on the left unit eigenvector entries.

The entries of the right unit eigenvector do change, because now a larger fraction of the population in the second age class survives to the third, increasing v_3 and consequently decreasing v_1 and v_2. □

One aspect of the sensitivity analysis of Leslie matrices may cause some concern. The p_i are survival probabilities so satisfy $0 \leq p_i \leq 1$, while the f_i are fecundities, so can be quite large (think insects), in which case the dominant eigenvalue may have higher sensitivity to the survival probabilities for no reason other than the different scales of the p_i and the f_i. To address this issue we calculate the *proportional sensitivity*, also called the *elasticity*. The idea is to measure the change in λ as a fraction of λ, divided by the change in m_{ij} as a fraction of m_{ij}. Then we see

$$e_{ij} = \frac{\partial\lambda/\lambda}{\partial m_{ij}/m_{ij}} = \frac{m_{ij}}{\lambda}\frac{\partial\lambda}{\partial m_{ij}} = \frac{m_{ij}}{\lambda}\frac{w_i v_j}{\vec{w}\vec{v}} \tag{14.31}$$

where we have used Eq. (14.29) for the last equality.

One reason elasticities are easier to interpret than sensitivities is that the elasticities sum to 1, so the elasticity e_{ij} is the fraction of the variation of λ that depends on m_{ij}.

The proof is interesting because it is indirect. It's difficult to see where even to start if we tried to show directly that

$$\sum_{i,j} e_{ij} = \sum_{i,j} \frac{m_{ij}}{\lambda}\frac{w_i v_j}{\vec{w}\vec{v}} = 1 \quad \text{or equivalently} \quad \sum_{i,j} m_{ij}\frac{w_i v_j}{\vec{w}\vec{v}} = \lambda$$

So we'll take a detour.

A function $f(x_1, \dots, x_n)$ is *homogeneous* of degree s if for all $c > 0$

$$f(cx_1, \dots, cx_n) = c^s f(x_1, \dots, x_n)$$

We'll use the property of homogeneous functions given in Lemma 14.4.1.

Lemma 14.4.1. (*Euler's theorem*): If f is homogeneous of degree s, then

$$\sum_{i=1}^{n} x_i \frac{\partial f}{\partial x_i} = s f(x_1, \dots, x_n)$$

Proof. Compute $\partial/\partial c$ of both sides of $f(cx_1, \dots, cx_n) = c^s f(x_1, \dots, x_n)$:

$$\frac{\partial}{\partial c} f(cx_1, \dots, cx_n) = \frac{\partial}{\partial c} c^s f(x_1, \dots, x_n)$$

$$\frac{\partial f}{\partial x_1}\frac{\partial cx_1}{\partial c} + \cdots + \frac{\partial f}{\partial x_n}\frac{\partial cx_n}{\partial c} = sc^{s-1} f(x_1, \dots, x_n)$$

$$\frac{\partial f}{\partial x_1} x_1 + \cdots + \frac{\partial f}{\partial x_n} x_n = sc^{s-1} f(x_1, \dots, x_n)$$

where we used the chain rule to get from the left-hand side of the first equation to that of the second equation. Euler's theorem follows by taking $c = 1$. □

Next we show that as a function of the matrix entries, the eigenvalues are homogeneous of degree 1. Let's write the eigenvector equation $M\vec{v} = \lambda\vec{v}$ in coordinates

$$\begin{bmatrix} m_{11} & \dots & m_{1n} \\ \dots & & \dots \\ m_{n1} & \dots & m_{nn} \end{bmatrix} \begin{bmatrix} v_1 \\ \dots \\ v_n \end{bmatrix} = \lambda \begin{bmatrix} v_1 \\ \dots \\ v_n \end{bmatrix}$$

Each row gives an equation:

$$m_{11}v_1 + \cdots m_{1n}v_n = \lambda v_1$$

$$\dots$$

$$m_{n1}v_1 + \cdots m_{nn}v_n = \lambda v_n$$

Multiply both sides of each equation by c

$$cm_{11}v_1 + \cdots cm_{1n}v_n = c\lambda v_1$$

$$\dots$$

$$cm_{n1}v_1 + \cdots cm_{nn}v_n = c\lambda v_n$$

Rewriting these as a matrix equation

$$\begin{bmatrix} cm_{11} & \dots & cm_{1n} \\ \dots & & \dots \\ cm_{n1} & \dots & cm_{nn} \end{bmatrix} \begin{bmatrix} v_1 \\ \dots \\ v_n \end{bmatrix} = c\lambda \begin{bmatrix} v_1 \\ \dots \\ v_n \end{bmatrix}$$

Recognizing that each eigenvalue depends on all the matrix entries,

$$\lambda = \lambda(m_{11}, \dots, m_{nn}),$$

this last matrix equation tells us that

$$\lambda(cm_{11}, \dots, cm_{nn}) = c\lambda(m_{11}, \dots, m_{nn})$$

That is, the eigenvalues are homogeneous of degree 1. Now apply Euler's theorem with $f = \lambda$ and $x_i = m_{ij}$:

$$\sum_{i,j} m_{ij} \frac{\partial \lambda}{\partial m_{ij}} = \lambda, \text{ so } \sum_{i,j} \frac{m_{ij}}{\lambda} \frac{\partial \lambda}{\partial m_{ij}} = 1 \text{ and } \sum_{i,j} e_{ij} = 1$$

No step of this calculation was difficult. But why would we think about homogeneous functions, and why would we think of Euler's theorem? Not all of math can be derived easily by simple calculations. As with every other field, in math to know some tricks is useful.

The elasticities for Example 14.4.2 then are

$$[e_{ij}] = \left[\frac{m_{ij}}{\lambda} \frac{w_i v_j}{\vec{w}\vec{v}} \right] = \begin{bmatrix} 0.049 & 0.477 & 0 \\ 0.475 & 0 & 0 \\ 0 & 0 & 0 \end{bmatrix} \tag{14.32}$$

Okay, these elasticities don't sum to 1, but remember, we've written them only to three digits so that the e_{ij} sum to 1.001 is the result of rounding. We notice several things when comparing the sensitivities and the elasticities of Example 14.4.2. First, the highest elasticity is for m_{12}, while the highest sensitivity is for m_{21}, though the elasticities of m_{21} and m_{12} are very close. Three matrix entries have higher sensitivities than m_{12}. So indeed the highest sensitivity and highest elasticity can occur for different matrix entries.

Next from Eq. (14.32) we see that if $m_{ij} = 0$, then right away $e_{ij} = 0$. Now this makes sense if a matrix entry is truly 0, if the first or last age classes never produce offspring, for example, or if movement from the last class to the second class is forbidden. If we stratify the population by age, the latter obviously is true, but maybe not if we stratify the population by size. On the other hand, if a matrix entry is 0 because we haven't yet measured the event it represents, then it may be reasonable to have a positive sensitivity for a matrix entry that is 0. We need to be a bit circumspect when interpreting $e_{ij} = 0$.

In addition to some non-zero sensitivities becoming zero elasticities, we notice other relatively large changes that have simple explanations. For example, m_{11} has sensitivity 0.524 and elasticity 0.049. That the sensitivity of this entry is so much higher than its elasticity follows from the small magnitude of m_{11}. Considering examples of this kind helps us build intuition for interpreting the elasticities of the dominant eigenvalue of a transition matrix.

We've introduced some complicated concepts in this section and the development may have tied them up in a fairly intricate way. So we'll end by summarizing the main points.

- A structured population $\vec{N}(t)$ grows according to $\vec{N}(t+1) = M\vec{N}(t)$, λ_1 is the dominant eigenvalue of M, and $\vec{w}$ and $\vec{v}$ are the unit left and right eigenvectors of λ_1.
- The entries of $\vec{v}$ are the eventual distribution of the stable population among its several classes.
- The entries of $\vec{w}$ give the reproductive value of the corresponding state, that is, the proportion of the eventual population descending from a single individual in that state.
- If some entries of M depend on a variable x, then $\dfrac{d\lambda_1}{dx} = \dfrac{\vec{w}(dM/dx)\vec{v}}{\vec{w}\vec{v}}$.

- A special case is the sensitivity of λ_1: $\dfrac{d\lambda_1}{dm_{ij}} = \dfrac{w_i v_j}{\vec{w}\vec{v}}$.
- The proportional sensitivity, or elasticity, is $e_{ij} = \dfrac{m_{ij}}{\lambda_1}\dfrac{w_i v_j}{\vec{w}\vec{v}}$.

Practice Problems

14.4.1. For a 4-state population that grows according to the matrix

$$M = \begin{bmatrix} 0 & 2 & 1 & 0 \\ 0.9 & 0 & 0 & 0 \\ 0 & 0.8 & 0 & 0 \\ 0 & 0 & 0.7 & 0 \end{bmatrix}$$

(a) find the eventual growth rate of the population,
(b) find the stable distribution of the population, and
(c) find the reproductive value of each state.

14.4.2. For a 3-state population that grows according to the matrix

$$M = \begin{bmatrix} 0.1 & 1.5 & 0 \\ 0.7 & 0 & 0 \\ 0 & 1 & 0 \end{bmatrix}$$

(a) find the eventual growth rate of the population,
(b) find the sensitivity of this growth rate to each matrix entry,
(c) find the elasticity of this growth rate to each matrix entry, and
(d) comment on the differences between the sensitivities and elasticities.

Practice Problem Solutions

14.4.1. The characteristic polynomial of M is $\lambda^4 - 1.8\lambda^2 - 0.72\lambda$, but even after factoring out a λ the resulting cubic does not have a simple solution, so we must be satisfied with a numerical approximation. The eigenvalues are about $1.509, 0, -0.451$, and -1.058.
(a) The eventual growth rate is the dominant eigenvalue, $\lambda_1 \approx 1.509$.
(b) The stable distribution of the population is given by the entries of the unit right eigenvector of λ_1: $\vec{v} = \langle 0.486,\ 0.290,\ 0.153,\ 0.071\rangle^{\text{tr}}$.
(c) The reproductive value of each state is the corresponding entry of the unit left eigenvector of λ_1: $\vec{w} = \langle 0.300,\ 0.502,\ 0.198,\ 0\rangle$.

14.4.2. (a) The characteristic polynomial is $-\lambda^3 + 0.1\lambda^2 + 1.05\lambda$; the numerical approximations are $\lambda = 1.076, 0$, and -0.976, so the eventual growth rate is the dominant eigenvalue $\lambda_1 \approx 1.076$.

(b) The sensitivities are given by Eq. (14.29): $\partial\lambda/\partial m_{ij} = w_i v_j/(\vec{w}\vec{v})$. To compute these, we need $\vec{v} = \langle 0.443, 0.288, 0.268\rangle^{tr}$, the unit right eigenvector of λ_1, and $\vec{w} = \langle 0.418, 0.582, 0\rangle$, the unit left eigenvector. With these, we compute

$$\left[\frac{\partial\lambda}{\partial m_{ij}}\right] = \left[\frac{w_i v_j}{\vec{w}\vec{v}}\right] = \begin{bmatrix} 0.524 & 0.341 & 0.317 \\ 0.731 & 0.477 & 0 \\ 0 & 0 & 0 \end{bmatrix}$$

(c) The elasticities are given by Eq. (14.31)

$$\left[e_{ij}\right] = \left[\frac{m_{ij}}{\lambda}\frac{w_i v_j}{\vec{w}\vec{v}}\right] = \begin{bmatrix} 0.049 & 0.476 & 0 \\ 0.476 & 0 & 0 \\ 0 & 0 & 0 \end{bmatrix}$$

(d) The most obvious difference is that $e_{13} = e_{22} = 0$ while $s_{13} \neq 0$ and $s_{22} \neq 0$. This one is clear: the elasticities are 0 because $m_{13} = m_{22} = 0$. More interesting is the difference between s_{11} and e_{11}. The elasticity is low because m_{11} is small, while the sensitivity is larger because a small change in the reproductive rate for class 0 cascades to produce additional offspring in class 1.

Exercises

14.4.1. For the populations that grow according to each of these matrices

$$\begin{bmatrix} 0 & 2 & 2 & 1 \\ 0.9 & 0 & 0 & 0 \\ 0 & 0.8 & 0 & 0 \\ 0 & 0 & 0.7 & 0 \end{bmatrix} \begin{bmatrix} 0 & 2 & 2 & 1 \\ 0.9 & 0 & 0 & 0 \\ 0 & 0.8 & 0 & 0 \\ 0 & 0 & 0.1 & 0 \end{bmatrix} \begin{bmatrix} 0 & 0.8 & 2 & 1 \\ 0.9 & 0 & 0 & 0 \\ 0 & 0.8 & 0 & 0 \\ 0 & 0 & 0.1 & 0 \end{bmatrix}$$

$$\begin{bmatrix} 0 & 0.8 & 2 & 0.5 \\ 0.8 & 0 & 0 & 0 \\ 0 & 0.7 & 0 & 0 \\ 0 & 0 & 0.5 & 0 \end{bmatrix} \begin{bmatrix} 0 & 0.8 & 2 & 0.5 \\ 0.6 & 0 & 0 & 0 \\ 0 & 0.7 & 0 & 0 \\ 0 & 0 & 0.5 & 0 \end{bmatrix} \begin{bmatrix} 0 & 0.6 & 1.1 & 0.5 \\ 0.6 & 0 & 0 & 0 \\ 0 & 0.7 & 0 & 0 \\ 0 & 0 & 0.5 & 0 \end{bmatrix}$$

(a) find the eventual growth rate of the population,
(b) find the stable distribution of the population, and
(c) find the reproductive value of each state.

14.4.2. For populations governed by these transition matrices

$$\begin{bmatrix} 0 & 1-x & x \\ x & 0 & 0 \\ 1-x & x & 1-x \end{bmatrix} \begin{bmatrix} 0 & 1-x & 1-x \\ x & 0 & x \\ 1-x & x & 0 \end{bmatrix} \begin{bmatrix} 0 & 1-x & 1-x \\ x & 0 & 0 \\ 1-x & x & x \end{bmatrix}$$

(a) for each of the three population classes, find the value of x that gives the largest value for the stable population in that class,
(b) for each of the three population classes, find the value of x that gives the smallest value for the stable population in that class, and
(c) describe the effect of x on the reproductive value of each population class.

14.4.3. For the populations that grow according to each of these matrices

$$\begin{bmatrix} 0.1 & 0.5 & 0.7 \\ 0.9 & 0.2 & 0.2 \\ 0 & 0.3 & 0.1 \end{bmatrix} \begin{bmatrix} 0.1 & 0 & 0.7 \\ 0.9 & 0.5 & 0.2 \\ 0 & 0.5 & 0.1 \end{bmatrix} \begin{bmatrix} 0.5 & 0.5 & 0 \\ 0.5 & 0.5 & 0.2 \\ 0 & 0 & 0.8 \end{bmatrix} \begin{bmatrix} 0.5 & 1.5 & 1 \\ 0.9 & 0 & 0 \\ 0 & 0.7 & 0 \end{bmatrix}$$

(a) find the eventual growth rate of the population,
(b) find the sensitivity of this growth rate to each matrix entry,
(c) find the elasticity of this growth rate to each matrix entry, and
(d) comment on the differences between the sensitivities and elasticities.

Chapter 15 A tiny bit of genetics

Modern genetics has transformed biology and has had an immense impact on medicine. Think of gene therapies, for example. We've said a bit about genetics already in Sects. 13.4 and 13.5, and we'll say more in Chapters 16, 18, and 19. Not surprisingly, the relationship between medicine and genetics is a two-way street. Genetics enables and refines therapies, and medical questions direct some genetics research. This is a vast topic, because of the complexity of the issues and also the number of scientists who study genetics.

When I was a kid, my parents got an encyclopedia. Each week at the grocery store, you could buy the next volume for $1.99 if you'd bought at least twenty dollars' worth of groceries. That encyclopedia was a treasure to me. When not at school, I always carried around a volume. Since then, I've loved encyclopedias. So it breaks my heart that I can't write an encyclopedia of mathematical genetics. Instead, in this chapter we'll explore a few parts of genetics where mathematical models can provide some insight. This is just a quick sketch of a few examples from among many, many more.

15.1 THE HARDY-WEINBERG LAW AND SELECTION EQUATIONS

Suppose at a particular chromosomal *locus* (position), genes come in two *alleles* (types), say A and a. The *genotype* refers to which pair of alleles, AA, aa, or Aa, occurs at that locus. The genotypes AA and aa are called *homozygous*; Aa is *heterozygous*. Usually we do not distinguish Aa from aA, though to get a proper count of offspring genotypes we do distinguish which allele is contributed by which parent. The *phenotype* is the trait—traits if the gene is pleiotropic—manifested as a result of the genotype.

Until around 1914, most biologists believed that rare alleles would leave the population. In this section we'll describe the theorem of G. H. Hardy and Wilhelm Weinberg, which states that under some reasonable, but far from universal, conditions, the genotype frequencies achieve a stable fixed point in one generation. The hypotheses of the theorem were imposed to simplify the proof details, almost to the point of transparency. Some of these conditions can be relaxed, at the expense of requiring more generations for the genotype frequencies to stabilize.

First, we describe some of the basic mechanics of inheritance through simple examples.

If A represents brown eyes and a blue eyes, we say blue eyes are a *recessive* trait because while the genotype aa has a phenotype with blue eyes, the genotypes AA and Aa have phenotypes with brown eyes. We say brown eyes are *dominant*.

Recall Gregor Mendel's mid-19th-century experiments with pea plants. (Incidentally, Karl Sigmund [331] reports that Mendel was a student of mathematics at the University of Vienna. Mendel became a monk in order to be able to continue his scientific investigations. He did not receive a teaching certificate because he twice failed his botany examination. I mention this for those who appreciate irony.) With respect to seed color, pure lines (homozygotes) come in two types, red seeds and yellow seeds. Mendel observed that crossing a pure red and a pure yellow gives yellow offspring. Crossing two of these yellow offspring gives both red and yellow offspring, in the ratio 1:3. How can this be explained?

Suppose seed color is determined by a single chromosomal locus, at which the gene has two alleles, r and y. Mendel's observations can be explained in this way.

- Pure lines have genotypes rr and yy, with corresponding phenotypes red and yellow.
- The first crossbred generation has genotype ry and phenotype yellow.
- For the second crossbred generation, with equal probability each parent contributes r or y, giving genotypes rr with probability $1/4$, yy with probability $1/4$,

and ry with probability 1/2. The corresponding phenotypes are red and yellow in the ratio 1:3.

We see the proportion in the second generation is explained by the observation that y is dominant, r recessive.

In what ways can this observation be generalized?

Suppose two alleles A_1 and A_2 can occupy a locus. Then three genotypes A_1A_1, A_1A_2, and A_2A_2 can occur. Say N_{11}, N_{12}, and N_{22} are the number of individuals in the population with these genotypes, so $N_{11} + N_{12} + N_{22} = N$, the total population. The *genotype frequencies* are

$$f_{11} = N_{11}/N, \quad f_{12} = N_{12}/N, \quad f_{22} = N_{22}/N$$

Each individual carries two genes at this locus, so the total number of these genes is $2N$. The total number of A_1 alleles in the population is $2N_{11} + N_{12}$ because the genotype A_1A_1 contributes two A_1 alleles and the genotype A_1A_2 contributes one A_1 allele. Similarly, the total number of A_2 genes is $N_{12} + 2N_{22}$.

The *allele frequencies* are

$$\text{freq}(A_1) = \frac{2N_{11} + N_{12}}{2N} = f_{11} + \frac{f_{12}}{2} = p$$
$$\text{freq}(A_2) = \frac{N_{12} + 2N_{22}}{2N} = \frac{f_{12}}{2} + f_{22} = 1 - p$$

Recall that for counting we distinguish the gene pair (A_1, A_2) from (A_2, A_1). The first member of the pair is the allele contributed by the mother, the second is the allele contributed by the father. Both have the genotype A_1A_2.

We assume *random unions*: genes are inherited independently of one another. Then the probabilities that offspring have a given gene pair are

$$P(A_1, A_1) = p^2 \quad P(A_1, A_2) = P(A_2, A_1) = p(1-p) \quad P(A_2, A_2) = (1-p)^2$$

by the multiplication rule for independent events (II.6). From this we see the offspring genotype frequencies are

$$\text{freq}(A_1A_1) = p^2 \quad \text{freq}(A_1A_2) = 2p(1-p) \quad \text{freq}(A_2A_2) = (1-p)^2$$

and the offspring allele frequencies are

$$\text{freq}(A_1) = p^2 + \frac{2p(1-p)}{2} = p^2 + p(1-p) = p$$
$$\text{freq}(A_2) = \frac{2p(1-p)}{2} + (1-p)^2 = p(1-p) + (1-p)^2 = 1 - p$$

Combining these results gives

Theorem 15.1.1. *The Hardy-Weinberg law.* For large populations that reproduce by random unions with a single locus having two alleles with no reproductive advantage of either phenotype,
(1) allele frequencies (p and $1-p$) are constant from generation to generation, and
(2) genotype frequencies (p^2, $2p(1-p)$, and $(1-p)^2$) are constant from the first offspring generation onward.

The Hardy-Weinberg law holds under more general circumstances, for example, if random unions are replaced with with random mating (random selection of individuals rather than of gametes), or if the locus supports N alleles instead of only 2. For sex-linked alleles, the calculation is more complicated, and the law must be modified so that the genotype frequencies p^2, $2p(1-p)$, and $(1-p)^2$ occur only in the limit after many generations. A good derivation of this is presented in Sect. 2.4 of [174].

Now we'll build a cartoon of selection in a large population that has non-overlapping generations, that reproduces by random unions, and in which selection acts on one chromosomal locus by modifying the probabilities of survival to reproductive age.

Suppose that locus has two alleles A_1 and A_2, and that in generation n they occur with relative frequencies $p_1(n)$ and $p_2(n) = 1 - p_1(n)$.

Under random unions, a child born in generation n has allele pair (A_i, A_j) (different from (A_j, A_i) for counting) with probability $p_i(n)p_j(n)$. Suppose

$$w_{ij} = P(\text{an individual with } (A_i, A_j) \text{ survives to reproductive age})$$

Because the allele pairs (A_i, A_j) and (A_j, A_i) have the same genotype, $w_{ij} = w_{ji}$. Suppose $N(n)$ is the number of newborns in generation n. The population is large, so about $p_i(n)p_j(n)N(n)$ have the allele pair (A_i, A_j) and about $w_{ij}p_i(n)p_j(n)N(n)$ survive to reproductive age. The total number that survive to reproductive age is

$$\sum_{i,j=1}^{2} w_{ij}p_i(n)p_j(n)N(n) = \left(w_{11}p_1(n)^2 + 2w_{12}p_1(n)p_2(n) + w_{22}p_2(n)^2\right)N(n)$$

Denote by $p_{ij}(n+1)$ the frequency of the allele pair (A_i, A_j) in generation $n+1$. The frequency is the number of offspring in generation $n+1$ with (A_i, A_j), divided by the total number of offspring in generation $n+1$. That is,

$$p_{ij}(n+1) = \frac{w_{ij}p_i(n)p_j(n)}{w_{11}p_1(n)^2 + 2w_{12}p_1(n)p_2(n) + w_{22}p_2(n)^2}$$

where we've canceled the $N(n)$ factors from the numerator and denominator. Note $p_{ij}(t+1) = p_{ji}(t+1)$ because $w_{ij} = w_{ji}$.

Let $p_i(n+1)$ denote the frequency of the allele A_i in generation $n+1$. Then

$$p_i(n+1) = \frac{1}{2}\Big(p_{i1}(n+1) + p_{i2}(n+1)\Big) + \frac{1}{2}\Big(p_{1i}(n+1) + p_{2i}(n+1)\Big)$$

because A_i is contributed by the mother in (A_i, A_j) with probability $1/2$, and by the father in (A_j, A_i) with probability $1/2$. Because $p_{ij}(n+1) = p_{ji}(n+1)$, we see $p_i(n+1) = p_{i1}(n+1) + p_{i2}(n+1)$ and we have the *discrete selection equation*

$$p_i(n+1) = p_i(n) \cdot \frac{w_{i1}p_1(n) + w_{i2}p_2(n)}{w_{11}p_1(n)^2 + 2w_{12}p_1(n)p_2(n) + w_{22}p_2(n)^2} \tag{15.1}$$

The denominator of Eq. (15.1) is the *average fitness*,

$$\langle f(n) \rangle = w_{11}p_1(n)^2 + 2w_{12}p_1(n)p_2(n) + w_{22}p_2(n)^2 \tag{15.2}$$

or *average selective value*, of the population: w_{ij} is the selective value of (A_i, A_j), and $p_i(n)p_j(n)$ is the probability that (A_i, A_j) occurs. That this sum is the average, or expected value, follows from Eq. (11.14).

Under evolution governed by Eq. (15.1), the average fitness of the population never decreases. This is shown by

Theorem 15.1.2. *The fundamental theorem of selection.* When the allele frequencies $(p_1(n), p_2(n))$ evolve to $(p_1(n+1), p_2(n+1))$ by Eq. (15.1), the average fitness behaves this way:

$$\langle f(n+1) \rangle \geq \langle f(n) \rangle$$

with equality if and only if $(p_1(n), p_2(n))$ is a fixed point. That is, average fitness increases to a local maximum.

The proof of Theorem 15.1.2 involves some fairly intricate combinatorics using several inequalities, including the arithmetic mean-geometric mean, Jensen's, and the Cauchy-Schwarz inequalities. These are interesting math, but they would involve a longish detour. You can find a proof in Sect. 3.4 of [174], where the argument is presented for m alleles at the locus, rather than just 2. The m-allele version of Eq. (15.1) is

$$p_i(n+1) = p_i(n) \frac{\displaystyle\sum_{j=1}^{m} w_{ij}p_j(n)}{\displaystyle\sum_{a=1}^{m}\sum_{b=1}^{m} w_{ab}p_a(n)p_b(n)} \tag{15.3}$$

In situations where fitness depends on more than one locus, that is, for almost every realistic situation, the selection theorem doesn't apply. Many factors interact in complex ways; the selection theorem is a simple cartoon that shows a general direction.

Even though the proof is tricky, in the two-allele case some results are easy to see. Because $p_2(n) = 1 - p_1(n)$ the average fitness defined by Eq. (15.2) can be expressed in terms of $p_1(n)$ alone:

$$\langle f(n)\rangle = (w_{11} - 2w_{12} + w_{22})p_1(n)^2 + 2(w_{12} - w_{22})p_1(n) + w_{22} \tag{15.4}$$

and with a bit of algebra the selection equation (15.1) can be rearranged into the form

$$p_1(n+1) - p_1(n) = \frac{p_1(n)(1-p_1(n))}{2\langle f(n)\rangle} \frac{d\langle f(n)\rangle}{dp_1(n)} \tag{15.5}$$

A probability is a fixed point if $p_1(n+1) = p_1(n)$. From the right-hand side of Eq. (15.5) we see the fixed points are $p_1(n) = 0$ and $p_1(n) = 1$ (the allele population at this locus is homogeneous), and $p_1(n) = p_*$, the critical point of $\langle f(n)\rangle$, if p_* lies between 0 and 1 (a heterogeneous allele population at this locus). With the form of $\langle f(n)\rangle$ of Eq. (15.4) we see that

$$p_* = \frac{w_{22} - w_{12}}{w_{11} - 2w_{12} + w_{22}} = \frac{w_{22} - w_{12}}{(w_{11} - w_{12}) + (w_{22} - w_{12})} \tag{15.6}$$

The allele frequencies will evolve toward one of the homogeneous states if the critical point does not lie in $(0,1)$. If the critical point does lie in $(0,1)$ and the second derivative

$$\left.\frac{d^2\langle f(n)\rangle}{dp_1(n)^2}\right|_{p_*} = 2((w_{11} - w_{12}) + (w_{22} - w_{12}))$$

is positive, the critical point is a fitness local minimum and the population evolves to a homogeneous state. If the second derivative is negative, the critical point p_* is a local max and the population evolves to this heterogeneous state, a mixture of the alleles.

If we abandon the assumption of non-overlapping generations, and if the population is large, the relative frequency x_i of the allele A_i evolves according to a differential equation, the continuous selection equation, which can be viewed as an appropriate limit of Eq. (15.3). See Chapter 23 of [174] for a complete treatment. Here we'll sketch a few of the main steps, in part to highlight a comparison of the discrete (non-overlapping generations) and continuous (overlapping generations) models.

For a large population of (variable) size N with births and deaths continuously and randomly distributed, at one locus suppose the gene has alleles

$A_1,\ldots,A_n$. A population of size N has $2N$ alleles, N_i copies of allele A_i, so the frequency of this allele is $x_i = N_i/2N$. Further, suppose the population is in Hardy-Weinberg equilibrium, so $x_i x_j = N_i N_j/(4N^2)$ is the frequency of the pair (A_i,A_j). Denote by b_{ij} and d_{ij} the birth and death rates of (A_i,A_j) individuals. Then the fitness of these individuals can be taken to be $f_{ij} = b_{ij} - d_{ij}$. Because the allele A_i occurs in both (A_i,A_j) and (A_j,A_i),

$$\frac{dN_i}{dt} = \sum_{j=1}^{n} f_{ij}\frac{N_i N_j}{4N^2}N + \sum_{j=1}^{n} f_{ji}\frac{N_j N_i}{4N^2}N = \frac{N_i}{2N}\sum_{j=1}^{n} f_{ij}N_j \tag{15.7}$$

where we've used $f_{ji} = f_{ij}$, a consequence of the functional identity of (A_i,A_j) and (A_j,A_i). Because each individual has two alleles at this locus, the rate of change of the population N is $1/2$ the sum of the N_i'. Then with a bit of algebra we see

$$\frac{dN}{dt} = \frac{1}{2}\sum_{i=1}^{n}\frac{dN_i}{dt} = N\sum_{i,j=1}^{n} f_{ij}x_i x_j \tag{15.8}$$

To find an expression for x_i', apply the quotient rule, substitute with Eqs. (15.7) and (15.8), and simplify. Then for $i = 1,\ldots,N$,

$$\frac{dx_i}{dt} = \left(\frac{N_i}{2N}\right)' = \frac{NN_i' - N_i N'}{2N^2} = x_i\left(\sum_a f_{ia}x_a - \sum_{b,c} f_{bc}x_b x_c\right) \tag{15.9}$$

These are the components of the continuous selection equation. The average fitness is $\langle f\rangle = \sum_{i,j} f_{ij}x_i x_j$, and we see the familiar Darwinian notion that the per capita growth rate of x_i is the difference between the fitness $\sum_j f_{ij}x_j$ of A_i and the average fitness $\langle f\rangle$ of the population.

The average fitness $\langle f\rangle$ is a Liapunov function for Eq. (15.9). Here's a sketch. Remember that $f_{ij} = f_{ji}$.

$$\begin{aligned}
\frac{d}{dt}\langle f\rangle &= \frac{d}{dt}\left(\sum_{i,j} f_{ij}x_i x_j\right) = \sum_{i,j} f_{ij}(x_i' x_j + x_i x_j') = 2\sum_{i,j} f_{ij}x_i' x_j \\
&= 2\sum_{i,j} f_{ij}x_i\left(\sum_a f_{ia}x_a - \sum_{b,c} f_{bc}x_b x_c\right)x_j \qquad \text{by Eq. (15.9)} \\
&= 2\sum_i x_i\left(\sum_a f_{ia}x_a\right)\left(\sum_j f_{ij}x_j\right) - 2\left(\sum_{i,j} f_{ij}x_i x_j\right)\left(\sum_{b,c} f_{bc}x_b x_c\right) \\
&= 2\sum_i x_i\left(\sum_j f_{ij}x_j\right)^2 - 2\left(\sum_{i,j} f_{ij}x_i x_j\right)^2
\end{aligned} \tag{15.10}$$

With some more algebra (try the $n = 2$ case, and remember that then $x_1 + x_2 = 1$), this last expression can be rearranged to give

$$\frac{d}{dt}\langle f\rangle = \sum_i x_i\left(\sum_a f_{ia}x_a - \sum_{b,c} f_{bc}x_b x_c\right)^2 \tag{15.11}$$

and we see that $\langle f\rangle' \geq 0$. Moreover, $\langle f\rangle' = 0$ if and only if for each i

$$x_i = 0 \qquad \text{or} \qquad \sum_a f_{ia}x_a - \sum_{b,c} f_{bc}x_b x_c = 0$$

From the continuous selection equation (15.9) we see that $\langle f\rangle' = 0$ exactly at the fixed points.

Compare the discrete selection equation (15.3) and the continuous selection equation (15.9). Why does the average fitness appear in the denominator of (15.3) and yet is subtracted in (15.9)? Here's a hint: $e^{a-b} = e^a/e^b$.

Also, the right side of Eq. (15.10) is 2 times the variance of $\sum_j f_{ij}x_j$, the fitness of A_i. With a bit more work we can show that the rate of increase of the average fitness is proportional to the variance of the fitness. Small variance means slow progress. Yet again we see that monocultures are not good news.

None of these models capture all the subtle dynamics of a biological population. Remember that one point of these models is to find how much of their complex behaviors can be captured by simple models.

Fitness, that is, reproductive success, is the force that drives evolution, but evolution is far more complex than this simple statement suggests.

For example, Richard Prum [300] points out that Darwin posited *two* selection principles, one based on survival fitness, the other on aesthetics. Mostly, Prum's evidence involves birds, little surprise because he is an ornithologist. Although it is unlikely that aesthetic selection functions on primitive creatures—do female Tardigrades care how male Tardigrades look?—we do need to be aware that this is a factor in the evolution of higher animals. Still, I don't expect "Survival of the fittest" to be replaced by "Survival of the cutest."

Then, too, the evolution of a species does not occur against a fixed background of all other species. Everything evolves together; *coevolution* is a more apt term. Evolutionary arms races abound. Every addition to the predator's arsenal is matched by a new defense mechanism for the prey.

Evolutionary phase transitions pose a different problem. In going from cell to organ, or from organ to organism, some cooperative behavior may be needed. The good of the organ or of the organism may force cells to behaviors suboptimal for the cell. Apoptosis of infected cells does not help the cell, but it does help the organism. These transitions correspond to shifts in organizational level. Richard Michod's book *Darwinian Dynamics* [235] is a good source. We are led to this question: does evolution exhibit scaling?

Mostly, we'll focus on fitness, but with these variations in mind.

Practice Problems

15.1.1. For a population in Hardy-Weinberg equilibrium, suppose a trait resulting from the genotype *aa* occurs in 20% of the population. Compute the frequencies of the alleles and genotypes for this trait. Assume two alleles, A dominant and a recessive.

15.1.2. What allele frequency will produce three times as many recessive homozygotes as heterozygotes?

15.1.3. Suppose a parental generation of moths have allele frequencies for color $B = 0.64$ (black pigment) and $W = 0.36$ (no black pigment). Due to some selective pressure, moths with the different genotypes have different probabilities of survival to reproductive age. The probabilities for the genotypes are

$$P_{sur}(BB) = 0.75 \quad P_{sur}(BW) = 0.6 \quad P_{sur}(WW) = 0.9$$

Calculate the allele frequencies of the next generation at maturity.

Practice Problem Solutions

15.1.1. Let p denote the frequency of the recessive allele a and q the frequency of the dominant allele A. Then the frequency of *aa* is p^2, so $p^2 = 0.2$ and $p = \sqrt{0.2} \approx 0.447$. Then $q = 1 - p \approx 0.553$. These are the allele frequencies. The genotype frequencies are

$$AA : q^2 \approx 0.306 \qquad Aa : 2pq \approx 0.494 \qquad aa : p^2 \approx 0.200$$

15.1.2. The frequency of *aa* is p^2 and that of *Aa* is $2pq = 2p(1-p)$. In order for the frequency of *aa* to be three times that of *Aa*, we see

$$p^2 = 3 \cdot 2p(1-p) = 6p - 6p^2$$

That is, $p = 6/7$ and so $q = 1/7$.

15.1.3. Using the allele frequencies of the parental generation, we calculate the genotype frequencies of the offspring at birth.

$$BB : 0.64^2 = 0.410 \quad BW : 2 \cdot 0.64 \cdot 0.36 = 0.461 \quad WW : 0.36^2 = 0.130$$

To get the genotype frequencies at maturity, we multiply the genotype frequencies by the probabilities of each genotype surviving to maturity, and divide by the sum of the products to normalize.

$$BB: 0.410 \cdot 0.75 = 0.307 \quad BW: 0.461 \cdot 0.6 = 0.277$$

$$WW: 0.130 \cdot 0.9 = 0.117 \quad \text{sum} = 0.307 + 0.277 + 0.117 = 0.701$$

So at maturity

$$BB: \frac{0.307}{0.701} \approx 0.438 \quad BW: \frac{0.277}{0.701} \approx 0.395 \quad WW: \frac{0.117}{0.701} = 0.167$$

Now to get the new allele frequencies of the offspring generation,

$$\text{freq}(B) = \frac{2 \cdot 0.438 + 0.395}{2} \approx 0.636$$

$$\text{freq}(W) = \frac{2 \cdot 0.167 + 0.395}{2} \approx 0.364$$

Exercises

15.1.1. Suppose 40% of a large population expresses a recessive phenotype. What percent of the population is heterozygous?

15.1.2. Suppose 60% of a large population possesses a dominant allele. Find the frequency of dominant homozygotes, of recessive homozygotes, and of heterozygotes.

15.1.3. Suppose 30% of a population carries at least one copy of the recessive allele. What is the frequency of the dominant phenotype? What is the frequency of the recessive phenotype?

15.1.4. Suppose 30% of a large population is homozygous recessive. Find the frequency of homozygous dominant, and of heterozygous.

15.1.5. Cystic fibrosis is recessive, affecting about 0.0004 Caucasian babies in the U.S. Find the frequency of the recessive allele, the dominant allele, and of heterozygotes (carriers).

15.1.6. Suppose at the start of a generation, 16% of the population has the genotype aa, where a is recessive and A is dominant. Find a relation between the probabilities $P_{sur}(AA)$, $P_{sur}(Aa)$, and $P_{sur}(aa)$ of survival to adulthood so that when this generation reaches adulthood $P(AA) = P(Aa) = P(aa)$.

15.1.7. Contrary to the classical model of genetic variation, in which geneticists believed natural populations have little genetic variation, the experimental technique gel electrophoresis revealed that populations actually maintain a large amount of variation over time. This led some population geneticists [196, 197] to propose the neutral mutation hypothesis, which asserts that the amount of genetic variation always is large, but much of it is neutral. For a particular polygenic locus (one that supports multiple alleles), three alleles are viable, A, B, and C, with frequencies 0.25, 0.4, and 0.35. The adulthood survival probabilities are

0.8 for AA, 0.7 for AB, 0.25 for AC, 0.98 for BB, 0.05 for BC, and 0.75 for CC.

Find the allele pair frequencies in adulthood of this generation.

15.1.8. Sickle cell anemia is a disease caused by the homogeneous inheritance of sickle cell alleles that produce a mutated form of hemoglobin, either Hb-S or Hb-C, instead of the wild type, Hb-A. While the inheritance of a homogeneous pair can be lethal, individuals with a heterogeneous combination (called *sickle cell trait*) are less likely to die from malaria because the immune system destroys sickled cells, and the parasites with them. Thus, the sickle cell alleles have a selective advantage in regions where malaria is present, most notably in parts of Africa. Jacques Simpore and coworkers [334, 335] surveyed 12,019 students in Ouagadougou, Burkina Faso, and found the genotypes listed in the table at the end of this problem. Note that while the "number of students" column entries sum to 12,019, the entries of the frequency column, each number divided by 12,019, do not quite sum to 1. This is numerical noise, a consequence of the fact that the largest number is more than two orders of magnitude larger than the smallest. Calculate the allele frequencies. (We'll mention the connection between sickle cell disease and malaria again in Sect. 19.1.) Do you see why we can't compute the allele by dividing, say, the number of AA genotypes by the total number of students?

Hb genotype	Number of students	genotype frequency
AA	8097	0.674
AC	2623	0.218
AS	980	0.082
SC	100	0.008
CC	194	0.016
SS	25	0.002

15.1.9. For a two-allele locus with average fitness

$$\langle f(n)\rangle = w_{11}p_1(n)^2 + 2w_{12}p_1(n)p_2(n) + w_{22}p_2(n)^2$$

(a) express $\langle f(n) \rangle$ in terms of the coefficients w_{ij} and $p_1(n)$.
(b) Compute the average fitness of homozygote states $p_1(n) = 0$ and $p_2(n) = 0$.
(c) Plot the graph of $\langle f(n) \rangle$ for $0 \leq p_1(v) \leq 1$, for these parameter values:
$w_{11} = 1$, $w_{12} = 1$, and $w_{22} = 2$
$w_{11} = 2$, $w_{12} = 1$, and $w_{22} = 1$
$w_{11} = 1$, $w_{12} = 2$, and $w_{22} = 1$
(d) For which of these does the maximum average fitness occur at a heterozygote point?
(e) Explain the plot when $w_{11} = w_{12} = w_{22} = 1$.

15.1.10. For a three-allele locus, show the average fitness is

$$\begin{aligned}\langle f(n) \rangle &= w_{11}p_1^2 + 2w_{12}p_1p_2 + 2w_{13}p_1(1 - p_1 - p_2) + w_{22}p_2^2 \\ &\quad + 2w_{23}p_2(1 - p_1 - p_2) + w_{33}(1 - p_1 - p_2)^2\end{aligned}$$

Take $w_{11} = 3$, $w_{12} = w_{13} = 1$, $w_{22} = w_{23} = 2$, and $w_{33} = 1$. Plot $\langle f \rangle$ for $0 \leq p_1 \leq 1$ and $0 \leq p_2 \leq 1 - p_1$.

15.2 PRICE'S EQUATION AND MULTILEVEL SELECTION

In his 1930 book *The Genetical Theory of Natural Selection* [116], Ronald Fisher presented his *fundamental theorem of natural selection*: the rate of increase of a population's average fitness is proportional to the variance of its fitness. From Sect. 11.6 of Volume 1 recall the variance of a random variable X is $\sigma^2(X) = \mathbb{E}(X^2) - (\mathbb{E}(X))^2$. Denoting by w the fitness of population members and by $\mathbb{E}(w)$ the expected value of the fitness, Fisher's theorem can be written as:

$$\Delta\mathbb{E}(w) \propto \sigma^2(w) \tag{15.12}$$

Fisher's argument to support this result was not so easy to follow, and his theorem has been misinterpreted as stating that average fitness always increases. The problem with this interpretation is that in some instances the average fitness is known to decrease.

Sometimes natural selection is understood as operating only on individuals. After all, individuals reproduce and send their genes into the future. Yet examples abound of *biological altruism*, where an individual accepts a reduction of its fitness to increase the fitness of others. If the others are relatives of the altruistic individual, this is called *kin selection*. Although Darwin discussed the concept, William Hamilton did the first mathematical study [156]. Hamilton found that natural selection can amplify genes that lead to biological altruism if

$$rB > C \tag{15.13}$$

where r is the relatedness parameter, B is the fitness benefit to the recipient of an altruistic act, and C is the fitness cost to the individual who performs an altruistic act. This is called *Hamilton's altruism equation.* In its simplest form, r represents the probability that a randomly selected allele is shared, so for parent and child $r = 1/2$, for parent and grandchild $r = 1/4$, and so on. Hamilton interpreted r as a regression coefficient, emphasizing the importance of genotype association above common ancestry. Negative values of r indicate that two individuals share fewer alleles than two randomly selected individuals from the population. Negative values have been used to explain the evolution of some spiteful behaviors: negative r and B can exceed C in Eq. (15.13). However, likely some people, and by extension some animals, are jerks regardless of genetics. How else to explain U.S. politics in 2020?

Trained as a chemist but with very wide interests, George Price became curious about Hamilton's altruism equation (15.13), and about altruism in general. Along the way, he showed that the common interpretation of Fisher's fundamental theorem (15.12) is incomplete. All this and more is contained in Price's equation [293, 295]. Much has been written on Price's equation. Steven Frank's "George Price's contributions to evolutionary genetics" [121] is a good source. We'll state and sketch a derivation of Price's equation, show that Fisher's fundamental theorem is a consequence of Price's equation, and show that it is important for understanding multilevel selection.

We'll need a tool from probability, the covariance $\text{cov}(X, Y)$ of two random variables X and Y, defined by

$$\text{cov}(X, Y) = \mathbb{E}((X - \mathbb{E}(X))(Y - \mathbb{E}(Y))) = \mathbb{E}(XY) - \mathbb{E}(X)\mathbb{E}(Y) \qquad (15.14)$$

where the second equality follows from the linearity of the expected value, $\mathbb{E}(X + Y) = \mathbb{E}(X) + \mathbb{E}(Y)$ and $\mathbb{E}(kX) = k\mathbb{E}(X)$. The product of $X - \mathbb{E}(X)$ and $Y - \mathbb{E}(Y)$ is positive if X and Y are both greater than their expected values, or are both less than their expected values. So a positive covariance means that on average X and Y lie on the same side of their expected values.

Suppose a population of size N is partitioned into subpopulations of sizes $n_1, \dots, n_k$, and in the ith subpopulation a trait z takes on a value z_i. Then $q_i = n_i/N$ is the frequency of trait z_i in the population and $\mathbb{E}(z) = \sum_i q_i z_i$ is the expected value of the trait z in this population.

Now consider the next generation, whose characteristics are indicated by $'$. Price used q_i' to denote the frequency in the next generation of the descendants of group i in the current generation. (A more expected interpretation would be the frequency in the next generation of those individuals with trait z_i. Price did approach things differently.) When we denote by w_i the fitness of elements in class i in the current generation, then $q_i' = q_i w_i / \mathbb{E}(w)$, where $\mathbb{E}(w)$ is the expected value of the fitness of the current generation.

Continuing with his notational choices, z_i' denotes the average of the trait z for the descendants of subpopulation i. Then the change in the trait for the descendants of i is $\Delta z_i = z_i' - z_i$ and the change in the expected value of the trait is

$$\Delta\mathbb{E}(z) = \sum_i q_i' z_i' - \sum_i q_i z_i \tag{15.15}$$

We see

$$\begin{aligned}
\Delta\mathbb{E}(z) &= \sum_i \left(\frac{q_i w_i}{\mathbb{E}(w)} (z_i + \Delta z_i) \right) - \sum_i q_i z_i \\
&= \sum_i q_i \left(\frac{w_i}{\mathbb{E}(w)} - 1 \right) z_i + \sum_i \frac{q_i w_i}{\mathbb{E}(w)} \Delta z_i \\
&= \sum_i q_i \left(\frac{w_i - \mathbb{E}(w)}{\mathbb{E}(w)} \right) z_i + \sum_i \frac{q_i w_i}{\mathbb{E}(w)} \Delta z_i \\
&= \frac{1}{\mathbb{E}(w)} \left(\sum_i q_i (w_i z_i) - \sum_i q_i (z_i \mathbb{E}(w)) + \sum_i q_i (w_i \Delta z_i) \right) \\
&= \frac{1}{\mathbb{E}(w)} \left(\mathbb{E}(w_i z_i) - \mathbb{E}(z)\mathbb{E}(w) + \mathbb{E}(w_i \Delta z_i) \right)
\end{aligned}$$

Now multiply both sides by $\mathbb{E}(w)$ and use the definition of covariance (15.14). This gives Price's equation:

$$\mathbb{E}(w)\Delta\mathbb{E}(z) = \mathrm{cov}(w_i, z_i) + \mathbb{E}(w_i \Delta z_i) \tag{15.16}$$

The covariance term of Eq. (15.16) is the change in the trait that results from differential fitness (reproductive success) of that value of the trait. The expected value term on the right-hand side of Eq. (15.16) is the change in the character values between an individual and its descendants.

Price pointed out [294] that in his derivation of his fundamental theorem (15.12), Fisher recognized that the total change in fitness has two sources, selection and "environment." Fisher was concerned with how fitness changes by selection in a fixed environment. This corresponds to the covariance term of Price's equation. Or put another way, Fisher worked in a setting where the second term of Price's equation is 0. That is, $\mathbb{E}(w)\Delta\mathbb{E}(z) = \mathrm{cov}(w, z)$, where we have i subscripts because we'll need them only when we explore the second term of Price's equation. Recall that z can be any trait, so take z to be the fitness w. Then

$$\mathbb{E}(w)\Delta\mathbb{E}(w) = \mathrm{cov}(w, w) = \mathbb{E}(w^2) - \mathbb{E}(w)^2 = \sigma^2(w)$$

We have recovered Fisher's theorem (15.12) and found that the proportionality constant is $\mathbb{E}(w)$. The common problems with applications of Fisher's theorem follow from the assumption that it refers to all changes in fitness. The various special conditions imposed in applications of the theorem minimized the environmental effects on fitness change. With this interpretation, the proof and applications of Fisher's theorem are clear.

Price's equation gives insight into the derivation and interpretation of Hamilton's altruism equation (15.13). This is a bit more complicated than we can get into here, but details can be found in [121, 301]. So his equation did help Price understand biological altruism.

Finally, to see how Price's equation can be applied to multilevel selection, note that the product inside the expected value on the right-hand side of Eq. (15.16) looks a lot like the product on the left-hand side. So if each subpopulation i is further subdivided into sub-subpopulations ij, we can iterate Price's equation

$$\begin{aligned}\mathbb{E}(w)\Delta\mathbb{E}(z) &= \text{cov}(w_i, z_i) + \mathbb{E}(w_i \Delta z_i)\\ &= \text{cov}(w_i, z_i) + \mathbb{E}(\text{cov}(w_{ij}, z_{ij}) + \mathbb{E}(w_{ij}\Delta z_{ij}))\end{aligned}$$

This can be continued for additional levels, should we desire.

How are we to understand when cells can behave in ways that reduce their own fitness in order to benefit that of the group (organ) to which they belong? When does selection work at the level of organs, of organisms, of collections of individuals, of populations? Sonya Bahar's wonderful book *The Essential Tension: Competition, Cooperation and Multilevel Selection in Evolution* [16] presents the history of this issue from many perspectives. See especially Chapter 13, but the entire book is a treat. A key observation is that "the transition from one level of selection to another [...] can be viewed as a shift from competition to conflict suppression." Other authors, Michod [235] for example, present this idea; the discussion in [16] is the most nuanced I've read.

In math, when you make some progress in the study of objects, a natural next step is to study the maps, the dynamics, between these objects. In genetics once a genome is sequenced, we can study the maps, the dynamics, provided by gene activation and inactivation networks. Maybe in evolution we can study the maps of organization on multiple levels. Do these have similar features? Will fractal geometry provide a useful meta-organization? Or maybe the complexities are too subtle for us now. The hard work of science, the thousands of hours spent staring at code or chalkboards or into microscopes, is rewarded by a few glimpses of a transcendent beauty.

Now a few words about George Price himself. As recounted in [159], Price derived his equation in his search to understand how altruism can evolve through natural selection. Price's life was complicated, and included work on the Manhattan Project and at Bell Labs and IBM. Then he moved to London to study evolution. Here he derived his equation. Eventually Price began to notice coincidences in his life. For example, when on his way to work he passed a rotating clock and thermometer on a bank, far more often he saw the clock first. These and other events he took to be signals from god. He gave away all his possessions, and took in homeless people until his grant ran out and he was evicted from his apartment. He lived on the streets for a while, became depressed, punctured his carotid artery with tailor's scissors and died. How could such a brilliant person go so far off the rails?

15.3 FITNESS LANDSCAPES

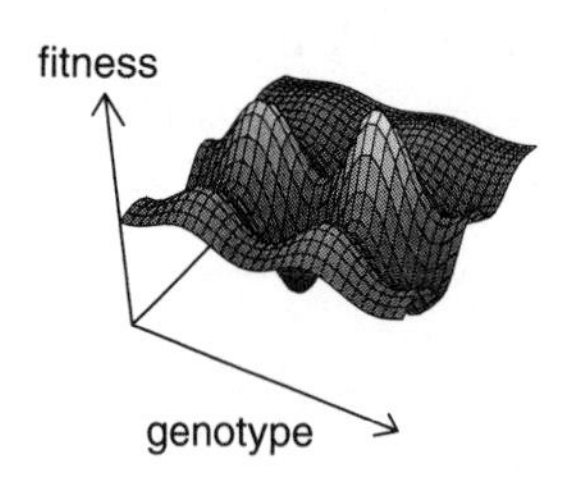

Figure 15.1. A fitness landacape.

Fitness landscapes are Sewall Wright's [391, 392] elegant geometric construction to visualize evolutionary progress. Nobel laureate Manfred Eigen and chemist Peter Schuster defined the landscape as height over genotypes. The term "fitness landscape" is used in two ways: as a map, mediated by phenotype, between genotype and fitness of individuals having that genotype, and as a map between parameters characterizing the genetic structure of the population and the average fitness of the population. Fig. 15.1 illustrates the first, fitness on the vertical axis, genotypes on the horizontal plane.

Because fitness landscapes are an instance of the evocative notion of a geometric representation of evolutionary dynamics, references abound: whole books [132, 273, 310], chapters [1, 258], and many papers [22, 34, 118, 119, 135, 153, 230, 306, 309, 311, 362, 377, 378], for example.

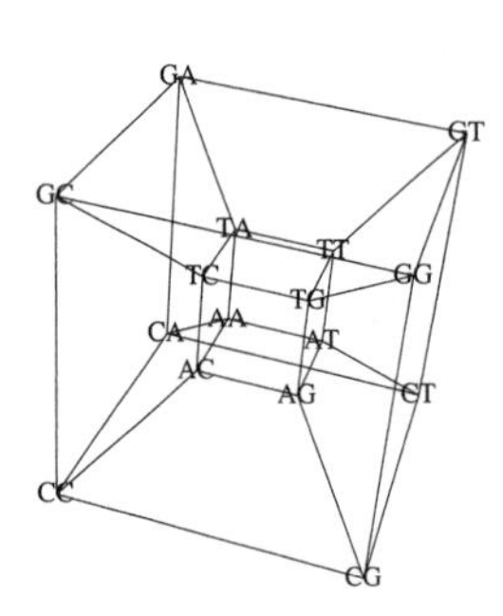

Figure 15.2. Hypercube vertices

For the moment, assume that we can compute the fitness of each genotype. Then certainly the relative values of these fitnesses do not depend on how the genotypes are arranged, on which are near one another. But the geometry of the landscape does depend on these relative positions.

One arrangement uses the *Hamming distance*, the number of positions at which two equal-length sequences differ. For example, the Hamming distance between *CTGGATCCA* and *CTCGATTCA* is 2. We arrange the genotypes by their Hamming distances. This might sound straightforward, but really it is a bit tricky. To achieve this for length-2 sequences, the genotypes must be placed with care on the vertices of a hypercube. One placement is shown in Fig. 15.2.

Even this geometry needs an addendum. Consider the base of the outer cube, that is, the square with corners labeled CC, CG, CA, and CT. Each of these pairs differ in only one position, so the Hamming distance between each pair is 1. This is represented by the edges CC-CG, CG-CT, CT-CA, and CA-CC. To be complete, we also should draw edges CC-CT and CG-CA, but these would make the picture too complicated.

Similar additions must be made to all the faces of the hypercube. To the square with corners CC, AC, TC, and GC (this may not look like a square, but a projection from 4 dimensions to 2 must induce some distortions) we should add the edges CC-TC and AC-GC.

Another way to arrange genotypes, one closer to biological reality, is to define two sequences as close if one can be gotten from the other by mutation—point mutation, insertion, or deletion, for example— operations that can occur in one step in biological settings. We won't pursue this here, just mention that some [258] arrange genotypes by this approach.

A length-N genotype has 4^N possible sequences, which can be represented as the corners of a $2N$-dimensional hypercube. The smallest genes have about 200 bases, the largest about 2,000,000, so we are dealing with a hypercube of between 400 and 4,000,000 dimensions. This has two immediate consequences. One is that any visualization beyond the most schematic is out of the question. The second consequence is that the space is too large for the computation of even such standard parameters as the average fitness. We need other measures, necessarily local.

Evolution, however, does not need other measures, or any measure, really. Evolution is massively parallel; each individual tests the fitness of its genome. The more fit, the more likely to survive and reproduce that genome. Over the generations, the average fitness of the population increases. On the fitness landscape, the population resembles a herd of mountain goats wandering to a peak.

Once we have arranged the genotypes and calculated the fitnesses, sometimes we can model movement across the landscape with the gradient of the fitness function. If the genome space is large enough, and the landscape is not too rugged, we can approximate the landscape by a smooth function and apply the tools of calculus. In Sect. 17.1 we'll study the directional derivative and the gradient, mostly as tools to integrate vector functions. But also the gradient can be used to find the direction of steepest ascent. Here we'll state the result; details are in Sect. 17.1. Though this construction can be adapted to any dimension, we'll introduce it for graphs $z = f(x,y)$. For any unit vector $\vec{u} = \langle u_1, u_2 \rangle$ the *directional derivative* of f in the direction $\vec{u}$ is

$$D_{\vec{u}}f(x,y) = \left\langle \frac{\partial f}{\partial x}, \frac{\partial f}{\partial y} \right\rangle \cdot \vec{u} = \nabla f \cdot \vec{u} \tag{15.17}$$

where the vector whose components are the partial derivatives of f is called the *gradient* of f, denoted ∇f. (Remember, details are in 17.1.) Computations are easy. For example, with $f(x,y) = x^2 + xy^3$ and $\vec{u} = \langle \sqrt{3}/2, 1/2 \rangle$, we find $D_{\vec{u}}f = (2x + y^3)\sqrt{3}/2 + 3xy^2/2$.

Our use of the directional derivative is based on the interpretation of the dot product of Eq. (13.52):

$$D_{\vec{u}}f = \|\nabla f\| \cos(\theta) \tag{15.18}$$

where θ is the angle between ∇f and $\vec{u}$, and we have used $\|\vec{u}\| = 1$. Here's the interpretation: if we have a model $z = f(x,y)$ for the fitness landscape, then at any point (x,y) the direction of fastest increase of z, the *direction of steepest ascent*, is the direction of $\nabla f(x,y)$. Without computational effort, although perhaps with some visualization challenges, this approach to the fastest increase of fitness can be applied in any number of dimensions.

Also from Eq. (15.17) we see that the direction of fastest decrease is opposite the direction of fastest increase, and perpendicular to these are the directions of no change.

In the second interpretation, the collection of genotypes of the whole population is represented by a point in the horizontal plane. A different point represents a different collection. Selection pressure drives the population to increase its average fitness, moving the population to a local maximum. This is the *hill-climbing problem*: local motion of population can reach a local maximum, but then any additional change in the distribution of population genotypes reduces the average fitness. How does the population move to a higher local maximum average fitness?

This is an active field of study, with many subtle aspects. How is fitness assessed? What about in a collection of changing populations? Gregory Sorkin [339] and others [311, 378] have studied rugged fitness landscapes; in some circumstances these are fractal. There are many opportunities to study different metrics on the fitness landscape, the dynamics of evolutionary and coevolutionary processes, curvature of the landscape, and curvature of the paths on teh landscape.

For smooth landscapes, nearby sequences have similar fitness; for rugged (fractal) landscapes such as the left image of Fig. 15.3, nearby sequences can have very different fitnesses. Pleiotropic genes are a possible source of this high

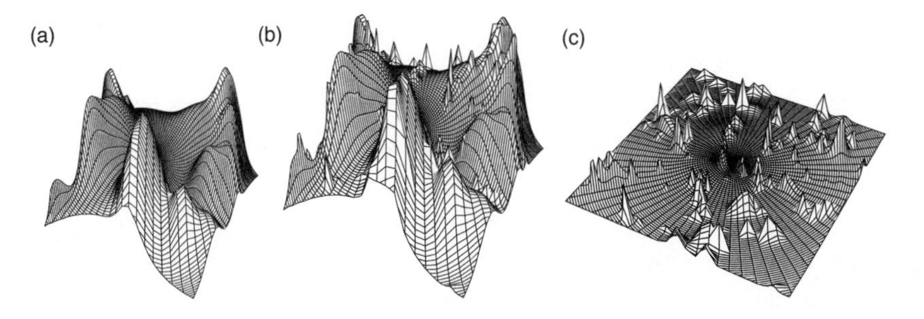

Figure 15.3. Left: a fractal fitness landscape. Middle: the largest-scale landscape features. Right: the difference between the first and second images.

variation: a single genotype mutation can alter many phenotypic factors and this may have a large impact on fitness.

If we can't get a view of the entire landscape, how can we tell that the landscape is fractal? How can we tell that the whole shape is made of pieces similar to the whole when we can't see the whole shape? We can use a more local characterization: the more closely we look, the more uphill paths we should find. Sorkin [339, 340] defines a fitness landscape as a fractal of type H if

$$\langle (f(\vec{x}) - f(\vec{y}))^2 \rangle \sim d^{2H}$$

where the average, indicated by the pointy brackets, is taken over all sequences $\vec{x}$ and $\vec{y}$ a distance d apart.

Many fitness landscapes are fractal in Sorkin's sense. For example, evidence of the fractality of RNA landscapes is presented in [1, 378].

Of course, evolution does not occur on a fixed landscape, because all the other species also are evolving simultaneously. More appropriate is a coevolutionary landscape, mountain climbing in an active earthquake zone. If we could understand the dynamics of the landscape, say for the influenza virus, we might make more appropriate vaccines for the current flu season. This is a daunting task, but the current development of bioinformatics may give a glimpse of hope.

Practice Problems

15.3.1. Consider binary sequences of length 3 and suppose the fitness function is $f(i,j,k) = 3 + i + j - k$.

(a) View the entries of the sequence ijk as x-, y-, and z-coordinates. Then represent these sequences as corners of a cube.

(b) Write the Hamming distance matrix D for these sequences. Note these distances are the smallest number of cube edges that separate the corresponding cube corners. Use the binary order for the rows and columns of this matrix. That is, the left-most column is 000, next is 001, ..., the right-most is 111. Also write the adjacency matrix A. The entry A_{ij} is 1 if the sequence that labels row i differs by one entry from the sequence that labels column j, and otherwise the entry is 0.

(c) Find the sequence with maximum fitness.

(d) Find two paths from 000 to the maximum-fitness sequence. Be sure that every step along each path increases fitness.

15.3.2. Suppose $f(x,y) = e^{-(x-1)^2-(y-1)^2}$ for $0 \le x \le 2$, $0 \le y \le 2$. Use ∇f to find a local maximum. Show it is a global maximum.

Practice Problem Solutions

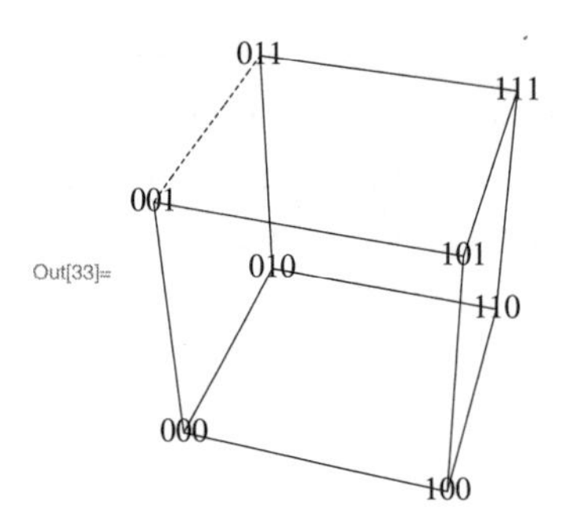

Figure 15.4. Cube vertices

15.3.1. (a) The cube plot is shown in Fig. 15.4. The dashed edge is explained in (b).

(b) To find the distance matrix, for example note that the entry in the second row and the fourth column is the Hamming distance between 001 and 011. Because these sequences differ only in the second entry, the distance is 1. The single edge that connects the corresponding vertices in the cube is dashed. The distance matrix D and the adjacency matrix A are

$$D = \begin{bmatrix} 0 & 1 & 1 & 2 & 1 & 2 & 2 & 3 \\ 1 & 0 & 2 & 1 & 2 & 1 & 3 & 2 \\ 1 & 2 & 0 & 1 & 2 & 3 & 1 & 2 \\ 2 & 1 & 1 & 0 & 3 & 2 & 2 & 1 \\ 1 & 2 & 2 & 3 & 0 & 1 & 1 & 2 \\ 2 & 1 & 3 & 2 & 1 & 0 & 2 & 1 \\ 2 & 3 & 1 & 2 & 1 & 2 & 0 & 1 \\ 3 & 2 & 2 & 1 & 2 & 1 & 1 & 0 \end{bmatrix} \qquad A = \begin{bmatrix} 0 & 1 & 1 & 0 & 1 & 0 & 0 & 0 \\ 1 & 0 & 0 & 1 & 0 & 1 & 0 & 0 \\ 1 & 0 & 0 & 1 & 0 & 0 & 1 & 0 \\ 0 & 1 & 1 & 0 & 0 & 0 & 0 & 1 \\ 1 & 0 & 0 & 0 & 0 & 1 & 1 & 0 \\ 0 & 1 & 0 & 0 & 1 & 0 & 0 & 1 \\ 0 & 0 & 1 & 0 & 1 & 0 & 0 & 1 \\ 0 & 0 & 0 & 1 & 0 & 1 & 1 & 0 \end{bmatrix}$$

(c) Because $f(i,j,k) = 3 + i + j - k$, the maximum fitness is obtained by the sequence with the largest values of i and j, and the smallest value of k. That is, 110 is the sequence with the maximum fitness, and the maximum fitness is $f(1,1,0) = 5$.

(d) The two paths are $000 \to 100 \to 110$ and $000 \to 010 \to 110$. To check that the fitness only increases along these paths, compute $f(0,0,0)=3, f(1,0,0)=4, f(1,1,0)=5$; and $f(0,0,0)=3, f(0,1,0)=4, f(1,1,0)=5$.

15.3.2. The fitness function $f(x,y)=e^{-(x-1)^2-(y-1)^2}$ is plotted in Fig. 15.5. What appears to be the global maximum is near the middle, but in order for a visual approach to work, we must be able to plot the landscape and interpret what we've plotted. Not so difficult for 2-dimensional plots, but more challenging for 300-dimensional plots.

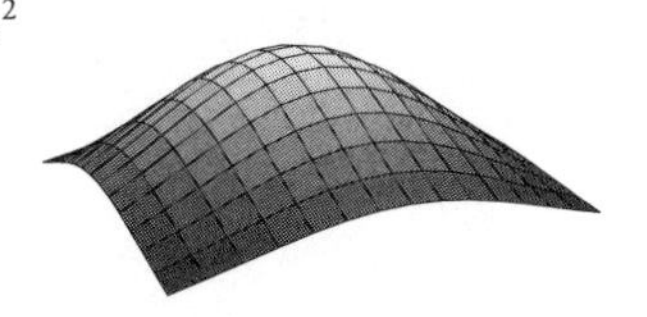

Figure 15.5. A smooth fitness function.

The landscape S defined by $z=f(x,y)$ also is defined by $g(x,y,z)=0$, where $g(x,y,z)=f(x,y)-z$. For any path $r(t)=(x(t),y(t),z(t))$ in S, $g(r(t))=0$. Then by the chain rule,

$$0=\frac{\partial g}{\partial x}x'+\frac{\partial g}{\partial y}y'+\frac{\partial g}{\partial z}z'=\left\langle\frac{\partial f}{\partial x},\frac{\partial f}{\partial y},-1\right\rangle\cdot\langle x',y',z'\rangle$$

Now $\langle x',y',z'\rangle$ is tangent to the landscape S. At a local maximum, all these tangent vectors are horizontal. In order for this to hold for all possible paths, we must have $\nabla f=\langle\partial f/\partial x,\partial f/\partial y\rangle=\langle 0,0\rangle$. Points (x,y) where $\nabla f(x,y)=\langle 0,0\rangle$ are called critical points. Certainly this can be generalized to fitness functions $f:\mathbb{R}^n\to\mathbb{R}$ for all $n\geq 1$.

But these critical points also could be local minima or saddle points, and in higher dimensions there are still other local geometries. How can we tell which critical points are local maxima?

There's a version of the familiar second-derivative test that also recalls our study of Liapunov functions. Suppose the third partial derivatives of f are continuous and that $(x_1,\dots,x_n)$ is a critical point of f. The *Hessian* of f at this critical point is the function

$$Hf(x_1,\dots,x_n)(u_1,\dots,u_n)=\frac{1}{2}\sum_{i,j=1}^{n}\frac{\partial^2 f}{\partial x_i\partial x_j}(x_1,\dots,x_n)u_1\cdots u_n \qquad (15.19)$$

(This matrix is named for the German mathematician Ludwig Hesse, and not, as I had hoped, for the German novelist Hermann Hesse.) If the Hessian is a positive definite function of $u_1,\dots,u_n$, then the critical point $(x_1,\dots,x_n)$ is a strict local minimum; if the Hessian is negative definite then the critical point is a strict local maximum. The proof is relatively involved and uses a version of Taylor's theorem for several variables. You can find a proof in any good vector calculus book. Theorem 4 of Sect. 4.2 of [226] is a clear exposition.

For the fitness function of this example, we compute

$$\frac{\partial f}{\partial x} = -2f(x,y)(x-1) \quad \text{and} \quad \frac{\partial f}{\partial y} = -2f(x,y)(y-1)$$

so the only critical point is $(x,y) = (1,1)$. To find the Hessian, first compute the second partials,

$$\frac{\partial^2 f}{\partial x^2} = f(x,y)(4x^2 - 8x + 2), \quad \frac{\partial^2 f}{\partial y^2} = f(x,y)(4y^2 - 8y + 2),$$

$$\text{and } \frac{\partial^2 f}{\partial x \partial y} = f(x,y)4(x-1)(y-1)$$

Next, at the critical point $(x,y) = (1,1)$,

$$\frac{\partial^2 f}{\partial x^2}(1,1) = -2, \quad \frac{\partial^2 f}{\partial y^2}(1,1) = -2, \quad \frac{\partial^2 f}{\partial x \partial y}(1,1) = 0$$

Then the Hessian is

$$Hf(1,1)(h_1,h_2) = -h_1^2 - h_2^2$$

This is negative definite, so the point $(1,1)$ is a local strict maximum.

In this case, we can argue directly from the form of f. Note that $e^{-(x-1)^2-(y-1)^2} = e^{-(x-1)^2}e^{-(y-1)^2}$. Elementary curve-sketching shows that these factors have global maxima at $x = 1$ and at $y = 1$, respectively. Consequently, the global maximum fitness occurs at $(x,y) = (1,1)$.

Exercises

Exercises 15.3.1–15.3.3 use the sequence space of Practice Problem 15.3.1. For each fitness function find the maximum fitness and two paths of non-decreasing fitness from 000 to a maximum sequence. Each step of each path should involve the change of a single sequence element.

15.3.1. $f(i,j,k) = 3 + i + j + k - ijk$.

15.3.2. $f(i,j,k) = 3 + ij + jk + ik$.

15.3.3. $f(i,j,k) = 3 + ij + jk + ik - ijk$.

15.3.4. For length-4 binary sequences with the Hamming distance, suppose the fitness function is $f(i,j,k,l) = 3 + 2ij + kl - il$.

(a) Find the maximum value of the fitness and the points at which that value occurs.

(b) Find a path from 0000 to 1100, one mutation for each step of the path, so that the fitness does not decrease along the path.
(c) Is there a path from 0000 to 1100 where the fitness increases at each step along the path?

15.3.5. Consider the landscape determined by the fitness function $f(x,y) = 4 - (x-2)^2 + (y-1)^2$ for $1 \le x \le 3$ and $0 \le y \le 2$.
(a) Show that the global maximum of f occurs at $(x,y) = (2,1)$.
(b) Find the path of steepest ascent from $(1,1)$ to $(2,1)$.
(c) Find the path of steepest ascent from $(1,0)$ to $(2,1)$.

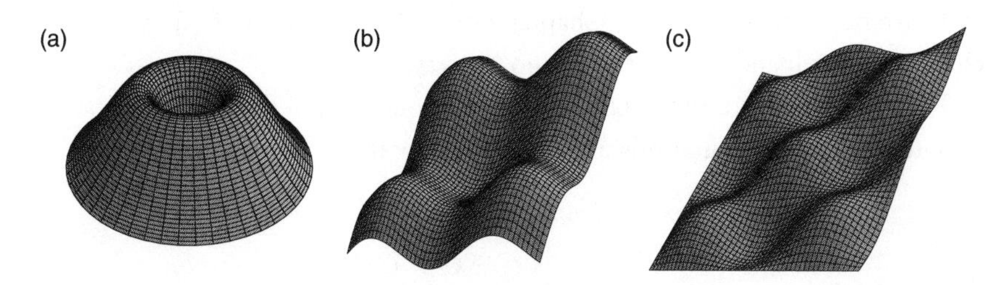

Figure 15.6. Fitness landscapes for Exercises 15.3.6 (left), 15.3.7 (center), and 15.3.8 (right).

15.3.6. For the left fitness landscape of Fig. 15.6, describe the eventual distribution of a population that begins somewhere low on the landscape.

15.3.7. For the center fitness landscape of Fig. 15.6, once the population reaches one of the fitness peaks, can short mutational steps move to another peak? Can the population split into subgroups that occupy different fitness peaks?

15.3.8. For the right fitness landscape of Fig. 15.6, once the population reaches one of the fitness peaks, can short mutational steps move to another peak? Can the population split into subgroups that occupy different fitness peaks?

15.3.9. In an environment that involves competition with many high mutation rate species, describe how the fitness landscape could change so rapidly that a low mutation rate species could be stuck in a suboptimal adaptation.

15.3.10. Find a paper on fitness landscapes and gene regulatory networks. What insights are provided by the fitness landscape approach?

15.4 THE QUASISPECIES EQUATION

The fitness landscapes provide a natural context to describe *quasispecies*, collections of nearby sequences generated from wild types (natural form) by mutation and selection. This approach was introduced by Manfred Eigen and Peter Schuster [105, 106] and has been very useful for understanding the evolution of viruses. Some RNA viruses with high mutation rates (HIV, for example) are better understood as a cloud of genotypes that explore a large region of their fitness landscape and adapt rapidly to landscape changes such as the introduction of antiviral drugs. A subtle point is the error threshold, the maximum mutation rate that allows both adaptation to local fitness peaks and genetic variation sufficient to explore landscape changes. For further background and current examples, see [91], Chapter 3 of [258], Chapter 11 of [1], Schuster's chapter of [277], and the papers [49, 90, 92, 325, 326, 384].

First, we'll look at a tiny quasispecies: this sequence and the two single point mutated sequences (mutations underlined) below it.

$$A\ T\ C\ A\ G\ G\ A\ C\ T\ C\ A$$
$$A\ T\ C\ \underline{G}\ G\ G\ A\ C\ T\ C\ A$$
$$A\ T\ C\ A\ G\ G\ A\ \underline{A}\ T\ C\ A$$

We consider a simple example, of base-2 sequences of length L. Order them in the obvious way and number them accordingly:

number	sequence
0	$\langle 0,0,\dots,0,0,0\rangle$
1	$\langle 0,0,\dots,0,0,1\rangle$
2	$\langle 0,0,\dots,0,1,0\rangle$
3	$\langle 0,0,\dots,0,1,1\rangle$
...	...
$n = 2^L - 1$	$\langle 1,1,\dots,1,1,1\rangle$

With this ordering, let's say x_i is the relative frequency of organisms with genome i, so $x_0 + x_1 + \cdots + x_n = 1$. Say f_i is the fitness of the genome i, that is, the reproductive rate of i. Recall that the average fitness of the population is $\langle f\rangle = x_0 f_0 + x_1 f_1 + \cdots + x_n f_n$. Finally, say q_{ij} is the probability that genome j mutates into genome i. How does each i population change? The j population reproduces at a rate f_j, and of these, q_{ij} mutate to i, so $\sum_j q_{ij} f_j x_j$ appears to represent the growth of population i. However, we have not yet taken Darwinian survival into account. The growth of the i population due to the average fitness of the entire population is $\langle f\rangle x_i$. If $\sum_j q_{ij} f_j x_j > \langle f\rangle x_i$, the i population grows; if $\sum_j q_{ij} f_j x_j <$

$\langle f\rangle x_i$, the i population drops. The rate of growth or decline is proportional to the difference $\sum_j q_{ij} f_j x_j - \langle f\rangle x_i$. This is the *quasispecies equation*

$$\frac{dx_i}{dt} = \left(\sum_{j=0}^{n} q_{ij} f_j x_j\right) - \langle f\rangle x_i \tag{15.20}$$

Note that $\sum_i dx_i/dt = 0$ (see Exercise 15.4.1), that is, the quasispecies equation describes the evolution of the total population fraction in each state i. If $\langle f\rangle > 0$, we expect the total population to grow. To represent this growth, define

$$u(t) = \int_0^t \langle f(s)\rangle \, ds \qquad \text{and} \quad y_i(t) = x_i(t)e^{u(t)}$$

Then

$$\begin{aligned}
\frac{dy_i}{dt} &= \frac{dx_i}{dt}e^u + x_i e^u \frac{du}{dt} = e^u\left(\frac{dx_i}{dt} + x_i\langle f\rangle\right) \\
&= e^u\left(\left(\left(\sum_{j=0}^{n} q_{ij} f_j x_j\right) - \langle f\rangle x_i\right) + x_i\langle f\rangle\right) = e^u \sum_{j=0}^{n} q_{ij} f_j x_j \\
&= \sum_{j=0}^{n} q_{ij} f_j x_j e^u = \sum_{j=0}^{n} q_{ij} f_j y_j
\end{aligned} \tag{15.21}$$

where the second equality is a consequence of the fundamental theorem of calculus, $du/dt = (d/dt)\int_0^t \langle f(s)\rangle \, ds = \langle f\rangle$, and the third equality is an application of the quasispecies equation (15.20). Eq. (15.21) is a linear equation:

$$\frac{d\vec{y}}{dt} = \begin{bmatrix} y_1 \\ \dots \\ y_n \end{bmatrix}' = \begin{bmatrix} q_{11}f_1 & \dots & q_{1n}f_n \\ & \dots & \\ q_{n1}f_1 & \dots & q_{nn}f_n \end{bmatrix} \begin{bmatrix} y_1 \\ \dots \\ y_n \end{bmatrix} = W\vec{y} \tag{15.22}$$

The matrix W is called the *mutation-selection matrix*. We can apply the methods of Sects. 8.5 and 13.1 to study the long-term behavior of $y_i(t)$.

We can rewrite the equation $\vec{y}\,' = W\vec{y}$ as $\vec{x}\,'e^u + \vec{x}e^u u' = W\vec{x}e^u$, so

$$\vec{x}\,' = (W - u')\vec{x} = (W - \langle f\rangle)\vec{x} \tag{15.23}$$

At a fixed point, $\vec{x}\,' = \vec{0}$, so Eq. (15.23) becomes $W\vec{x} = \langle f\rangle\vec{x}$. That is, the fixed point of Eq. (15.23) is the unit eigenvector of W with eigenvalue $\langle f\rangle$.

The quasispecies equation describes the evolution of the population across the fitness landscape over the sequence space. Roughly, the gradient of the fitness landscape drives this evolution.

More precisely, we say *adaptation* occurs when a quasispecies locates a peak in the fitness landscape and stays at or near the peak. A high mutation rate expedites exploring a large part of sequence space, but makes difficult the task of staying on the peak. Can we find the *error threshold*, the maximum mutation rate compatible with adaptation?

We approach this question through continuing the simple example of binary sequences of length L. Suppose p is the probability of a *point mutation*, a mutation at a single location in the sequence. Suppose that this probability is the same for all locations, and that a mutation at one location is independent of a mutation at another location. With these assumptions, q_{ij}, the probability that when sequence j reproduces it mutates to sequence i, is

$$q_{ij} = p^{h(i,j)}(1-p)^{L-h(i,j)} \tag{15.24}$$

where $h(i,j)$ is the Hamming distance between sequences i and j, that is, the number of positions on which sequences i and j differ. Now suppose sequence 0, that is, $\langle 0,0,0,\dots,0\rangle$ is the wild type and has fitness $f_0 > 1$, and all other sequences are mutants, with fitness $f_i = 1$. This is an immense oversimplification, but our goal is to study in broadest strokes the effect of a wild type with fitness higher than those of point mutants. Then the probability that a wild type reproduces a wild type is $q_{00} = (1-p)^L = q$; the probability that a wild type produces a point mutant is $q_{10} = 1-q$. Finally, assume we can neglect *back mutation*, mutants mutating to revert to wild type, so $q_{01} = 0$ and consequently $q_{11} = 1$. (However, back mutations are observed: a dwarf Alberta spruce can revert to wild type and grow long gangly branches from its top that eventually will overwhelm the tree.) Say x_0 is the frequency of the wild type, and $x_1 = 1 - x_0$ is the frequency of all the point mutants. Then the quasispecies equation (15.20) becomes

$$\frac{dx_0}{dt} = q_{00}f_0x_0 + q_{01}f_1x_1 - \langle f\rangle x_0 = qf_0x_0 - \langle f\rangle x_0$$

$$\frac{dx_1}{dt} = q_{10}f_0x_0 + q_{11}f_1x_1 - \langle f\rangle x_1 = (1-q)f_0x_0 + x_1 - \langle f\rangle x_1$$

In the equation for dx_0/dt substitute in the value of the average fitness $\langle f\rangle = f_0x_0 + f_1x_1 = f_0x_0 + x_1$ and rearrange to obtain

$$\frac{dx_0}{dt} = x_0(qf_0 - 1 - x_0(f_0 - 1)) = g(x_0)$$

The fixed points are the solutions of $g(x_0) = 0$, that is,

$$x_0 = 0 \text{ and } x_0 = \frac{qf_0 - 1}{f_0 - 1}$$

To test the stability of these fixed points, compute dg/dx_0 and evaluate it at $x_0 = 0$ and at $x_0 = (qf_0 - 1)/(f_0 - 1)$. First,

$$\frac{dg}{dx_0} = qf_0 - 1 - 2x_0(f_0 - 1)$$

From this we see the fixed point $x_0 = 0$ (the wild type disappears, no adaptation) is stable if $qf_0 < 1$. The fixed point $x_0 = (qf_0 - 1)/(f_0 - 1)$ is stable if $qf_0 > 1$. Thus, adaptation is compatible with mutation if $qf_0 > 1$, that is, if $(1-p)^L f_0 > 1$, and so

$$L \cdot \ln(1-p) + \ln(f_0) > 0$$

From the Taylor series (Sect. 10.8)

$$\ln(1-p) = -p - \frac{p^2}{2} - \frac{p^3}{3} - \frac{p^4}{4} - \cdots$$

we see that for small positive p, $\ln(1-p) \approx -p$, and so adaptation is compatible with mutation if

$$p < \frac{\ln(f_0)}{L}$$

If the fitness advantage of the wild type is not too large and not too small, $\ln(f_0) \approx 1$. Then the error threshold, the maximum mutation rate compatible with adaptation, is

$$p_c < \frac{1}{L} = \frac{1}{\text{genome length}}$$

The mutation rate compatible with adaptation to a fitness peak is related to the narrowness of that peak. Only the highest peaks matter if the mutation rate is low. For higher mutation rates, broader peaks, even if they are lower, may give a higher average fitness for the whole quasispecies.

There now is some experimental evidence that some RNA viruses have evolved to a mutation rate near the error threshold. The review [193] reports on experiments where viruses are treated with the mutagen ribavirin that pushes their mutation rate over the error threshold, after which the deleterious mutations accumulate and the virus dies out. So far as we can tell, in a billion years evolution has explored all manner of adaptive niches we have discovered. Which niches has evolution found that we have not? Yet.

Practice Problems

15.4.1. Consider binary sequences of length 10. Find the smallest N for which point mutations on $\leq N$ sites allow exploration of $1/4$ of the sequence space.

15.4.2. Find the equilibrium quasispecies distribution if the fitnesses are $f_1 = 0.8$, $f_2 = 1.1$, $f_3 = 0.6$ and the mutation-selection matrix is $[w_{ij}] = [q_{ij} f_i]$ where

$$[q_{ij}] = \begin{bmatrix} 0.5 & 0.2 & 0.3 \\ 0.4 & 0.6 & 0.3 \\ 0.1 & 0.2 & 0.4 \end{bmatrix}$$

Find the average fitness of this equilibrium quasispecies distribution.

15.4.3. For a point mutation rate $p = 10^{-3}$, find the length L of a genome that has a 50% probability of replication without error.

Practice Problem Solutions

15.4.1. This sequence space has 2^{10} elements, so 1/4 of the sequence space has $2^{10}/4 = 2^8$ elements. Consequently, $N = 8$ is the smallest number of point mutations needed to explore 1/4 of the sequence space. But the problem isn't well-posed. Which quarter of the sequence space do we want the mutations to visit? The $N = 8$ calculation is based on the assumption that two of the locations in the sequence need not be mutated, that, for example, only the first eight locations need to vary. But in general, the quarter of sequence space to explore could be spread over all locations, so all we can say is $8 \leq N \leq 10$.

15.4.2. Apply Eq. (15.23) to the mutation-selection matrix

$$W = \begin{bmatrix} 0.4 & 0.28 & 0.18 \\ 0.32 & 0.84 & 0.18 \\ 0.08 & 0.28 & 0.24 \end{bmatrix}$$

The eigenvalues of W are $\lambda = 1.075, 0.203 \pm 0.052i$. The unit eigenvector for $\lambda = 1.075$ is $\vec{x} = \langle 0.272, 0.526, 0.202 \rangle$.

The average fitness $\langle f \rangle$ is the eigenvalue $\langle f \rangle = 1.075$. To check this, we'll compute the average fitness by the definition $\langle f \rangle = f_1 x_1 + f_2 x_2 + f_3 x_3$ with the fitness values f_i posed in the exercise and the population proportions x_i that make up the unit eigenvector. This gives $\langle f \rangle = 1.075$, so the check succeeds.

15.4.3. If p denotes the probability of a point mutation, then $1 - p$ is the probability of no point mutation at a site. We assume that individual mutations are independent of one another, so the probability of no mutation in a string of length L is $(1 - p)^L = q$. Given p and q, we find $L = \log(q)/\log(1 - p)$. Then the values $p = 10^{-3}$ and $q = 0.5$ give $L \approx 692.8$. That is, we can be about 50% certain that a genome of length 693 is copied without error.

Exercises

15.4.1. Apply the quasispecies equation (15.20) to show $\sum_i dx_i/dt = 0$.

15.4.2. Show that $\sum_i y_i = e^u$ and that $\left(\sum_i y_i\right)' = \left(\sum_i y_i\right)\langle f \rangle$.

15.4.3. For a genome of length $L = 10000$, what point mutation rate p gives a 99% probability of errorless reproduction?

15.4.4. Keep the point mutation rate p constant. If the length L of the genome doubles, how does the probability q of error-free reproduction change?

15.4.5. With the $[q_{ij}]$ matrix of Practice Problem 15.4.2 suppose $f_1 = 0.9$ and $f_2 = 1.1$. Find the smallest value of f_3 (accurate to two places to the right of the decimal) for which the average fitness of the equilibrium quasispecies distribution is 1.0.

15.4.6. If every value of the fitness f_i is doubled and the $[q_{ij}]$ matrix is unaltered, how does the average fitness of the equilibrium quasispecies distribution change?

15.4.7. Suppose the genome splits into two subsets, neither of which can mutate into the other. Specifically, suppose the $[q_{ij}]$ matrix has this form,

$$[q_{ij}] = \begin{bmatrix} q_{00} & q_{01} & 0 & 0 \\ q_{01} & q_{11} & 0 & 0 \\ 0 & 0 & q_{22} & q_{23} \\ 0 & 0 & q_{23} & q_{33} \end{bmatrix}$$

and the fitnesses satisfy $f_0 = f_1$ and $f_2 = f_3$. Express the average fitness of the equilibrium quasispecies distribution in terms of those of the upper left 2×2 submatrix and of the lower right 2×2 submatrix of the mutation-selection matrix.

15.4.8. Recall that $y_i(t) = x_i(t)e^{u(t)}$, and that in Exercise 15.4.2 we showed that $y = \sum_i y_i = e^u$. Deduce that $x_i = y_i/y$ so y_i represents the absolute number of individuals with genome i. Further, show that the total population, y, has growth rate $\langle f \rangle$, the average fitness of the population.

15.4.9. Suppose in the equilibrium quasispecies distribution $x_0, \dots, x_n$ each $x_i > 0$. Show that none of the genomes will disappear in a finite time.

15.4.10. Find a paper on the error threshold and RNA virus evolution. What would happen if the mutation rate were half the error threshold? Twice the error threshold?

15.5 BIOINFORMATICS

Originally bioinformatics referred to information processing by biological systems. Now it mostly refers to (computer) manipulation and analysis of biological sequence data, DNA and protein sequences, for example. See [35, 111, 205, 245, 393] and [185, 186] for pedagogical implications. The development of next-generation (high throughput) sequencing in the 1990s led to the abundant availability of massive data strings. The human genome, for example, is made of almost 3 1/4 billion base pairs. Any efficient search or subsequence extraction must use clever algorithms.

We'll describe the most commonly used tool, BLAST, the *basic local alignment search tool.* Developed by Stephen Altschul, Warren Gish, Webb Miller, Eugene Myers, and David Lipman [10], BLAST is an algorithm, realized in software available at [44].

BLAST finds regions of similarity between amino acid sequences in proteins or nucleotide sequences in DNA or RNA. This information can help deduce evolutionary and functional relationships between sequences. For DNA and RNA sequences, the alphabet consists of four letters, A, T, G, and C for DNA; A, U, G, and C for RNA. For protein sequences, the alphabet consists of twenty letters, A, R, N, D, C, Q, E, G, H, I, L, K, M, F, P, S, T, W, Y, and V, representing the twenty amino acids. (See Sect. A.22.) We'll build an example with DNA sequences, and comment on the differences in the search for protein sequences.

Suppose we have a new DNA sequence $\vec{x} = x_1 \dots x_n$ and we want to match subsequences of $\vec{x}$ with subsequences of a database $\vec{y}_1, \dots, \vec{y}_N$. We want to find subsequences of some of the $\vec{y}_i$ that match, at least fairly closely, subsequences of $\vec{x}$. Why are we interested in "fairly closely"? One answer is that if organisms whose genomes contain $\vec{x}$ are related evolutionarily to organisms whose genomes contain $\vec{y}_i$, mutations could connect the subsequences.

The first step in a BLAST analysis is to remove low-complexity regions of $\vec{x}$. By low-complexity we mean an extended string of a repeated pattern. These can give misleading high comparison scores for subsequences. These regions are removed with the SEG program for protein sequences, and with the DUST program for DNA sequences.

Next, make a list of subsequences, usually length-3 for protein sequences, length-11 for DNA sequences. This is the time to set up an example we'll follow through the sketch of BLAST. Suppose

$$\vec{x} = TGAATTCAAGTTTGGTGCAAA$$

(This is the beginning of the sequence of instructions to build amylase.) For this illustration, rather than length-11 let's use length-6 subsequences. These are

$$TGAATT,\ GAATTC,\ AATTCA,\ ATTCAA,\ \dots,\ TGCAAA$$

Compare each of these subsequences with with all possible sequences that have the same length. For length-3 protein sequences the number of possible sequences is $20^3 = 8000$; for length-11 DNA sequences the number is $4^{11} = 4194304$. Seems like a lot, but it's not bad with a fast computer.

What do we mean by "compare"? First, we'll need the *substitution matrix*. For proteins this is a 20×20 array, actually, a family of arrays called BLOSUM (BLOcks SUbstitution Matrix) indexed by the degree of similarity of the sequences. The most commonly used is BLOSUM62. For DNA sequences the substitution matrix is is

$$
\begin{array}{c} \\ A \\ T \\ G \\ C \end{array}
\begin{array}{c} \begin{array}{cccc} A & T & G & C \end{array} \\ \begin{bmatrix} 5 & -4 & -4 & -4 \\ -4 & 5 & -4 & -4 \\ -4 & -4 & 5 & -4 \\ -4 & -4 & -4 & 5 \end{bmatrix} \end{array}
$$

The matrix defines a kind of distance between DNA symbols. For example, $d(A,A) = 5$ and $d(A,T) = -4$. To see how this is used, take the subsequence $CAAGTT$ from $\vec{x}$ and compare it with the sequence $CACGTA$, one of the (in this case) 4^6 sequences that have the same length as $CAAGTT$. Then compared with $CAAGTT$, the *score* of $CACGTA$ is

$$d(C,C) + d(A,A) + d(A,C) + d(G,G) + d(T,T) + d(T,A) = 12$$

Discard those sequences with scores under a threshold τ, usually taken to be 10. (This is a common choice, but a bit arbitrary.)

For each sequence z that has not been discarded, search through $\vec{y}_1, \ldots, \vec{y}_N$ and mark all occurrences of z. Repeat this for every length-6 subsequence of $\vec{x}$. All the subsequences of the $\vec{y}_i$ identified in this way are the *seed sequences*. BLAST uses these to look for longer subsequences of the $\vec{y}_i$ that match longer subsequences of $\vec{x}$ with a score above the threshold τ. We'll illustrate this process as we continue the example.

Take the subsequence $CAAGTT$ from $\vec{x}$ and match it with the subsequence $CACGTA$ of, say, $\vec{y}_1$. We plot $\vec{x}$ and (part of) $\vec{y}_1$ with these subsequences underlined and aligned.

$$
\begin{aligned}
\vec{x} &= TGAATT\underline{CAAGTT}TGGTGCAAA \\
\vec{y}_1 &= \quad \ldots CAT\underline{CACGTA}TGT\ldots
\end{aligned}
$$

Extend both subsequences one place to the left and to the right. The corresponding entries of both extended subsequences match, so the score increases to 22. Extend again. The left entries differ, the right agree, so again the score increases, now to 23. Extend again. Now both entries disagree so the score decreases to 15. BLAST will continue to extend the subsequences until the score begins to decrease. In this example, the best match is *ATCACGTATG* to the subsequence *TTCAAGTTTG* of $\vec{x}$.

Then the significance of each of these matches is evaluated with a distribution called the Gumbel extreme value distribution. The test parameter E is the expected number of times a random sequence will obtain a higher score. Finally, BLAST reports out all subsequences with E lower than a previously set threshold value.

In this brief description we've omitted some important points, for example how to deal with gaps in the matches. These could occur as the genomes evolve, through insertion or deletion. Nevertheless, this sketch illustrates the basic mechanism of BLAST. In addition, it helps us understand one of the main points of bioinformatics: biological data sets now are so very large that we can no longer say "search for matches" without devoting some cleverness to exactly how this search can be carried out. Biology provides math and computer science and statistics with strong motivation to study big data.

Two very similar sequences are called *homologues*. Often these have evolved from a common ancestor, and have similar 3-dimensional structures and biochemical functions. Rough estimates of the degree of similarity needed to call two sequences homologous are available, but the E value computed by the BLAST algorithm is a more reliable indicator. Suppose BLAST matches the sequence of an unfamiliar protein to that of a well-understood protein and that the match has a very low E value. Then likely the new protein and the familiar protein share many conformational and functional characteristics.

BLAST has several variations and was preceded by another search program, FASTA. Because BLAST uses heuristics, it may not find optimal matches. The Smith-Waterman algorithm [337] is more accurate, but slower and more demanding of computer memory. Likely new algorithms and new approaches will be developed as even larger data sets and more subtle search goals arise.

We sketch an application of BLAST in Sect. 19.2. Drug design is another use of bioinformatics, targeted therapies is yet another. As more computing resources become available—and we expect they will if Moore's law (computer speed and memory doubles about every two years) continues to hold—new questions and new categories of analysis will appear.

Exercises

Because bioinformatics is about the analysis and interpretation of large data sets, rather than a manual calculation, we'll do some web-based investigations. We'll give URLs and describe webpages as they appear now. Both may change by the time you read this. Google can help you find new URLs. Read the webpages to navigate site changes.

These exercises are general suggestions. The point is to get you to explore how to work with BLAST.

15.5.1. Find a nucleotide sequence and save it as a text-only file. It must be a string of C, A, T, and G without comma separations.

15.5.2. Open the BLAST website [44]. Select Nucleotide BLAST. Either paste the sequence you found in Exercise 15.5.1 into the text field, or click the Browse button and upload your text file. Scroll down the page to the Choose Search Set box and select a collection for comparison. Report what you find.

15.5.3. Rerun the search of Exercise 15.5.2 with the Query subrange limited to a portion of the query sequence. How do the search results change?

15.5.4. Find a protein sequence and save it as a text-only file. It must be a string of A, R, N, D, C, Q, E, G, H, I, L, K, M, F, P, S, T, W, Y, and V without comma separations.

15.5.5. Open the BLAST website [44]. Select Protein BLAST. Either paste the sequence you found in Exercise 15.5.4 into the text field, or click the Browse button and upload your text file. Scroll down the page to the Choose Search Set box and select a collection for comparison. Report what you find.

15.5.6. Rerun the search of Exercise 15.5.5 with the Query subrange limited to a portion of the query sequence. How do the search results change?

15.6 SOME OTHER TOPICS

The impact of genetics on medicine is immense, and in fact the intersection of genetics, medicine, and math also is substantial. So far we've covered a few topics, familiar to me, in this intersection. But there is so much more. Here we'll mention briefly a few others, just because I can't imagine not writing a tiny bit about each. But do look at this intersection yourself. I expect you'll find wonders, unknown to me, that will knock off your socks.

Some background in immunology is in Sect. A.15; some background in molecular genetics is in Sect. A.22.

15.6.1 Evolutionary game theory

Evolutionary game theory is a way to model fitnesses that are not constant. Modern game theory was introduced by John von Neumann [366], and presented in encyclopedic detail by von Neumann and Oskar Morgenstern [367]. Their purpose was to develop a mathematical method to study human social and economic interactions. Before we look at applications to evolution, we'll give a simple example, the prisoner's dilemma.

Two people, x and y, have been accused of a crime, arrested, and held in separate rooms. Both are asked to confess. If both confess (we say they *cooperate*, C), each will spend a week in jail. We say the *payoff* is $(x,y) = (-1,-1)$. If one confesses and the other does not (we say the other *defects*, D), the one who confesses spends 10 weeks in jail while the who does not goes free. If neither confesses, both spend 5 weeks in jail. We can represent these outcomes with a *payoff matrix*. Label the rows with the behavior of x, row 1 for cooperate, row 2 for defect. Label the columns with the behavior of y. The matrix entries are the payoffs (x,y). The payoff matrix for this prisoner's dilemma is

$$\begin{bmatrix} (-1,-1) & (-10,0) \\ (0,-10) & (-5,-5) \end{bmatrix} \quad \text{or} \quad \begin{bmatrix} -1 & -10 \\ 0 & -5 \end{bmatrix}$$

where in the second matrix the payoffs are for the row player: C against C receives -1, C against D receives -10, D against C receives 0, and D against D receives -5.

Here's the dilemma. To x the situation looks like this: if I defect and y cooperates, I get off free; if I defect and y defects, I get 5 weeks. If I cooperate and y cooperates, I get 1 week; if I cooperate and y defects, I get 10 weeks. Whatever y does, I get a better outcome if I defect.

Seems clear enough, but y can make the same argument and decide to defect, so both x and y will get 5 weeks. If we assume both x and y play rationally, both will spend 5 weeks in jail, not a very good outcome. The strategy of defection for both players is a *Nash equilibrium*: neither player can improve their payoff by changing their strategy.

This game may seem to be a bit less interesting than tic-tac-toe, but I'll mention that at least two books [289, 304] and one novel [292] have been written about the prisoner's dilemma.

Repeated play with memory of previous plays is a source of cooperation. A popular, and detailed, exposition of the evolution of cooperative behavior is given in [260]. For repeated play the notion of strategy must be extended a bit. Suppose in a single iteration of the game there are n strategies, $S_1, \dots, S_n$. These are called *pure strategies*. The prisoner's dilemma has two pure strategies:

cooperate and defect. Now a strategy for player x is a vector $\vec{x} = \langle x_1, \ldots, x_n \rangle$ that gives the conditions under which x plays S_i, or the probability that x plays S_i. These are called *mixed strategies*. An example of a mixed strategy for the prisoner's dilemma is *tit-for-tat*: in the next iteration you do whatever your opponent did—cooperate or defect—in the current iteration.

John Maynard Smith and George Price, the same Price who gave us Price's equation of Sect. 15.2, applied game theory to questions of animal conflict [229, 228]. Specifically, they addressed the puzzle that most animal conflict that is competitive for territory or mates is ritualized and results in few injuries. Of course, predator-prey conflict often results in considerable injuries. The puzzle is why these conflicts are limited. From the perspective of the fitness (reproductive success) of the individual, it makes more sense to kill your opponent and inherit their mates thus giving you more opportunities to spread your genes. From the point of view of the population, genetic variation is an advantage and so the presence of several strong males is desirable. But selection is driven by the fitness of individuals, not populations. Individuals reproduce; populations do not. The resolution of the puzzle is based on the fact that fitness of an individual cannot be calculated in isolation because it is influenced by how the individual interacts with other members of the population. Then the map to game theory is achieved by comparing phenotypic characteristics and behaviors to strategies and by comparing fitness to the payoff. Through this lens we see that limited aggression can benefit both individuals and populations. A good survey of evolutionary game theory is [178].

What medical significance can we find in these population models? Recall in Sect. 13.4 we applied some of the techniques of population biology with the body as the ecosystem, viruses as prey, and immune system elements as predators. Game theory can explain cooperative behavior in populations of animals, including people. This approach can be adapted to understand cooperative behavior of cells.

We'll sketch one application, an interpretation of multiple myeloma phenotype as the result of an evolutionary game between normal and malignant cells [86, 87, 232]. Normally bone resorption mediated by osteoclasts (OC) is balanced by bone formation mediated by osteoblasts (OB). Multiple myeloma cells (MM) perturb the OB-OC balance toward OB. Label the rows and columns OC, OB, and MM. Then we can view the interactions of these populations as a three-player game with the payoff matrix on the left

$$\begin{bmatrix} 0 & a & b \\ c & 0 & -d \\ e & 0 & 0 \end{bmatrix} \quad \text{or rescaled to} \quad [p_{ij}] = \begin{bmatrix} 0 & 1 & \beta \\ 1 & 0 & -\delta \\ \beta & 0 & 0 \end{bmatrix} \tag{15.25}$$

where the constants a, b, c, d, and e are positive. In the right matrix $\beta = e/c$ and $\delta = de/bc$. How are these matrices related? Multiply every entry of the first column of the left matrix by $1/c$, every entry of the second column by $1/a$, and every entry of the third column by e/bc. This is an example of a projective transformation, and although it will change the locations of the fixed points, it will not change the type of fixed point. See [175], for example.

David Dingli and coauthors [86, 87] used an understanding of evolutionary game theory to deduce these two matrices exhibit the same dynamics. The identification of the dynamically important parameters β and δ depended on model reformulation based on evolutionary game theory.

Denote by x_1, x_2, and x_3 the relative frequencies of the OC, OB, and MM cells. Then the fitness f_i of cell type i and the average fitness $\langle f \rangle$ are given by

$$f_i = \sum_{j=1}^{3} p_{ij} x_j \qquad \text{and} \qquad \langle f \rangle = \sum_{i=1}^{3} \sum_{j=1}^{3} p_{ij} x_i x_j$$

The per capita reproduction rate of each x_i is the difference between the fitness f_i and the average fitness $\langle f \rangle$. If cell type i has higher fitness than the average, Darwinian selection says the relative frequency x_i should grow. That is,

$$\frac{dx_i}{dt} = x_i \sum_{j=1}^{3} (f_i - \langle f \rangle) \tag{15.26}$$

This should remind us of the quasispecies equation (15.20). With the right matrix of (15.25), the differential equations (15.26) become

$$\begin{aligned} x_1' &= x_1 x_2 + \beta x_1 x_3 - 2x_1^2 x_2 - 2\beta x_1^2 x_3 + \delta x_1 x_2 x_3 \\ x_2' &= x_1 x_2 - \delta x_2 x_3 - 2x_1 x_2^2 - 2\beta x_1 x_2 x_3 + \delta x_2^2 x_3 \\ x_3' &= \beta x_1 x_3 - 2x_1 x_2 x_3 - 2\beta x_1 x_3^2 + \delta x_2 x_3^2 \end{aligned} \tag{15.27}$$

Because x_1, x_2, and x_3 are relative frequencies, we have $x_1 + x_2 + x_3 = 1$. We can visualize this space as the filled-in triangle with vertices $(1,0,0)$, $(0,1,0)$, and $(0,0,1)$ in 3-dimensional space. At any time the population of OB, OC, and MM cells is represented by a point on this triangle. The trajectories of the differential equations (15.27) show how the relative frequencies change with time.

Label one vertex OB, one OC, and one MM. If the population is at vertex OB, then all cells are OB and none are OC and MM. With the methods of Sect. 9.2, extended to higher dimensions in Chapter 13, we find the fixed points by simultaneously setting to 0 the right-hand sides of Eq. (15.27) and solving; the stability is determined by the eigenvalues of the derivative matrix of Eq. (15.27), evaluated at each fixed point. After some algebra we find the vertices and test their stability. The number and stability of the fixed points depend on the parameters β and δ. Fig. 15.7 shows the three typical arrangements.

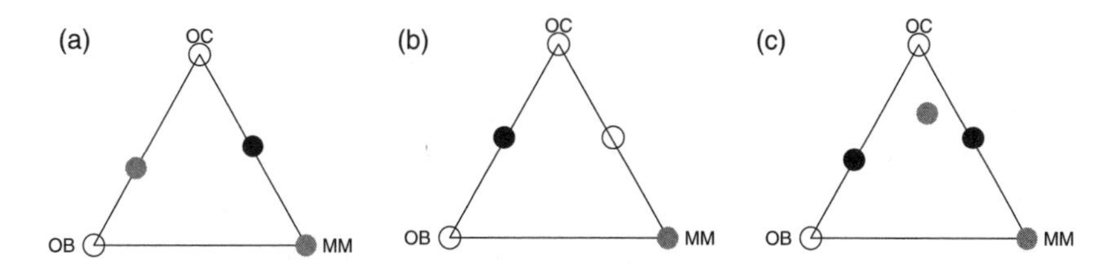

Figure 15.7. Fixed points for $\beta > 1$ (left); $\beta < 1$ and $\beta + \delta < 1$ (center); and $\beta < 1$ and $\beta + \delta > 1$ (right). Circles indicate unstable fixed points, gray discs indicate saddle points, and black discs inducate stable fixed points.

Untreated multiple myeloma patients most often have $\beta > 1$. In the left triangle of Fig. 15.7 we see that the only stable fixed point is along the OC-MM edge—bad for the patient. In the center triangle we see that if $\beta < 1$ and $\beta + \delta < 1$, the only stable fixed point is along the OB-OC edge and corresponds to the elimination of MM. The condition $\beta + \delta < 1$ requires a considerable reduction of β, or equivalently, a reduction of e far below c. A more modest reduction can put the patient in the circumstance of the third triangle of Fig. 15.7 where the addition of a saddle point in the triangle interior opens another therapeutic path. If therapy can move the current state of the system near the unstable direction of this saddle point that leads to the stable equilibrium on the OB-OC edge, then natural selection can force the MM population into extinction.

High-dose chemotherapy or stem cell transplants can lower the tumor load, allow some bone repair, and prolong life up to a point, but always the disease recurs. Several clinical trials show promise. Treatment with anakinra (an Interleukin 1, IL-1, receptor antagonist) and antisense therapy (yes, this is a real thing—ask our learned friend Google) against *MIP*-1α (macrophage inflammatory protein), reduce β and slow the progression of smoldering to active multiple myeloma. Similarly, an antibody directed against *Dkk*-1, important in inflammation-induced bone loss, can reduce δ; this also improves the patient's survival time. The model predicts that an adequate reduction of both β and δ may lead to the elimination of the MM cells by natural selection.

The effectiveness of some therapies exploits natural selection pressure to drive the pathological genotype to extinction. Careful mathematical analysis, including evolutionary game theory, has revealed some new pathways for therapies. Other examples are [33, 131, 194, 265, 268].

15.6.2 Epigenetics

For some time now examples have accumulated of inheritable alterations of gene activity without changes to the DNA sequence. Epigenetics is the study of the mechanisms of these changes. (Epigenetics is distinct from the discredited Lamarckism. Google will tell you what this is if you don't know.) These mechanisms include methylation, phosphorylation, and histone modification in chromatin. Epigenetic effects, largely consequences of unfavorable environmental factors, can perturb these mechanisms from their normal functions and so contribute to the cause of many ailments.

We must mention that Conrad Waddington developed the epigenetic landscape model, analogous to fitness landscape models, to help visualize the effects of epigenetics. See [15] for a discussion.

We'll focus mostly on methylation. The addition of a methyl group to the $5'$ end of cytosine (see Sect. A.22) in a CpG (cytosine-phosphate-guanosine) dinucleotide prevents transcription factors from binding and thus silences that gene. (The process is more complex that this and involves chromatin proteins. Details are in [184].) Generally the CG pair is underrepresented in DNA sequences, but is abundant in *CpG islands*, regions of length between 500 and 4000 bases, often close to promoter regions.

Cancer The reviews [113, 184] are good references. In every type of human cancer the promoter regions near CpG islands are hypermethylated and the transcriptional activity of these genes is blocked. Hypermethylation is at least as common as mutation of tumor suppressor genes (TSG, see Sect. 16.5), and hypermethylation silences some genes thought to be TSGs. For example, in between 10% and 15% of women with non-familial breast cancer, the TSG *BRCA*1 is hypermethylated. Methylation changes the absorption wavelength of cytosine into the range of sunlight, promoting skin cancer. And methylated cytosine is a preferred binding site for many carcinogens. Probably this is enough to convince you that epigenetic hypermethylation is bad news, but we'll mention another. In Sect. 16.5 we discuss Alfred Knudson's two-hit hypothesis, which states that both alleles of a TSG must be inactivated in order to promote tumor growth. Several studies have observed tumors where one TSG is inactivated by mutation and the other by hypermethylation.

Few things in biology are straightforward, so while hypermethylation contributes to the formation of many tumors, in tumor cells hypomethylation (wherein the fraction of methylated cytosine in tumor cells is lower than in wild type cells) is more common [84]. Now evidence suggests that hypomethylation promotes chromosomal instability and oncogene activation. Both of these concepts are discussed in Sect. 16.5, but even their names suggest they are dangerous.

Even now, some of the basic biology of cancers remains unsettled. The standard view is that cancer is monoclonal, the result of an initial mutation. In an alternate model [113], cancer is polyclonal and arises from epigenetically disrupted stem cells. The story is far from finished, but at this point most researchers agree that cancer has an epigenetic component.

Systemic lupus erythematosus (SLE) is an inflammatory autoimmune disease whose symptoms include malar rash, painful and swollen joints, muscle pain, fever, and fatigue. SLE patients exhibit autoantibodies against cell nuclear, cytoplasmic, and surface antigens. Relatively symptom-free periods are punctuated by flares of some severity. SLE increases the risk of cardiovascular disease and for some patients can reduce lifespan considerably.

Recent studies support the importance of epigenetic factors in other autoimmune diseases, including rheumatoid arthritis, scleroderma, type 1 diabetes, and multiple sclerosis [161]. So far, SLE exhibits the best evidence of an epigenetic influence.

Identical twin studies show that in fewer than 60% [26], and in as few as 25% in some studies [161], of the pairs in which one twin has SLE, the other twin does, too. Because identical twins are monozygous, these findings point to non-genetic factors in the development of SLE. Natural candidates are epigenetic modifications triggered by environmental agents, and some of the genes deregulated in SLE are known to be deregulated by epigenetic mechanisms. For example, T cells from patients with active SLE have hypomethylated DNA, and this has a direct effect on autoreactivity. Histone modifications are another epigenetic effect that contributes to SLE. SLE patients exhibit hypomethylated helper T cells and also reduced expression of CD5 (a T cell marker) on B cells, which promotes autoreactivity. Also, human endogenous retroviral elements (HERV, about 8% of our genomes, inserted millions of years ago by retrovirus infections), usually inactivated by methylation, exhibit reduced methylation in SLE patients. The resulting perturbation of gene expression likely contributes to autoimmune diseases. Application of de-methylating drugs (hydralazine or procainamide) induces SLE symptoms and so supports the hypothesis that DNA hypomethylation is an important component in the development of SLE.

Histone modification and microRNA (MiRNA) overexpression also are almost certainly involved in autoimmune diseases, but the processes are complex and not yet so well understood. These, too, provide avenues for epigenetic influence on autoimmune diseases.

As described in [128], epigenetic effects can be reversed by some drugs. This opens the possibility for treatment even though (so far) the genetic mutations cannot be reversed.

Schizophrenia is a mental disorder affecting about 1% of the population. Its symptoms include unusual behavior and speech, hallucinations, impaired memory, and a different perception of reality. Its causes are thought to have both genetic and environmental (uncovered through epidemiological studies) components. The existence of genetic components is supported by identical twin studies that show if one twin has schizophrenia, the probability that the other twin will also have schizophrenia is 53%. For dizygotic twins the probability drops to 15%. Can the environmental components have epigenetic factors?

In the review [314], the authors discuss evidence for the role of DNA methylation and histone modifications in the development of schizophrenia through regulation of CNS (central nervous system) gene activation.

DNA methylation is important for synaptic plasticity and the formation of associative memories. And remarkably (at least, it surprised me) there is evidence that postnatal care can affect DNA methylation. DNA methylation also plays a role in the dysfunction of neurons that produce the neurotransmitter GABA (γ-aminobutyric acid) and altered GABA activity is responsible for at least some of the clinical features of schizophrenia. Recent studies have found about 100 loci that exhibit *CpG* methylation in schizophrenia patients. In addition, evidence supports epigenetic influences on serotonin, dopamine, and BDNF (brain-derived neurotrophic factor), important in some aspects of schizophrenia. Though not (yet) as strongly supported by evidence as is DNA methylation, some studies suggest that histone modification may be another route for epigenetic influence of schizophrenia.

Epigenetic effects can be reversed, though this is tricky: methylation is good in some loci, bad in others. The targeting must be precise. Still, there is some evidence that chromatin-modifying drugs may be effective in the treatment of schizophrenia. This disease torments 70,000,000 people worldwide. The investigation of a possible treatment is worth some effort.

Finally, we mention the cleverly titled paper (the title references Hegel's dialectic) [283] in which Arturas Petronis argues that a Hegelian synthesis of thesis (genetic causes) and antithesis (epigenetic causes) will give the best outcome for the study of schizophrenia and perhaps other complex diseases. He makes the important point that a substantial portion of phenotypic variation may be due to random fluctuations in epigenetic inputs.

So, there are probable epigenetic effects in cancers, lupus, and schizophrenia. Of course, this is nowhere near a complete list. But maybe you wonder where's the math?

Epigenetics and math Instead of yet another collection of differential equations, here we'll give a very brief sketch of the use of epigenetic information—specifically, histone modifications—to refine models of gene regulatory networks [56]. This is a problem of clear interest, and of clear difficulty. We won't give many details, just show how epigenetic information can be used to build models. The approach of Haifen Chen and coworkers is based on Bayesian inference, a technique we encountered first in Sect. 11.12. But here and in previous work [58, 354, 394, 396, 401] the goal is to deduce the structure of a network, which unlike a simple causal diagram, has many loops to express positive and negative feedback that are known features of gene regulation. Dynamical Bayesian networks [247] are a natural tool in this setting.

Although still there is no clear picture of how epigenetic factors impact the regulatory relations between individual genes, the study [151] suggests that genes in the same regulatory pathways exhibit similar epigenetic features.

The Chen group worked with the gene regulatory network of yeast. The phenotypic changes studied were alterations in gene expression caused by histone modifications, reported in [287]. The work of [56] is an extension of [401], where histone modification patterns were the prior distribution in Bayesian networks used to refine the gene regulatory network model. We'll present a simplified sketch, to emphasize the main points. Each vertex of the network corresponds to a gene, each (directed) edge to a conditional dependence between the genes. So the gene regulatory network can be represented by a directed graph G. The gene expression levels are sampled at times $j = 1, \dots, T$. (The "dynamical" of dynamical Bayesian networks captures how the expression levels vary with time, an essential ingredient to study feedback loops.) The entries $x_i(j)$ of the vectors $\vec{x}(j)$ are the level of expression of gene i at time j. Then the probability $P(\vec{x}(j)|G)$ of observing gene expression levels $\vec{x}(j)$ at time j is a function of the conditional probabilities $P(x_i(j) = a|x_k(j-1) = b)$, where a and b run over all possible levels of expression, and the probability is 0 if the graph G has no edge from x_k to x_i.

Next, apply Bayes' theorem, Eq. (11.10) to the relation between gene expression levels and regulatory network graphs:

$$P(G|\vec{x}(j)) = \frac{P(\vec{x}(j)|G)P(G)}{\sum_H P(\vec{x}(j)|H)P(H)}$$

where the sum is over all possible regulatory network graphs. For even a modest number of genes, the number of network graphs is immense. Summing over all possibilities takes too long, so in practice the sum is taken over a random sample of possible graphs. Then the gene regulatory network is the G that maximizes the posterior probability $P(G|\vec{x}(j))$.

Chen's group showed that the use of epigenetic data offers a significant improvement in the accuracy of the gene regulatory network.

Don't be fooled by the small number of equations in this section. This construction uses a lot of math, most of it invisible because of the brevity of this sketch. But the math is there. Math is everywhere.

15.6.3 Targeted therapies

In the early years of medical oncology, the most common agents were primitive chemotherapeutics. These are poisons (the reason the technique is called "cytotoxic chemotherapy"), effective to the extent that they are because many cancer cells grow and divide more quickly than non-cancer cells, so the cancer cells absorb more poison than other cells. Chemotherapy poisons everything; the oncologist's art is to balance the dosage so the cancer cells die while (most of) the non-cancer cells don't.

In the early 1960s my father's older sister Ruth was diagnosed with Hodgkin's lymphoma. She was treated with Mustargen, an early chemotherapy agent, but she died after several months of misery. I was twelve years old then. My aunt Ruthie and I had spent a lot of time together. In her garden we investigated the veins of leaves, the roughness of dirt clumps, the movements of insects. She taught me the types of clouds, and which could bring rain. In the evening sky she pointed out Venus, a whole planet, almost as large as Earth. That I became a scientist is due more directly to Aunt Ruthie than it is to anyone else. When she got sick, I went crazy, as crazy as a little kid can go.

And I had a dream: I would work very hard, day and night, go into medical research, and find a cure for Hodgkin's lymphoma. Then I'd rush to Ruthie's hospital room, administer the cure, and save my aunt. Only a child could have such a fantasy. But this added some urgency to the curiosity Ruthie had awakened in me. I knew I couldn't save her, but maybe I could help others. Turns out, I'm not smart enough. But maybe you are. I hope you are. Every year new treatments are invented, new ideas explored. We'll end this section and this chapter with a brief description of a new class of therapies.

Targeted therapies have several forms. Some are chemicals ("biologics") that interfere with a protein or enzyme, essential for cancer growth, that has a mutation found in cancer cells but not in normal cells. Others are enzymes, artificial antigens, that attach to cancer cell surfaces and mark them for destruction by agents of the immune system. This approach belongs to the general category of immunotherapy.

Bruton's tyrosine kinase (BTK) inhibitors are a class of targeted therapy drugs. They include ibrutinib (PCI-32765) and ARQ 531. BTK inhibitors target B cell receptor signal complexes in patients with B cell malignancies, including CLL, chronic lymphocytic leukemia. Most often CLL is an indolent cancer, progressing so slowly that no intervention is required in a patient's lifetime. But some mutations (13q and 17p deletions, for example) make CLL much more dangerous. For patients with one of these CLL variants, currently ibrutinib often is the first treatment. Eventually some mutant cancer cells develop resistance to ibrutinib and selection pressure amplifies a clone of that mutant strain. Bad news for the patient. Currently in clinical trials, ARQ 531 is effective against ibrutinib-resistant CLL clones, for a while.

The development of targeted therapies is fluid, and offers treatments that attack cancer cells and little else. We'll explore one example [248] more closely.

The *cancer stem cell* hypothesis [46, 307], supported by a considerable body of evidence, is that cancerous cells are produced by a small population of cancer stem cells (CSC) that can self-renew and also differentiate to form other cancer cells. The success of traditional therapies is measured by the reduction of tumor size, but if some CSC remain after treatment, the cancer will recur. CSC have been identified in many cancers. The model of [248] concerns head and neck squamous cell carcinomas [296]. In [200] the cytokine (signal molecule) Interleukin 6 (IL-6) is identified as essential for the growth of these tumors. The study [248] models treatment with anti-IL-6 antibodies, tocilizumab (TCZ), for example. TCZ binds to IL-6R, the natural receptor of IL-6, on tumor cells and so suppresses the formation of IL-6/IL-6R complexes. The receptor IL-6R is overexpressed on CSC, so TCZ therapy is a promising approach.

The model is a collection of fourteen differential equations, more than we care to present here. The model variables are the populations of CSC, progenitor and differentiated tumor cells; fmol (femtomoles) of IL-6; fmol of IL-6R and bound complexes of IL-6/IL-6R on CSC, on progenitor cells, and on differentiated cells; and fmol of free anti-IL-6 antibodies in the tumor and bound complexes of anti-IL-6/IL-6R on CSC, on progenitor cells, and on differentiated cells. The model includes molecular signaling dynamics, CSC self-renewal and differentiation, and the effects of TCZ on the CSC population. It predicts time-dependence of the fraction of bound IL-6 receptors and the impact on tumor growth.

The model predicts that even small doses of TCZ are adequate to produce a 25% reduction in tumor volume. This reduction is a consequence of simple competition: TCZ outcompetes IL-6 for available IL-6R. Because IL-6 signals are important to the survival of tumor cells and to the self-renewal of CSC, the disruption of these signal pathways increases the tumor cell death rate and reduces the CSC self-renewal probability. Work in progress includes extensions of the model to combinations of TCZ and traditional chemotherapy agents.

Many of the model predictions are consistent with experimental data. With adequate validation, model experiments can explore vast arrays of therapy combinations and treatment schedules. To paraphrase a quote of Daisetsu Teitaro Suzuki ("As many Buddhists, so many Buddhisms," or so I remember,) I offer *as many cancer patients, so many cancers*. Targeted therapies allow treatments tailored to the patient. But in the absence of an underlying predictive theory—and we appear to be very far from any such theory—we need tons of data. Predictions from reliable models are an essential piece of this process. As Colleen Clancy of the UC Davis College of Biological Sciences says, "Math instead of mice."

Chapter 16 Markov chains in biology

Most biological processes involve some degree of randomness. In the SIR model of Sect. 3.2, the parameter ϵ is the probability that an encounter between an infected and a susceptible will result in the infection of the susceptible, and the principle of mass action estimates the average likelihood of an encounter between a susceptible and an infected. Individual encounter numbers vary with behavior: a retired susceptible who rarely leaves home is less likely to encounter an infected than is an elementary school teacher, for example.

In the age-structured population models of Sect. 14.3, the reproductive rate and survival rate of an age class represent averages. The results for an individual depend on probabilistic factors: encountering a suitable mate, locating food, avoiding predators, surviving diseases.

In this chapter we'll move to much smaller size scales and investigate some Markov chain models in genetics and in transmembrane ion channels. Along the way we'll learn what we mean by the exponential of a matrix, and how to solve some matrix differential equations. We'll see that a major contributor of membrane channel randomness is thermal noise, heat knocking molecules around. Often heat has no memory: which state we enter next depends on the current state and not on the history leading to the current state. In these and similar circumstances, Markov chains are suitable models.

A wonderful expression of the random nature of heat was made by Albert Einstein's friend, the mathematician Marcel Grossmann. Grossmann noted that when he sat on a recently vacated chair, the warmth left by the previous occupant bothered him, until he learned in a physics class that heat is impersonal.

Impersonal, but not unimportant.

16.1 GENETIC DRIFT

A common application of Markov chains is the study of genetic drift. We'll investigate the Wright-Fisher model of haploid (cells with unpaired chromosomes) allele frequency change [116, 391]. With a minor modification, Wright-Fisher can analyze diploid (cells with paired chromosomes) allele frequency change.

Suppose in a population of constant size N with non-overlapping generations we concentrate on a locus with two alleles, A and a, both neutral in that neither confers a survival advantage. In addition, we assume no mutation occurs during the time we observe the population. In each generation a parent from this population is selected randomly with replacement. That is, the same member can be selected as parent more than once. Because there are only two outcomes (A and a) selected randomly, and because in each generation the probability of A is constant (this is where we use the fact that a parent can be selected more than once), the Wright-Fisher model uses the binomial distribution to compute the transition probabilities. Here's how.

In generation t suppose the population has exactly i instances of the allele A, and consequently $N-i$ instances of a. We'll signify this by writing $|A(t)|=i$ and $|a(t)|=N-i$. Then the probability that $|A(t+1)|=j$ is

$$p_{j,i} = Pr(|A(t+1)|=j \mid |A(t)|=i) = \binom{N}{j}\left(\frac{i}{N}\right)^j\left(1-\frac{i}{N}\right)^{N-j} \qquad (16.1)$$

because in the current generation the probability of selecting allele A is i/N, so the probability of selecting exactly j copies of A, and consequently exactly $N-j$ copies of a, is $(i/N)^j(1-i/N)^{N-j}$. The number of arrangements of the j copies of A in a population of size N is the binomial coefficient N choose j.

We can use properties of the binomial distribution derived in Sect. 11.7 to deduce some simple aspects of allele frequency dynamics in the Wright-Fisher model. First, if $|A(t)|=i$, by Eq. (11.29) the expected value of $|A(t+1)|$ is

$$\mathbb{E}(|A(t+1)|) = Np = N\frac{i}{N} = i \qquad (16.2)$$

so on average the number of copies of allele A does not change. This does *not* mean that the number of copies of A is unchanged, because we can apply Eq. (11.32) to deduce

$$\sigma^2(|A(t+1)|) = Np(1-p) = i\left(1 - \frac{i}{N}\right) \tag{16.3}$$

This variance is a signature of genetic drift. From the middle expression of Eq. (16.3) we see that if we fix a proportion $p = i/N$ of copies of the allele A, the variance increases with N. And from the right expression of Eq. (16.3) we see that if we fix the number i of copies of A, the variance increases with N. That a larger population would exhibit more variance is hardly a surprise; these formulas for $\sigma^2(|A(t+1)|)$ show how the variance depends on the population size.

A common measure of the degree of genetic spread is the *heterozygosity*, $H(t)$, the probability of finding both alleles when we randomly select two population members without replacement. At time t the population has $|A(t)|$ copies of the allele A, so the probability of randomly selecting a copy of A is

$$P(\text{selecting } A) = \frac{|A(t)|}{N}$$

This leaves a population of size $N-1$ that contains $|A(t)|-1$ copies of A. In this population, the probability of selecting a is

$$\begin{aligned}&P(\text{selecting } a \text{ if one } A \text{ has been removed from the population})\\ &= \frac{(N-1)-(|A(t)|-1)}{N-1} = \frac{N-|A(t)|}{N-1}\end{aligned}$$

Except for the effect of the first sample on the populations, an effect already taken into account, these samples are independent of one another. Then the probability of selecting both A and a is given by the multiplication rule for independent events, Eq. (11.6),

$$H(t) = \frac{|A(t)|}{N} \cdot \frac{N-|A(t)|}{N-1} = \frac{N}{N-1}p(t)(1-p(t)) \tag{16.4}$$

where $p(t) = |A(t)|/N$ is the probability of randomly selecting A in generation t.

Wright-Fisher dynamics imply that the expected value of $H(t)$ decreases with successive generations:

Proposition 16.1.1. In the Wright-Fisher model,

$$\mathbb{E}(H(t)) = \left(1 - \frac{1}{N}\right)^t \mathbb{E}(H(0)) \tag{16.5}$$

Proof. In order to sample both alleles from two randomly selected population members in generation t, the selected alleles must have different ancestors (recall each population member carries only one allele and there is no mutation) and these ancestors must have different alleles. The second selected allele has $N-1$ choices of ancestor different from that of the first allele, so the probability that the two selected alleles have different ancestors is $(N-1)/N = 1-(1/N)$. The probability that these ancestors (in generation $t-1$) have different alleles is $\mathbb{E}(H(t-1))$. Combining these two observations, we have

$$\mathbb{E}(H(t)) = \left(1 - \frac{1}{N}\right)\mathbb{E}(H(t-1))$$

Applying the same reasoning to $\mathbb{E}(H(t-1))$ and so on back to $\mathbb{E}(H(1))$, we obtain Eq. (16.5). □

A simple, and not at all surprising, consequence of Eq. (16.5) is that the heterozygosity decreases more slowly for large populations. Regardless of the population size, Wright-Fisher dynamics imply that (without mutation) genetic variation will vanish if we wait long enough.

So how long is "long enough"? We'd like to compute the expected number of generations to *fixation* of the allele A or of the allele a, that is, find the average t that gives $|A(t)| = N$ or $|a(t)| = N$. Without mutation or immigration, once one allele disappears from the population, it won't return. To find fixation time, consider a large ensemble of populations, all of size N. Then the distribution of allele frequencies is an $(N+1)$-dimensional vector $\vec{\theta}(t)$ with components $\theta_i(t)$ equal to the proportion of ensemble populations at time t having exactly i copies of allele A. The evolution of the distribution of allele frequencies is governed by

$$\vec{\theta}(t+1) = P\vec{\theta}(t) \tag{16.6}$$

where

$$P = \begin{bmatrix} 1 & p_{0,1} & p_{0,2} & \cdots & p_{0,N-1} & 0 \\ 0 & p_{1,1} & p_{1,2} & \cdots & p_{1,N-1} & 0 \\ 0 & p_{2,1} & p_{2,2} & \cdots & p_{2,N-1} & 0 \\ & & & \cdots & & \\ 0 & p_{N-1,1} & p_{N-1,2} & \cdots & p_{N-1,N-1} & 0 \\ 0 & p_{N,1} & p_{N,2} & \cdots & p_{N,N-1} & 1 \end{bmatrix} \tag{16.7}$$

Recall that $p_{i,j}$ denotes the probability that a population with j copies of A becomes a population with i copies of A. The first column takes this form because a population with 0 copies of A must stay a population with 0 copies of A. To understand the first row, note for example that $p_{0,1}\theta_1(t)$ is the proportion of the generation t ensemble having 1 copy of A, times the probability that a population with 1 copy of A loses that copy. Analogous unpacking of the other terms explains the form of each row.

That this is a stochastic matrix is straightforward: consider the second column for example. We see that

$$p_{0,1} + p_{1,1} + p_{2,1} + \cdots + p_{N-1,1} + p_{N,1} = 1$$

because a population with exactly 1 copy of A must evolve into a population with 0, 1, 2, ..., $N-1$, or N copies of A.

The states $|A(t)| = 0$ and $|A(t)| = N$ are absorbing states, a concept introduced in Sect. 14.1. In that language, the average number of generations until fixation of the allele A or a is the average time to reach one of the absorbing states $|A(t)| = N$ or $|A(t)| = 0$. We'll employ the method developed in Sect. A.13, most easily understood through a simple example expressed by the matrix P. The first step is to reorder the rows and columns so the absorbing states occupy the first two rows and columns of the reordered matrix P'.

$$P = \begin{bmatrix} 1 & 0.1 & 0 & 0 & 0 \\ 0 & 0.3 & 0.4 & 0.2 & 0 \\ 0 & 0.3 & 0.3 & 0.5 & 0 \\ 0 & 0.3 & 0.2 & 0.3 & 0 \\ 0 & 0 & 0.1 & 0 & 1 \end{bmatrix} \text{ gives } P' = \begin{bmatrix} 1 & 0 & 0.1 & 0 & 0 \\ 0 & 1 & 0 & 0.1 & 0 \\ 0 & 0 & 0.3 & 0.4 & 0.2 \\ 0 & 0 & 0.3 & 0.3 & 0.5 \\ 0 & 0 & 0.3 & 0.2 & 0.3 \end{bmatrix} \qquad (16.8)$$

The correspondence of the row (and column) entries of P and P' is

$$P \to P': \quad 1 \to 1, \quad 2 \to 3, \quad 3 \to 4, \quad 4 \to 5, \quad 5 \to 2$$

So, for example, $0.5 = P_{3,4} = P'_{4,5}$.

Now decompose the matrix P' into four blocks, Q, R, S, and 0. The matrix Q gives the transition probabilities between transient (i.e., non-absorbing) states, R the transition probabilities from transient states to absorbing states, S the transition probabilities among absorbing states, and 0 is a matrix of all 0s because there are no transitions from absorbing states to transient states. For this example we have

$$P' = \begin{bmatrix} S & R \\ 0 & Q \end{bmatrix} \text{ where } S = \begin{bmatrix} 1 & 0 \\ 0 & 1 \end{bmatrix}, \ R = \begin{bmatrix} 0.1 & 0 & 0 \\ 0 & 0.1 & 0 \end{bmatrix},$$

$$Q = \begin{bmatrix} 0.3 & 0.4 & 0.2 \\ 0.3 & 0.3 & 0.5 \\ 0.3 & 0.2 & 0.3 \end{bmatrix}, \text{ and } 0 = \begin{bmatrix} 0 & 0 \\ 0 & 0 \\ 0 & 0 \end{bmatrix}$$

In matrix P the states 2, 3, and 4 (states 3, 4, and 5 in P') are called transient because eventually the system described by this Markov chain will exit these states, never to return; the system will become fixed in state 1 or 5. In P' we see that Q is the transition matrix among the transient states. In Prop. A.13.1 we see that $Q^n \to 0$ as $n \to \infty$, that is, eventually the transient states empty out. This is no surprise. A consequence (Prop. A.13.2) is that $I - Q$ is invertible. Denote $(I - Q)^{-1}$ by N. Then Prop. A.13.4 shows that the expected number of steps before reaching one of the absorbing states when starting from transient state i is the ith entry of $\begin{bmatrix} 1 & 1 & 1 \end{bmatrix} N$. For the example we have

$$N = \begin{bmatrix} 5.2 & 4.267 & 4.533 \\ 4.8 & 5.733 & 5.467 \\ 3.6 & 3.467 & 4.933 \end{bmatrix} \text{ and } \begin{bmatrix} 1 & 1 & 1 \end{bmatrix} N = \begin{bmatrix} 13.6 & 13.47 & 14.93 \end{bmatrix}$$

That is, if initially we have 1, 2, or 3 copies of allele A, the expected time to fixation is 13.6, 13.5, and 14.9 iterates.

Another natural problem is to find the probability of being absorbed in a population of all a or a population of all A, given that we begin with 1, 2, or 3 copies of A. This is calculated in Prop. A.13.5: the probability of ending in the absorbing state i when starting from the initial state j is $[RN]_{i,j}$. In the example, we have

$$RN = \begin{bmatrix} 0.1 & 0 & 0 \\ 0 & 0.1 & 0 \end{bmatrix} \begin{bmatrix} 5.2 & 4.267 & 4.533 \\ 4.8 & 5.733 & 5.467 \\ 3.6 & 3.467 & 4.933 \end{bmatrix} = \begin{bmatrix} 0.52 & 0.427 & 0.453 \\ 0.48 & 0.573 & 0.547 \end{bmatrix}$$

So we see that starting from a single copy of A the system ends in all as with probability 0.52 and all As with probability 0.48. The highest probability of ending in all a occurs when starting from 1 copy of A (no surprise here), while the highest probability of ending in all A occurs when starting from 2 copies of A, not from 3 copies of A, and this is a bit surprising.

Though the Wright-Fisher model is quite simple, it does capture part of the dynamics of genetic drift. Some of the simplifying assumptions can be relaxed to more realistic conditions, without making much difference in the outcomes. Simple models often are effective for capturing the main features of a system. Wright-Fisher is an example.

Other sections of this chapter involve more complex models, but the simple Markov chain calculations we've done here will be useful in these settings. Its wide applicability is one of the most appealing features of mathematics.

Practice Problems

These are the matrices for the practice problems.

$$P_1 = \begin{bmatrix} 1 & 0 & 0.1 & 0 & 0 \\ 0 & 0.2 & 0.2 & 0.2 & 0 \\ 0 & 0.3 & 0.3 & 0.3 & 0 \\ 0 & 0.4 & 0.4 & 0.4 & 0 \\ 0 & 0.1 & 0 & 0.1 & 1 \end{bmatrix} \qquad P_2 = \begin{bmatrix} 1 & 0 & p & 0 \\ 0 & q & 0 & 0 \\ 0 & 1-2q & 1-p & 0 \\ 0 & q & 0 & 1 \end{bmatrix}$$

16.1.1. For the Wright-Fisher model determined by P_1,
(a) find the expected number of steps until fixation starting from 1, from 2, and from 3 copies of the allele A.
(b) From which initial state do we find a higher probability of ending with 0 copies of A?

16.1.2. For the Wright-Fisher model determined by P_2, with $0 < p < 1$ and $0 < q < 1/2$,
(a) find the expected number of steps until fixation starting from 1 and from 2 copies of the allele A. Call these $es_1(p,q)$ and $es_2(p,q)$.
(b) Set $q = 0.3$ and plot es_1 and es_2 for $0.1 \leq p \leq 0.9$. Find a numerical approximation for the point of intersection of these curves.
(c) Find the probability of fixing at $|A| = 0$ and at $|A| = 3$ when starting from $|A| = 1$. Find the value of q for which the probability of reaching each absorbing state is $1/2$.
(d) Find the probability of fixing at $|A| = 0$ and at $|A| = 3$ when starting from $|A| = 2$. Explain your result based on the form of the transition matrix P_2.

Practice Problem Solutions

16.1.1. First, convert P_1 to P_1' following the example of Eq. (16.8). This gives

$$P_1' = \begin{bmatrix} 1 & 0 & 0 & 0.5 & 0 \\ 0 & 1 & 0.1 & 0 & 0.1 \\ 0 & 0 & 0.2 & 0.2 & 0.2 \\ 0 & 0 & 0.3 & 0.3 & 0.3 \\ 0 & 0 & 0.4 & 0.4 & 0.4 \end{bmatrix} \quad R = \begin{bmatrix} 0 & 0.1 & 0 \\ 0.1 & 0 & 0.1 \end{bmatrix} \quad Q = \begin{bmatrix} 0.2 & 0.2 & 0.2 \\ 0.3 & 0.3 & 0.3 \\ 0.4 & 0.4 & 0.4 \end{bmatrix}$$

To answer (a) and (b) we'll apply Props. A.13.4 and A.13.5, and for these we need the matrices $N = (I - Q)^{-1}$ and RN.

$$N = \begin{bmatrix} 3 & 2 & 2 \\ 3 & 4 & 3 \\ 4 & 4 & 5 \end{bmatrix} \text{ and } RN = \begin{bmatrix} 0.3 & 0.4 & 0.3 \\ 0.7 & 0.6 & 0.7 \end{bmatrix}$$

(a) By Prop. A.13.4 we see that the expected number of steps to fixation from an initial distribution of $|A| = 1$, $|A| = 2$, and $|A| = 3$ is the first, second, and third entries of

$$\begin{bmatrix} 1 & 1 & 1 \end{bmatrix} N = \begin{bmatrix} 10 & 10 & 10 \end{bmatrix}$$

so 10 steps regardless of the (transient) initial condition. In general, this isn't expected, but perhaps it is a result of the constant rows in the Q matrix. Test this with an experiment of your design.
(b) To answer this question, we must be careful with interpreting the value of $|A|$ associated with each row. In matrix P_1, the first row corresponds to entering the $|A| = 0$ absorbing state, the fifth row to entering the $|A| = 4$ absorbing state. Then in matrices R and RN the first row corresponds to $|A| = 0$, the second row to $|A| = 4$. Then to fix on $|A| = 0$, we want to enter the first row of RN. Because $(RN)_{1,2} = 0.4 > 0.3 = (RN)_{1,1} = (RN)_{1,3}$, the higher probability of fixing with $|A| = 0$ is to start with $|A| = 2$.

16.1.2. First, convert P_2 to P_2' following the example of Eq. (16.8). This gives

$$P_2' = \begin{bmatrix} 1 & 0 & 0 & p \\ 0 & 1 & q & 0 \\ 0 & 0 & q & 0 \\ 0 & 0 & 1-2q & 1-p \end{bmatrix} \quad R = \begin{bmatrix} 0 & p \\ q & 0 \end{bmatrix} \quad Q = \begin{bmatrix} q & 0 \\ 1-2q & 1-p \end{bmatrix}$$

Then

$$N = (I - Q)^{-1} = \begin{bmatrix} \dfrac{1}{1-q} & 0 \\ \dfrac{1-2q}{p(1-q)} & \dfrac{1}{p} \end{bmatrix} \quad \begin{bmatrix} 1 & 1 \end{bmatrix} N = \begin{bmatrix} \dfrac{1+p-2q}{p(1-q)} & \dfrac{1}{p} \end{bmatrix}$$

and

$$RN = \begin{bmatrix} \dfrac{1-2q}{1-q} & 1 \\ \dfrac{q}{1-q} & 0 \end{bmatrix}$$

(a) By Prop. A.13.4 the expected number of steps to fixation are the entries of $\begin{bmatrix} 1 & 1 \end{bmatrix} N$. That is,

$$es_1(p, q) = \frac{1+p-2q}{p(1-q)} \qquad es_2(p, q) = \frac{1}{p}$$

(b) In Fig. 16.1 we plot es_1, labeled 1, and es_2, labeled 2. The curves intersect where $(1+p-0.6)/(0.7p) = 1/p$, that is, $p = 0.3$. Then $es_1(0.3,0.3) = es_2(0.3,0.3) = 10/3$.

(c) We'll use Prop. A.13.5. Consequently, the probability of fixing at $|A| = 0$ and at $|A| = 3$ when starting from $|A| = 1$ are the matrix entries $(RN)_{1,1}$ and $(RN)_{1,2}$. These are

$$(RN)_{1,1} = \frac{1-2q}{1-q} \quad \text{and} \quad (RN)_{1,2} = \frac{q}{1-q}$$

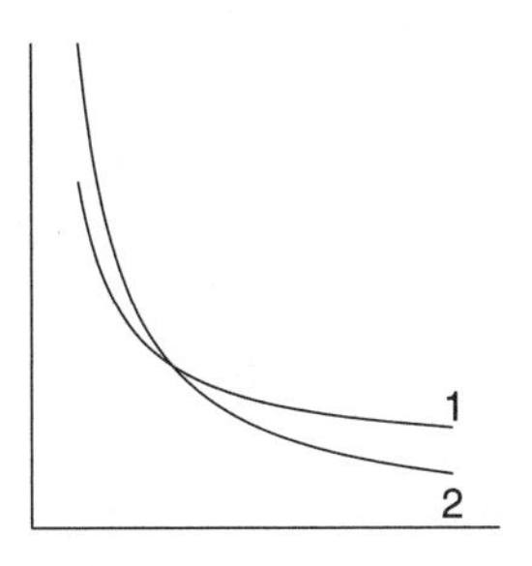

Figure 16.1. Plots of es_1 and es_2.

Solving $(RN)_{1,1} = 1/2$ gives $q = 1/3$. Not surprisingly, solving $(RN)_{1,2} = 1/2$ also gives $q = 1/3$. For $q < 1/3$, starting from $|A| = 1$ more likely leads to $|A| = 0$; for $1/3 < q < 1/2$, starting from $|A| = 1$ more likely leads to $|A| = 3$.

(d) The probability of fixing at $|A| = 0$ and at $|A| = 3$ when starting from $|A| = 2$ are the matrix entries $(RN)_{2,1} = 1$ and $(RN)_{2,2} = 0$. This last isn't a surprise: from $|A| = 2$ we can go to $|A| = 2$ and to $|A| = 0$, and nowhere else. There is no path from $|A| = 2$ to $|A| = 3$.

Exercises

Here are the matrices for Exercises 16.1.1 and 16.1.2.

$$P_3 = \begin{bmatrix} 1 & 0 & 0 & 0.1 & 0 \\ 0 & 0.2 & 0.3 & 0.2 & 0 \\ 0 & 0.2 & 0.4 & 0.3 & 0 \\ 0 & 0.5 & 0.3 & 0.4 & 0 \\ 0 & 0.1 & 0 & 0 & 1 \end{bmatrix} \qquad P_4 = \begin{bmatrix} 1 & 0 & 0 & 0.1 & 0 \\ 0 & 0.3 & 0.3 & 0.2 & 0 \\ 0 & 0.2 & 0.4 & 0.3 & 0 \\ 0 & 0.5 & 0.3 & 0.3 & 0 \\ 0 & 0 & 0 & 0.1 & 1 \end{bmatrix}$$

16.1.1. For the Wright-Fisher model determined by P_3,
(a) find the expected number of steps until fixation starting from 1, from 2, and from 3 copies of the allele A.
(b) From which initial state do we find a higher probability of ending with 0 copies of A?
(c) From which initial state do we find a higher probability of ending with 0 copies of a?

16.1.2. Repeat Exercise 16.1.1 using P_4.

Here are the matrices for Exercises 16.1.3 and 16.1.4. For P_5 take $0 \le b \le 0.4$; for P_6 take $0 \le b \le 0.3$.

$$P_5 = \begin{bmatrix} 1 & 0 & 0 & 0.1 & 0 \\ 0 & 0.2 & 0.3+b & 0.2 & 0 \\ 0 & 0.2 & 0.4-b & 0.3 & 0 \\ 0 & 0.5 & 0.3 & 0.4 & 0 \\ 0 & 0.1 & 0 & 0 & 1 \end{bmatrix} \quad P_6 = \begin{bmatrix} 1 & 0 & 0.1 & 0 & 0 \\ 0 & 0.3+b & 0.2 & 0.3 & 0 \\ 0 & 0.2 & 0.4 & 0.3+b & 0 \\ 0 & 0.5-b & 0.3 & 0.3-b & 0 \\ 0 & 0 & 0 & 0.1 & 1 \end{bmatrix}$$

16.1.3. For the Wright-Fisher model determined by P_5, find the number of initial copies of A, $1 \leq |A| \leq 3$, that give the maximum expected number of steps to fixation. Does this maximum increase or decrease as b increases?

16.1.4. For the Wright-Fisher model determined by P_6,
(a) compute the probability of fixing on $|A| = 0$ starting from $|A| = 1$.
(b) Compute the probability of fixing on $|A| = 4$ starting from $|A| = 1$.
(c) Plot these probabilities as functions of b for $0 \leq b \leq 0.3$.
(d) Find a numerical approximation for the value of b for which these probabilities are equal.

16.1.5. In the Wright-Fisher model, find the relation between N and $|A(t)|$ that gives $\sigma^2(|A(t+1)|) = 0.01|A(t)|$.

These are the matrices for Exercises 16.1.6–16.1.9. For P_7, $0 \leq p+q \leq 1$; for P_8, $0 \leq p \leq 0.8$; for P_9, $0 \leq p+q \leq 1$; and for P_{10}, $0 \leq p \leq 1$ and $0 \leq q \leq 1$.

$$P_7 = \begin{bmatrix} 1 & p & 0 \\ 0 & 1-p-q & 0 \\ 0 & q & 1 \end{bmatrix} \quad P_8 = \begin{bmatrix} 1 & 0.1 & 0.1 & 0 \\ 0 & 0.8-p & p & 0 \\ 0 & p & 0.8-p & 0 \\ 0 & 0.1 & 0.1 & 1 \end{bmatrix}$$

$$P_9 = \begin{bmatrix} 1 & p & 0 & 0 \\ 0 & q & 1-p-q & 0 \\ 0 & 1-p-q & p & 0 \\ 0 & 0 & q & 1 \end{bmatrix} \quad P_{10} = \begin{bmatrix} 1 & p & 0 & 0 \\ 0 & 0 & 1-q & 0 \\ 0 & 1-p & 0 & 0 \\ 0 & 0 & q & 1 \end{bmatrix}$$

16.1.6. (a) Draw the transition graph for P_7.
(b) Find the expected number of steps until fixation.
(c) Find the probability of being absorbed in $|A| = 0$ and in $|A| = 1$.

16.1.7. (a) Draw the transition graph for P_8.
(b) Find the expected number of steps until fixation starting from $|A(0)| = 1$ and from $|A(0)| = 2$.

(c) Find the probability of being absorbed in $|A| = 0$ and in $|A| = 3$ starting from $|A(0)| = 1$ and from $|A(0)| = 2$. Do you need to apply Prop. A.13.5 to answer this?

16.1.8. (a) Draw the transition graph for P_9.
(b) Find the expected number of steps until fixation starting from $|A(0)| = 1$ and from $|A(0)| = 2$.
(c) Find the probability of being absorbed in $|A| = 0$ starting from $|A(0)| = 1$ and from $|A(0)| = 2$. Which is larger?
(d) Plot $[RN]_{1,1}$ and $[RN]_{1,2}$ for $0 \le p \le 0.9$, $q = 0.1$, then for $0 \le p \le 0.5$, $q = 0.5$, and then for $p = 0.1$, $0 \le q \le 0.9$. What conclusions can you draw?

16.1.9. (a) Draw the transition graph for P_{10}.
(b) Find the expected number of steps until fixation starting from $|A(0)| = 1$ and from $|A(0)| = 2$.
(c) Find the relation between p and q that gives a value of ≥ 10 for the expected number of steps to fixation starting from $|A(0)| = 1$.
(d) Find the probability of being fixed at $|A| = 0$ starting from $|A(0)| = 1$ and from $|A(0)| = 2$. Which is larger?

These are the matrices for Exercise 16.1.10. Here $0 \le a \le 0.3$ and $0 \le b \le 0.3$.

$$P_{11} = \begin{bmatrix} 1 & a & a & a & 0 \\ 0 & 0.3-a & 0.3-a & 0.3-a & 0 \\ 0 & 0.3-b & 0.3-b & 0.3-b & 0 \\ 0 & 0.4 & 0.4 & 0.4 & 0 \\ 0 & b & b & b & 1 \end{bmatrix}$$

$$P_{12} = \begin{bmatrix} 1 & b & a & a & 0 \\ 0 & 0.3-a & 0.3-a & 0.3-a & 0 \\ 0 & 0.3-b & 0.3-b & 0.3-b & 0 \\ 0 & 0.4 & 0.4 & 0.4 & 0 \\ 0 & a & b & b & 1 \end{bmatrix}$$

16.1.10. (a) For P_{11} compute the probability of fixing at $|A| = 0$ and the probability of fixing at $|A| = 4$ from an initial distribution of states evenly spread among $|A(0)| = 1$, $|A(0)| = 2$, and $|A(0)| = 3$. If $a = 2b$, is the probability of fixing at $|A| = 0$ twice that of fixing at $|A| = 4$?
(b) For P_{12} compute the probability of fixing at $|A| = 0$ and the probability of fixing at $|A| = 4$ from an initial distribution of states evenly spread among $|A(0)| = 1$, $|A(0)| = 2$, and $|A(0)| = 3$. If $a = 2b$, is the probability of fixing at $|A| = 0$ twice that of fixing at $|A| = 4$?

16.2 MATRIX DIFFERENTIAL EQUATIONS

In Sects. 16.3 and 16.4 we study ion channels. There we'll find that the probabilities for transitions between channel conformational states vary with time. For the state transition matrix by $A(t)$, we'll encounter differential equations of the form

$$\frac{d}{dt}A(t) = BA(t) \tag{16.9}$$

where $A(t)$ and B are matrices and the entries of B are constants. In this section we'll focus on small matrices with integer entries so we can find simple solutions. Real applications are much messier.

First, recall from Example 4.6.3 that the solution of $dP/dt = \lambda P$ is $P(t) = e^{\lambda t}P(0)$. To apply this to the matrix differential equation (16.9), we must exponentiate a matrix. The Taylor series for e^t,

$$e^t = 1 + t + \frac{t^2}{2!} + \frac{t^3}{3!} + \frac{t^4}{4!} + \cdots$$

provides the idea. For an $n \times n$ matrix B, we'll define

$$e^B = I + B + \frac{1}{2!}B^2 + \frac{1}{3!}B^3 + \frac{1}{4!}B^4 + \cdots \tag{16.10}$$

Convergence of this series is addressed in Sect. A.12.1. Let's see how to compute the terms of the exponential of a matrix.

Example 16.2.1. *Exponentiating a matrix.* Find e^B where $B = \begin{bmatrix} 1 & 0 \\ 2 & 1 \end{bmatrix}$.

To apply the definition of Eq. (16.10), we'll need enough powers of B to see the pattern.

$$B^2 = \begin{bmatrix} 1 & 0 \\ 4 & 1 \end{bmatrix}, \; B^3 = \begin{bmatrix} 1 & 0 \\ 6 & 1 \end{bmatrix}, \; B^4 = \begin{bmatrix} 1 & 0 \\ 8 & 1 \end{bmatrix}, \; \ldots\text{and } B^n = \begin{bmatrix} 1 & 0 \\ 2n & 1 \end{bmatrix}$$

Then

$$e^B = \begin{bmatrix} 1 & 0 \\ 0 & 1 \end{bmatrix} + \begin{bmatrix} 1 & 0 \\ 2 & 1 \end{bmatrix} + \frac{1}{2!}\begin{bmatrix} 1 & 0 \\ 4 & 1 \end{bmatrix} + \frac{1}{3!}\begin{bmatrix} 1 & 0 \\ 6 & 1 \end{bmatrix} + \frac{1}{4!}\begin{bmatrix} 1 & 0 \\ 8 & 1 \end{bmatrix} + \cdots$$

So we see

$$(e^B)_{11} = (e^B)_{22} = 1 + 1 + \frac{1}{2!} + \frac{1}{3!} + \frac{1}{4!} + \cdots = e^1 = e, \quad (e^B)_{10} = 0$$

$$(e^B)_{21} = 0 + 2 + \frac{4}{2!} + \frac{6}{3!} + \frac{8}{4!} + \cdots = 2\left(1 + \frac{2}{2!} + \frac{3}{3!} + \frac{4}{4!} + \cdots\right)$$

$$= 2\left(1 + 1 + \frac{1}{2!} + \frac{1}{3!} + \cdots\right) = 2e$$

From this we deduce that

$$e^B = \begin{bmatrix} e & 0 \\ 2e & e \end{bmatrix}$$

To construct the series for each term of the matrix is straightforward, though possibly tedious if we can't easily find a pattern for the entries of the powers of B. More difficult is to find simple expressions for the sums of the series. Here, as with most things, practice is our friend. □

From this example you might guess $e^{[B_{ij}]} = [B_{ij}e]$, but do you really think it's always that simple? Of course not. For example, you can adapt the calculation above to show that

$$\text{for } B = \begin{bmatrix} 1 & 1 \\ 1 & 1 \end{bmatrix} \text{ we have } e^B = \frac{1}{2}\begin{bmatrix} 1+e^2 & -1+e^2 \\ -1+e^2 & 1+e^2 \end{bmatrix}$$

This is a good check that you understand the method, and it's a reminder of the skill, developed in Sect. 10.9, of interpreting some infinite series as polynomials in e.

Now we'll see how to use the exponential of a matrix to solve a matrix differential equation.

Example 16.2.2. *Matrix differential equation solutions.* Solve

$$\frac{d}{dt}A(t) = BA(t) \quad \text{where } B = \begin{bmatrix} 1 & 0 \\ 2 & 1 \end{bmatrix} \text{ and } A(0) = \begin{bmatrix} 1 & 1 \\ 1 & 0 \end{bmatrix} \tag{16.11}$$

Reasoning by analogy, which has worked for us before and which is an acceptable approach because we always can check our solution, we guess that the solution of Eq. (16.11) is $A(t) = e^{Bt}A(0)$ where the $A(0)$ factor is included because when $t = 0$ we have $e^{Bt} = I$.

Following the argument of Example 16.2.1 we have

$$(e^{Bt})_{11} = (e^{Bt})_{22} = 1 + t + \frac{t^2}{2!} + \frac{t^3}{3!} + \frac{t^4}{4!} + \cdots = e^t, \quad (e^{Bt})_{10} = 0$$

$$(e^{Bt})_{21} = 0 + 2t + \frac{4t^2}{2!} + \frac{6t^3}{3!} + \frac{8t^4}{4!} + \cdots = 2t\left(1 + \frac{2t}{2!} + \frac{3t^2}{3!} + \frac{4t^3}{4!} + \cdots\right)$$

$$= 2t\left(1 + t + \frac{t^2}{2!} + \frac{t^3}{3!} + \cdots\right) = 2te^t$$

That is, we think the solution is

$$A(t) = \begin{bmatrix} e^t & 0 \\ 2te^t & e^t \end{bmatrix} A(0) = \begin{bmatrix} e^t & 0 \\ 2te^t & e^t \end{bmatrix}\begin{bmatrix} 1 & 1 \\ 1 & 0 \end{bmatrix} = \begin{bmatrix} e^t & e^t \\ 2te^t + e^t & 2te^t \end{bmatrix}$$

Let's check that this is a solution of Eq. (16.11). That it satisfies the initial condition is clear from the first equality in the line above. Next, we'll check that the proposed solution satisfies the differential equation. First, we'll compute the derivative $A'(t)$, and then we'll compute $BA(t)$. If $A(t)$ is a solution, these will be equal.

$$A(t)' = \begin{bmatrix} e^t & e^t \\ 2te^t + e^t & 2te^t \end{bmatrix}' = \begin{bmatrix} e^t & e^t \\ 2te^t + 2e^t + e^t & 2te^t + 2e^t \end{bmatrix}$$

Now let's compare this expression for $A'(t)$ with the right side of $dA/dt = BA$, keeping in mind that in general matrix multiplication is not commutative. (A lot of the physical world isn't commutative: think of the two operations "put on your socks" and "put on your shoes.")

$$BA(t) = \begin{bmatrix} 1 & 0 \\ 2 & 1 \end{bmatrix} \begin{bmatrix} e^t & e^t \\ 2te^t + e^t & 2te^t \end{bmatrix} = \begin{bmatrix} e^t & e^t \\ 2e^t + 2te^t + e^t & 2e^t + 2te^t \end{bmatrix} \qquad \square$$

Summarizing, the matrix differential equation

$$\frac{d}{dt}A(t) = BA(t) \tag{16.12}$$

has solution

$$A(t) = e^{Bt}A(0) \tag{16.13}$$

Practice Problems

16.2.1. Solve $dA(t)/dt = BA(t)$ for $B = \begin{bmatrix} 2 & 1 \\ 2 & 1 \end{bmatrix}$ and $A(0) = \begin{bmatrix} 1 & 1 \\ 2 & 1 \end{bmatrix}$.

16.2.2. Solve $dA(t)/dt = BA(t)$ for $B = \begin{bmatrix} 0 & 1 \\ 1 & 1 \end{bmatrix}$ and $A(0) = \begin{bmatrix} 0 & 1 \\ 1 & 2 \end{bmatrix}$.

Practice Problem Solutions

16.2.1. From Eq. (16.13) we know the solution has the form $e^{Bt}C$. To find e^{Bt} we first need powers of B:

$$B^2 = \begin{bmatrix} 6 & 3 \\ 6 & 3 \end{bmatrix}, B^3 = \begin{bmatrix} 18 & 9 \\ 18 & 9 \end{bmatrix}, B^4 = \begin{bmatrix} 54 & 27 \\ 54 & 27 \end{bmatrix}, \dots, B^n = \begin{bmatrix} 2\cdot 3^{n-1} & 3^{n-1} \\ 2\cdot 3^{n-1} & 3^{n-1} \end{bmatrix}$$

Now we need to find a pattern in

$$e^{Bt} = I + Bt + \frac{1}{2!}B^2t^2 + \frac{1}{3!}B^3t^3 + \frac{1}{4!}B^4t^4 + \cdots$$

Substituting in the values of B^n we find

$$(e^{Bt})_{11} = 1 + 2t + 6t^2/2! + 18t^3/3! + 54t^4/4! + \cdots$$
$$(e^{Bt})_{12} = t + 3t^2/2! + 9t^3/3! + 27t^4/4! + \cdots$$
$$(e^{Bt})_{21} = 2t + 6t^2/2! + 18t^3/3! + 54t^4/4! + \cdots$$
$$(e^{Bt})_{22} = 1 + t + 3t^2/2! + 9t^3/3! + 27t^4/4! + \cdots$$

Can we find patterns and recognize these matrix entries as exponentials? We'll apply some of the skills we developed in Sect. 10.9.

$$\begin{aligned}(e^{Bt})_{11} &= 1 + 2t + \frac{6t^2}{2!} + \frac{18t^3}{3!} + \frac{54t^4}{4!} + \cdots \\ &= 1 + 2\left(t + \frac{3t^2}{2!} + \frac{3^2t^3}{3!} + \frac{3^3t^4}{4!} + \cdots\right) \\ &= 1 + \frac{2}{3}\left(3t + \frac{(3t)^2}{2!} + \frac{(3t)^3}{3!} + \frac{(3t)^4}{4!} + \cdots\right) \\ &= 1 - \frac{2}{3} + \frac{2}{3}\left(1 + 3t + \frac{(3t)^2}{2!} + \frac{(3t)^3}{3!} + \frac{(3t)^4}{4!} + \cdots\right) = \frac{1}{3} + \frac{2}{3}e^{3t}\end{aligned}$$

Here we factored out $1/3$ in the third equality to match the exponents of 3 with those of t. In the fourth equality we've added 1 inside the brackets to make the bracketed terms the Taylor series for e^{3t}. Because we add 1 inside the brackets, we must subtract $2/3$ outside the brackets. Similar calculations give the other elements of e^{Bt}, and we have

$$e^{Bt} = \begin{bmatrix} \frac{1}{3} + \frac{2}{3}e^{3t} & -\frac{1}{3} + \frac{1}{3}e^{3t} \\ -\frac{2}{3} + \frac{2}{3}e^{3t} & \frac{2}{3} + \frac{1}{3}e^{3t} \end{bmatrix}$$

From here we find the general solution $A(t) = e^{Bt}A(0)$. That is,

$$A(t) = \begin{bmatrix} -\frac{1}{3} + \frac{4}{3}e^{3t} & e^{3t} \\ \frac{2}{3} + \frac{4}{3}e^{3t} & e^{3t} \end{bmatrix}$$

We won't check that this $A(t)$ satisfies the differential equation, but you might try the exercise. Use the solution-check portion of Example 16.2.2 as a template. Watching the unwanted terms subtract out can be very satisfying.

16.2.2. Again we use Eq. (16.13), so we know the solution has the form $e^{Bt}A(0)$. To find e^{Bt} we first need powers of B:

$$B^2 = \begin{bmatrix} 1 & 1 \\ 1 & 2 \end{bmatrix}, \; B^3 = \begin{bmatrix} 1 & 2 \\ 2 & 3 \end{bmatrix}, \; B^4 = \begin{bmatrix} 2 & 3 \\ 3 & 5 \end{bmatrix}, \; B^5 = \begin{bmatrix} 3 & 5 \\ 5 & 8 \end{bmatrix}, \; B^6 = \begin{bmatrix} 5 & 8 \\ 8 & 13 \end{bmatrix},$$

Now we need to find a pattern in

$$e^{Bt} = I + Bt + \frac{1}{2!}B^2t^2 + \frac{1}{3!}B^3t^3 + \frac{1}{4!}B^4t^4 + \cdots$$

Here are the first few terms in the series expansion of each matrix entry

$$(e^{Bt})_{11} = 1 + \frac{t^2}{2!} + \frac{t^3}{3!} + \frac{2t^4}{4!} + \frac{3t^5}{5!} + \frac{5t^6}{6!} + \dots$$

$$(e^{Bt})_{12} = (e^{Bt})_{21} = t + \frac{t^2}{2!} + \frac{2t^3}{3!} + \frac{3t^4}{4!} + \frac{5t^5}{5!} + \frac{8t^6}{6!} + \dots$$

$$(e^{Bt})_{22} = 1 + t + \frac{2t^2}{2!} + \frac{3t^3}{3!} + \frac{5t^4}{4!} + \frac{8t^5}{5!} + \frac{13t^6}{6!} + \cdots$$

Starting with the t^2 term of $(e^{Bt})_{11}$, and with the first term of the other three $(e^{Bt})_{ij}$, the coefficients are successive Fibonacci numbers, the sequence

$$f_1 = 1,\ f_2 = 1,\ f_3 = 2,\ f_4 = 3,\ f_5 = 5,\ f_6 = 8, f_7 = 13,\ \dots$$

that is, $f_1 = f_2 = 1$ and for $n \geq 3$, $f_n = f_{n-1} + f_{n-2}$. Although this sequence was known in India for over 1500 years prior to discovery by its namesake, in the west it appeared first in *Liber Abaci*, written in 1202 by Leonardo of Pisa, now known as Fibonacci. The sequence is a very early model of rabbit populations. Suppose you have a pair of newborn rabbits, one female and one male, who at age two months become reproductively active, producing one pair of newborns (always one female and one male) every month. If no rabbits die, how many rabbits do you have at the end of a year?

Perhaps surprisingly, there is a formula for f_n that depends only on n:

$$f_n = \frac{\varphi_+^n - \varphi_-^n}{\sqrt{5}} \tag{16.14}$$

where $\varphi_+ = (1+\sqrt{5})/2$ and $\varphi_- = (1-\sqrt{5})/2$. If this is unfamiliar, you can verify that this formula reproduces as many Fibonacci numbers as you wish (if you want a derivation, Google is your friend).

We'll use this property of the Fibonacci numbers to compute $(e^{Bt})_{11}$

$$\begin{aligned}
(e^{Bt})_{11} &= 1 + f_1\frac{t^2}{2} + f_2\frac{t^3}{3!} + f_3\frac{t^4}{4!} + \cdots \\
&= 1 + \frac{\varphi_+ - \varphi_-}{\sqrt{5}}\frac{t^2}{2} + \frac{\varphi_+^2 - \varphi_-^2}{\sqrt{5}}\frac{t^3}{3!} + \frac{\varphi_+^3 - \varphi_-^3}{\sqrt{5}}\frac{t^4}{4!} + \cdots \\
&= 1 + \frac{1}{\varphi_+\sqrt{5}}\left(\frac{(\varphi_+ t)^2}{2} + \frac{(\varphi_+ t)^3}{3!} + \cdots\right) \qquad \text{(a)}
\end{aligned}$$

$$
\begin{aligned}
&- \frac{1}{\varphi_- \sqrt{5}} \left(\frac{(\varphi_- t)^2}{2} + \frac{(\varphi_- t)^3}{3!} + \cdots \right) && \text{(b)} \\
&= 1 + \frac{1}{\varphi_+ \sqrt{5}} e^{\varphi_+ t} - \frac{1}{\varphi_- \sqrt{5}} e^{\varphi_- t} && \text{(c)} \\
&- \frac{1}{\varphi_+ \sqrt{5}} \left(1 + \varphi_+ t \right) - \frac{-1}{\varphi_- \sqrt{5}} \left(1 + \varphi_- t \right) && \text{(d)} \\
&= \frac{1}{\varphi_+ \sqrt{5}} e^{\varphi_+ t} - \frac{1}{\varphi_- \sqrt{5}} e^{\varphi_- t}
\end{aligned}
$$

To convert the bracketed terms in (a) and (b) into the exponentials of (c) we add in the terms subtracted out in (d). Substituting in the expressions for φ_+ and φ_- and simplifying, the terms in (d) sum to -1, which explains the last equality in the sequence.

Similar calculations give

$$
(e^{Bt})_{12} = (e^{Bt})_{21} = \frac{1}{\sqrt{5}} e^{\varphi_+ t} - \frac{1}{\sqrt{5}} e^{\varphi_- t}, \quad (e^{Bt})_{22} = \frac{\varphi_+}{\sqrt{5}} e^{\varphi_+ t} - \frac{\varphi_-}{\sqrt{5}} e^{\varphi_- t}
$$

Substitute in the expressions for the terms of e^{Bt} and multiply by $A(0)$. This shows that $A(t)$ is

$$
\begin{bmatrix}
\dfrac{e^{\varphi_+} - e^{\varphi_-}}{\sqrt{5}} & \dfrac{(5+3\sqrt{5})e^{\varphi_+} + (5-3\sqrt{5})e^{\varphi_-}}{10} \\
\dfrac{(1+\sqrt{5})e^{\varphi_+} + (-1+\sqrt{5})e^{\varphi_-}}{2\sqrt{5}} & \dfrac{(2+\sqrt{5})e^{\varphi_+} + (-2+\sqrt{5})e^{\varphi_-}}{\sqrt{5}}
\end{bmatrix}
$$

This last problem points out something interesting, or discouraging. Finding the series expansion for the terms of e^{Bt} is straightforward, if we can find a pattern in the powers of B. But finding more compact expressions for these terms can be much trickier. Some special cases are straightforward (See Exercise 16.2.5, for example.) Here we relied upon recognizing that the numerators of the coefficients are Fibonacci numbers (these are well known, so this isn't such a stretch), then knowing Eq. (16.14) and figuring out how to use it. Sometimes we might not be able to go beyond the series expansions. It just depends on what we know and what we recognize. But then, much of science works this way.

Exercises

16.2.1. Find e^B for these B. Hint: $(2^3 - 1)/3! = 2^3/3! - 1/3!$.

$$
\text{(a)} \begin{bmatrix} 0 & 1 \\ 1 & 0 \end{bmatrix} \quad \text{(b)} \begin{bmatrix} 1 & 2 \\ 0 & 1 \end{bmatrix} \quad \text{(c)} \begin{bmatrix} 1 & 1 \\ 2 & 2 \end{bmatrix} \quad \text{(d)} \begin{bmatrix} 1 & 3 \\ 1 & 3 \end{bmatrix} \quad \text{(e)} \begin{bmatrix} 1 & 4 \\ 1 & 4 \end{bmatrix} \quad \text{(f)} \begin{bmatrix} 1 & 0 \\ 1 & 2 \end{bmatrix}
$$

16.2.2. Solve $dA(t)/dt = BA(t)$ for these B with $A(0) = \begin{bmatrix} 1 & 1 \\ 0 & 2 \end{bmatrix}$.

$$\text{(a)} \begin{bmatrix} 1 & 3 \\ 0 & 1 \end{bmatrix} \text{(b)} \begin{bmatrix} 1 & 0 \\ 4 & 1 \end{bmatrix} \text{(c)} \begin{bmatrix} 1 & 1 \\ 1 & 1 \end{bmatrix} \text{(d)} \begin{bmatrix} 1 & 2 \\ 1 & 2 \end{bmatrix} \text{(e)} \begin{bmatrix} 3 & 3 \\ 1 & 1 \end{bmatrix} \text{(f)} \begin{bmatrix} 1 & 1 \\ 3 & 3 \end{bmatrix}$$

16.2.3. Find e^B for these B.

$$\text{(a)} \begin{bmatrix} 1 & 0 & 1 \\ 0 & 1 & 1 \\ 0 & 0 & 1 \end{bmatrix} \text{(b)} \begin{bmatrix} 1 & 0 & 1 \\ 0 & 1 & 0 \\ 1 & 0 & 1 \end{bmatrix} \text{(c)} \begin{bmatrix} 1 & 1 & 0 \\ 0 & 1 & 0 \\ 0 & 1 & 1 \end{bmatrix} \text{(d)} \begin{bmatrix} 1 & 0 & 1 \\ 1 & 0 & 0 \\ 1 & 0 & 1 \end{bmatrix}$$

16.2.4. Solve $\dfrac{dA(t)}{dt} = BA(t)$ for these B with $A(0) = \begin{bmatrix} 1 & 1 & 0 \\ 1 & 1 & 1 \\ 0 & 0 & 2 \end{bmatrix}$. Recall that $1+2+\cdots+n = n(n+1)/2$. Also, if Fibonacci numbers appear as coefficients, don't forget the method introduced in Practice Problem 16.2.2.

$$\text{(a)} \begin{bmatrix} 1 & 1 & 0 \\ 0 & 1 & 1 \\ 0 & 0 & 1 \end{bmatrix} \text{(b)} \begin{bmatrix} 1 & 1 & 0 \\ 0 & 1 & 0 \\ 0 & 0 & 1 \end{bmatrix} \text{(c)} \begin{bmatrix} 1 & 0 & 1 \\ 0 & 1 & 0 \\ 1 & 0 & 0 \end{bmatrix} \text{(d)} \begin{bmatrix} 1 & 0 & 1 \\ 0 & 1 & 1 \\ 1 & 0 & 1 \end{bmatrix}$$

16.2.5. Compute e^B for these matrices.

$$B = \begin{bmatrix} a & 0 \\ 0 & b \end{bmatrix} \quad B = \begin{bmatrix} a & 0 & 0 \\ 0 & b & 0 \\ 0 & 0 & c \end{bmatrix}$$

Do you see a pattern?

16.2.6. For the matrices B of Exercises 16.2.1 (b) and (c) and 16.2.3 (b) and (c), compute $\det(e^B)$ and $e^{\text{tr}(B)}$. What do you see? This result is the reason that e^B is invertible for any square matrix B, the matrix version of why $e^x \neq 0$ for any real x.

16.2.7. Suppose A and B are $n \times n$ matrices that commute, that is, $AB = BA$.
(a) Write the series expansion for e^A, e^B, and e^{A+B}.
(b) Multiply the series expansions for e^A and e^B. In this product, group together the terms of the same total power of A and B. (For example, A^3, A^2B, AB^2, and B^3 are the cubic terms.)
(c) Deduce that $e^{A+B} = e^A e^B$ if A and B commute.

16.2.8. For the matrices

$$A = \begin{bmatrix} 1 & 0 \\ 1 & 1 \end{bmatrix} \qquad B = \begin{bmatrix} 0 & 1 \\ 1 & 1 \end{bmatrix}$$

(a) verify that $AB \neq BA$.
(b) Compute e^A and e^B.
(c) Compute e^{A+B}.
(d) Show $e^A e^B \neq e^{A+B}$.
Hints: Some of these calculations are in the practice problems and earlier exercises. In order to show that two matrices are not equal, it suffices to show that they differ in at least one corresponding entry. So to do part (d), you need not compute the whole product $e^A e^B$. If the $(1,1)$ entry of $e^A e^B$ differs from the $(1,1)$ entry of e^{A+B}, you're finished.

16.2.9. Suppose $A' = BA$. If we double every entry of $A(0)$, do we double every entry of $A(t)$ for all $t \geq 0$? Give a reason if you think this is true; find a counterexample if you think it is false.

16.2.10. For the differential equation $A' = BA$, suppose $A_1(t)$ is the solution for the initial condition $A(0)$ and $A_2(t)$ is the solution for the initial condition $A(0)^{\text{tr}}$, the transpose of $A(0)$. Does it follow that $A_2(t) = A_1(t)^{\text{tr}}$ for all $t \geq 0$? Give a reason if you think this is true; find a counterexample if you think it is false.

If you want more problems, one point requires some care. Writing these problems is easy; it's a different matter to find a B matrix that gives recognizable patterns of coefficients of e^{Bt}. But you might try this: start with A, compute A', then find the matrix B that satisfies $A' = BA$. If A is invertible, which we can guarantee because we select A, then $B = A'A^{-1}$. But before you think this is simple, remember that B must be a constant matrix.

16.3 ION CHANNEL DYNAMICS

Now we'll use Markov chains to model the dynamics of ion flow through transmembrane channels. Specifically, we'll study the *dwell times*, successive times the channel remains open or closed before switching. The next state of the channel is determined by its present state and some environmental factors. The current flow through a single channel (yes, this can be measured directly by the patch clamp technique, described briefly along with some background membrane biology in Sect. A.14) is sampled at high frequency to make likely that at most one transition occurs between successive measurements.

Whether a channel is open or closed depends on the configuration of channel molecules, which are arranged in a local energy minimum until environmental changes or thermal fluctuations push them to another local energy minimum. The probability of switching configurations depends on only the height of the energy barrier separating the minima, and not on the sequence of configurations that lead to the current configuration. This is the justification for the use of Markov chains to model the transitions.

The transition matrix is $P = [p_{ij}]$ where $p_{ij} = P(S_j \to S_i)$. Because transitions depend on only the current state, successive transitions are independent of one another. Then by Eq. (11.6), the multiplication rule for independent events,

$$P(S_k \to S_j \to S_i) = P(S_k \to S_j) \cdot P(S_j \to S_i) = p_{jk}p_{ij} = p_{ij}p_{jk}$$

Summing over all intermediate states S_j,

$$P(S_k \to S_i \text{ in 2 steps}) = p_{i1}p_{1k} + p_{i2}p_{2k} + \cdots + p_{in}p_{nk} = (P^2)_{ik}$$

where n is the number of states. This might remind you of Eq. (14.7).

If the channel is in state S_j, the probability of remaining in S_j for one time step, that is, being in S_j for *two* consecutive time steps, is p_{jj}. The probability of remaining in S_j for two time steps (that is, being in S_j for three consecutive time steps) is p_{jj}^2, not $(P^2)_{jj}$, because remaining in S_j corresponds to the single path $S_j \to S_j \to S_j$, and not to the sum over all paths $S_j \to S_k \to S_j$, which is $(P^2)_{jj}$.

The probability of switching from S_j to any other state in one time step is $1 - p_{jj}$.

The probability of staying in S_j for k time steps and then switching out of S_j, that is, the probability of having a dwell time of k in S_j is

$$p_{jj}^{k-1}(1 - p_{jj})$$

So we see that the probability of residing in any state for k time steps decreases exponentially with k. Because typically $p_{jj}^{k-1} \neq p_{ii}^{k-1}$ for $i \neq j$, the residence time histograms for the membrane current in a single-channel patch clamp should be fit by a sum of exponentials, one for each channel configuration. These exponentials reveal themselves on different time scales. For example, one configuration contributes most of the short open time events, another most of the long open time events.

Some early explorations of membrane channel opening and closing involved the nAChR (nicotinic acetylcholine receptor) in a neuromuscular junction of frog muscles. Experimenters studied the effects of *agonists*, chemicals that bind to cell membrane receptors and influence the configuration of a channel. They found

1. for low agonist concentration, the membrane current varies as (agonist concentration)^2 [192], and
2. dwell times exhibit bursts [78].

Observation (1) suggests that increasing the channel opening probability requires a pair of agonists. Observation (2) suggests that the dwell times are not well fit by a single exponential.

In [78] David Colquhoun and Bert Sakmann report their measurements of the dwell times, varying the membrane potential and the concentration and type of agonist. They found the closed state dwell times are fit by three exponentials and the open state dwell times by two. Building on [54, 190], Colquhoun and Sakmann's model has closed states with 0, 1, or 2 agonists bound, and open states with 1 or 2 agonists bound. We'll use a notation different from theirs: for us the states $O1$ and $O2$ represent open channels with 1 and 2 bound agonists; the states $C0$, $C1$, and $C2$ represent closed channels with 0, 1, and 2 bound agonists. Because there is no evidence that a channel with 0 bound agonists can open, the state $O0$ is not included in the model. Fig. 16.2 is the transition graph of the model.

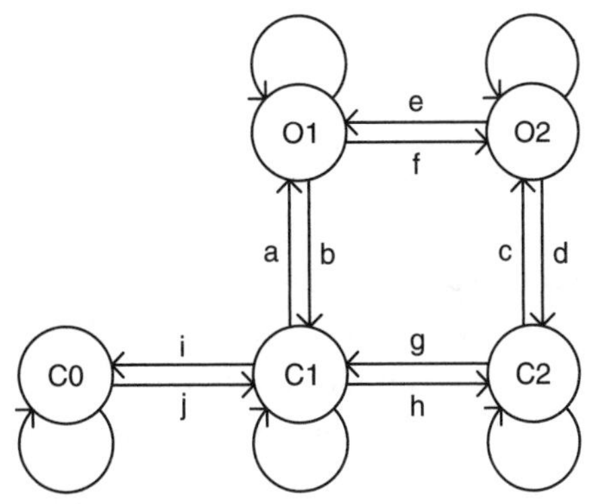

Figure 16.2. The Colquhoun-Sakmann transition graph.

The binding of agonists doesn't open or close channels but can alter the probability that a channel will open or close. Patch clamp measurements give the rate of transitions, represented by the matrix B of Eq. (16.15). States 1 through 5 are $O1$, $O2$, $C2$, $C1$, and $C0$. Let's investigate the entries in the first column of B. The entry B_{21} is the rate of moving from state 1 to state 2, which depends on a rate constant f and χ, the agonist concentration, because this transition involves adding an agonist. The entry B_{41} is the rate of going from $O1$ to $C1$, that is, the rate, b, of an open channel with one bound agonist closing. Because transitions from 1 to 3 and to 5 cannot occur (for example, going from open with one bound agonist to closed with two bound agonists requires two transitions), $B_{31} = B_{51} = 0$. Then $B_{11} = -(b + f\chi)$ is the rate of leaving state 1. The other entries of B are explained similarly, but we'll mention B_{12}, the rate of going from state 2 to state 1. Specifically, why don't we need a factor of the agonist concentration χ? That's because for this transition the channel sheds an agonist. To add an agonist the channel must encounter an agonist, which depends on χ; to shed an agonist, the concentration χ is not a significant factor.

$$B = \begin{bmatrix} -(b+f\chi) & e & 0 & a & 0 \\ f\chi & -(e+d) & c & 0 & 0 \\ 0 & d & -(c+g) & h\chi & 0 \\ b & 0 & g & -(a+h\chi+i) & j\chi \\ 0 & 0 & 0 & i & -j\chi \end{bmatrix} \tag{16.15}$$

To see the relation between B and the Markov chain transition matrix P, we'll argue by analogy. Recall that if r is the per capita growth rate of x, then $dx/dt = rx$. The corresponding matrix equation is

$$\frac{dP}{dt} = BP$$

We learned how to solve matrix differential equations in Sect. 16.2, specifically, Eqs. (16.12) and (16.13):

$$P(t) = e^{Bt}P(0) \tag{16.16}$$

where

$$e^{Bt} = I + Bt + \frac{B^2t^2}{2!} + \frac{B^3t^3}{3!} + \cdots \tag{16.17}$$

In Sect. 16.2 we were able to find patterns for the entries of B^n and from these a simple expression for e^{Bt}. Simple patterns do not present themselves for the matrix B of the Colquhoun-Sakmann model. Compute B^2 and B^3 if you wish. And maybe you'll find a pattern I didn't.

Usually we're stuck with using the first few terms of the series expansion Eq. (16.17) to approximate the transition matrix.

Now suppose a membrane contains of N channels, opening and closing independently of one another. Measure the membrane current M times, obtaining values $I_1, \dots, I_M$. The expected value and the variance are

$$\mathbb{E}(I) = \frac{1}{M}\sum_j I_j \qquad \sigma^2 = \left(\frac{1}{M}\sum_j I_j^2\right) - \mathbb{E}(I)^2$$

Now suppose p is the probability that a single channel will be open when it is measured, and suppose i is the current that flows through an open channel. Then in M measurements, the channel should be open about $k = pM$ times, and so the expected current through an individual channel is $\mathbb{E}(i) = pi$. If k channels are open, then $i_j = i$ for k measurements and 0 otherwise. Then the variance in M measurements of a single channel is

$$\sigma^2(i) = \frac{1}{M}\sum_j i_j^2 - \mathbb{E}(i)^2 = \frac{1}{M}ki^2 - (pi)^2 = \frac{1}{M}pMi^2 - (pi)^2 = p(1-p)i^2$$

If each of the N channels of the membrane behave this way, the expected value of the membrane current is

$$\mathbb{E}(I) = \mathbb{E}(i_1 + \cdots + i_N) = \mathbb{E}(i_1) + \cdots + \mathbb{E}(i_N) = Npi$$

(we've used Prop. 11.6.1(c)) and because individual channels are independent, the variance is

$$\begin{aligned}\sigma^2(I) &= \sigma^2\left(\sum_j i_j\right) = \sum_j \sigma^2(i_j) \quad \text{by Prop. 11.6.2}\\ &= \sum_j p(1-p)i^2 = Np(1-p)i^2 = Npi(1-p)i = \mathbb{E}(I)(1-p)i\end{aligned}$$

That is, the current flow through a single open channel is

$$i = \frac{\sigma^2(I)}{\mathbb{E}(I)(1-p)} \tag{16.18}$$

At first glance this may not seem impressive, but look more closely. The current, i, through a single channel is determined by the expected value and the variance of the membrane current, macroscopic parameters, measurable long before the patch clamp was developed. There is the probability, p, that a single channel will be open. True, we can't measure this without the patch clamp, but with the voltage clamp, available since the late 1940s, we can set the membrane potential so most channels are closed at any time. Then $1-p \approx 1$ and so i can be calculated in terms of I. This calculation was done in 1973 by C. Anderson and C. Stevens [11], before the patch clamp was developed in the late 1970s. Stephen Ellner and John Guckenheimer [107], page 92, described it this way:

> This was science at its best: data plus models plus theory led to predictions that were confirmed by new experiments that directly tested the predictions.

Practice Problems

16.3.1. For a channel with two open states (states 1 and 2) and two closed states (states 3 and 4) suppose the state transition rates are

$$B = \begin{bmatrix} -0.2 & 0.1 & 0.1 & 0 \\ 0.1 & -0.2 & 0 & 0 \\ 0.1 & 0.1 & -0.2 & 0.1 \\ 0 & 0 & 0.1 & -0.1 \end{bmatrix}$$

(a) For the approximation $P = e^{Bt} \approx I + tB$ with $t = 0.1$, show that $I + tB$ is a stochastic matrix. Draw the transition graph and deduce that $(I + tB)^3$ is a positive matrix, and consequently the dominant eigenvalue of $I + tB$ is $\lambda = 1$.
(b) For the approximation $P = e^{Bt} \approx I + tB$ with $t = 0.1$, find the unit right eigenvector of the eigenvalue $\lambda = 1$.
(c) For the approximation $P = e^{Bt} \approx I + tB + (t^2/2)B^2$ with $t = 0.1$, draw the transition graph and find the unit right eigenvector of the eigenvalue $\lambda = 1$.
(d) Interpret the long-term state distribution using this eigenvector.
(e) Comment on the appropriateness of the linear approximation $e^{Bt} \approx I + tB$.

16.3.2. Denote by $\mathbb{E}(I)$ and $\sigma^2(I)$ the expected value and variance of the membrane current I. Does the current in an individual channel increase or decrease with $\sigma^2(I)$, with $\mathbb{E}(I)$, and with p?

Practice Problem Solutions

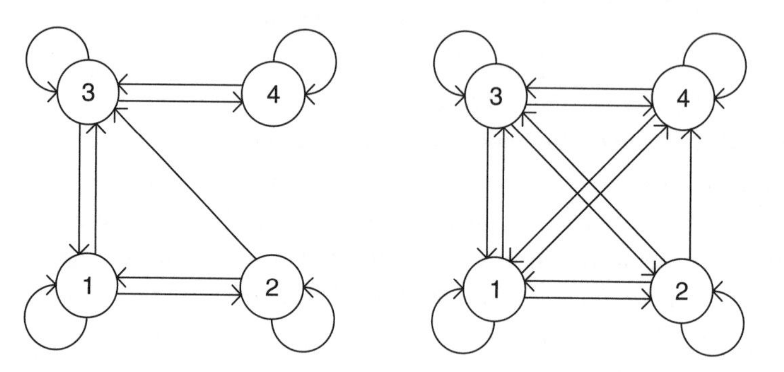

Figure 16.3. Left: transition graph for $I + tB$. Right: $I + tB + (t^2/2)B^2$.

16.3.1. For $t = 0.1$ the matrices $I + tB$ and $I + tB + (t^2/2)B^2$ are

$$I + tB = \begin{bmatrix} 0.98 & 0.01 & 0.01 & 0 \\ 0.01 & 0.98 & 0 & 0 \\ 0.01 & 0.01 & 0.98 & 0.01 \\ 0 & 0 & 0.01 & 0.99 \end{bmatrix}$$

$$I + tB + \frac{t^2}{2}B^2 = \begin{bmatrix} 0.9803 & 0.00985 & 0.0098 & 0.00005 \\ 0.0098 & 0.98025 & 0.00005 & 0 \\ 0.00985 & 0.00985 & 0.9803 & 0.00985 \\ 0.00005 & 0.00005 & 0.00985 & 0.9901 \end{bmatrix}$$

(a) All entries of $I + tB$ with $t = 0.1$ are non-negative, and the entries of each column sum to 1. The transition graph is the left graph of Fig. 16.3. From this graph we see that the missing transitions are $1 \to 4$, $2 \to 4$, $3 \to 2$, $4 \to 1$, and $4 \to 2$. We can achieve each of these transitions through longer paths:

$$1 \to 3 \to 4,\ 2 \to 3 \to 4,\ 3 \to 1 \to 2,\ 4 \to 3 \to 1, \text{ and } 4 \to 3 \to 1 \to 2$$

We see that we can get from 4 to 2 in three steps. The loops $1 \to 1$, $2 \to 2$, $3 \to 3$, and $4 \to 4$ allow us to get from any state to any state in three steps. For example $1 \to 1 \to 1 \to 2$. Consequently, every entry of $(I + tB)^3$ is positive. By the Perron-Frobenius theorem, the largest eigenvalue of $I + tB$ is 1.
(b) The unit right eigenvector of $\lambda = 1$ is the transpose of

$$\begin{bmatrix} 2/9 & 1/9 & 1/3 & 1/3 \end{bmatrix}$$

(c) The transition graph for $I + tB + (t^2/2)B^2$ is the right graph of Fig. 16.3. The unit right eigenvector of $\lambda = 1$ is the vector we found in (b).
(d) The right unit eigenvector of the eigenvalue $\lambda = 1$ gives the stable distribution of the channel among its states. We see that when the channel is open, it is twice as likely to be in state 2 as in state 1, that when the channel is closed it is equally likely to be in states 3 and 4, and that the channel is twice as likely to be closed as open.
(e) Although the matrices $I + tB$ and $I + tB + (t^2/2)B^2$ are a bit different, these differences are small. More importantly, their right unit eigenvectors are identical, so both make the same prediction about the long-term distribution among the four states. The linear approximation seems adequate.

16.3.2. A straightforward approach is to compute the partial derivatives using Eq. (16.18)

$$\frac{\partial i}{\partial \sigma^2(I)} = \frac{1}{\mathbb{E}(I)(1-p)} > 0 \qquad \frac{\partial i}{\partial \mathbb{E}(I)} = -\frac{\sigma^2(I)}{\mathbb{E}^2(I)(1-p)} < 0$$

$$\frac{\partial i}{\partial p} = \frac{\sigma^2(I)}{\mathbb{E}(I)(1-p)^2} > 0$$

So we see that i decreases with $\mathbb{E}(I)$ and increases with $\sigma^2(I)$ and p.

Exercises

Use these matrices for Exercises 16.3.1 and 16.3.2.

$$B_1 = \begin{bmatrix} -0.2 & 0.1 & 0.1 & 0 \\ 0.1 & -0.3 & 0.1 & 0 \\ 0.1 & 0.1 & -0.3 & 0.1 \\ 0 & 0.1 & 0.1 & -0.1 \end{bmatrix} \quad B_2 = \begin{bmatrix} -0.2 & 0.2 & 0.2 & 0 \\ 0.1 & -0.3 & 0 & 0 \\ 0.1 & 0.1 & -0.3 & 0.1 \\ 0 & 0 & 0.1 & -0.1 \end{bmatrix}$$

16.3.1. (a) For $t = 0.1$ find the unit right eigenvector of $\lambda = 1$ for $I + tB_1$.
(b) Repeat (a) for $t = 0.2$ and for $t = 1$.
(c) Describe the effect of t on the time spent among the four states.

16.3.2. (a) For $t = 0.1$ find the unit right eigenvector of $\lambda = 1$ for $I + tB_2$.
(b) Repeat (a) for $t = 0.2$ and for $t = 1$.
(c) Describe the effect of t on the time spent among the four states.

Use these matrices for Exercises 16.3.3 and 16.3.4.

$$B_3 = \begin{bmatrix} -0.1 & 0 & 0 & \alpha \\ 0.1 & -0.1 & 0 & \beta \\ 0 & 0.1 & -0.1 & \gamma \\ 0 & 0 & 0.1 & -0.1 \end{bmatrix} \quad B_4 = \begin{bmatrix} -0.2 & 0 & 0 & 0 \\ 0 & -0.2 & 0 & 0.1 \\ 0 & 0.1 & -0.1 & 0.1 \\ 0.2 & 0.1 & 0.1 & -0.2 \end{bmatrix}$$

16.3.3. Find the unit right eigenvector of $\lambda = 1$ for $I + 0.1B_3$ when $(\alpha, \beta, \gamma) =$ (a) $(0.1, 0, 0)$, (b) $(0, 0.1, 0)$, and (c) $(0, 0, 0.1)$.
(d) Explain the final distributions. Sketch the transition graph of $I + 0.1B_3$.

16.3.4. For $P = I + 0.1B_4$ compute the expected number of steps to reach one of the absorbing states. Recall Prop. A.13.4. Comment on the likelihood that evolution would produce transient states in membrane channels.

Use these matrices for Exercises 16.3.5 and 16.3.6.

$$B_5 = \begin{bmatrix} -0.3 & 0.1 & 0.1 & 0.1 \\ 0.1 & -0.2 & 0 & 0.1 \\ 0.1 & 0 & -0.2 & 0.1 \\ 0.1 & 0.1 & 0.1 & -0.3 \end{bmatrix} \quad B_6 = \begin{bmatrix} -0.1 & 0 & 0 & 0.1 \\ 0 & -0.1 & 0.1 & 0 \\ 0 & 0.1 & -0.1 & 0 \\ 0.1 & 0 & 0 & -0.1 \end{bmatrix}$$

16.3.5. (a) Find the unit right eigenvector of $\lambda = 1$ for $P = I + 0.1B_5$. Find the long-term distribution of the channel among these states.
(b) Change the $(1, 3)$ entry from 0.1 to 0.15 and the $(4, 3)$ entry from 0.1 to 0.05. Find the right unit eigenvector of $\lambda = 1$, and comment on the long-term distribution of the channel among these states.
(c) In B_5 change the $(2, 4)$ entry from 0.1 to 0. Show the largest eigenvalue is not 1. Why doesn't this contradict the Perron-Frobenius theorem? Okay, it's a theorem so it can't be contradicted. The better question is, why doesn't the Perron-Frobenius theorem apply in this situation?

16.3.6. (a) Find the eigenvalues of $P = I + 0.1B_6$. Note that $\lambda = 1$ occurs twice. Find the right unit eigenvectors for each value of $\lambda = 1$. Why can't the Perron-Frobenius theorem be applied to this matrix?
(b) Change the $(1,1)$ entry to -0.2 and the $(4,1)$ entry to 0.2. How does this change the probability of the channel winding up in each state?
(c) In B_6 change the $(1,4)$ and $(2,3)$ entries to 0.09 and change the $(2,4)$ and $(4,3)$ entries to 0.001. Find the right unit eigenvector of $\lambda = 1$. Why does this eigenvalue occur only once? How has the probability of the channel being in each state changed from the answer in (a)?

16.3.7. Suppose we measure the expected value $\mathbb{E}(I)$ of the total current through a membrane. In Practice Problem 16.3.2 we saw how to estimate the single-channel current i by holding p close to 0 with a voltage clamp, described in Sect. A.14 but maybe you can guess what it does. Using this, and setting p close to 1 with a voltage clamp, show how to estimate N, the number of membrane channels. How can we use the results of the $p \approx 0$ voltage clamp as part of a calculation with a $p \approx 1$ voltage clamp?

16.3.8. From the calculation of Practice Problem 16.3.2, we see that the expected current through an open channel increases with the variance of the membrane current. Find a simple explanation of this effect. (Not necessarily the true explanation. Just find something simple.)

Use these matrices for Exercises 16.3.9 and 16.3.10.

$$P_1 = \begin{bmatrix} 0.9 & 0.2 & 0 \\ 0.1 & 0.6 & 0.1 \\ 0 & 0.2 & 0.9 \end{bmatrix} \qquad P_2 = \begin{bmatrix} 0.5 & 0.2 & 0.2 \\ 0.4 & 0.6 & 0.2 \\ 0.1 & 0.2 & 0.6 \end{bmatrix}$$

16.3.9. Here we consider the channel with transition matrix P_1. States 1 and 2 are closed; state 3 is open.
(a) Show the stable distribution in states 1, 2, and 3 is $2/5$, $1/5$, and $2/5$.
(b) Collapse the system into two states by combining the two closed states to a single state. Find the transition matrix for this 2-state system. Hint: among the closed states, the probability of being in state 1 is $P(1|C) = (2/5)/(2/5+1/5) = 2/3$. And $P(C \to C) =$

$$P(1|C)\cdot(P(1\to 1)+P(1\to 2)) + P(2|C)\cdot(P(2\to 1)+P(2\to 2))$$

(c) Find the stable distribution of this 2-state system.

16.3.10. Here we consider the channel with transition matrix P_2. States 1 and 2 are closed; state 3 is open.
(a) Find the stable distribution.
(b) Find the transition matrix for the 2-state system formed by combining the two closed states into a single closed state.
(c) Find the stable distribution of this 2-state system.
(d) In this and the previous exercise, show that the sum of the probabilities of being in closed states found in (a) equals the probability of being in a closed state in (c). Do you think this always is the case? (Then how can we tell the difference? If it truly is a single closed state, the occupancy histogram is well fit by a single exponential; two states are better fit by two exponentials.)

16.4 THE CLANCY-RUDY CARDIAC ARRHYTHMIA MODEL

In this section we'll study a model of cardiac sodium channels developed to identify a mutation responsible for type 3 long QT syndrome (LQT3). Of the 15 known LQT types, LQT3 is most likely to result in lethal cardiac events [280]. To understand LQT3, we'll start with some basics of ECG plots [5].

An electrocardiogram (ECG or EKG) is a plot of time (horizontal axis) against voltage (vertical axis) measured by electrodes attached across the patient's body. Around 1895, Dutch physiologist Willem Einthoven developed a device to measure cardiac currents, assigned the letters P, Q, R, S, and T to components of the ECG waveform, and described how some cardiovascular diseases alter the standard waveform. For this work he was awarded the 1924 Nobel Prize in Physiology or Medicine.

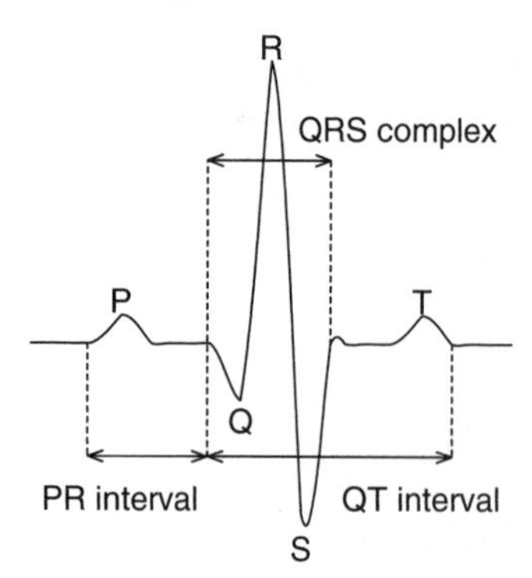

Figure 16.4. ECG components.

The resting potential of the interior of a cardiac cell is negative relative to the exterior of the cell. *Depolarization* of a cell occurs when sodium and calcium channels open, allowing these positively charged ions to flow into the cell and thus making the membrane potential less negative. Depolarization causes the cell to contract. The increased positive charge inside the cell causes the potassium channels to open, initiating *repolarization.* The cell interior returns to its resting potential and the cell relaxes. Throughout this process, sodium-potassium pumps in the cell membrane adjust the concentrations of these ions to prepare for the next depolarization.

As we mentioned in Sect. 2.5, usually, a heartbeat is triggered by an electrical impulse from the sinoatrial (SA) node, a group of cells in the wall of the right atrium. The SA node is the heart's principal pacemaker. These impulses depolarize both atria, causing them to contract and pump blood from atria to ventricles.

In the ECG:

- Atrial depolarization appears as a P wave. With about a 100 msec delay, allowing the atria to drain into the ventricles, the SA impulse causes the atrioventricular node (AV) to send an impulse to the ventricles, which causes the ventricles to depolarize and contract.
- Ventricular depolarization appears as the QRS complex. Then the ventricles and the atria repolarize and relax, ready for the next heartbeat.
- Ventricular repolarization appears as the T wave. Atrial repolarization occurs at the same time, but is masked by the larger ventricular repolarization wave.

The SCN5A gene includes instructions for the construction of proteins in sodium channels of cardiac cells. The LQT3 mutation of this gene has been found in patients suffering from Brugada syndrome, progressive familial heart block, LQ syndrome type 3, and sick sinus syndrome [134]. In [68] Colleen Clancy and Yoram Rudy built a Markov chain model of sodium channels to study the effect of the ΔKPQ mutation that influences the fast inactivation of the sodium channel, the response to depolarization.

The sodium channel, which is voltage-gated, consists of four domains, each containing six transmembrane spanning segments. The fourth segment of each domain is the voltage sensor. The linker between domains III and IV is responsible for fast inactivation of the channel by blocking the channel opening inside the cell. The ΔKPQ mutation is a deletion of the amino acids Lys-1505, Pro-1506, and Gln-1507 in the III-IV linker, so this mutation disables fast inactivation of the sodium channel. The clinical manifestation is a significant prolongation of the QT interval.

In Fig. 16.5 we show the transition graphs of the Clancy-Rudy model for the wild type and the ΔKPQ mutant sodium channel. Double-headed arrows stand for two arrows, one in each direction. To further simplify the diagram, the arrow from each state to itself has been omitted.

The wild type model has three closed states C_1, C_2, and C_3. (The closed time histograms are not well fit by one or two exponentials, but are by three, signaling three closed states, one for each configuration.) In addition, the wild type model has an open state O, and fast and slow inactivation states IF and IS.

The mutant model has two gating modes, *background* and *burst*. In background mode, the mutant channel behaves similarly to the wild type channel,

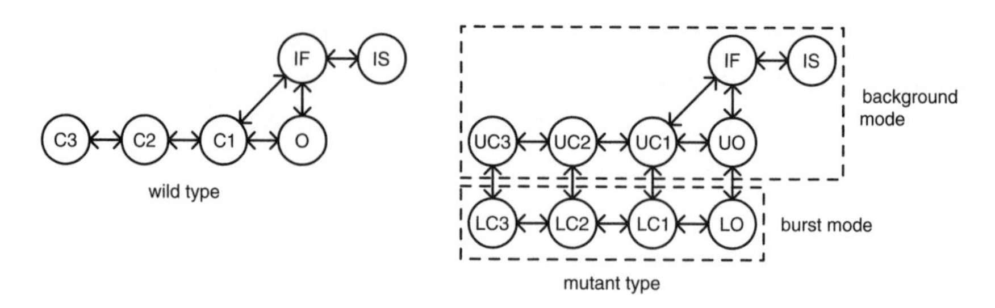

Figure 16.5. The Clancy-Rudy Markov models for the wild type sodium channel (left) and the ΔKPQ mutant channel (right).

except that the mutant channels activate more quickly. The strong coupling between the activated and inactivated states implies that the higher probability of activation leads to faster inactivation in the mutant channel. Also, the current in mutant channels decays faster than that in wild type channels. Similarly, the higher rate constants in mutant channels mean these recover from inactivation more rapidly than do wild type channels. This leads to a dispersal of mutant background channel re-openings.

In the mutant burst mode, the channel cannot inactivate. Transition rates between burst modes and background modes are very low, so it is unlikely for a channel to enter the burst mode, but once it does, it spends a lot of time switching between $LC3$, $LC2$, $LC1$, and LO. The frequent openings are the bursts giving this mode its name. Together with the dispersal of background channel re-openings, these bursts produce a late sodium current.

Model validation is established by the agreement of experimental and simulated data for the wild type and for both the background and burst modes of the mutant type. To test whether the ΔKPQ mutation can cause arrhythmias, Clancy and Rudy needed to see the effect of this sodium channel mutation on a cardiac cell. This they did by inserting their mutant sodium channel in the Luo-Rudy [218, 219, 399] model of a cardiac ventricular cell. Comparing a Luo-Rudy cell with wild type sodium channels to one with ΔKPQ mutant sodium channels, Clancy and Rudy found that the mutant cells exhibit faster depolarization and longer-lived action potential and consequently an elongated QT interval. Also, they observed an abnormal depolarization waveform called *early after-depolarization*, also a source of arrhythmia.

Because most individuals are heterozygous for LQT3, transcription of both wild type and ΔKPQ mutant SCN5A likely will occur. Clancy and Rudy ran simulations of the Luo-Rudy ventricular cell with different fractions of mutant sodium channels. They found that the action potential duration increases with the fraction of mutant channels, and that when half the sodium channels have the ΔKPQ mutation, they observed early after-depolarizations caused by frequent abortive action potentials before normal repolarization is complete. Clancy and Rudy concluded that this mutation can cause LQT3.

This was the first step in using a mathematical model of an ion channel to link channel mutations to cardiac arrhythmias. The work continues. See, for example, [65, 66, 69, 70, 71, 213, 312, 341, 355], among others. In [72] Clancy, Rudy, and Zheng Zhu apply their Markov chain model of cardiac cell sodium channels to test the relative therapeutic effects of mexiletine and lidocaine on removing early after-depolarization seen in LQT3. In [73] Lindsay Clegg and Feilim Mac Gabhann survey applications of computational models to drug development. These are fascinating ways for math and medicine to talk with one another.

In [67] Clancy and Robert Kass develop a Markov chain model of the neuronal sodium channel $Na_V1.1$ to study ways gene defects can alter normal channel gating. Some forms of epilepsy are the result of ion channel mutations; Clancy and Kass's goal is to understand the molecular mechanisms that underlie the epilepsy and use this to develop targeted drug interventions.

An advantage of Clancy's approach, which she calls "Math instead of mice: computational approaches to reveal mechanisms of excitable diseases," is that many experimental parameters can be varied by modifying the model code. In addition to cardiac tissue, Clancy and her colleagues model the hippocampus in order to study the pathological synchronization of neural nets that lead to epileptic seizures.

A 2011 report [144] in *Scientific American* described work by Clancy and her colleagues [239] to code a virtual heart, an aggregate of thousands of model ventricular cells with sodium channels modified to represent arrhythmia. Experimental data on the effects of anti-arrhythmia drugs guided the model integration of virtual anti-arrhythmics. The drugs under study, flecainide and lidocaine, both reduced tachycardia in single heart cells, both in vitro and in simulations. But in virtual heart simulations, flecainide produced side effects that doubled the likelihood of fatal arrhythmia, an occurrence observed in clinical trials run before the development of these computer models. The heart is a complex network of cells working together. Little surprise that the network can exhibit dynamics unseen in an individual cell. Soon, virtual heart trials may precede human trials.

This is quite a way to go from a model of sodium channels to study arrhythmia. I think we can agree that math and medicine aren't just talking now. They're dancing.

Much more modestly, in the practice problems and exercises we'll explore some simple variations of the Clancy-Rudy model, both wild type and mutant, and investigate the effects of changes in the transition probabilities.

Practice Problems

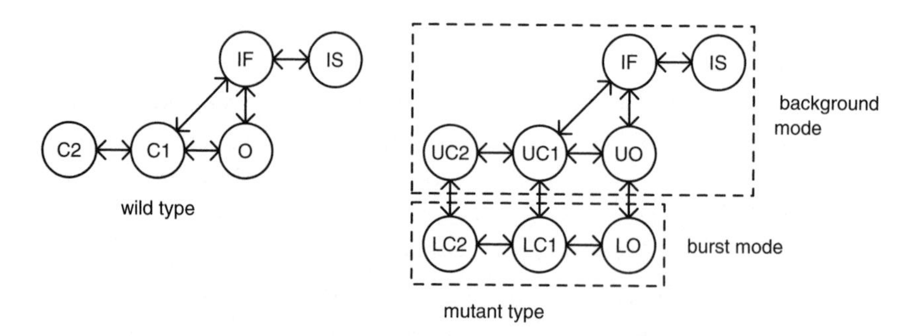

Figure 16.6. Practice Problem 16.4.1 and 16.4.2 graphs. Loops from a state to itself are omitted to simplify the diagram.

16.4.1. For the wild type channel of Fig. 16.6, suppose these are the non-zero transition probabilities:

$$
\begin{array}{lll}
P(C2 \to C2) = 0.8 & P(C2 \to C1) = 0.2 & P(C1 \to C2) = 0.5 \\
P(C1 \to C1) = 0.2 & P(C1 \to O) = 0.2 & P(C1 \to IF) = 0.1 \\
P(O \to C1) = 0.2 & P(O \to O) = 0.7 & P(O \to IF) = 0.1 \\
P(IF \to C1) = 0.2 & P(IF \to O) = 0.1 & P(IF \to IF) = 0.6 \\
P(IF \to IS) = 0.1 & P(IS \to IF) = 0.1 & P(IS \to IS) = 0.9
\end{array}
$$

(a) Write the transition matrix P. Verify that P^3 is a positive matrix. Can you deduce this directly from the transition graph?
(b) Find the unit eigenvector of the eigenvalue $\lambda = 1$. Find the fraction of time this channel will be open, closed, and inactivated.
(c) Write a simulation based on the transition matrix P. Compare the residence times of the simulation with the fractions obtained in (b).

16.4.2. Now consider the mutant type channel of Fig. 16.6 with this transition matrix, where the rows and columns are labeled in this order: $LC2$, $LC1$, LO, $UC2$, $UC1$, UO, IF, and IS.

$$P = \begin{bmatrix} 0.79 & 0.5 & 0 & 0.01 & 0 & 0 & 0 & 0 \\ 0.2 & 0.29 & 0.2 & 0 & 0.01 & 0 & 0 & 0 \\ 0 & 0.2 & 0.79 & 0 & 0 & 0.01 & 0 & 0 \\ 0.01 & 0 & 0 & 0.79 & 0.5 & 0 & 0 & 0 \\ 0 & 0.01 & 0 & 0.2 & 0.19 & 0.2 & 0.2 & 0 \\ 0 & 0 & 0.01 & 0 & 0.2 & 0.69 & 0.1 & 0 \\ 0 & 0 & 0 & 0 & 0.1 & 0.1 & 0.6 & 0.1 \\ 0 & 0 & 0 & 0 & 0 & 0.1 & 0.1 & 0.9 \end{bmatrix}$$

So, for example, $P_{32} = 0.1 = P(LC1 \to LO)$. (In the Clancy-Rudy model, these transition probabilities are voltage dependent. For example, Clancy and Rudy set $P_{76} = 9.178e^{-v/20.3}$. Our practice problems and exercises are intended to explore the connection between network topology and the distribution of times spent in each state, so we ignore voltage dependence.)

(a) Find an n for which P^n is a positive matrix.

(b) Find the fraction of time that this channel spends in each of these eight states.

(c) Compare this distribution with that of Practice Problem 16.4.1. In particular, comment on the effect of the relatively small transition probabilities between the L states and the U states.

Practice Problem Solutions

16.4.1. (a) The transition matrix P is on the left; P^3 is on the right. The rows and columns are ordered this way: $C2$, $C1$, O, IF, and IS.

$$\begin{bmatrix} 0.8 & 0.5 & 0 & 0 & 0 \\ 0.2 & 0.2 & 0.2 & 0.2 & 0 \\ 0 & 0.2 & 0.7 & 0.1 & 0 \\ 0 & 0.1 & 0.1 & 0.6 & 0.1 \\ 0 & 0 & 0 & 0.1 & 0.9 \end{bmatrix} \quad \begin{bmatrix} 0.692 & 0.5 & 0.18 & 0.17 & 0.01 \\ 0.2 & 0.198 & 0.198 & 0.17 & 0.036 \\ 0.07 & 0.183 & 0.433 & 0.195 & 0.026 \\ 0.036 & 0.1 & 0.165 & 0.29 & 0.175 \\ 0.002 & 0.019 & 0.024 & 0.175 & 0.753 \end{bmatrix}$$

From the transition graph we see that the states $C2$ and IS are the most widely separated; the shortest path between them is $C2 \to C1 \to IF \to IS$. The loop at each state means we can go from any state to any other state in 3 transitions. For example, $C3 \to C3 \to C3 \to C1$.

(b) The unit eigenvector of the eigenvalue $\lambda = 1$ is (decimal and rational)

$$\vec{v} = \begin{bmatrix} 0.444444\ldots \\ 0.177777\ldots \\ 0.155555\ldots \\ 0.111111\ldots \\ 0.111111\ldots \end{bmatrix} = \begin{bmatrix} 4/9 \\ 8/45 \\ 7/45 \\ 1/9 \\ 1/9 \end{bmatrix}$$

Some of these equivalents are clear. If $0.155555\cdots = 7/45$ isn't clear, we can see this with a geometric series

$$0.155555\cdots = \frac{1}{10} + \frac{5}{10^2} + \frac{5}{10^3} + \frac{5}{10^4} + \cdots$$
$$= \frac{1}{10} + \frac{5}{10^2}\left(1 + \frac{1}{10} + \frac{1}{10^2} + \cdots\right) = \frac{1}{10} + \frac{5}{10^2}\left(\frac{1}{1-1/10}\right) = \frac{7}{45}$$

We see that on average the channel will be inactive about $2/9$ of the time, open about $7/45$ of the time, and closed about $28/45$ of the time.
(c) Using the code in Sect. B.21, we find these fractions,

$$0.44554,\ 0.17937,\ 0.15735,\ 0.10688,\ 0.11086$$

which are pretty close to the results from (b).

16.4.2. (a) In the transition graph we see that the most distant states are $LC2$ and IS, joined by a path of length 4. Because there is a loop from each state to itself (in the transition matrix this is exhibited by each diagonal entry being non-zero) we see that we can get from any state to any other state in 4 steps. Consequently, every entry of P^4 is positive and so by the Perron-Frobenius theorem the dominant eigenvalue is $\lambda = 1$.
(b) The unit eigenvector of $\lambda = 1$ is $\vec{v}^{\,tr}$, where

$$\vec{v} \approx \begin{bmatrix} 0.2434 & 0.0972 & 0.0968 & 0.2496 & 0.1000 & 0.0878 & 0.0626 & 0.0626 \end{bmatrix}$$

(c) The relatively low transition probabilities for $UC2 \leftrightarrow LC2$, $UC1 \leftrightarrow LC1$, and $UO \leftrightarrow LO$ mean that transitions between the burst mode and the background mode are unlikely. On short time scales, the burst mode and background mode are uncoupled. The first three states of $\vec{v}$ make up the burst mode, the last six the background mode. Let's approximately normalize these subsets by dividing by the sum of their entries:

$$\vec{w}_a = \begin{bmatrix} v_1/va \\ v_2/va \\ v_3/va \end{bmatrix} = \begin{bmatrix} 0.4494 \\ 0.4385 \\ 0.1121 \end{bmatrix} \quad \text{and} \quad \vec{w}_b = \begin{bmatrix} v_4/vb \\ v_5/vb \\ v_6/vb \\ v_7/vb \\ v_8/vb \end{bmatrix} = \begin{bmatrix} 0.4437 \\ 0.1777 \\ 0.1561 \\ 0.1113 \\ 0.1113 \end{bmatrix}$$

where $va = v_1 + v_2 + v_3$ and $vb = v_4 + v_5 + v_6 + v_7 + v_8$. We see that $\vec{w}_b$ and $\vec{v}$ of Practice Problem Solution 16.4.1(b) are similar.

With what should we compare $\vec{w}_a$? Decoupling the burst mode from the background mode, we obtain

$$\begin{bmatrix} 0.8 & 0.5 & 0 \\ 0.2 & 0.3 & 0.2 \\ 0 & 0.2 & 0.8 \end{bmatrix} \text{ with } \lambda = 1 \text{ and unit eigenvector } \vec{v}_a = \begin{bmatrix} 0.5555 \\ 0.2222 \\ 0.2222 \end{bmatrix}$$

and we see that $\vec{v}_a$ is not so close to $\vec{w}_a$. At least for this example we have an oddly asymmetrical situation: the fraction of the total background time spent in each background state is close to the stable population of the uncoupled background matrix, but even this weak coupling between the burst and background states perturbs the fraction of total burst time spent in each burst state. Also note that about 44% of the time the channel is in the burst mode, so this mutation-generated mode will have a considerable effect on the behavior of the channel.

In the exercises we'll investigate the effects of modifying the transition probabilities of the mutant channel of Practice Problem 16.4.2, and the rate at which the simulation of Practice Problem 16.4.1 (c) converges to the unit eigenvector populations.

Exercises

In Exercises 16.4.1 through 16.4.5 we'll investigate the effect of weakening the connection between the burst mode and the background mode. That is, in the matrix P of Practice Problem 16.4.2 we'll change P_{11}, P_{41}, P_{22}, P_{52}, P_{33}, P_{63}, P_{14}, P_{44}, P_{25}, P_{55}, P_{36}, and P_{66}.

16.4.1. Change P_{41}, P_{52}, P_{63}, P_{14}, P_{25}, and P_{36} to 0.001 and make the corresponding increases to $P_{11}, \dots, P_{66}$. How does the unit eigenvector of $\lambda = 1$ change from that found in Practice Problem 16.4.2?

16.4.2. Return to the matrix P of Practice Problem 16.4.2. Change P_{52}, P_{63}, P_{25}, and P_{36} to 0 and make the corresponding changes to P_{22}, P_{33}, P_{55}, and P_{66}. How does the unit eigenvector of $\lambda = 1$ change from that found in Practice Problem 16.4.2?

16.4.3. Return to the matrix P of Practice Problem 16.4.2. Change P_{52}, P_{63}, P_{14} and P_{25} to 0 and make the corresponding changes to P_{22}, P_{33}, P_{44}, and P_{55}. How does the unit eigenvector of $\lambda = 1$ change from that found in Practice Problem 16.4.2?

16.4.4. Return to the matrix P of Practice Problem 16.4.2. Change P_{41}, P_{52}, P_{63}, P_{14}, P_{25}, and P_{36} to 0.1 and make the corresponding increases to $P_{11}, \dots, P_{66}$.

How does the unit eigenvector of $\lambda = 1$ change from that found in Practice Problem 16.4.2?

16.4.5. Interpret the results of problems 16.4.1 through 16.4.4 to comment on the effects of varying the coupling between the burst mode and the background mode of the mutant channel model.

16.4.6. Return to the matrix P of Practice Problem 16.4.2. Set the state self-couplings (the diagonals of the transition matrix) close to 1 and all the other non-zero entries to 0.01. That is, $P_{11} = P_{33} = P_{44} = 0.98$, $P_{22} = P_{66} = P_{77} = 0.97$, $P_{55} = 0.96$, $P_{88} = 0.99$, and $P_{21} = P_{41} = P_{12} = P_{32} = P_{52} = P_{23} = P_{63} = P_{14} = P_{54} = P_{25} = P_{45} = P_{65} = P_{75} = P_{36} = P_{56} = P_{76} = P_{57} = P_{67} = P_{87} = P_{78} = 0.01$. How does the unit eigenvector of $\lambda = 1$ change from that found in Practice Problem 16.4.2? Interpret your result.

16.4.7. For any value of a, $0 < a < 0.24$, find the unit eigenvalue of $\lambda = 1$ for this transition matrix.

$$P = \begin{bmatrix} 1-2a & a & 0 & a & 0 & 0 & 0 & 0 \\ a & 1-3a & a & 0 & a & 0 & 0 & 0 \\ 0 & a & 1-2a & 0 & 0 & a & 0 & 0 \\ a & 0 & 0 & 1-2a & a & 0 & 0 & 0 \\ 0 & a & 0 & a & 1-4a & a & a & 0 \\ 0 & 0 & a & 0 & a & 1-3a & a & 0 \\ 0 & 0 & 0 & 0 & a & a & 1-3a & a \\ 0 & 0 & 0 & 0 & 0 & 0 & a & 1-a \end{bmatrix}$$

Interpret the result. How sensitive is this result to equal probabilities for all off-diagonal allowed transitions? For example, what happens if you change P_{12} to 0 and P_{22} to $1-2a$?

16.4.8. Return to the model of Practice Problem 16.4.1 (c). Here we'll investigate how the state distribution generated by the simulation depends on the duration of the simulation. Before we start, we must take care of one tricky point. In Exercise 16.4.9 we'll explore the effect of different random number strings. Here, we're interested in the effect of the duration of the simulation, not in the particular random number sequence. In the code of Sect. B.21, immediately before the line $s = Random[Integer, \{1, 5\}]$; insert the line $SeedRandom[1]$;.

(a) Find the fraction of time in each of the five states for $num = 500, 1000, 5000, 10000, 50000$, and 100000.

(b) Comment on the effects of simulation duration.

16.4.9. Still with the model of Practice Problem 16.4.1 (c), here we'll investigate how the state distribution generated by the simulation depends on the random number sequence. With $num = 100000$, find the fraction of time in each of the five states for *SeedRandom*[2], *SeedRandom*[3], *SeedRandom*[4], *SeedRandom*[5], and *SeedRandom*[314], and for others if you wish. Comment on the effects of the random number seed.

16.4.10. Modify the code of Practice Problem 16.4.1 (c) to the model of Practice Problem 16.4.2. With $num = 100000$, compare the simulated populations of the eight states with the eigenvector calculation of Practice Problem 16.4.2.

16.5 TUMOR SUPPRESSOR GENES

We end this chapter with an application of Markov chains to a model for how colon cancer can arise. Why choose this cancer? The work of Bert Vogelstein and Kenneth Kinzler [364] reveals much about the evolutionary pathways of colon cancer, so colon cancer is a good laboratory to study evolutionary aspects of cancer.

In 1971 Alfred Knudson [199] proposed the notion of a tumor suppressor gene (TSG). Based on this idea, Knudson and Suresh Moolgavkar [238] constructed a mathematical model of the development of cancer. Normal cells have two TSG alleles, both active. Knudson's *two-hit hypothesis* is that both alleles must be inactivated in order to increase the reproductive rate of the cell, a circumstance that can lead to cancer. The first allele inactivation is a result of a point mutation; the second could occur from a point mutation, or loss of heterozygosity (LOH) [316], that is, the loss of part of the chromosome that contains the unmutated TSG allele followed by the replacement of it with a copy of the chromosome part that contains the inactivated allele.

Every day the epithelium of the colon sheds a large number of cells. These are replaced, and the corresponding large number of cell divisions is ripe ground for mutations, a major way for cancer to emerge. Evolution has discovered a clever way to reduce this risk: the colon wall contains about 10^7 crypts, roughly cylindrical protrusions into the colon wall. Each crypt has a fixed-size population of a few thousand cells, and the base of each crypt consists of several stem cells. These stem cells divide and produce differentiated cells that migrate up the crypt wall, dividing as they go, while the cells at the top of the crypt die and are pushed out. Mutations that occur in the dividing differentiated cells have little chance of developing cancer before these cells die. Only the stem cells are adequately long-lived to have a noticeable chance to become cancerous if mutated.

The arrangement of the tissues of an organ into a large number of small compartments is an effective way to reduce the probability that mutations in the organ's cells can lead to cancer. In fact, this pattern is used in every organ that exhibits rapid cell division. Was this geometry discovered independently by evolution for each of these organs, or did evolution invent and spread this metapattern?

We describe a Markov chain model, presented by Martin Nowak in Sect. 12.4 of [258], of the inactivation of both alleles of the TSG in the cells of a colon crypt. Nowak divides the population of crypt cells into three classes:

0 cells having two active TSG alleles
1 cells having one active and one inactive TSG allele
2 cells having two inactive TSG alleles

The issue we'll address is this: what is the expected number of time steps for a single cell with two inactive TSG alleles to arise? To keep the population size constant, at each time step one cell is selected randomly for removal (death), and another is selected randomly to reproduce, with a possibility of mutation. Say

u_1 is the mutation rate to inactivate a first TSG allele
u_2 is the mutation rate to inactivate the second TSG allele

Usually $u_1 < u_2$ because as we have seen, two inactivation mechanisms, point mutation and loss of heterozygosity, are available to the second allele. The Markov chain has $N+2$ states. For $i = 0, \ldots, N$, state i consists of i cells of class 1 and $N-i$ cells of class 0. State $N+1$ contains a cell in class 2. Recall that $P_{i,j} = P(j \to i)$. The non-zero transition probabilities are

$$P_{i-1,i} = \frac{i}{N}\frac{N-i}{N}(1-u_1) \qquad \text{for } i = 1, \ldots, N \tag{16.19}$$

$$P_{i,i} = \frac{i}{N}\left(\frac{i}{N}(1-u_2) + \frac{N-i}{N}u_1\right) + \left(\frac{N-i}{N}\right)^2(1-u_1) \qquad \text{for } i = 0, \ldots, N \tag{16.20}$$

$$P_{i+1,i} = \frac{N-i}{N}\left(\frac{i}{N}(1-u_2) + \frac{N-i}{N}u_1\right) \qquad \text{for } i = 0, \ldots, N-1 \tag{16.21}$$

$$P_{N+1,i} = \frac{i}{N}u_2 \qquad \text{for } i = 0, \ldots, N \tag{16.22}$$

$$P_{N+1,N+1} = 1 \tag{16.23}$$

The elements $P_{i-1,i}$ are on the superdiagonal of the transition matrix, the $P_{i,i}$ are on the diagonal, the $P_{i+1,i}$ are on the subdiagonal, and $P_{N+1,i}$ and $P_{N+1,N+1}$

are on the last row of the transition matrix. How are we to understand these matrix elements?

An $i \to i-1$ transition occurs if a class 1 cell is eliminated and a class 0 cell reproduces a class 0 cell. The probability of selecting a class 1 cell for elimination is i/N. The probability of selecting a class 0 cell for reproduction is $(N-i)/N$. The probability that a class 0 cell reproduces without inactivating a TSG allele is $1-u_1$. These steps are independent of one another, so by generalizing Eq (11.6) to three events, we see their product gives $P_{i-1,i}$.

An $i \to i$ transition occurs by three combinations of events, each of which we call a *pathway.* A class 1 cell can be eliminated with probability i/N. Then one pathway to keep the number of class 1 cells equal to i consists of selecting of a class 1 cell for reproduction (prob is i/N) and letting it reproduce without inactivating the second TSG allele (prob is $1-u_2$). The product of these three factors is is the probability of one pathway by which i is kept constant. Another pathway consists of removing a type 1 cell (prob is i/N), selecting a type 0 cell for reproduction (prob is $(N-i)/N$), and reproducing it with a mutation that inactivates a first TSG allele (prob is u_1). The product of these three factors is the probability of a second pathway that keeps i constant. The third pathway consists of removing a class 0 cell (prob is $(N-i)/N$), selecting a class 0 cell for reproduction (prob is $(N-i)/N$), and reproducing it without a mutation that inactivates a first TSG allele (prob is $1-u_1$). The product of these three factors is the probability of the third pathway that keeps i constant. These three pathways are mutually exclusive, so their sum (Eq. (11.5)) gives $P_{i,i}$.

Two pathways give an $i \to i+1$ transition. First, remove a class 0 cell (prob is $(N-i)/N$), select a class 1 cell for reproduction (prob is i/N), and reproduce it without a mutation that inactivates the second TSG allele (prob is $1-u_2$). The second pathway consists of removing a class 0 cell (prob is $(N-i)/N$), selecting a class 0 cell for reproduction (prob is $(N-i)/N$), and reproduceing it with a mutation that inactivates a first TSG allele (prob is u_1). Adding the probabilities of these pathways gives $P_{i+1,i}$.

An $i \to N+1$ transition occurs in only one way: select a class 1 cell to reproduce (prob is i/N), and reproduce it with a mutation that inactivates the second TSG allele (prob is u_2). Because the goal is to produce a class 2 cell, which cell is eliminated does not alter the calculation.

Finally, once the system enters state $N+1$, it stays in state $N+1$, an absorbing state.

Now that we have this model, we ask a question: what is the expected number of time steps before the system enters state $N+1$, that is, before the appearance of a class 2 cell? In Prop. A.13.4 of Sect. A.13 we have a way to calculate the expected number of steps to reach an absorbing state. This involves inverting an $(N+1) \times (N+1)$ matrix. Because N is several thousand for a typical crypt, we'd like to avoid inverting this matrix if we can.

We want to find the expected time for a system in state 0 (all cells have both TSG alleles active) to state $N+1$ (at least one cell has both TSG alleles inactive). For all i, $0 \leq i \leq N$, write $T_i = \mathbb{E}(i \to N+1)$, the expected time to go from state i to state $N+1$. Then by Eq. (11.14),

$$T_i = 1 \cdot p_1(i) + 2 \cdot p_2(i) + 3 \cdot p_3(i) + 4 \cdot p_4(i) + \cdots$$

where $p_k(i)$ is the probability of making a transition from state i to state $N+1$ in k time steps. We'll find a relation between the T_i that we can solve, at least in principle, for T_0.

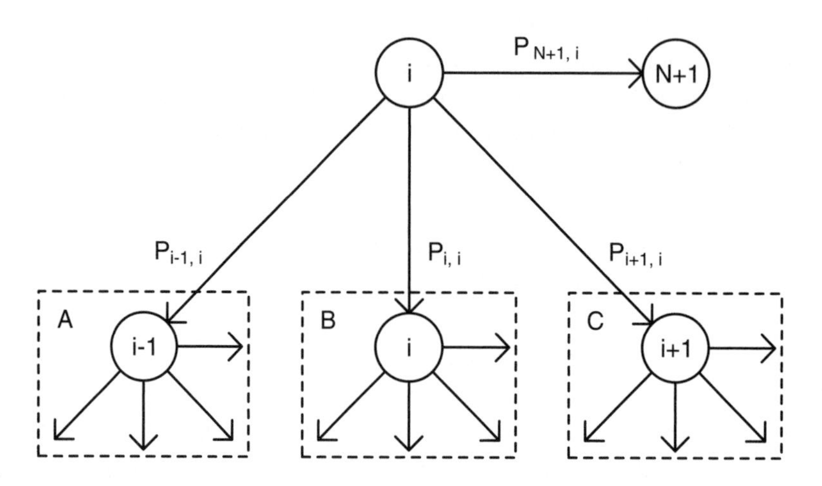

Figure 16.7. Steps in computing T_i.

The general relation for $1 \leq i \leq N-1$ is sketched in Fig 16.7. From state i the system can

(1) move into state $N+1$ with probability $P_{N+1,i}$,
(2) move into state $i+1$ with probability $P_{i+1,i}$,
(3) remain in state i with probability $P_{i,i}$, or
(4) move into state $i-1$ with probability $P_{i-1,i}$.

In case (1), with probability $P_{N+1,i}$ the system enters state $N+1$ in one time step, contributing $1 \cdot P_{N+1,i}$ to the expected value T_i.

In case (2), with probability $P_{i-1,i}$ the system enters state $i-1$ in one time step, contributing $1 + T_{i-1}$, in the dashed box A, to T_i.

In case (3), with probability $P_{i,i}$ the system remains in state i in the next time step, contributing $1+T_i$, in the dashed box B, to T_i.

In case (4), with probability $P_{i+1,i}$ the system enters state $i+1$ in one time step, contributing $1+T_{i+1}$, in the dashed box C, to T_i.

Combining these we obtain the relation

$$\begin{aligned}T_i &= P_{i-1,i}(1+T_{i-1}) + P_{i,i}(1+T_i) + P_{i+1,i}(1+T_{i+1}) + P_{N+1,i} \\ &= (P_{i-1,i} + P_{i,i} + P_{i+1,i} + P_{N+1,i}) + P_{i-1,i}T_{i-1} + P_{i,i}T_i + P_{i+1,i}T_{i+1} \\ &= 1 + P_{i-1,i}T_{i-1} + P_{i,i}T_i + P_{i+1,i}T_{i+1} \qquad (16.24)\end{aligned}$$

where we have used the fact that

$$P_{i-1,i} + P_{i,i} + P_{i+1,i} + P_{N+1,i} = 1$$

because from state i the only allowed transitions are to $i-1$, i, $i+1$, and $N+1$.

Some modifications are needed for T_0 and T_N.

First, T_0. From state 0 the only allowed transitions are to state 0 (a class 0 cell is removed and another class 0 cell is reproduced without a mutation that inactivates a first TSG allele) and to state 1 (a class 0 cell is removed and another class 0 cell is reproduced with a mutation that inactivates a first TSG allele). To enter state 2 requires two mutations in a cell, not possible in one time step when we start from a population of cells all of class 0. The appropriate modifications of Fig. 16.7 give

$$T_0 = P_{0,0}(1+T_0) + P_{1,0}(1+T_1) = 1 + P_{0,0}T_0 + P_{1,0}T_1 \qquad (16.25)$$

where we've used $P_{0,0} + P_{1,0} = 1$.

Now for T_N. From state N the only allowed transitions are to state N and $N+1$. There is no transition to state $N-1$ because in state N all N cells have one inactive TSG allele, and mutations that reverse the inactivation of the a TSG allele are vanishingly rare. Then

$$T_N = P_{N,N}(1+T_N) + P_{N+1,N} = 1 + P_{N,N}T_N \qquad (16.26)$$

where we've used $P_{N,N} + P_{N+1,N} = 1$. Now from Eqs. (16.22) and (16.20), $P_{N+1,N} = u_2$ and $P_{N,N} = 1-u_2$. Solving Eq. (16.26) we find

$$T_N = \frac{1}{u_2}$$

The system Eqs. (16.25), (16.24), and (16.26) can be solved for $\mathbb{E}(T_0)$, giving the expected time for the onset of cancer. The system parameters are u_1, u_2, and N. Of course, for large N solving this system might not be much simpler than inverting the matrix of Prop. A.13.4.

The evolutionary aspects of cancer are not restricted to TSG. When activated, the scarily named oncogenes [348] can disable the apoptosis (programmed cell death) mechanism of the cell, allowing it to survive and divide, even if the cell has been damaged. So far, several dozen human oncogenes have been identified. While both TSG alleles must be deactivated in order for the TSG to cause trouble, activating a single allele of an oncogene is enough to make it dangerous.

In addition, chromosomal instability (CIN) [356, 361] can amplify these effects. During cell division, several hundred genes manage the correct duplication of chromosomes and their distribution into both daughter cells. Mutations of these genes can lead to chromosomal instability, where all or part of some chromosomes are deleted or duplicated during cell division. CIN can lead to the LOH replacement of the second point mutation of a TSG. Christoph Lengauer, Kenneth Kinzler, and Bert Vogelstein [203, 204] calculated that the rate of LOH in CIN cells is about 10^{-2} per chromosome per cell division, while for non-CIN cells the rate is closer to 10^{-7}. That is, CIN is responsible for a significant increase in the speed of inactivation of the second TSG allele.

The evolutionary aspects of cancer still are revealing surprises. The complexity of cancers reminds us of the baroque intricacy of the immune system, briefly sketched in Sect. A.15. Elegance and simplicity appear to have little evolutionary advantage. Redundancy and variation carry the day.

Practice Problems

16.5.1. (a) For $N = 2$ the system Eqs. (16.25), (16.24), and (16.26) reduces to three equations. Write those equations.
(b) Solve these equations to find expressions for T_0, T_1, and T_2 in terms of u_1 and u_2.
(c) For $u_1 = 0.05, 0.1$, and 0.2, plot T_0, T_1, and T_2 for $u_1 \leq u_2 \leq 0.95$.
These are unrealistic numbers; we use them to illustrate the relation between time, measured in number of cell divisions, until the appearance of a cell with both TSG alleles inactivated, and the probability of these mutations.

16.5.2. (a) For $N = 3$ write the transition matrix determined by Eqs. (16.19)–(16.23).
(b) Take $u_1 = 0.1$ and $u_2 = 0.2$ and use the method of Prop. A.13.4 to compute the expected number of cell divisions to enter the absorbing state 4 when starting from state 0.

Practice Problem Solutions

16.5.1. (a) For $N=2$, Eqs. (16.25), (16.24), and (16.26) are

$$T_0 = 1 + P_{0,0}T_0 + P_{1,0}T_1$$
$$T_1 = 1 + P_{0,1}T_0 + P_{1,1}T_1 + P_{2,1}T_2 \quad \text{and} \quad T_2 = 1 + P_{1,2}T_1 + P_{2,2}T_2$$

(b) Substituting in the transition probabilities, we find

$$T_0 = 1 + (1-u_1)T_0 + u_1 T_1$$
$$T_1 = 1 + \frac{1-u_1}{4}T_0 + \frac{2-u_2}{4}T_1 + \frac{1-u_2+u_1}{2}T_2$$
$$T_2 = 1 + (1-u_2)T_2$$

The solutions are

$$T_0 = \frac{u_1 + u_1^2 + 2u_2 + 3u_1u_2 + u_2^2}{u_1u_2(1+u_1+u_2)}$$
$$T_1 = \frac{u_1 + u_1^2 + u_2 + 2u_1u_2}{u_1u_2(1+u_1+u_2)} \quad \text{and} \quad T_2 = \frac{1}{u_2}$$

(c) The graphs are shown in Fig. 16.8.

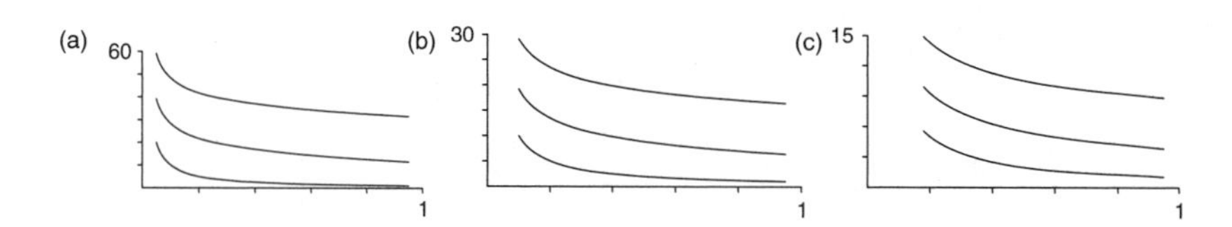

Figure 16.8. Plots of T_0 (upper), T_1 (middle), and T_2 (lower) for $u_1 = 0.05$ (left), 0.1 (middle), and 0.2 (right), for Practice Problem 16.5.1.

16.5.2. (a) First, recall that for $N=3$ the rows and columns are numbered 0, 1, 2, 3, and 4. Next, for example, by Eq. (16.19) with $i=1$ we see that

$$P_{0,1} = \frac{1}{3}\frac{2}{3}\left(1-u_1\right) = \frac{2(1-u_1)}{9}$$

Then we find the transition matrix is

$$P = \begin{bmatrix} 1-u_1 & \dfrac{2(1-u_1)}{9} & 0 & 0 & 0 \\ u_1 & \dfrac{5-2u_1-u_2}{9} & \dfrac{2(1-u_1)}{9} & 0 & 0 \\ 0 & \dfrac{2(1+2u_1-u_2)}{9} & \dfrac{5+u_1-4u_2}{9} & 0 & 0 \\ 0 & 0 & \dfrac{2+u_1-2u_2}{9} & 1-u_2 & 0 \\ 0 & \dfrac{u_2}{3} & \dfrac{2u_2}{3} & u_2 & 1 \end{bmatrix}$$

As long as we have this matrix, let's check that it's stochastic. Because $0 \leq u_1 \leq 1$ and $0 \leq u_2 \leq 1$, each matrix entry is non-negative. And with a little algebra, we see that the sum of the entries of each column is 1.

(b) To put this in the form to extract the matrix Q, needed for Prop. A.13.4, we renumber the rows and columns this way: $0 \to 1$, $1 \to 2$, $2 \to 3$, $3 \to 4$, and $4 \to 0$. This gives

$$Q = \begin{bmatrix} 1-u_1 & \dfrac{2(1-u_1)}{9} & 0 & 0 \\ u_1 & \dfrac{5-2u_1-u_2}{9} & \dfrac{2(1-u_1)}{9} & 0 \\ 0 & \dfrac{2(1+2u_1-u_2)}{9} & \dfrac{5+u_1-4u_2}{9} & 0 \\ 0 & 0 & u_2 & 1-u_2 \end{bmatrix}$$

For $u_1 = 0.1$ and $u_2 = 0.2$,

$$N = (I-Q)^{-1} \approx \begin{bmatrix} 19.814 & 9.814 & 3.759 & 0 \\ 4.907 & 4.907 & 1.879 & 0 \\ 2.088 & 2.088 & 2.715 & 0 \\ 2.088 & 2.088 & 2.715 & 5 \end{bmatrix}$$

Then the expected number of time steps (cell divisions) to go from state 0 (all TSG have both alleles activated) to the absorbing state 4 (at least one TSG has both alleles inactivated) is the first entry of $\begin{bmatrix} 1 & 1 & 1 & 1 \end{bmatrix} N$. That is, 28.898.

Exercises

16.5.1. Continuing the calculations of Practice Problem 16.5.2, find the expected number of cell divisions to go from all TSG with both alleles active to at least one TSG with both alleles inactive for these probabilities:

$(u_1, u_2) =$ (a) $(0.01, 0.2)$, (b) $(0.001, 0.2)$, (c) $(0.1, 0.3)$, (d) $(0.1, 0.4)$.
(e) Based on these results, do you think the expected number of cell divisions is an increasing or a decreasing function of u_1? Of u_2?

16.5.2. Continuing the calculations of Practice Problem 16.5.2, hold u_2 at 0.1 and let u_1 get smaller and smaller. Take $u_1 = 0.01$, $u_1 = 0.001$, $u_1 = 0.0001$, $u_1 = 0.00001$, and $u_1 = 0.000001$. Does the expected number of cell divisions to go from state 0 to state 4 appear to increase without bound, or does it appear to approach a limit?

16.5.3. One last continuation of Practice Problem 16.5.2.
(a) Hold $u_1 = u_2/2$ and let u_2 get smaller and smaller. Take $u_2 = 0.1$, $u_2 = 0.01$, $u_2 = 0.001$, $u_2 = 0.0001$, $u_2 = 0.00001$, and $u_2 = 0.000001$. Does the expected number of cell divisions to go from state 0 to state 4 appear to increase without bound, or does it appear to approach a limit?
(b) Hold $u_1 = u_2/10$ and let u_2 get smaller and smaller. Does the expected number of cell divisions to go from state 0 to state 4 appear to increase without bound, or does it appear to approach a limit?

16.5.4. (a) For $N = 3$ the system Eqs. (16.25), (16.24), and (16.26) reduces to four equations. Write those equations.
(b) Solve these equations for $T_0, \dots, T_3$ when $u_1 = 0.1$ and $u_2 = 0.2$.
(c) Solve these equations for $T_0, \dots, T_3$ when $u_1 = 0.01$ and $u_2 = 0.2$.

16.5.5. Continuing Exercise 16.5.4, find $T_0, \dots, T_3$ when $u_2 = 0.2$ and $u_1 =$ (a) 0.001, (b) 0.0001, (c) 0.00001, and (d) 0.000001.
(e) Do these results suggest T_0, T_1, and T_2 increase linearly as $1/u_1$?

16.5.6. Still continuing, find $T_0, \dots, T_3$ when $u_2 = \sqrt{u_1}$ and $u_1 =$ (a) 0.1, (b) 0.01, (c) 0.001, (d) 0.0001, (e) 0.00001, and (f) 0.000001.
(g) Do these results suggest T_0, T_1, and T_2 increase linearly as $1/u_1$?

16.5.7. (a) For $N = 4$ write the transition matrix determined by Eqs. (16.19)–(16.23).
(b) Write the matrix Q used in the decomposition of the transition matrix in Sect. A.13.

16.5.8. Using the Q matrix of Exercise 16.5.7, find the expected number of time steps from state 0 to state 5 for $u_2 = 0.1$ and $u_1 =$

(a) 0.01, (b) 0.001, (c) 0.0001, (d) 0.00001, and (e) 0.000001.
(f) Compare these results with those of Exercise 16.5.2.

16.5.9. Using the Q matrix of Exercise 16.5.7, find the expected number of time steps from state 0 to state 5 for $u_1 = 0.00001$ and $u_2 =$
(a) 0.1, (b) 0.01, (c) 0.001, and (d) 0.0001.
(e) Compare the rate of increase of the expected number of time steps as u_2 decreases to the rate as u_1 decreases found in Exercise 16.5.8.

16.5.10. Find what you can about current thinking on whether CIN arises early or late during the development of colon cancer.

Chapter 17 Some vector calculus

Vector calculus is the next step in calculus, usually undertaken after techniques of integration and infinite series. Very roughly, vector calculus is the calculus of vector fields, the study of how to differentiate and integrate vector fields. A complete treatment of this subject would fill another book, indeed has filled many other books. These are the basic concepts.

1. Three kinds of derivatives involve vector fields. The gradient is a vector field that is a derivative of a (scalar) function, the divergence is a (scalar) function that is a derivative of a vector field, and the curl is a vector field that is a derivative of a vector field.
2. Integrals of vector fields along curves are called *line integrals*.
3. If the vector field is a gradient of a function, the line integral of the vector field along a curve is the difference of the function at the curve endpoints. This is the fundamental theorem of line integrals (FTLI).
4. The line integral of a vector field along a closed curve in the plane is equal to the integral of the derivative of the vector field over the region bounded by the curve. This is *Green's theorem*.
5. Integrals of vector fields over surfaces are *surface integrals*.

6. *Stokes' theorem* equates the integral of the curl of a vector field over a surface with boundary to the line integral of the vector field along that boundary curve.
7. Gauss' theorem equates the integral of a vector field over a closed surface to the integral of the divergence of the vector field over the region bounded by the surface.

(Grammar states that the possessive of a singular noun that ends in *s* is formed by adding 's. The names should be *Stokes's theorem* and *Gauss's theorem*. But long tradition is to use the names *Stokes' theorem* and *Gauss' theorem*.) All parts have applications in the biomedical sciences. The gradient gives a simple way to understand the evolutionary movement of populations across fitness landscapes, the movements of ions across transmembrane channels, and many things in between. Line integrals give a way to preclude closed trajectories from some regions, and also a way to define a number—the index of a vector field of a fixed point—that restricts combinations of fixed points in regions. Gauss' theorem leads to the diffusion equation, important to model the flow of drugs through a body, diseases through a population, and much more.

Also, we'll mention that vector calculus is the right way to formulate the laws of electromagnetism and of fluid mechanics, and that vector calculus generalizes to differential geometry the language of Einstein's general theory of relativity. And from a mathematical point of view, we should mention that div, curl, and grad are versions of the same derivative, and Gauss' theorem, Stokes' theorem, and the fundamental theorem of line integrals all are versions of one theorem. This is a beautiful branch of mathematics. We'll see only a tiny glimpse.

17.1 GRADIENT FIELDS

From Sect. 15.3 recall the *gradient field* of a differentiable function f,

$$\nabla f = \left\langle \frac{\partial f}{\partial x}, \frac{\partial f}{\partial y} \right\rangle = \frac{\partial f}{\partial x}\vec{i} + \frac{\partial f}{\partial y}\vec{j} \tag{17.1}$$

where $\vec{i} = \langle 1,0 \rangle$ and $\vec{j} = \langle 0,1 \rangle$. This definition of ∇f in the plane generalizes to higher dimensions in the obvious way: $\nabla f = \langle \partial f/\partial x, \partial f/\partial y, \partial f/\partial z \rangle$ if $f = f(x,y,z)$, and so on. Calculations are straightforward; applications are interesting.

First, suppose $\vec{c}(t) = \langle x(t), y(t) \rangle$ is a *level curve* for the function f. That is, for all t,

$$k = f(\vec{c}(t)) = f(x(t), y(t)) \tag{17.2}$$

for some constant k. The name "level curve" comes from topography, where a level curve on a map is the collection of all locations on the map at which the altitude takes a given value, represented by the constant k on Eq. (17.2). In our more general setting, the function f takes the role of altitude. Then differentiating both sides of Eq. (17.2) and applying the chain rule (Eq. (9.4)) to the right-hand side in the second equality,

$$0 = \frac{d}{dt} f(x(t), y(t)) = \frac{\partial f}{\partial x} x' + \frac{\partial f}{\partial y} y' = \nabla f \cdot \langle x', y' \rangle \tag{17.3}$$

Recall the dot product (Eq. (13.51)) of vectors $\langle a, b \rangle$ and $\langle c, d \rangle$ is

$$\langle a, b \rangle \cdot \langle c, d \rangle = ac + bd \tag{17.4}$$

We'll use Fig. 17.1 to derive the useful geometric interpretation (Eq. (13.52)) of the dot product by application of the law of cosines to the triangle with sides $\langle a, b \rangle$, $\langle c, d \rangle$, and $\langle a - c, b - d \rangle$, the third side being opposite the angle θ between the other two sides. Recall the (Euclidean) length of a vector $\langle a, b \rangle$ is $\| \langle a, b \rangle \| = \sqrt{a^2 + b^2}$. This is just an application of the Pythagorean theorem. Then from the law of cosines, which is the Pythagorean theorem generalized to triangles that need not have a right angle, we see

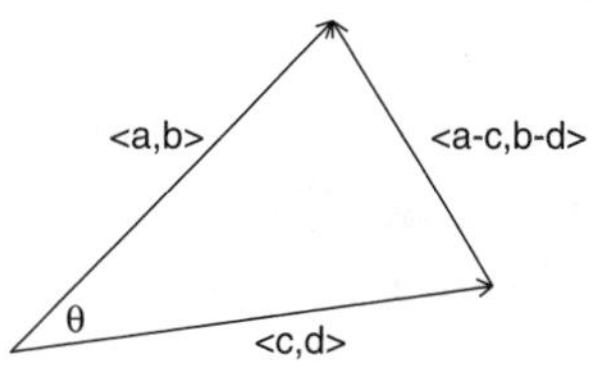

Figure 17.1. Dot product geometry

$$\begin{aligned}
\| \langle a - c, b - d \rangle \|^2 &= \| \langle a, b \rangle \|^2 + \| \langle c, d \rangle \|^2 - 2 \| \langle a, b \rangle \| \| \langle c, d \rangle \| \cos(\theta) \\
(a - c)^2 + (b - d)^2 &= a^2 + b^2 + c^2 + d^2 - 2 \| \langle a, b \rangle \| \| \langle c, d \rangle \| \cos(\theta) \\
-2ac - 2bd &= -2 \| \langle a, b \rangle \| \| \langle c, d \rangle \| \cos(\theta)
\end{aligned}$$

Combined with Eq. (17.4) this shows

$$\langle a, b \rangle \cdot \langle c, d \rangle = \| \langle a, b \rangle \| \| \langle c, d \rangle \| \cos(\theta) \tag{17.5}$$

In particular, so long as neither vector has length 0, if the dot product is 0, then the vectors are perpendicular. Consequently, Eq. (17.3) shows that the gradient field ∇f is perpendicular to $\langle x', y' \rangle$, the tangent vector to the level curve. We'll see that the direction of the gradient vector is the direction of greatest increase of the function f. Generalized to higher dimensions, the gradient of the fitness function gives the direction of greatest increase on the fitness landscape.

To relate the gradient to the direction of greatest increase, we need to compute the increase in every direction. This we can do with the directional derivative. First, recall that we defined $\partial f / \partial x$ as the rate of change of f in the x-direction and $\partial f / \partial y$ as the rate of change of f in the y-direction:

$$\frac{\partial f}{\partial x} = \lim_{t\to 0}\frac{f(x+t,y)-f(x,y)}{t} = \lim_{t\to 0}\frac{f((x,y)+t(1,0))-f(x,y)}{t}$$
$$\frac{\partial f}{\partial y} = \lim_{t\to 0}\frac{f(x,y+t)-f(x,y)}{t} \lim_{t\to 0}\frac{f((x,y)+t(0,1))-f(x,y)}{t}$$

So for any vector $\langle u,v\rangle$ we might think that the rate of change of f in the direction of $\langle u,v\rangle$ is

$$\lim_{t\to 0}\frac{f((x,y)+t(u,v))-f(x,y)}{t} \tag{17.6}$$

This is pretty close, but not quite there. The problem is that the vectors $\langle u,v\rangle$ and $\langle 2u,2v\rangle$ are in the same direction, but the limit above can give different values for these two vectors. For example, take $f(x,y)=x^2+xy$ and compute the limit for the vectors $\langle 1,1\rangle$ and $\langle 2,2\rangle$. To avoid this problem, we'll use only unit vectors (vectors of Euclidean length 1) when computing the derivative in the direction of a vector. That is, for any unit vector $\vec{w}=\langle u,v\rangle$, the directional derivative of f in the direction of $\vec{w}$ is

$$D_{\vec{w}}f(x,y) = \lim_{t\to 0}\frac{f((x,y)+t(u,v))-f(x,y)}{t} \qquad \text{where } \|\vec{w}\|=1 \tag{17.7}$$

Note that Eq. (17.7) is the same equation as (17.6), but here we emphasize that the direction vector must be a unit vector.

This is fine, until we remember that we don't compute partial derivatives with the limit definition if we possibly can avoid it, or unless asked to do so in a homework problem. Can we compute directional derivatives without using limits? Sure. Here's how.

If f is a differentiable function and $\vec{w}=\langle u,v\rangle$ is a unit vector, then

$$D_{\vec{w}}f(x,y) = \nabla f(x,y)\cdot\vec{w} = \left(\left.\frac{\partial f}{\partial x}\right|_{(x,y)}\right)u + \left(\left.\frac{\partial f}{\partial y}\right|_{(x,y)}\right)v \tag{17.8}$$

To see this, take the straight line path $c(t)=(x,y)+t(u,v)$ where $\vec{w}=\langle u,v\rangle$ is a unit vector. Then starting with Eq. (17.7),

$$\begin{aligned} D_{\vec{w}}f(x,y) &= \lim_{t\to 0}\frac{f((x,y)+t(u,v))-f(x,y)}{t} = \lim_{t\to 0}\frac{f(c(t))-f(c(0))}{t} \\ &= \frac{d}{dt}f(c(t)) = \left(\left.\frac{\partial f}{\partial x}\right|_{(x,y)}\right)\frac{d\,tu}{dt} + \left(\left.\frac{\partial f}{\partial y}\right|_{(x,y)}\right)\frac{d\,tv}{dt} \\ &= \left(\left.\frac{\partial f}{\partial x}\right|_{(x,y)}\right)u + \left(\left.\frac{\partial f}{\partial y}\right|_{(x,y)}\right)v = \nabla f\cdot\vec{w} \end{aligned}$$

where the fourth equality follows by the chain rule.

Using the geometric interpretation of the dot product Eq. (17.5) and recalling $\|w\| = 1$,

$$D_{\vec{w}}f(x,y) = \|\nabla f\| \cos(\theta) \tag{17.9}$$

where θ is the angle between ∇f and $\vec{w}$.

Now Eq. (17.9) offers an elaboration of our earlier observation that the gradient is perpendicular to the level curves of the function f. Recalling that $\cos(0) = 1$, $\cos(\pm\pi/2) = 0$, and $\cos(\pi) = -1$, we see that

- f increases most rapidly in the direction of ∇f,
- f remains unchanged in both directions perpendicular to ∇f, and
- f decreases most rapidly in the direction $-\nabla f$, that is, in the direction opposite to that of ∇f.

So we see that the gradient allows us to recognize some of the geometry of functions. In addition, it will turn out to be very useful in evaluating some line integrals.

Practice Problems

17.1.1. For $f(x,y) = x^2 + xy^2$, find the lines or curves along which ∇f consists of horizontal vectors.

17.1.2. Compute the directional derivative of $f(x,y) = x^2 - 3xy^2 + y^3$ at the point $(x,y) = (3,2)$ in the direction of $\langle 2,1 \rangle$.

17.1.3. Find the direction of most rapid increase of the function $f(x,y) = x^3y^4$ at the point $(x,y) = (2,1)$. Find the magnitude of that increase.

Practice Problem Solutions

17.1.1. For $f(x,y) = x^2 + xy^2$, we see $\nabla f = \langle 2x + y^2, 2xy \rangle$. The vector field consists of horizontal vectors when the vertical component is 0, that is, when $2xy = 0$. This happens when $x = 0$ or when $y = 0$, so ∇f consists of horizontal vectors along the x- and y-axes.

17.1.2. We'll use Eq. (17.8) to compute the directional derivative. But first we need to find a unit vector in the direction of $\langle 2,1 \rangle$. This is easy: $\|\langle 2,1 \rangle\| = \sqrt{2^2 + 1^2} = \sqrt{5}$, so the unit vector in this direction is

$$\vec{w} = \langle u, v \rangle = \langle 2/\sqrt{5}, 1/\sqrt{5} \rangle$$

Next, $\partial f/\partial x = 2x - 3y^2$ and $\partial f/\partial y = -6xy + 3y^2$. Combining this information in Eq. (17.8) we find

$$D_{\vec{w}} f(3,2) = -6 \cdot 2/\sqrt{5} + -24 \cdot 1/\sqrt{5} = -36/\sqrt{5}$$

17.1.3. From Eq. (17.9) we have seen that the direction of most rapid increase at $(2,1)$ is the direction of the gradient ∇f at $(2,1)$. Now $\partial f/\partial x = 3x^2y^4$ and $\partial f/\partial y = 4x^3y^3$. Then by Eq. (17.1),

$$\nabla f(2,1) = \langle (\partial f/\partial x)|_{(2,1)}, (\partial f/\partial y)|_{(2,1)} \rangle = \langle 12, 32 \rangle$$

Because we just asked for the direction, this vector will do. No need to find the unit vector in this direction.

Also from Eq. (17.9) we see that the magnitude of greatest increase is

$$\|\nabla f(2,1)\| = \|\langle 12, 32 \rangle\| = 4\sqrt{73}$$

Exercises

17.1.1. Find the gradients of these functions at every point (x,y) of their domains. Some of these are a bit messy. Think of them as an opportunity to polish your differentiation skills.

(a) $x^3y^2 + \sqrt{xy}$ (b) $x^3y/(y^4+1)$ (c) $\sin(x^2 - y^3)$
(d) $\ln(x^2 + y^4 + 1)$ (e) $x^2e^{x^3-y^2}$ (f) $(x + \sin(y))/(y^2 + \cos^2(x))$
(g) $\tan(x + \tan(y))$ (h) $e^{x^2/(2+y^4)}$ (i) $(x^2 + y^3 + \sin(xy))^{1/3}$

17.1.2. Find the directional derivatives of these functions at the point $(x,y) = (1,1)$ and in the direction $\langle -1, 2 \rangle$.

(a) $x^2 - y^2$ (b) $x^2 - y^3$ (c) $x^2 - xy^2 + y^3$
(d) $e^{x^2-y^2}$ (e) $\sin(x^2 - y^3)$ (f) $\cos(x^2 - y^3)$
(g) $xy^3 + x^2y^2 + x^3y$ (h) $(x^2 + y^4)^{1/2}$ (i) $\tan(x^2 - y^3)$

17.1.3. Find the directions of greatest increase, greatest decrease, and no change for the functions of Exercise 17.1.1, at the point $(1,1)$. Your answers need not be unit vectors. Numerics suffice for (f), (g), and (i).

17.1.4. Find the lines or curves along which ∇f is horizontal; find the lines or curves along which ∇f is vertical. Be careful: saying ∇f is horizontal means that the $\vec{j}$-component of ∇f is 0 and the $\vec{i}$-component is non-zero.

(a) $x^2y - xy^2$ (b) x^2y^4 (c) $x(x^2/3 + y^2)$ (d) $x^3 - 3xy + y^3$

17.1.5. (a) Suppose $\vec{F}(x,y) = g(x,y)\vec{i} + h(x,y)\vec{j}$ with g and h continuously differentiable. If $\vec{F} = \nabla f$, show that $\partial g/\partial y = \partial h/\partial x$. Hint: Clairaut's theorem states that if f has continuous second partial derivatives, then $\partial^2 f/\partial x\partial y = \partial^2 f/\partial y\partial x$.
(b) Show that $\vec{F} = x^2 y\vec{i} + x^2 y^3\vec{j}$ is not the gradient of any function f.

17.2 LINE INTEGRALS

In your introduction to integral calculus you learned that the area under the graph $y = f(x)$ between $x = a$ and $x = b$ is the definite integral $\int_a^b f(x)dx$. The left side of Fig. 17.2 illustrates this.

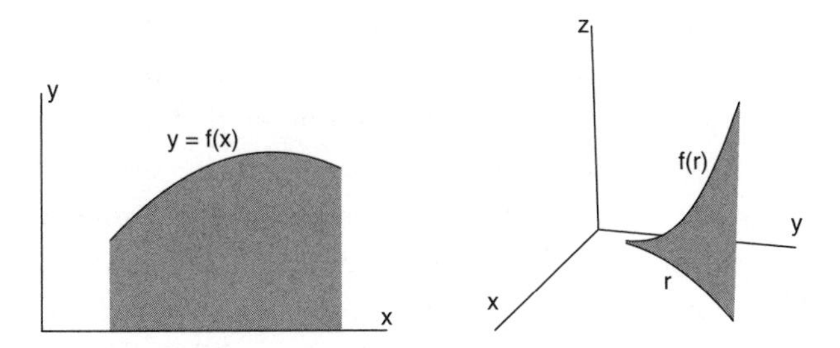

Figure 17.2. Left: integral as area over a straight line. Right: integral as area over a curve.

We should be able to calculate the area if the base is a curve C in the xy-plane rather than a segment of the x-axis. But what kind of integral? We can't expect the area to depend only on the endpoints of the curve. How much the curve bends, how long it is, must figure into the area.

Here's how to deal with area over curves. Parameterize the curve C as $\vec{r}(t) = \langle x(t), y(t)\rangle$ for $a \le t \le b$. Thinking of $\vec{r}(t)$ as describing position along C at time t, the length of the tangent vector

$$\|\vec{r}'(t)\| = \|\langle x'(t), y'(t)\rangle\| = \sqrt{(x'(t))^2 + (y'(t))^2}$$

is the speed of this path along the curve C. Integrating speed as a function of time gives distance traveled, so the *arclength* of the curve is

$$s = \int_a^b \|\vec{r}'(t)\|\, dt \tag{17.10}$$

Then to find the area of the surface pictured on the right of Fig. 17.2 we must adjust the area above each bit of the curve by the length of its tangent vector, standing in for the little bit of arclength at the base of the surface. We call this a *line integral*,

$$\int_C f\, ds = \int_a^b f(\vec{r}(t))\|\vec{r}'(t)\|\, dt \tag{17.11}$$

We use the differential *ds* to indicate that the integral is with respect to arclength along the curve. This is the way to evaluate line integrals of functions along curves. (*Scalar* functions are functions that take on real-number values; integrating *vector* functions is covered next.) All these notions—parameterization of a curve, the length of the tangent vector—work just as well in higher dimensions.

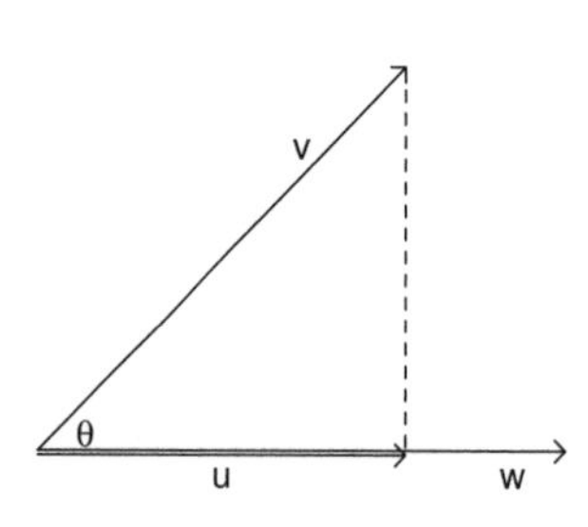

Figure 17.3. Projection of the vector $\vec{v}$ in the direction of the vector $\vec{w}$.

While this is a useful extension of the familiar notion of integral, our main purpose is to figure out how to integrate vector fields along curves. To do this we need one more tool: a way to project one vector in the direction of another. Fig. 17.3 shows how to project a vector $\vec{v}$ in the direction of the vector $\vec{w}$. The geometry is simple: drop a perpendicular to $\vec{w}$ from the end of $\vec{v}$. This marks off $\vec{u}$, the projection of $\vec{v}$ to $\vec{w}$, also called the *component* of $\vec{v}$ in the direction of $\vec{w}$. We'd like a simpler, more formulaic way to calculate this projection. From the figure

$$\cos(\theta) = \|\vec{u}\|/\|\vec{v}\|, \quad \text{and by Eq. (17.5)} \quad \vec{w}\cdot\vec{v} = \|\vec{w}\|\,\|\vec{v}\|\cos(\theta)$$

Eliminating $\cos(\theta)$ we find

$$\|\vec{u}\| = \frac{\vec{w}\cdot\vec{v}}{\|\vec{w}\|} = \frac{\vec{w}}{\|\vec{w}\|}\cdot\vec{v} \tag{17.12}$$

Finally, we're ready to understand the line integral of a vector field. For motivation we'll use the definition of work. From high school physics we recall that for an object that moves in a straight line, a constant force applied in the direction of motion does the amount of work given by

$$\text{work} = \text{force} \times \text{distance}$$

Only the component of the force in the direction of motion does any work. For curved paths and varying force, the work is the line integral of the component of the force in the direction of motion. This is a reason for the definition of the line integral of a vector field $\vec{F}$:

$$\int_C \vec{F}\cdot d\vec{r} = \int_C \vec{F}(\vec{r})\cdot \frac{\vec{r}'}{\|\vec{r}'\|}\, ds$$
$$= \int_a^b \vec{F}(\vec{r}(t))\cdot \frac{\vec{r}'(t)}{\|\vec{r}'(t)\|}\|\vec{r}'(t)\|\, dt = \int_a^b \vec{F}(\vec{r}(t))\cdot \vec{r}'(t)\, dt \qquad (17.13)$$

The first equality is the definition: the line integral of the component of $\vec{F}$ in the direction of $\vec{w}$. Being a dot product, the component is a scalar function, so the second equality is the arclength correction (Eq. (17.11)) for the line integral of a scalar function. Now let's work out an example of the line integral of a vector field along a curve.

Example 17.2.1. *Line integral, direct evaluation.* Suppose $\vec{F} = \langle x^2, y^3\rangle$ and C is the quarter circle from $(1,0)$ to $(0,1)$. To evaluate the line integral, parameterize C by $\vec{r}(t) = \langle \cos(t), \sin(t)\rangle$ for $0 \le t \le \pi/2$. Then

$$\int_C \vec{F}\cdot d\vec{r} = \int_0^{\pi/2} \langle \cos^2(t), \sin^3(t)\rangle \cdot \langle -\sin(t), \cos(t)\rangle\, dt$$
$$= \int_0^{\pi/2} -\cos^2(t)\sin(t) + \sin^3(t)\cos(t)\, dt$$
$$= \left(\frac{\cos^3(t)}{3} + \frac{\sin(t)^4}{4}\right)\Bigg|_0^{\pi/2} = -\frac{1}{3} + \frac{1}{4} = -\frac{1}{12}$$

where the penultimate equality was obtained by the substitution of $u = \cos(t)$ in the first term and $u = \sin(t)$ in the second. □

We began our development of line integrals with the observation that surely the integral depends on the path, not just on the path endpoints. So let's integrate this vector field over another path that starts at $(1,0)$ and ends at $(0,1)$. The straight line segment C' parameterized by $\vec{r}(t) = \langle 1-t, t\rangle$ will do. Then

$$\int_{C'} \vec{F}\cdot d\vec{r} = \int_0^1 \langle (1-t)^2, t^3\rangle \cdot \langle -1, 1\rangle dt = \int_0^1 (t^3 - t^2 + 2t - 1) dt$$
$$= \left(\frac{t^4}{4} - \frac{t^3}{3} + t^2 - 1\right)\Bigg|_0^1 = -\frac{1}{12}$$

Shouldn't we have gotten a different answer? Maybe that our choice of C' gave the same line integral is just luck. Maybe another path would give a different answer. Maybe most different paths would give different answers. It turns out this isn't true. For this vector field $\vec{F}$, the line integral along *every* path from $(1,0)$ to $(0,1)$ gives the same answer. If the path starts at $(1,0)$, makes the voyage to Arcturus, then to Deneb, then back to $(0,1)$, the line integral still will be $-1/12$. Why is this true for this vector field? Is it true for all vector fields? Let's see.

For a vector field $\vec{F}$, a function f with $\vec{F} = \nabla f$ is called a *potential function* for $\vec{F}$. Think of gravitational force and gravitational potential. Continuing with this analogy, we say a vector field $\vec{F}$ is *conservative* if $\vec{F} = \nabla f$ for some function f. We'll see the motivation for the term "conservative" in Cor. 17.2.3.

Next we'll revisit the fundamental theorem of calculus (FTC), $\int_a^b f'(t)dt = f(b) - f(a)$. In order to formulate this for line integrals, we'll need to have a vector field that is the derivative of a function. But we already know one kind of vector that is a derivative: the vector field ∇f is the derivative of the function f. With this kind of derivative there is a vector version of FTC, the *fundamental theorem of line integrals (FTLI)*.

Theorem 17.2.1. If C is any path from (a, b) to (c, d),

$$\int_C \nabla f \cdot d\vec{r} = f(c, d) - f(a, b) \tag{17.14}$$

Proof. For a parameterization $\vec{r}(t) = \langle x(t), y(t) \rangle$ of C with $\vec{r}(0) = (a, b)$ and $\vec{r}(1) = (c, d)$, we see

$$\begin{aligned}
\int_C \nabla f \cdot d\vec{r} &= \int_0^1 \nabla f(\vec{r}(t)) \cdot \vec{r}\,'(t)dt \\
&= \int_0^1 \left\langle \frac{\partial f(\vec{r}(t))}{\partial x}, \frac{\partial f(\vec{r}(t))}{\partial y} \right\rangle \cdot \left\langle \frac{dx}{dt}, \frac{dy}{dt} \right\rangle dt \\
&= \int_0^1 \frac{\partial f(\vec{r}(t))}{\partial x}\frac{dx}{dt} + \frac{\partial f(\vec{r}(t))}{\partial y}\frac{dy}{dt} dt \\
&= \int_0^1 \frac{df(\vec{r}(t))}{dt} dt = f(\vec{r}(1)) - f(\vec{r}(0)) = f(c, d) - f(a, b)
\end{aligned}$$

We apply the chain rule for the fourth equality and FTC for the fifth. □

Note that FTLI says that the line integral of ∇f along a path C depends on only the value of f at the endpoints of C, and not on any particulars of the path. Because we'll use this notion often, we'll give it a name, *path independence*, and record it in the next corollary. With path independence we can replace a line integral along a complicated path with a line integral along a simpler path, provided both paths have the same initial point and the same endpoint.

Corollary 17.2.1. If $\vec{F}$ is conservative, that is, $\vec{F} = \nabla f$ for some f, and the paths C_1 and C_2 have the same initial point and have the same endpoint, then

$$\int_{C_1} \vec{F} \cdot d\vec{r} = \int_{C_2} \vec{F} \cdot d\vec{r}.$$

Proof. This is a straightforward application of FTLI, Thm. 17.2.1. □

Now we'll show that the converse of Cor. 17.2.1 holds on a suitable region R. That is, if the line integrals of a vector field are path independent, then the vector field is conservative. To make a precise statement, we need two more geometrical concepts. Both can be formulated in far more general settings, but we're content with subsets of the plane. A set A is *open* if for every point (x,y) of A, there is a positive number r for which $D_r(x,y)$, the disc with radius r and center (x,y), lies entirely inside A. For example, the disc $D_1 = \{(x,y) : x^2+y^2 < 1\}$ is open, while the disc $D_2 = \{(x,y) : x^2+y^2 \leq 1\}$ is not. To see the latter, note that the point $(1,0)$ belongs to D_2, and every disc with center $(1,0)$, no matter how small its radius, contains points inside D_2 and also points outside D_2. Consequently, D_2 is not open. We'll leave to you the task of showing that D_1 is open to you.

The second concept is that of a *connected* space. Because we are dealing with subsets of the plane, we can use the simpler concept of *path-connected*. A set A is path-connected if every pair of points in A can be joined by a path lying entirely in A. (In fact, for subsets of any Euclidean space $\mathbb{R}^n$, connected and path-connected are equivalent concepts.)

Corollary 17.2.2. Suppose the line integrals $\int_C \vec{F} \cdot d\vec{r}$ are path independent for all paths C in an open connected region R. Then there is a function f defined on R for which $\nabla f = \vec{F}$.

Proof. We'll prove $\vec{F} = \langle P, Q\rangle$ is conservative by constructing a function f defined on R for which $\partial f/\partial x = P$ and $\partial f/\partial y = Q$. Start by picking a point p in the region R, shown in Fig. 17.4 as a rectangle with dashed edges indicating that the sides of the rectangle do not belong to R, a necessary condition for R to be open.

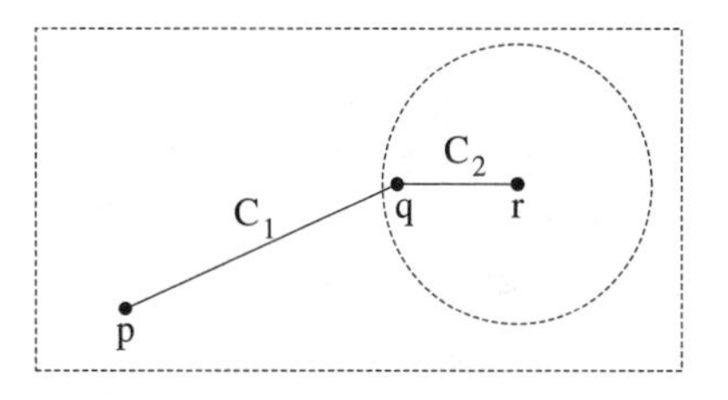

Figure 17.4. Constructing a potential function.

Now given any point $r = (x,y)$ in R, because R is open we know that some small disc centered on (x,y) lies entirely inside R. Take any point $q = (x_1,y)$ in this disc, with $x_1 < x$. Certainly, the straight horizontal line, C_2, between q and r lies in R. Because R is connected, there is a path, C_1, between p and q. Here's our definition of the function f (see the notation comment at the section end):

$$f(x,y) = \int_{C_1} P\,dx + Q dy + \int_{C_2} P\,dx + Q dy \tag{17.15}$$

How does this make sense? Wouldn't the value of f at (x,y) depend on the paths C_1 and C_2? No, that's where path independence comes in. Eq. (17.15) is a perfectly good way to define the function f.

Now we'll show that $\partial f/\partial x = P$:

$$\begin{aligned}\frac{\partial f}{\partial x} &= \frac{\partial}{\partial x}\int_{C_1} P\,dx + Q\,dy + \frac{\partial}{\partial x}\int_{C_2} P\,dx + Q\,dy \\ &= 0 + \frac{\partial}{\partial x}\int_{C_2} P\,dx + Q\,dy = \frac{\partial}{\partial x}\int_{C_2} P\,dx = \frac{\partial}{\partial x}\int_{x_1}^{x} P(t,y)\,dt = P(x,y)\end{aligned}$$

where the second equality follows because C_1 doesn't depend on x, the third because $dy = 0$ on C_2, and the fifth by FTC.

Replacing $q = (x_1, y)$ by a point (x, y_1) with $y_1 < y$ and making the appropriate replacements for C_1 and C_2, by a similar argument we show that $\partial f/\partial y = Q$. So we see that path independence of line integrals (on open, connected sets) implies the vector field is conservative. □

Recall we say the path C is closed if the initial point and the endpoint of C are the same point. We'll restrict our attention to simple closed curves, closed curves that do not cross themselves: 0s, not 8s. For simplicity we'll call these curves *loops*. Because closed curves that do cross themselves can be viewed as the union of a collection of loops, this restriction doesn't (much) limit the closed curves we can consider. Rather, it allows us to spell out results without having to account for complicated cases.

While it may seem obvious that a loop in the plane separates the plane into two pieces, one bounded (the inside of the curve, a region of finite area) and one unbounded (the outside of the curve, a region of infinite area), that this is true is the *Jordan curve theorem*, a result difficult to prove using elementary methods. The first proof, quite long, was published in 1887 by Camille Jordan. A short proof can be written using techniques from algebraic topology, but then learning algebraic topology takes some effort.

Integrating a conservative vector field around a loop is especially easy. Just apply FTLI along with the closed path condition $(a,b) = (c,d)$. That is,

Corollary 17.2.3. If $\vec{F} = \nabla f$ for some f, the path C is closed, and f has continuous partial derivatives throughout the region bounded by C, then

$$\oint_C \vec{F} \cdot d\vec{r} = 0$$

The notation $\oint$ is used to indicate a line integral around a closed path.

Think of the trade-off between kinetic and potential energy in the motion of a pendulum without friction. The path of the pendulum is repeated exactly; total energy, the sum of kinetic and potential, is conserved. The motion is a consequence of gravitational force, conservative because it is the gradient of gravitational potential.

Combining Cors. 17.2.2 and 17.2.3 we see that if $\int_C \vec{F}\cdot d\vec{r}$ is path independent for all paths C in a region R, then $\oint_C \vec{F}\cdot d\vec{r} = 0$ for all closed paths C in R. Now we'll show the converse is true, namely

Corollary 17.2.4. If $\oint_C \vec{F}\cdot d\vec{r} = 0$ for all closed paths C in a region R, then $\int_C \vec{F}\cdot d\vec{r}$ is path independent for all paths C in R.

Proof. Suppose both paths C_1 and C_2 start at a point p and end at a point q. Let $-C_2$ denote the path C_2 but going backwards, from q to p. At each point of the curve, the tangent vector $d\vec{r}$ along $-C_2$ is the negative of the tangent vector along C_2. Consequently, $\int_{-C_2} \vec{F}\cdot d\vec{r} = -\int_{C_2} \vec{F}\cdot d\vec{r}$. Next note that $C_1 \cap -C_2$, that is, the path along C_1 from p to q, followed by the path along $-C_2$ from q to p, is a closed curve. So

$$0 = \oint_{C_1 \cup -C_2} \vec{F}\cdot d\vec{r} = \int_{C_1} \vec{F}\cdot d\vec{r} + \int_{-C_2} \vec{F}\cdot d\vec{r} = \int_{C_1} \vec{F}\cdot d\vec{r} - \int_{C_2} \vec{F}\cdot d\vec{r}$$

Because this is true for every pair of points p and q in R, and for every pair of paths C_1 and C_2 connecting p and q, the line integrals $\int_C \vec{F}\cdot d\vec{r}$ are path independent. □

Conservative vector fields lead to three questions.

1. Is every vector field the gradient of some function?
2. If some aren't gradients, how can we tell which are and which aren't?
3. If a vector field is a gradient of some function, can we find such a function?

Exercise 17.1.5 gives an approach to Questions 1 and 2. Write $\vec{F} = \langle P, Q\rangle$ and suppose $\vec{F} = \nabla f$. Then

$$P = \frac{\partial f}{\partial x} \quad \text{and} \quad Q = \frac{\partial f}{\partial y}$$

Assume the second partials of f are continuous. Apply Clairaut's theorem (the third equality in this string) to obtain

$$\frac{\partial P}{\partial y} = \frac{\partial}{\partial y}\frac{\partial f}{\partial x} = \frac{\partial^2 f}{\partial y \partial x} = \frac{\partial^2 f}{\partial x \partial y} = \frac{\partial}{\partial x}\frac{\partial f}{\partial y} = \frac{\partial Q}{\partial x}$$

That is, if $\vec{F} = \langle P, Q\rangle$ equals ∇f, then

$$\frac{\partial P}{\partial y} = \frac{\partial Q}{\partial x} \tag{17.16}$$

If Eq. (17.16) does not hold for a vector field $\vec{F} = \langle P, Q\rangle$, then that vector field is not the gradient of any function. For instance,

Example 17.2.2. *A vector field that is not a gradient.* Take $\vec{F} = \langle y^3, x^2\rangle$. Then

$$\frac{\partial P}{\partial y} = 3y^2 \quad \text{and} \quad \frac{\partial Q}{\partial x} = 2x$$

so certainly Eq. (17.16) does not hold for this vector field and consequently this vector field is not a gradient. □

This answers Question 1: there are vector fields, very many in fact, that are not gradients. And it gets part of the answer to Question 2: if Eq. (17.16) does not hold, then the vector field is not a gradient. But what about the other part of Question 2? If Eq. (17.16) does hold for a vector field, then is that vector field necessarily the gradient of some function? Surprisingly, the answer depends on the geometry of the domain, specifically on whether the domain of the vector field has holes. In Sect. 17.4 we'll develop a technique to answer this question. For the rest of this section we'll accept the fact, proved in Cor. 17.4.1, that if a vector field satisfies Eq. (17.16) and its domain has no holes, then the vector field is the gradient of a function.

Let's return to Example 17.2.1. For this vector field $\vec{F} = \langle x^2, y^3\rangle$ we have $P(x,y) = x^2$ and $Q(x,y) = y^3$. Then certainly

$$\frac{\partial P}{\partial y} = 0 \quad \text{and} \quad \frac{\partial Q}{\partial x} = 0$$

That is, this vector field satisfies Eq. (17.16). The domain of the vector field is all of the xy-plane, which has no holes. (How could the domain have holes? Suppose $P = 1/(x^2+y^2)$ and $Q = xy^2$. Then P is not defined at the origin, and the domain of the vector field $\langle P, Q\rangle$ is the plane minus the origin. The origin is a hole in the domain.) Consequently, Cor. 17.4.1 implies that there is a function f for which $\vec{F} = \nabla f$. With one slightly subtle point, finding f is not so hard to do. For any such f we know

$$\frac{\partial f}{\partial x} = P = x^2 \quad \text{and} \quad \frac{\partial f}{\partial y} = Q = y^3 \tag{17.17}$$

Let's start with the first. Because $\partial f/\partial x = x^2$, we know $f = \int x^2 dx$. Seems easy enough, but remember that f is a function of both x and y and we're integrating with respect to x only. In basic calculus we learned that $\int x^2 dx = x^3/3 + k$ for some constant k, because $(x^3/3+k)' = x^2$. So now when we integrate x^2 we get $x^3/3 + g(y)$ for some function $g(y)$, because

$$\frac{\partial}{\partial x}\left(\frac{x^3}{3} + g(y)\right) = x^2$$

That is, from $\partial f/\partial x = x^2$ we deduce that $f(x,y) = x^3/3 + g(y)$. Then $\partial f/\partial y = y^3$ becomes

$$y^3 = \frac{\partial}{\partial y}\left(\frac{x^3}{3} + g(y)\right) = g'(y)$$

and consequently

$$g = \int y^3 dy = \frac{y^4}{4} + k$$

where now k is an actual constant: it does not depend on the variable x. Combining these bits we find

$$f(x,y) = \frac{x^3}{3} + \frac{y^4}{4} + k$$

Of course, it's easy to check that $\nabla f = \vec{F}$.

Does it matter that from Eq. (17.17) we've started with $\partial f/\partial x = P$? No, not at all. Try starting with $\partial f/\partial y = Q$ and follow the analogous steps. You'll get exactly the same function f.

Once we know a potential function f, we can apply FTLI to evaluate any line integral of this vector field for any path C. For instance, suppose C is any path leading from $(1,0)$ to $(0,1)$. Then

$$\int_C \vec{F} \cdot d\vec{r} = \int_C \nabla f \cdot d\vec{r} = f(0,1) - f(1,0) = \frac{1}{4} - \frac{1}{3} = -\frac{1}{12}$$

What goes wrong with this approach if $\partial Q/\partial x \neq \partial P/\partial y$? An example will illustrate this. Take $\vec{F} = \langle y + x^2 y, x^3/3 + y^2 \rangle$. Then $\partial Q/\partial x = x^2$ and $\partial P/\partial y = 1 + x^2$. If there were a function f with $\partial f/\partial x = P$ and $\partial f/\partial y = Q$, then

$$f = \int \frac{\partial f}{\partial x}\, dx = \int (y + x^2 y)\, dx = xy + \frac{x^3}{3} y + g(y)$$

Then $\partial f/\partial y = x^3/3 + y^2$ becomes

$$\frac{x^3}{3} + y^2 = \frac{\partial}{\partial y}\left(xy + \frac{x^3}{3} y + g(y)\right) = x + \frac{x^3}{3} + g'(y)$$

and consequently $g'(y) = y^2 - x$, not possible because g depends only on y and not at all on x. This kind of thing goes wrong when we try to find a potential function for a vector field that does not satisfy Eq. (17.16).

In the next example we'll see that for the vector field of Example 17.2.2, which we recall does not satisfy Eq. (17.16), line integrals do depend on the choice of path between the endpoints.

Example 17.2.3. *Path dependence.* Integrate the vector field $\vec{F} = \langle y^3, x^2\rangle$ of Example 17.2.2 along C_1, the quarter circle from $(1,0)$ to $(0,1)$, and along C_2, the straight line segment from $(1,0)$ to $(0,1)$. The curve C_1 is parameterized by $\vec{r}(t) = \langle \cos(t), \sin(t)\rangle$ for $0 \le t \le \pi/2$. Then

$$\int_{C_1} \vec{F}\cdot d\vec{r} = \int_0^{\pi/2} \langle \sin^3(t), \cos^2(t)\rangle \cdot \langle -\sin(t), \cos(t)\rangle\, dt$$

$$= \int_0^{\pi/2} (-\sin^4(t) + \cos^3(t))\, dt$$

$$= \left(\frac{-3t}{8} + \frac{\sin(2t)}{4} - \frac{\sin(4t)}{32} + \sin(t) - \frac{\sin^3(t)}{3}\right)\Bigg|_0^{\pi/2} = \frac{2}{3} - \frac{3\pi}{16}$$

where we have used the double angle formulas $\sin^2(t) = (1-\cos(2t))/2$ and $\cos^2(t) = (1+\cos(2t))/2$ to integrate $\sin^4(t)$, and the identity $\cos^3(t) = (1-\sin^2(t))\cos(t)$ and simple substitution to integrate $\cos^3(t)$.

Next, the straight line path C_2 is parameterized by $\vec{r}(t) = \langle 1-t, t\rangle$ for $0 \le t \le 1$. Along this path the line integral is

$$\int_{C_2} \vec{F}\cdot d\vec{r} = \int_0^1 \langle t^3, (1-t)^2\rangle \cdot \langle -1, 1\rangle\, dt = \int_0^1 (-t^3 + t^2 - 2t + 1)\, dt = \frac{1}{12}$$

So for this vector field, the line integral does depend on the path. □

Now we'll give another application of Cor. 17.2.1, the path independence of line integrals of conservative vector fields.

Example 17.2.4. *Other paths.* Take $\vec{F} = \langle x^2, e^{y^2}\rangle$ and C the part of the parabola $y = x(1-x)$ between $x = 0$ and $x = 1$, parameterized by $\vec{r}(t) = \langle t, t(1-t)\rangle$ for $0 \le t \le 1$. The definition of a line integral gives

$$\int_C \vec{F}\cdot d\vec{r} = \int_0^1 \langle t^2, e^{(t(1-t))^2}\rangle \cdot \langle 1, 1-2t\rangle\, dt = \int_0^1 t^2 + e^{(t-t^2)^2}(1-2t)\, dt$$

Substitute $u = t - t^2$. The second term becomes $\int e^{u^2}\, du$, not an integral we'd like to try. But note that $\partial P/\partial y = \partial(x^2)/\partial y = 0$ and $\partial Q/\partial x = \partial(e^{y^2})/\partial x = 0$. Once again we assume, as we'll see in Sect. 17.4, that for a vector field that satisfies Eq. (17.16) and whose domain has no holes, the vector field is conservative so its line integrals are path independent and we can apply Cor. 17.2.1. The simplest path between the endpoints of C is the straight line C' parameterized by $\vec{r}(t) = \langle t, 0 \rangle$ for $0 \leq t \leq 1$. Then

$$\int_C \vec{F} \cdot d\vec{r} = \int_{C'} \vec{F} \cdot d\vec{r} = \int_0^1 \langle t^2, e^{0^2} \rangle \cdot \langle 1, 0 \rangle \, dt = \int_0^1 t^2 \, dt = \frac{1}{3}$$

Pretty easy, don't you think? □

Finally, one item about notation: writing the vector field as $\vec{F} = \langle P, Q \rangle$ and the parameterized path C as $\vec{r}(t) = \langle x(t), y(t) \rangle$, both the expressions

$$\int_C \vec{F} \cdot d\vec{r} \quad \text{and} \quad \int_C P dx + Q dy$$

represent the same line integral. Mathematicians usually prefer the first notation, physicists and engineers the second. Our facility with mathematical techniques should be invariant under change of notation.

In the next section we turn to non-vector calculus and develop a method for integrating a function over a region in the plane. This is the last ingredient we need for Green's theorem, a major tool for all the tricks in the remainder of this chapter. And, yes, we will see biological applications.

Practice Problems

17.2.1. Integrate $f(x, y) = x$ along the curve C defined by $y = x^2/2$ from $x = 0$ to $x = 1$.

17.2.2. Integrate $\vec{F} = \langle xy^2, -yx^2 \rangle$ along the curve C defined by $y = x^2$ from $x = 0$ to $x = 1$.

17.2.3. (a) For $\vec{F} = \langle y^2 - 3x^2, 2xy \rangle$ find a function f with $\nabla f = \vec{F}$.
(b) Evaluate the line integral $\int_C \vec{F} \cdot d\vec{r}$ where C is the path $y = x - x^2 - x^3 + 3x^5$ between $x = 0$ and $x = 1$.

17.2.4. Evaluate the line integral $\int_C P dx + Q dy$ where $P(x, y) = y \sin(x)$, $Q(x, y) = x \ln(1 + y^2)$, and the path C consists of two line segments, C_1, the vertical segment from $(0, 1)$ to $(0, 0)$, followed by C_2, the horizontal segment from $(0, 0)$ to $(1, 0)$.

Practice Problem Solutions

17.2.1. We can parameterize the curve by $\vec{r}(t) = \langle t, t^2/2 \rangle$ between $t = 0$ and $t = 1$. Then $\vec{r}'(t) = \langle 1, t \rangle$ and $\|\vec{r}'(t)\| = \sqrt{1 + t^2}$. Now we can evaluate the integral $\int_C f \, ds$

$$= \int_0^1 f(\vec{r}(t))\|\vec{r}'(t)\|\,dt = \int_0^1 t\sqrt{1+t^2}\,dt = \left.\frac{(1+t^2)^{3/2}}{3}\right|_0^1 = \frac{2^{3/2}-1}{3}$$

17.2.2. For this vector field $P = xy^2$ and $Q = -yx^2$, so $\partial P/\partial y = 2xy$ and $\partial Q/\partial x = -2xy$. We cannot use the fundamental theorem of line integrals or path independence, so we must evaluate the line integral directly. The curve C is parameterized by $\vec{r}(t) = \langle t, t^2\rangle$ for $0 \le t \le 1$.

$$\int_C \vec{F}\cdot d\vec{r} = \int_0^1 \langle t^5, -t^4\rangle \cdot \langle 1, 2t\rangle dt = \int_0^1 -t^5\,dt = -\frac{1}{6}$$

17.2.3. (a) Probably it's a good idea to check that this vector field satisfies Eq. (17.16) before trying to find f. For this vector field, $P = y^2 - 3x^2$ and $Q = 2xy$. Then $\partial P/\partial y = 2y$ and $\partial Q/\partial x = 2y$. The equation is satisfied, so let's find f.

From $\nabla f = \vec{F}$ we know $\partial f/\partial x = y^2 - 3x^2$ and $\partial f/\partial y = 2xy$. Let's start with the second equation, from which we obtain

$$f = \int 2xy\,dy = xy^2 + h(x)$$

Then the first equation gives

$$y^2 - 3x^2 = \frac{\partial f}{\partial x} = \frac{\partial}{\partial x}(xy^2 + h(x)) = y^2 + h'(x)$$

From this we conclude that $h'(x) = -3x^2$, so $h(x) = \int -3x^2\,dx = -x^3 + k$. Combining these we obtain $f = xy^2 - x^3 + k$.

(b) We'll use the fundamental theorem of line integrals so we're interested only in the endpoints of the curve C. These are $(0,0)$ and $(1,2)$. Then

$$\int_C \vec{F}\cdot d\vec{r} = f(1,2) - f(0,0) = (1\cdot 2^2 - 1^3 + k) - (0 - 0 + k) = 3$$

The detailed arithmetic near the end of the equation is included to point out that the value of the constant k does not alter the line integral. This is identical to a familiar result from introductory calculus: the constant added to an antiderivative does not figure in evaluating a definite integral.

17.2.4. First, it is clear from the definition of line integrals that

$$\int_C P\,dx + Q\,dy = \int_{C_1} P\,dx + Q\,dy + \int_{C_2} P\,dx + Q\,dy$$

Along C_1 we see $x = 0$ (C_1 is part of the y-axis) and $dx = 0$ (C_1 is vertical). Then

$$\int_{C_1} P\,dx + Q\,dy = \int_{C_1} x\ln(1+y^2)\,dy = 0$$

Similarly, along C_2 we see $y=0$ and $dy=0$. Then

$$\int_{C_2} P\,dx + Q\,dy = \int_{C_2} y\sin(x)\,dx = 0$$

Combining these results, we see that the original line integral is 0.

Exercises

17.2.1. Integrate these functions along the indicated curve C.
(a) $f(x,y)=x+\sqrt{y}$ and C is $y=x^2$ for $0 \le x \le 1$.
(b) $f(x,y)=x+2y$ and C is $x^2+y^2=1$ from $(1,0)$ to $(0,1)$.
(c) $f(x,y)=e^{\sqrt{y}}$ and C is given by $\vec{r}(t)=\langle t^2/2, t^2\rangle$ for $0 \le t \le 1$.
(d) $f(x,y)=x^3/y^2$ and C is given by $\vec{r}(t)=\langle t^2/2, t^3/3\rangle$ for $1 \le t \le 2$.
(e) $f(x,y)=xy$ and C is the curve $y=x^4/4$ for $0 \le x \le 1$.

17.2.2. Integrate these vector fields along the indicated curve C.
(a) $\vec{F}=\langle xy^2, x^2\rangle$ and C is $y=x^2$ for $0 \le x \le 1$.
(b) $\vec{F}=\langle x+y, y^2\rangle$ and C is $x^2+y^2=1$ from $(1,0)$ to $(-1,0)$.
(c) $\vec{F}=\langle x, x-y\rangle$ and C is given by $\vec{r}(t)=\langle t^3, t^2\rangle$ for $0 \le t \le 1$.
(d) $\vec{F}=\langle xy, x^2\rangle$ and C is $x=y^3$ for $0 \le y \le 1$.
(e) $\vec{F}=\langle x+y^2, x-y\rangle$ and C is $y=1+x^2$ for $-1 \le x \le 1$.

17.2.3. Integrate these vector fields by applying both path independence and FTLI. Verify $\partial P/\partial y = \partial Q/\partial x$.
(a) $\vec{F}=\langle x^3+xy^2, x^2y\rangle$ and C is $y=\sin(x)$ for $0 \le x \le 2\pi$.
(b) $\vec{F}=\langle \sin(x)+y^2, \cos(y)+2xy\rangle$ and C is $y=1-x^2$ for $-1 \le x \le 1$.
(c) $\vec{F}=\langle x^3+y, x+\sin(y^2)\rangle$ and C is $y=x-x^2$ for $0 \le x \le 1$.
(d) $\vec{F}=\langle yx^3, x^4/4+y^3\rangle$ and C is $y=x^3$ for $-1 \le x \le 1$.
(e) $\vec{F}=\langle y+\cos(x), x-\ln(1+y^2)\rangle$ and C is $y=x-x^3$ for $0 \le x \le 1$.

17.3 DOUBLE INTEGRALS

We know how to differentiate and integrate functions of a single variable, and we've learned to differentiate functions of two variables by partial derivatives: differentiate one variable at a time, treating the other variables as constant. More or less the same idea allows us to integrate functions of two (or more) variables. Integrate with respect to one variable at a time, holding the other variables constant during the integration.

This is easiest to illustrate if we're integrating over a rectangle R given by $a \le x \le b$ and $c \le y \le d$. Suppose $f(x,y)=x^2y+y^3$. Then

$$\begin{aligned}\iint_R f(x,y)\,dA &= \int_a^b \int_c^d x^2y+y^3\,dy\,dx = \int_a^b \left(\frac{x^2y^2}{2}+\frac{y^4}{4}\right)\Bigg|_c^d dx \\ &= \int_a^b \left(\frac{x^2(d^2-c^2)}{2}+\frac{d^4-c^4}{4}\right) dx = \left(\frac{x^3}{3}\frac{d^2-c^2}{2}+\frac{x(d^4-c^4)}{4}\right)\Bigg|_a^b \\ &= \frac{(b^3-a^3)(d^2-c^2)}{6}+\frac{(b-a)(d^4-c^4)}{4}\end{aligned}$$

In the first integral dA indicates integration with respect to a differential of area. This method is called *iterated integration* because we integrate with respect to one variable, then integrate with respect to the other.

In this calculation we integrated first with respect to y and then with respect to x. Does the order matter? A first step is to repeat this calculation, integrating first with respect to x, then with respect to y. If you want some practice with double integrals, try this. You'll get the same answer. Is this true for all functions? No. The conditions on f which guarantee that the order of integration is unimportant are spelled out in Fubini's theorem. The list of conditions is fairly involved, but Cauchy (years before Fubini) knew that equality of iterated integrals holds for continuous functions. Cauchy also found examples of unbounded functions for which changing the order of integration changes the answer. So long as we restrict our attention to continuous functions, we can change the order of integration whenever it's convenient.

Before we move on, we'll mention a point of notation. We do integration from the inside out. For example, first integrate with respect to y, so the integral with limits on y and the dy are closer to $f(x,y)$. Then we integrate with respect to x.

$$\int_{\text{limits on } x} \int_{\text{limits on } y} f(x,y)\,dy\,dx$$

In some other books you'll find integration done left to right

$$\int_{\text{limits on } x} \int_{\text{limits on } y} f(x,y)\,dx\,dy$$

That is, as written here, the left limits on x pair with the left variable, indicated by dx. After integrating with respect to the left variable, we move on to the right variable, here dy, paired with the right limits of integration.

If both sets of limits of integration are constants, only by knowing the author's convention can we tell if the integral is from inside to out or from left to right. For example, in

$$\int_1^2 \int_1^4 x^2+xy\,dx\,dy$$

our convention is to integrate from inside to out, so the limits are $1 \le x \le 4$ and $1 \le y \le 2$. The answer is $129/4$ if you're checking. If we were to integrate from left to right (which we aren't), the limits would be $1 \le x \le 2$ and $1 \le y \le 4$, giving $73/4$. Clearly, the value of the integral depends on which limits go with which variables. Spelling out the convention, inside to out or left to right, is a way to be sure we apply the intended limits to each variable. Usually it's easy to tell which choice an author has made, but inattentive readers can paint themselves into a very confusing corner.

Not every double integral is over a rectangle. For more general regions we need a bit of care in setting up the integrals. While there are more complicated regions, most can be broken into pieces that look like one of the examples in Fig. 17.5.

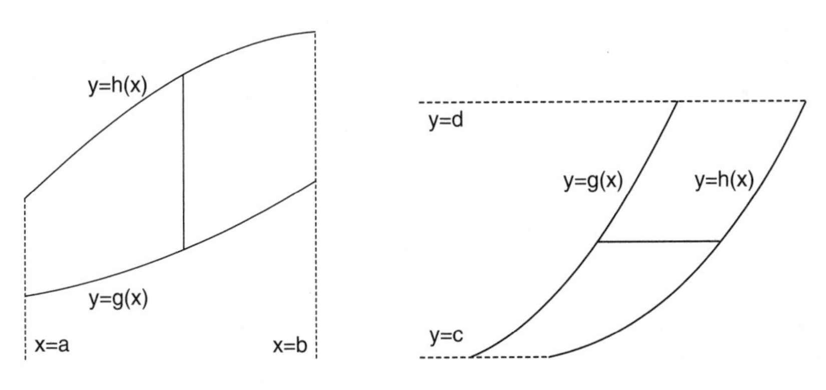

Figure 17.5. Two non-rectangular regions.

How can we describe region (i)? Simplest to see is that x lies between constant left and right bounds, $a \le x \le b$. The y-bounds are more complicated because they depend on x. For each x in the interval $a \le x \le b$, y lies in the interval $h(x) \le y \le g(x)$. Because we are integrating with respect to both x and y, when we've finished integrating we should see neither x nor y in the answer. That means for regions of the shape (i) we must integrate first with respect to y, then with respect to x. Using our convention of integrating from the inside out, this is

$$\iint_R f(x,y)\,dA = \int_a^b \int_{g(x)}^{h(x)} f(x,y)\,dy\,dx$$

To illustrate this point more concretely, suppose the region is $0 \le x \le 1$ and $x^2 \le y \le x$. To make the example as simple as possible, we'll take $f(x,y) = 1$.

Then the right way gives

$$\int_0^1 \int_{x^2}^{x} 1\, dy\, dy = \int_0^1 y\Big|_{x^2}^{x} dx = \int_0^1 (x - x^2)\, dx = \left(\frac{x^2}{2} - \frac{x^3}{3}\right)\Big|_0^1 = \frac{1}{6}$$

while the wrong way gives

$$\int_{x^2}^{x} \int_0^1 1\, dx\, dy = \int_{x^2}^{x} x\Big|_0^1 dy = \int_{x^2}^{x} 1\, dy = y\Big|_{x^2}^{x} = x^2 - x$$

The integral should be a number, so this makes no sense. If there are variable limits of integration, they need to be on the inside integral.

For shapes in the form of region (i), integration in this order sweeps through the region with vertical segments of varying length.

Region (ii) can be described by $c \leq y \leq d$ and $g(y) \leq x \leq h(y)$. Keeping in mind that the variable limits of integration must be on the inner integral, integrals over regions of this type have the form

$$\iint_R f(x,y) dA = \int_c^d \int_{g(y)}^{h(y)} f(x,y)\, dx\, dy$$

Integration in this order sweeps through the region with horizontal segments of varying length.

If a region can be described in both ways (rectangles for sure, but there are many others as well), then we can change the order of integration, though must be careful with the limits of integration. Here's an example.

Example 17.3.1. *Order of integration.* Suppose $f(x,y) = e^{y^2}$ and R is the triangular region $0 \leq x \leq 1, x \leq y \leq 1$, shown in Fig. 17.6. With this description of the region, the double integral of f over R is

$$\iint_R f(x,y) dA = \int_0^1 \int_x^1 e^{y^2}\, dy\, dx$$

Do you know an antiderivative of e^{y^2}? I don't. The best we can do is use the method of Sect. 10.8: find the Taylor series for e^{y^2}, integrate term by term, and go from there. More than a little complicated.

But suppose we change the order of variables in our description of the region R? The numerical limits on y are $0 \le y \le 1$, and for each y, the limits on x are $0 \le x \le y$. With this description, the double integral becomes

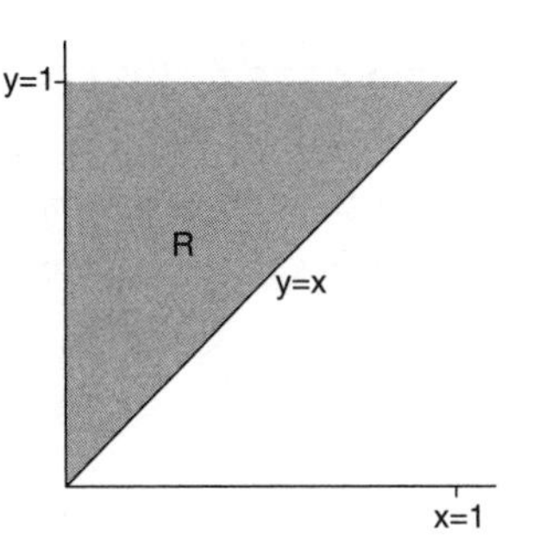

Figure 17.6. Triangular region.

$$\iint_R f(x,y)dA = \int_0^1 \int_0^y e^{y^2}\, dx\, dy =$$

$$\int_0^1 xe^{y^2}\Big|_0^y dy = \int_0^1 ye^{y^2}\, dy = \frac{e^{y^2}}{2}\Big|_0^1 = \frac{e-1}{2}$$

where the last integral is evaluated by the substitution $u = y^2$, so $du = 2ydy$. We see that switching the order of integration can make a difficult problem much simpler. The Taylor series approach may be of interest, but frustration is the more likely result. □

This example shows that changing the form of an integral can simplify its computation. Utilizing a symmetry of the problem can lead to significant economies, and because circular symmetry is common in nature, we'll learn how to do double integrals in polar coordinates. There is only one issue: how to express $dA = dx\, dy$ in terms of dr and $d\theta$.

In Cartesian coordinates dA is the area of a rectangle with side lengths dx and dy. But really, these are the distances between $x + dx$ and x, and between $y + dy$ and y. In polar coordinates, we look at the radial distance between $r + dr$ and r, and the angular distance between $\theta + d\theta$ and θ. The area dA is the shaded region in Fig. 17.7. The radial sides are straight lines with length dr, simple enough. The arcs of the circles have arclength $r\, d\theta$ (the shorter arc) and $(r + dr)d\theta$ (the longer arc). Here we use the fact that for a circle of radius r, an angle θ (measured in radians) determines an arc of length $r\theta$.

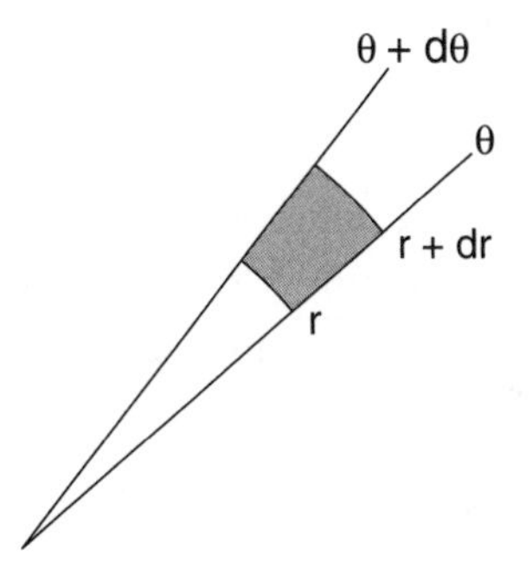

Figure 17.7. Polar area

Back to dA. We'll take the angular distance to be given by the average of the two arclengths, so in polar coordinates

$$dA = \frac{r + (r + dr)}{2}\, d\theta\, dr = r\, d\theta\, dr + \frac{1}{2}\, d\theta\, (dr)^2$$

Because $d\theta$ and dr are small, the last term is much smaller still and vanishes in the $dr \to 0$, $d\theta \to 0$ limit. So in polar coordinates,

$$dA = r\, d\theta\, dr \tag{17.18}$$

Let's see how to evaluate a double integral in polar coordinates.

Example 17.3.2. *Polar coordinate integration.* To Integrate $f(x,y) = x+y$ over the region R in the first quadrant and between the circles of radius 1 and 2, the Cartesian description of R must be broken into two subregions. In polar coordinates R is a rectangle:

$$1 \leq r \leq 2 \quad \text{and} \quad 0 \leq \theta \leq \pi/2$$

Then

$$\iint_R x+y\,dA = \int_1^2 \int_0^{\pi/2} (r\cos(\theta)+r\sin(\theta))r\,d\theta\;dr$$
$$= \int_1^2 r^2(-\sin(\theta)+\cos(\theta))\Big|_0^{\pi/2}\,dr = \int_1^2 2r^2\,dr = (2r^3/3)\Big|_1^2 = 14/3$$

This is not such a difficult calculation. □

If you don't want to be bothered with learning to integrate in polar coordinates, try to do this integral in Cartesian coordinates. That should be enough to convince you that we should use every symmetry the system possesses.

Practice Problems

17.3.1. Integrate $f(x,y) = x+2y$ over the region R bounded by $y=0$, $y=1$, $y=x$, and $x=2$.

17.3.2. Integrate $f(x,y) = x-y$ over the region R bounded by the line segments connecting $(1,0)$ to $(0,0)$ and connecting $(0,0)$ to $(\sqrt{2}/2, \sqrt{2}/2)$, and the curve $y=\sqrt{1-x^2}$ for $\sqrt{2}/2 \leq x \leq 1$.

17.3.3. Evaluate the integral $\displaystyle\int_0^1 \int_{\sqrt{y}}^1 \sin(x^3)\,dx\,dy$.

Practice Problem Solutions

17.3.1. The simplest description of R is $0 \leq y \leq 1$ and $y \leq x \leq 2$. Then

$$\iint_R x+2y\,dA = \int_0^1 \int_y^2 x+2y\,dx\,dy = \int_0^1 \left(\frac{x^2}{2}+2xy\right)\Big|_y^2\,dy$$
$$= \int_0^1 \left(-\frac{5y^2}{2}+4y+2\right)dy = \left(-\frac{5y^3}{6}+2y^2+2y\right)\Big|_0^1 = \frac{19}{6}$$

17.3.2. First note that the curve $y=\sqrt{1-x^2}$ for $\sqrt{2}/2 \leq x \leq 1$ is an eighth circle, and that the region R is the portion of the unit disc in the first quadrant and below the line $y=x$. The circular portion of the boundary suggests using polar coordinates. Then the region again is a rectangle $0 \leq r \leq 1$ and $0 \leq \theta \leq \pi/4$. The integral is

$$\iint_R x-y\,dA = \int_0^1 \int_0^{\pi/4} r\cos(\theta) - r\sin(\theta)\; r\,d\theta\,dr$$
$$= \int_0^1 r^2(\sin(\theta)+\cos(\theta))\Big|_0^{\pi/4} dr = \int_0^1 r^2(\sqrt{2}-1)\,dr = \frac{\sqrt{2}-1}{3}$$

17.3.3. Probably you don't know an antiderivative for $\sin(x^3)$. I certainly don't. Mathematica gives a complicated answer involving Gamma functions, but let's agree that we don't know this antiderivative. So let's sketch the region R and reverse the order of integration. As presented in Fig. 17.8, the region is $0 \leq y \leq 1$ and $\sqrt{y} \leq x \leq 1$. There's no trick here: we just read them from the limits of integration in the integral.

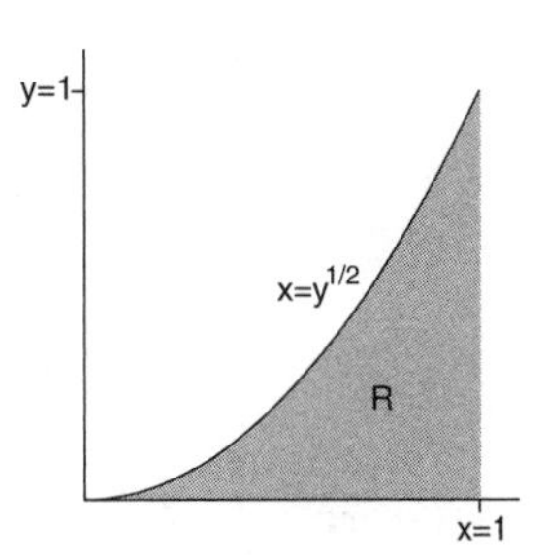

Figure 17.8. Integration region of 17.3.3.

To reverse the order of the variables, first note that $x=\sqrt{y}$ becomes $y=x^2$. Then we see that R also can be described as $0 \leq x \leq 1$ and $0 \leq y \leq x^2$. Then the integral becomes

$$\int_0^1 \int_{\sqrt{y}}^1 \sin(x^3)\,dx\,dy = \int_0^1 \int_0^{x^2} \sin(x^3)\,dy\,dx = \int_0^1 (y\sin(x^3))\Big|_0^{x^2} dx$$
$$= \int_0^1 x^2\sin(x^3)\,dx = -\frac{\cos(x^3)}{3}\Big|_0^1 = \frac{1}{3} - \frac{\cos(1)}{3}$$

Exercises

17.3.1. Evaluate $\iint_R f(x,y)\,dA$ over the regions R indicated.
(a) $f(x,y)=x+x^2y$ and R is above $y=x^2$ and below $y=1$.
(b) $f(x,y)=x+x^2y$ and R is above $y=x^2$ and below $y=2-x^2$.
(c) $f(x,y)=x+x^2y$ and R is between $x=y^2$ and $x=2-y^2$.
(d) $f(x,y)=xy^2$ and R is between $y=x^2$ and $y=x^3$ for $0 \leq x \leq 1$.
(e) $f(x,y)=xy^2$ and R is between $y=x^2$ and $y=x^3$ for $0 \leq x \leq 2$.

17.3.2. Evaluate $\iint_R f(x,y)\,dA$ over the regions R and sketch R.
(a) $f(x,y)=x+y^2$ and R is the region between $r=1$ and $r=2$.
(b) $f(x,y)=x-y$ and R is the region inside $r=1$ and below $y=x$.
(c) $f(x,y)=x+y^2$ and R is the region inside $(x-1)^2+y^2=1$. Hint: find a

polar representation for R by substitution of $x = r\cos(\theta)$ and $y = r\sin(\theta)$ into the equation for the circle, then solve for r.

(d) $f(x,y) = x + 2y$ and R is the region between $(x-1)^2 + y^2 = 1$ and $x^2 + (y-1)^2 = 1$.

(e) $f(x,y) = xy$ and R is the region inside $x^2 + y^2 = 4$ and to the right of $x = 1$.

17.3.3. Evaluate $\iint_R f(x,y)\, dA$ over the regions R indicated.

(a) $\displaystyle\int_0^1 \int_{\arcsin(y)}^{\pi/2} \cos(x)\, dx\, dy$ (b) $\displaystyle\int_1^e \int_{\ln(y)}^1 e^x\, dx dy$

(c) $\displaystyle\int_0^1 \int_0^{\arccos(y)} e^{\sin(x)}\, dx\, dy$ (d) $\displaystyle\int_0^1 \int_{x^2}^1 e^{\sqrt{y}}/y\, dy\, dx$

(e) $\displaystyle\int_0^1 \int_\theta^1 \sin(1+r^3) r\, dr\, d\theta$ (f) $\displaystyle\int_0^1 \int_y^1 x/(1+x^3)\, dx\, dy$

17.4 GREEN'S THEOREM

Green's theorem is a beautiful result, for many (including me) the first view of the subtle relations between geometry and calculus. Green's theorem shows that the line integral of a vector field around a closed curve equals the double integral of a sort of derivative of the vector field over the region bounded by the curve. With Green's theorem we'll develop conditions (Sect. 17.6) that guarantee a differential equation does not have trajectories that are closed curves. Also, with Green's theorem we can define a number, the index of a fixed point of a vector field (Sect. 17.7), that encapsulates information about the structure of trajectories near the fixed point. Surprisingly, we'll find a vector field which, when integrated around a curve, gives the *area* enclosed by the curve. That integration of the length of the tangent vector field of a closed curve gives the perimeter (length) of the curve we expect and understand. But it's a surprise that there's a vector field we can integrate along a closed curve to get the area of the region bounded by the curve. And there's a mechanical device, a *planimeter*, that calculates areas in this way.

So we'll state the theorem, sketch a proof, and see some applications, including the proof that $\partial Q/\partial x = \partial P/\partial y = 0$ implies $\vec{F} = \langle P, Q\rangle$ is conservative, that is, $\vec{F} = \nabla f$ for some function f. In Sect. 17.5 we'll find some of the interesting geometry involved in extensions and elaborations of Green's theorem.

Before stating the theorem, we need two concepts. First, recall a region is simply-connected if it has no holes. Another formulation is that any closed curve in a simply-connected region can be shrunk to a point without going outside the region. The left region shown in Fig. 17.9 is simply-connected. The middle region is not.

The second concept involves the orientation of the curve relative to the region it bounds. In the right image of Fig. 17.9, the curve C is *oriented positively*. This means that if the curve is drawn on the floor and you walk around the curve in the direction indicated, then the region is on your left.

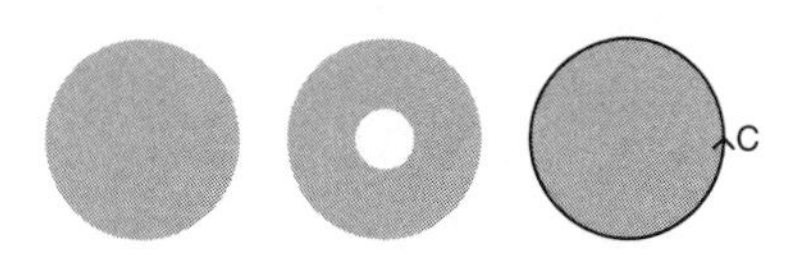

Figure 17.9. Three regions.

Theorem 17.4.1. *Green's theorem.* Suppose C is a closed curve bounding a simply-connected region R and C is oriented positively. If P and Q have continuous partial derivatives throughout R, then

$$\oint_C P\,dx + Q\,dy = \iint_R \frac{\partial Q}{\partial x} - \frac{\partial P}{\partial y}\,dA \tag{17.19}$$

We'll sketch the proof in a simple situation. The full proof involves no new ideas, just a lot of bookkeeping. We'll show

$$\oint_C P\,dx = -\iint_R \frac{\partial P}{\partial y}\,dA \tag{17.20}$$

A similar argument shows

$$\oint_C Q\,dy = \iint_R \frac{\partial Q}{\partial x}\,dA \tag{17.21}$$

We'll leave that step for you. Adding these two results gives Eq. (17.19).

Suppose the region R is as shown in Fig. 17.10, and the boundary curve C is split into a lower curve C_1 and an upper curve C_2. Split this way, both the lower and upper curves are graphs of functions, $y = g_1(x)$ for C_1 and $y = g_2(x)$ for C_2. For both these curves $a \leq x \leq b$, but notice that the orientation of C_1 agrees with increasing x, while the orientation of C_2 corresponds to decreasing x. With these orientations, the parameterization of C_1 is $\langle x, g_1(x)\rangle$ for x from a to b, and for C_2 we have $\langle x, g_2(x)\rangle$ for x from b to a. Now we can integrate Pdx around the curve C by integrating along each piece, where the orientation of each piece is consistent with that of C and using the parameterizations by g_1 and g_2.

$$\oint_C P(x,y)\,dx = \int_{C_1} P(x,y)\,dx + \int_{C_2} P(x,y)\,dx$$

$$= \int_a^b P(x, g_1(x))\,dx + \int_b^a P(x, g_2(x))\,dx$$

$$= \int_a^b P(x, g_1(x))\, dx - \int_a^b P(x, g_2(x))\, dx$$

$$= -\int_a^b (P(x, g_2(x)) - P(x, g_1(x)))\, dx$$

$$= -\int_a^b P(x,y)\Big|_{y=g_1(x)}^{y=g_2(x)}\, dx = -\int_a^b \left(\int_{g_1(x)}^{g_2(x)} \frac{\partial P}{\partial y}\, dy \right) dx = -\iint_R \frac{\partial P}{\partial y}\, dA$$

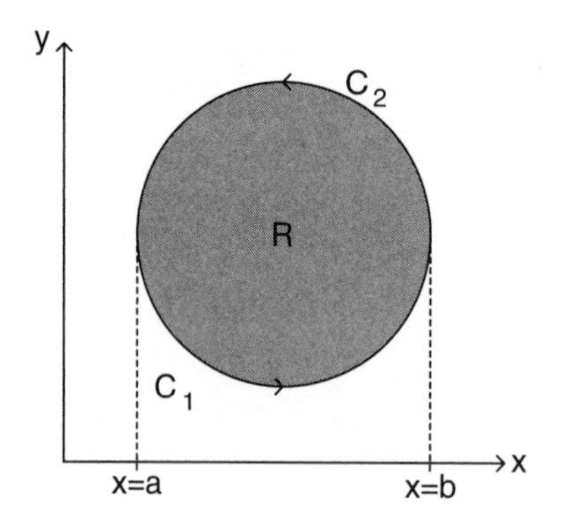

Figure 17.10. A setup for Green's theorem.

This shows Eq. (17.20). Break C into left and right parts. Then a similar argument proves Eq. (17.21). For a more complicated region, break it into simple regions, apply Green's theorem to each piece, and add the results. In Sect. 17.5 we'll see that the line integrals cancel along the interior edges formed when we break the complicated region into simple regions. □

Time for some examples. We'll use Green's theorem to evaluate some line integrals.

Example 17.4.1. *Green's theorem on a rectangle.* Suppose R is the filled-in triangle with vertices $(0,0)$, $(1,0)$, and $(1,2)$, so we can describe R by $0 \leq x \leq 1$ and $0 \leq y \leq 2x$. Suppose C is the boundary of R, oriented positively. To integrate the vector field $\vec{F} = \langle e^x, xy \rangle$ over C we could decompose C into three line segments and integrate $\vec{F}$ along each segment. Not challenging, but tedious. Or we could apply Green's theorem.

$$\oint_C e^x\, dx + xy\, dy = \iint_R \left(\frac{\partial}{\partial x} xy - \frac{\partial}{\partial y} e^x \right) dA = \int_0^1 \int_0^{2x} y\, dy\, dx$$

The inner integral evaluates to $2x^2$. Then the outer integral gives $2/3$. □

The next example better shows the power of Green's theorem.

Example 17.4.2. *Green's theorem to simplify the integrand.* Suppose R is the filled-in square with vertices $(0,0)$, $(1,0)$, $(1,1)$, and $(0,1)$, C is the boundary of R oriented positively, and the vector field is $\vec{F} = \langle e^{x^2} - y, x^2 + \ln(2 + \sin^2(y)) \rangle$. Some of the terms in P and Q look difficult to integrate directly, and in any case we'd have to do four line integrals, so let's use Green's theorem to evaluate this line integral.

$$\begin{aligned}
&\oint_C (e^{x^2} - y)\, dx + (x^2 + \ln(2+\sin^2(y)))\, dy \\
&= \iint_R \left(\frac{\partial}{\partial x}(x^2 + \ln(2+\sin^2(y))) - \frac{\partial}{\partial y}(e^{x^2} - y) \right) dA \\
&= \int_0^1 \int_0^1 (2x+1)\, dy\, dx = \int_0^1 (2x+1)y \Big|_0^1 dx = (x^2+x)\Big|_0^1 = 2
\end{aligned}$$

The partial derivatives kill the hard-to-integrate terms. □

Now a much simpler, yet surprising, example, the *planimeter*.

Example 17.4.3. *Green's theorem and areas.* Suppose R is any simply-connected region in the plane and C is its boundary curve, oriented positively. Take the vector field $\vec{F} = (1/2)\langle -y, x\rangle$. Then by Green's theorem

$$\frac{1}{2}\oint_C -y\, dx + x\, dy = \frac{1}{2}\iint_R \left(\frac{\partial}{\partial x}x - \frac{\partial}{\partial y}(-y) \right) dA = \iint_R dA = \text{area}(R)$$

That is, the area A enclosed by the simple closed curve C is

$$\text{area}(A) = \frac{1}{2}\oint_C -y\, dx + x\, dy \qquad (17.22)$$

Already this is interesting, that integrating this vector around the boundary of a region gives the area of the region. The real surprise is that there is a mechanical device, a *planimeter*, a wheel on an arm with a jointed elbow that can measure the area of a region by running the wheel around the boundary of the region. On the web you can find many pictures of mechanical planimeters, as well as apps to simulate the operation of a planimeter. As in so many other quests, in this Google is your true and trusted friend. □

Now we'll show that if Eq. (17.16) holds on a suitable region R, then the vector field $\vec{F} = \langle P, Q\rangle$ is conservative.

Corollary 17.4.1. Suppose P and Q have continuous partial derivatives throughout an open, connected, and simply-connected region R and that $\partial Q/\partial x = \partial P/\partial y$ in R. Then there is a function f defined on R for which $\nabla f = \langle P, Q\rangle$.

Proof. We'll apply Green's theorem to show that the line integral of $\vec{F} = \langle P, Q\rangle$ around any simple closed curve in R is 0. Then Cor. 17.2.4 shows that line integrals of $\vec{F}$ are path independent, and so Cor. 17.2.2 shows that $\vec{F}$ is conservative. Here's the application of Green's theorem. Recall that because the region R is simply-connected, the simple closed curve C bounds a region D in R. Then

$$\oint_C \vec{F} \cdot d\vec{r} = \oint_C P\,dx + Q\,dy = \iint_D \frac{\partial Q}{\partial x} - \frac{\partial P}{\partial y}\,dA = \iint_D 0\,dA = 0$$

and we're finished. □

In the next section we'll investigate Green's theorem for regions with more complicated geometries.

Practice Problems

17.4.1. Evaluate $\oint_C x^3\,dx + xy\,dy$ where C consists of the line segments from $(0,0)$ to $(1,0)$, from $(1,0)$ to $(0,1)$, and from $(0,1)$ to $(0,0)$.

17.4.2. Evaluate $\oint_C (\sin(y) - y^2)\,dx + (x^2 + x\cos(y))\,dy$, where C is the boundary of the unit square $0 \leq x \leq 1$, $0 \leq y \leq 1$, oriented positively.

17.4.3. Find the area enclosed by $x^{2/3} + y^{2/3} = 1$.

Practice Problem Solutions

17.4.1. Apply Green's theorem. The region R bounded by the curve C can be described by $0 \leq x \leq 1$ and $0 \leq y \leq 1 - x$. Then

$$\oint_C x^3\,dx + xy\,dy = \iint_R \frac{\partial}{\partial x}xy - \frac{\partial}{\partial y}x^3\,dA$$
$$= \int_0^1 \int_0^{1-x} y\,dy\,dx = \int_0^1 \frac{1}{2} + \frac{x^2}{2} - x\,dx = \frac{1}{6}$$

17.4.2. The region R is the unit square bounded by the segments of C.

$$\oint_C (\sin(y) - y^2)\,dx + (x^2 + x\cos(y))\,dy$$
$$= \iint_D \frac{\partial}{\partial x}(x^2 + x\cos(y)) - \frac{\partial}{\partial y}(\sin(y) - y^2)\,dA$$
$$= \iint_D (2x + \cos(y)) - (\cos(y) - 2y)\,dA = \int_0^1 \int_0^1 (2x + 2y)\,dx\,dy$$
$$= \int_0^1 (x^2 + 2yx)\Big|_0^1\,dy = \int_0^1 (1 + 2y)\,dy = (y + y^2)\Big|_0^1 = 2$$

17.4.3. We'll calculate the area using the formula Eq. (17.22), so we'll need a parameterization of the curve $x^{2/3}+y^{2/3}=1$. Except for the cube roots, this looks like $x^2+y^2=1$, which we parameterize by $x(t)=\cos(t)$ and $y(t)=\sin(t)$. So let's try $x(t)=\cos^3(t)$, $y(t)=\sin^3(t)$ for $0\le t\le 2\pi$. It's easy to check that this parameterization satisfies $x^{2/3}+y^{2/3}=1$. Then Eq. (17.22) gives the area for the region enclosed by the curve.

$$\frac{1}{2}\int_C x\,dy-y\,dx=\frac{1}{2}\int_0^{2\pi}\cos^3(t)\,d(\sin^3(t))-\sin^3(t)\,d(\cos^3(t))$$
$$=\frac{1}{2}\int_0^{2\pi}(\cos^3(t)3\sin^2(t)\cos(t)-\sin^3(t)3\cos^2(t)(-\sin(t))\,dt$$
$$=\frac{3}{2}\int_0^{2\pi}(\cos^4(t)\sin^2(t)+\sin^4(t)\cos^2(t))\,dt=\frac{3}{2}\int_0^{2\pi}(\cos(t)\sin(t))^2\,dt$$
$$=\frac{3}{2}\int_0^{2\pi}\left(\frac{\sin(2t)}{2}\right)^2dt=\frac{3}{8}\int_0^{2\pi}\frac{1-\cos(4t)}{2}\,dt=\frac{3\pi}{8}$$

This curve is called an *astroid.*

Exercises

17.4.1. Evaluate $\oint_C P\,dx+Q\,dy$ for these P, Q, and C.
(a) $P=Q=xy$, and C is the ellipse $x^2/4+y^2=1$, oriented positively.
(b) $P=x+y^2$, $Q=xy^3$, and C is the part of the parabola $y=x^2$ from $(0,0)$ to $(1,1)$, followed by the part of the line $y=x$ from $(1,1)$ to $(0,0)$.
(c) $P=y^2+\sin(x^3)$, $Q=x^2+e^{y^3}$, and C is the upper semicircle from $(1,0)$ to $(-1,0)$, followed by the portion of the x-axis from $(-1,0)$ to $(1,0)$.
(d) $P=y^2+x$, $Q=x^2+\ln(1+y^2)$, and C is the path of (c).
(e) $P=-yx^2+1/(1+x^4)$, $Q=xy^2+y^3$, and C is the circle $x^2+y^2=1$, oriented positively.

17.4.2. Evaluate $\oint_C P\,dx+Q\,dy$ for these P, Q, and C.
(a) $P=xy+\cos(x^2)$, $Q=e^y$, and C is the circle $x^2+y^2=1$, oriented positively.
(b) $P=-y+x^3$, $Q=x+y^3$, and C is the curve $x^2-2x+y^2=0$ for $0\le x\le 2$, oriented positively. Hint for identifying the curve: complete the square for the terms involving x.
(c) $P=xy+\sqrt{x}$, $Q=xy^2-\sqrt{y}$, and C is the boundary of the triangle $0\le x\le 1$, $0\le y\le x$, oriented positively.
(d) $P=y\cos(\pi x)$, $Q=x^2-y^2$, and C is the boundary of the unit square $0\le x\le 1$, $0\le y\le 1$, oriented positively.

(e) $P = -y\ln(1+x)+x^2$, $Q = y^2$, and C is the boundary of $0 \le x \le 1$, $0 \le y \le x$, oriented positively.

17.4.3. Find the area of each of these regions.
(a) The curve with polar representation $r = \sin(2\theta)$ for $0 \le \theta \le \pi/2$.
(b) The curve with polar representation $r = 3\sin(2\theta)$ for $0 \le \theta \le \pi/2$. Is this area 3 times that of the region in (a)?
(c) The curve with polar representation $r = \sin(3\theta)$ for $0 \le \theta \le \pi/3$.
(d) The region bounded by $\langle x(t), y(t)\rangle = \langle t - \sin(t), 1 - \cos(t)\rangle$ for $0 \le t \le 2\pi$, together with the x-axis. If the sign of your answer puzzles you, think about the direction around the curve determined by this parameterization.
(e) The region bounded by $x(t) = 2\cos(t) - \cos(2t)$ and $y(t) = 2\sin(t) - \sin(2t)$ for $0 \le t \le 2\pi$.

17.5 INTERESTING GEOMETRIES

Green's theorem makes two geometric requirements: the line integral is around a closed curve C, and that C bounds a simply-connected region R on which the vector field and its partial derivatives are continuous. First, we'll show that sometimes Green's theorem can be applied if the curve C isn't closed. Then we'll see that sometimes we can use Green's theorem even if the region it bounds isn't simply-connected, either because the vector field is not defined everywhere in the bounded region, or the geometry of the problem excludes some portion of the bounded region.

The curve isn't closed.

First, we may be able to use Green's theorem to integrate a vector field over a curve C_1 that is not closed. We can add another curve C_2 so that $C_1 \cup C_2$ is a closed curve, then apply Green's theorem to the region bounded by $C_1 \cup C_2$ and subtract the line integral along C_2. For this to be a useful approach, the line integral over C_2 must be simpler to evaluate than the line integral over C_1. Here's an example.

Example 17.5.1. *Closing the curve.* Suppose C_1 consists of three line segments, from $(1,0)$ to $(1,1)$, then to $(0,1)$, and finally to $(0,0)$. Certainly, C_1 is not a closed curve. For

$$P = x^3 - y^3 \quad \text{and} \quad Q = x^2 + e^{y^2}$$

evaluating $\int_{C_1} P\,dx + Q\,dy$ does not look inviting. (If you try it, you may find a lucky subtraction.) But because C_1 isn't a closed curve, we can't apply Green's theorem. However, suppose we follow C_1 with a path C_2, the line segment from $(0,0)$ to $(1,0)$. Then $C_1 \cup C_2$ is a closed curve and encloses the region R given by $0 \le x \le 1$ and $0 \le y \le 1$. Now we can apply Green's theorem.

$$\oint_{C_1 \cup C_2} (x^3 - y^3)dx + (x^2 + e^{y^2})dy = \iint_R \frac{\partial}{\partial x}(x^2 + e^{y^2}) - \frac{\partial}{\partial y}(x^3 - y^3)dA$$

$$= \int_0^1 \int_0^1 (2x + 3y^2)dy\, dx = \int_0^1 (2xy + y^3)\Big|_0^1 dx = \int_0^1 (2x+1)dx = 2$$

From this we have

$$\oint_{C_1 \cup C_2} (x^3 - y^3)dx + (x^2 + e^{y^2})dy = 2$$

$$\int_{C_1} (x^3 - y^3)dx + (x^2 + e^{y^2})dy + \int_{C_2} (x^3 - y^3)dx + (x^2 + e^{y^2})dy = 2$$

$$\int_{C_1} (x^3 - y^3)\, dx + (x^2 + e^{y^2})dy = 2 - \int_{C_2} (x^3 - y^3)dx + (x^2 + e^{y^2})dy$$

Along the path C_2, $y = 0$ and $dy = 0$, so the line integral along C_2 is

$$\int_{C_2} (x^3 - y^3)dx + (x^2 + e^{y^2})dy = \int_0^1 x^3\, dx = \frac{1}{4}$$

Combining these results, we see

$$\int_{C_1} (x^3 - y^3)dx + (x^2 + e^{y^2})dy = 2 - \frac{1}{4} = \frac{7}{4}$$

In order for this to be effective, we must be able to evaluate the double integral given by Green's theorem, and also the line integral along C_1. When we can do both steps, this is a good way to enlarge the collection of line integrals that can be evaluated using Green's theorem. □

The bounded region isn't simply-connected.

If C bounds a region that isn't simply-connected, a trick may allow us to apply Green's theorem. Split the region into pieces that are simply-connected, apply Green's theorem to each piece, and then add the results. An example will illustrate this trick.

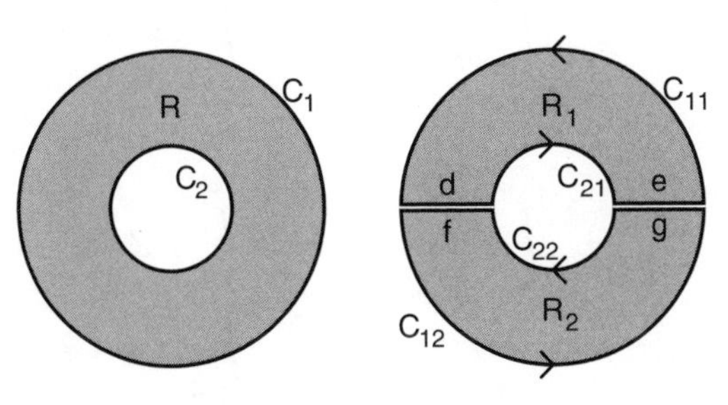

Figure 17.11. Decomposing a region.

The first image of Fig. 17.11 is a region R that is not simply-connected: notice the empty space in its middle. The boundary of R consists of two curves, C_1 and C_2. The second image shows how to break R into two simply-connected regions. Working out the justification will take a bit of time, but the result is simple. We'll state it now, so knowing the goal may help you understand some of the steps.

As long as the components of the boundary of R are oriented positively, that is, so R is on the left-hand side as we follow the boundary,

$$\oint_{C_1} P\,dx + Q\,dy + \oint_{C_2} P\,dx + Q\,dy = \iint_R \frac{\partial Q}{\partial x} - \frac{\partial P}{\partial y}\,dA \tag{17.23}$$

Now if C_2 is oriented clockwise, $-C_2$ is oriented counterclockwise and Eq. (17.23) becomes

$$\oint_{C_1} P\,dx + Q\,dy - \oint_{-C_2} P\,dx + Q\,dy = \iint_R \frac{\partial Q}{\partial x} - \frac{\partial P}{\partial y}\,dA$$

Solving for $\oint_{C_1} P\,dx + Q\,dy$ we obtain

$$\oint_{C_1} P\,dx + Q\,dy = \oint_{-C_2} P\,dx + Q\,dy + \iint_R \frac{\partial Q}{\partial x} - \frac{\partial P}{\partial y}\,dA \tag{17.24}$$

So we split R into two pieces, R_1 and R_2, both of which are simply-connected. We split the curve C_1 into two pieces, C_{11} and C_{12}, and split C_2 into C_{21} and C_{22}. To complete the boundary of R_1, we add two segments, d and e, oriented left to right. To complete the boundary of R_2, we add two segments, g and f, oriented right to left. One slightly subtle point: although they are drawn a bit offset in Fig. 17.11, the lines d and f are the same lines, but oriented in opposite directions. The same goes for the lines e and g. Consequently,

$$\int_d P\,dx + Q\,dy = -\int_f P\,dx + Q\,dy\,, \quad \int_e P\,dx + Q\,dy = -\int_g P\,dx + Q\,dy$$

Now apply Green's theorem to R_1 and to R_2:

$$\begin{aligned}\int_{R_1} \frac{\partial Q}{\partial x} - \frac{\partial P}{\partial y}\,dA &= \int_{C_{11}} P dx + Q dy + \int_d P\,dx + Q\,dy \\ &\quad + \int_{C_{21}} P\,dx + Q\,dy + \int_e P\,dx + Q\,dy \\ \int_{R_2} \frac{\partial Q}{\partial x} - \frac{\partial P}{\partial y}\,dA &= \int_{C_{12}} P\,dx + Q\,dy + \int_g P\,dx + Q\,dy \\ &\quad + \int_{C_{22}} P\,dx + Q\,dy + \int_f P\,dx + Q\,dy\end{aligned}$$

Adding these equations, we obtain Eq. (17.23). We'll see an example in the practice problems.

This can be extended in the obvious way to regions with more than one hole. So long as the region is subdivided into simply-connected pieces, which curves we use to subdivide the region doesn't matter. The next example shows how to apply Green's theorem when the vector field isn't defined throughout the region bounded by the curve.

Example 17.5.2. *The curve bounds a region with a singularity.* Evaluate

$$\oint_{C_1} \frac{-y}{x^2+y^2}\,dx + \frac{x}{x^2+y^2}\,dy$$

where C_1 is the ellipse $x^2/4+y^2/9=1$. We can parameterize the ellipse by $\vec{r}(t) = \langle 2\cos(t), 3\sin(t)\rangle$ for $0 \leq t \leq 2\pi$. If we try to evaluate the integral directly, we obtain

$$\oint_{C_1} \frac{-y}{x^2+y^2}\,dx + \frac{x}{x^2+y^2}\,dy = \int_0^{2\pi} \frac{6}{4\cos^2(t)+9\sin^2(t)}\,dt$$

This does not look very inviting. Suppose we try Green's theorem. The relevant partial derivatives are

$$\frac{\partial Q}{\partial x} = \frac{\partial}{\partial x}\frac{x}{x^2+y^2} = \frac{-x^2+y^2}{x^2+y^2} \text{ and } \frac{\partial P}{\partial y} = \frac{\partial}{\partial y}\frac{-y}{x^2+y^2} = \frac{-x^2+y^2}{x^2+y^2}$$

Then $\partial Q/\partial x - \partial P/\partial y = 0$. Good, right? Not so fast: P and Q aren't defined throughout the region bounded by C_1, so we can't apply Green's theorem, at least not to the whole region inside C_1.

Suppose C_2 is the circle $x^2+y^2=1$, oriented clockwise, and R is the region between C and C_1. Then Green's theorem, as formulated in Eq. (17.24), applied to C_1, C_2, and R gives

$$\begin{aligned}&\oint_{C_1} \frac{-y}{x^2+y^2}\,dx + \frac{x}{x^2+y^2}\,dy\\ &= \oint_{-C_2} \frac{-y}{x^2+y^2}\,dx + \frac{x}{x^2+y^2}\,dy + \iint_R \frac{\partial Q}{\partial x} - \frac{\partial P}{\partial y}\,dA\end{aligned}$$

Because $\partial Q/\partial x = \partial P/\partial y$ throughout R,

$$\oint_{C_1} \frac{-y}{x^2+y^2}\,dx + \frac{x}{x^2+y^2}\,dy = \oint_{-C_2} \frac{-y}{x^2+y^2}\,dx + \frac{x}{x^2+y^2}\,dy$$

where $-C_2$ is parameterized counterclockwise, by $\vec{r}(t) = \langle \cos(t), \sin t \rangle$ for $0 \leq t \leq 2\pi$. So we find

$$\oint_{C_1} \frac{-y}{x^2+y^2}\,dx + \frac{x}{x^2+y^2}\,dy = \oint_{-C_2} \frac{-y}{x^2+y^2}\,dx + \frac{x}{x^2+y^2}\,dy$$

$$= \int_0^{2\pi} \frac{-\sin(t)}{\cos^2(t)+\sin^2(t)}(-\sin(t))\,dt + \frac{\cos(t)}{\cos^2(t)+\sin^2(t)}\cos(t)\,dt = 2\pi$$

and we're finished. □

The more general point here is that with clever applications of geometry sometimes we can apply calculus techniques to circumstances outside their original hypotheses. Geometry is a wonderful way to view the world.

Practice Problems

17.5.1. Evaluate $\oint_{C_1} e^x\,dx + (xy + \ln(2+y^2))\,dy$ where C_1 is the line segments from $(1,0)$ to $(1,2)$ to $(0,0)$.

17.5.2. Evaluate the line integral

$$\oint_{C_1} \frac{-y}{x^2+y^2}\,dx + \left(x + \frac{x}{x^2+y^2}\right)dy$$

where C_1 is the ellipse $x^2/16 + y^2/9 = 1$ oriented counterclockwise.

Practice Problem Solutions

17.5.1. Here C_1 is not a closed curve, but if we add the curve C_2 from $(0,0)$ to $(1,0)$, we have a closed curve. So

$$\int_{C_1} e^x\,dx + (xy+\ln(2+y^2))\,dy + \int_{C_2} e^x\,dx + (xy+\ln(2+y^2))\,dy$$

$$= \oint_{C_1 \cup C_2} e^x\,dx + (xy+\ln(2+y^2))\,dy$$

The region R bounded by the curves C_1 and C_2 can be described by $0 \leq x \leq 1$ and $0 \leq y \leq 2x$. Then we can apply Green's theorem

$$\oint_{C_1 \cup C_2} e^x dx + (xy+\ln(2+y^2))dy = \iint_R \frac{\partial}{\partial x}(xy+\ln(2+y^2)) - \frac{\partial}{\partial y}e^x dA$$

$$= \int_0^1 \int_0^{2x} y\,dy\,dx = \int_0^1 2x^2\,dx = \frac{2}{3}$$

Then

$$\int_{C_1} e^x\,dx + (xy+\ln(2+y^2))\,dy = \frac{2}{3} - \int_{C_2} e^x\,dx + (xy+\ln(2+y^2))\,dy$$

The path C_2 is easily parameterized: $\vec{r}(t) = \langle t, 0\rangle$ for $0 \le t \le 1$. Then

$$\int_{C_2} e^x\,dx + (xy + \ln(2+y^2))\,dy = \int_0^1 e^t\,dt = e - 1$$

The last equality follows because $dy = 0$ along C_2.

Combining these calculations, we have

$$\int_{C_1} e^x\,dx + (xy + \ln(2+y^2))\,dy = \frac{2}{3} - (e-1) = \frac{5}{3} - e$$

17.5.2. The curve C_1 can be parameterized by $\vec{r}(t) = \langle 4\cos(t), 3\sin(t)\rangle$ for $0 \le t \le 2\pi$. Direct evaluation is discouraging, but we can't apply Green's theorem directly because P and Q are not defined at the origin. So let C_2 denote the circle $x^2 + y^2 = 1$, oriented clockwise, and apply Green's theorem to the region R between the ellipse and the circle as shown in Fig. 17.12. We'll use Green's theorem in the form of Eq. (17.24).

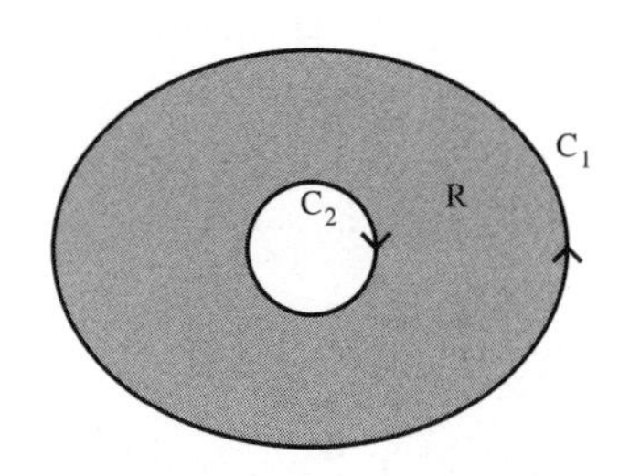

Figure 17.12. A region with a hole.

To compute the double integral, we need

$$\frac{\partial Q}{\partial x} = \frac{\partial}{\partial x}\left(x + \frac{x}{x^2+y^2}\right) = 1 + \frac{-x^2+y^2}{x^2+y^2} \quad \text{and}$$

$$\frac{\partial P}{\partial y} = \frac{\partial}{\partial y}\left(\frac{y}{x^2+y^2}\right) = \frac{-x^2+y^2}{x^2+y^2}$$

Then the double integral becomes

$$\iint_R \frac{\partial Q}{\partial x} - \frac{\partial P}{\partial y}\,dA = \iint_R \left(1 + \frac{-x^2+y^2}{x^2+y^2}\right) - \left(\frac{-x^2+y^2}{x^2+y^2}\right) dA = \iint_R 1\,dA$$

That is, the double integral is the area of the region R. Recall (or Google if you've forgotten) the area of an ellipse with semi-major axis a and semi-minor axis b is πab. For the ellipse $x^2/16 + y^2/9 = 1$, we have $a = 4$ and $b = 3$, so the area bounded by the ellipse is 12π. Now R is the region between the ellipse and the circle, so

$$\text{area}(R) = \text{area(ellipse)} - \text{area(circle)} = 12\pi - \pi = 11\pi$$

To complete the problem, we need the line integral around $-C_2$. Recall that $-C_2$ is oriented counterclockwise, so this curve can be parameterized by $\vec{r}(t) = \langle\cos(t), \sin(t)\rangle$ for $0 \le t \le 2\pi$. Then the line integral is

$$\oint_{-C_2} P\,dx + Q\,dy$$
$$= \int_0^{2\pi} \frac{-\sin(t)}{\cos^2(t)+\sin^2(t)}\,d\cos(t) + \left(\cos(t) + \frac{\cos(t)}{\cos^2(t)+\sin^2(t)}\right) d\sin(t)$$
$$= \int_0^{2\pi} (1+\cos^2(t))\,dt = \left(t + \frac{t}{2} + \frac{1}{4}\sin(2t)\right)\Big|_0^{2\pi} = 3\pi$$

Using Eq. (17.24) to combine this with the value of the double integral, we have

$$\oint_{C_1} \frac{-y}{x^2+y^2}\,dx + \left(x + \frac{x}{x^2+y^2}\right) dy = 3\pi + 11\pi = 14\pi$$

Exercises

17.5.1. Evaluate the line integral $\int_C P\,dx + Q\,dy$ for these P, Q, and C.
(a) $P = x^2 + xy^3$, $Q = x^2 - y^4$, and C consists of the line segments from $(1,0)$ to $(1,1)$, then from $(1,1)$ to $(0,1)$, then from $(0,1)$ to $(0,0)$.
(b) $P = yx^2 + x^4$, $Q = e^y + xy^2$, and C is the upper semicircle $x^2 + y^2 = 1, y \geq 0$ from $(1,0)$ to $(-1,0)$.
(c) $P = yx^2 + x^4$, $Q = e^y + xy^2$, and C is the quarter of the unit circle $x^2 + y^2 = 1, x \geq 0, y \geq 0$, from $(1,0)$ to $(0,1)$.
(d) $P = yx + \sin(x^3)$, $Q = xy^2 + y$, and C is the straight line segments from $(0,-1)$ to $(1,0)$, then from $(1,0)$ to $(0,1)$.
(e) $P = y + x$, $Q = 2x - \tan(y^4)$, and C is the part of the circle $(x-1)^2 + y^2 = 1$ with $y \geq 0$, from $(2,0)$ to $(0,0)$.

17.5.2. Evaluate the line integral $\int_C P\,dx + Q\,dy$ for these P, Q, and C.
(a) $P = y - y/(x^2+y^2)$, $Q = y + x/(x^2+y^2)$, and C is the ellipse $x^2/4 + y^2/9 = 1$, oriented positively.
(b) $P = -y/(x^2+y^2)$, $Q = y^2 + x/(x^2+y^2)$, and C is the ellipse $x^2/4 + y^2/9 = 1$, oriented positively.
(c) $P = -y/(x^2+y^2)$, $Q = xy + x/(x^2+y^2)$, and C is the ellipse $x^2/4 + y^2/9 = 1$, oriented positively.
(d) $P = -y/(x^2+y^2)$, $Q = x + x/(x^2+y^2)$, and C is the portion of the parabola $y = 4 - x^2$ for $-3 \leq x \leq 3$, along with the line segment between $(-3,-5)$ and $(3,-5)$, oriented positively.
(e) $P = -y/(x^2+y^2)$, $Q = x + x/(x^2+y^2)$, and C is the square with vertices $(\pm 2, \pm 2)$, oriented positively.

17.6 BENDIXSON'S CRITERION

Bendixson's criterion gives conditions that guarantee a differential equation has no closed trajectories in a simply-connected region. Moreover, if the differential equation does have periodic trajectories, it allows us to determine regions that do not contain the entire trajectory. The proof is a straightforward application of Green's theorem.

Theorem 17.6.1. *Bendixson's criterion.* Suppose P and Q gave continuous partial derivatives throughout a simply-connected region R. If $\partial P/\partial x + \partial Q/\partial y$ has the same sign throughout R, then the differential equation $x' = P, y' = Q$ has no periodic trajectory that lies entirely in R.

Proof. For any closed curve C contained in R, the curve C bounds a region R' contained in R, and R' is necessarily simply-connected. Then by Green's theorem

$$\oint_C -Q\,dx + P\,dy = \iint_{R'} \frac{\partial P}{\partial x} + \frac{\partial Q}{\partial y}\,dA \tag{17.25}$$

Because $\partial P/\partial x + \partial Q/\partial y$ does not change sign in R, the sum must be either positive throughout R or negative throughout R. Consequently, the double integral in Eq. (17.25) cannot be 0.

On the other hand, suppose $dx/dt = P$ and $dy/dt = Q$, which we'll rewrite as $dx = Pdt$ and $dy = Qdt$. If this differential equation has a closed trajectory C parameterized by $\vec{r}(t) = \langle x(t), y(t)\rangle$ for $0 \le t \le T$, where T is the time required to go once around C, then

$$\oint_C -Q\,dx + P\,dy = \int_0^T -Q(P\,dt) + P(Q\,dt) = \int_0^T (-QP + PQ)\,dt = 0$$

But the double integral of Eq. (17.25) cannot be 0. The only questionable step was when we supposed there is a closed trajectory C contained entirely within the simply-connected region R. No such trajectories exist. □

Now for a couple of applications.

Example 17.6.1. *No closed trajectories by Bendixson's criterion.* A biological application of Bendixson's criterion is a genetic toggle switch (Sect. 18.1). The term "toggle switch" may be unfamiliar to the generations raised on digital electronics. (They were welcome in the box of wires, resistors, capacitors, and inductors that was the electronics lab of my childhood.) Google can show you pictures that explain the concept.

A genetic toggle switch is a network of two types of repressor genes, each inhibiting the production of mRNA in the other. In the simplest version of such a network, x and y are the concentrations of the two types of repressor. In Sect. 18.1 we'll see the model is

$$x' = -x + \frac{a}{1+y^b} = P(x,y) \quad \text{and} \quad y' = -y + \frac{a}{1+x^b} = Q(x,y)$$

where a and b are constants.

Obviously, concentrations cannot be negative, so the domain R is the first quadrant, which is simply-connected. Now $\partial P/\partial x + \partial Q/\partial y = -2$, which clearly doesn't change signs. So by Bendixson's criterion, no trajectory of the genetic toggle switch is a closed trajectory. Not surprisingly, we'll see the dynamics depend on the values of a and b, but we'll see this later. Here we see the power of Bendixson's criterion, when its conditions are met. No derivative matrix, no eigenvalues, no trace-determinant, no trapping regions—just check if $\partial P/\partial x + \partial Q/\partial y$ has a constant sign on a simply-connected domain. □

Example 17.6.2. *No closed trajectories in some regions.* Recall the Lotka-Volterra predator-prey equations, Eqs. (7.4) and (7.5),

$$x' = -ax + bxy = P(x,y) \quad \text{and} \quad y' = cy - dxy = Q(x,y)$$

Here x and y are populations. As in the previous example, x and y are restricted to the first quadrant, which is simply-connected. To apply Bendixson's criterion we compute

$$\frac{\partial P}{\partial x} = -a + by \text{ and } \frac{\partial Q}{\partial y} = c - dx, \quad \text{so } \frac{\partial P}{\partial x} + \frac{\partial Q}{\partial y} = -a + c + by - dx$$

To find where $\partial P/\partial x + \partial Q/\partial y$ is positive and where it is negative, first find where it is 0. This occurs on the line $y = (d/b)x + (a-c)/b$. While it certainly is true that $\partial P/\partial x + \partial Q/\partial y$ changes sign in the first quadrant, so Bendixson's criterion does not preclude closed trajectories (as expected, because we know the Lotka-Volterra system does have closed trajectories), nevertheless we can deduce something about where closed trajectories cannot entirely lie.

Above the line $y = (d/b)x + (a-c)/b$ we see $\partial P/\partial x + \partial Q/\partial y$ is always positive, so no closed trajectory can lie entirely above the line. Similarly, no closed trajectory can lie entirely below the line. □

For the Lotka-Volterra system, we could deduce this from the directions of x' and y' in various regions of the first quadrant. Or because the fixed point $(c/d, a/b)$ lies on the line $y = (d/b)x + (a-c)/b$ and a closed trajectory must enclose a fixed point (Cor. 17.7.1), any closed trajectory must cross this line. For more complicated systems, the x' and y' directions can be harder to read. The regions where $\partial P/\partial x + \partial Q/\partial y$ has a constant sign can restrict the location of any closed trajectories.

Practice Problems

17.6.1. Show this system has no closed trajectory in the plane.

$$x' = x + y + x^3 - y^2 = P \quad \text{and} \quad y' = -2x + 2y + x^2y + y^3/3 = Q$$

17.6.2. Show this system has no closed trajectories in the plane, or identify regions in the plane that cannot contain closed trajectories.

$$x' = x^3/3 + xy^2 - y^3 = P \quad \text{and} \quad y' = -2xy + x^4 = Q$$

Practice Problem Solutions

17.6.1. From the values of P and Q we compute

$$\frac{\partial P}{\partial x} = 1 + 3x^2 \text{ and } \frac{\partial Q}{\partial y} = 2 + x^2 + y^2 \quad \text{so} \quad \frac{\partial P}{\partial x} + \frac{\partial Q}{\partial y} = 3 + 4x^2 + y^2$$

This is positive for all points (x, y) in the plane, which is simply-connected, so Bendixson's criterion precludes the existence of a closed trajectory for this system.

17.6.2. From the values of P and Q we compute

$$\frac{\partial P}{\partial x} = x^2 + y^2 \text{ and } \frac{\partial Q}{\partial y} = -2x \quad \text{so} \quad \frac{\partial P}{\partial x} + \frac{\partial Q}{\partial y} = x^2 - 2x + y^2$$

The sum of the partials is negative for x near 1 and y near 0, and is positive if either x or y is large. So Bendixson's criterion cannot forbid a closed trajectory. We'll identify where $\partial P/\partial x + \partial Q/\partial y = 0$ and use this to divide the plane into regions where the sum is positive and regions where the sum is negative. Completing the square gives

$$\frac{\partial P}{\partial x} + \frac{\partial Q}{\partial y} = x^2 - 2x + 1 + y^2 - 1 = (x-1)^2 + y^2 - 1$$

We see that inside the circle $(x-1)^2 + y^2 = 1$ the sum of the partials is negative, while outside that circle the sum is positive. The region inside the circle is simply-connected, so we can apply Bendixson's criterion and deduce that no closed trajectory can lie entirely within that circle. However, the region outside the circle is not simply-connected, so Bendixson's criterion cannot be applied to that region.

Exercises

17.6.1. For each of these P and Q, show the differential equation $x' = P, y' = Q$ has no closed trajectory in the plane, or find a region or regions in the plane that cannot contain closed trajectories.
(a) $P = -x^3 + y^3$, $Q = -yx^2 + x^2$
(b) $P = 2x + \cos(x) + y$, $Q = 3y - \sin(y) + \sin(x)$
(c) $P = x^3/3 + y^2$, $Q = -y^3/3 + x^3$
(d) $P = xy^2$, $Q = -yx^2 + x^3$
(e) $P = 3xy + x + y^3$, $Q = -2xy + x^3$

17.6.2. Consider the differential equation

$$x' = x + y - x(x^2 + y^2) \quad \text{and} \quad y' = -x + y - y(x^2 + y^2)$$

(a) Use Benixson's criterion to show that this differential equation has no closed trajectory inside the circle of radius $1/\sqrt{2}$.
(b) Use the methods of Sect. 9.6 to show that the differential equation has a limit cycle for $r > 1/\sqrt{2}$.

17.7 THE INDEX OF A FIXED POINT

Vector fields are closely related to differential equations, and now we know how to integrate a vector field along a curve. Bendixson's criterion is one application of line integrals to differential equations. In this section we'll see another.

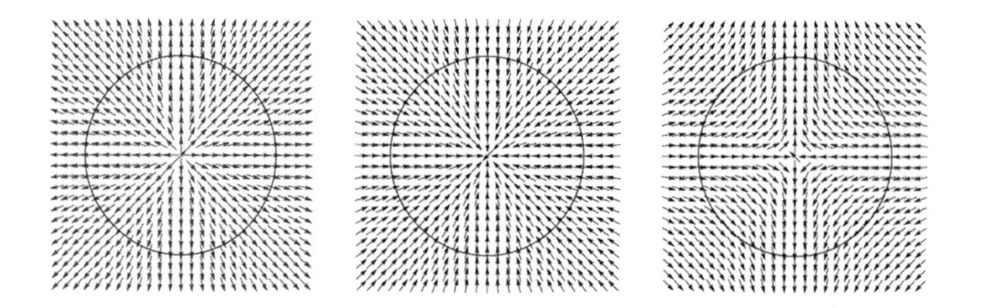

Figure 17.13. Visually computation of the index of vector fields around a curve. The vectors are the same length to make clearer their directions.

We'll use line integrals to define a number, the index, associated with a vector field and with a fixed point. We'll see that under some circumstances, only certain combinations of indices can occur. This, in turn, can limit the number and types of fixed points in a region. The index gives very little information about local structure of the vector field, but it does give some global information about the possible relative placement of fixed points and limit cycles.

First, we'll see the underlying geometry. At each point where it is defined, a vector field has a unique vector. Following a vector field once around a closed curve, the vectors must make a whole number of turns. Signs are important, so we'll make some orientation choices. We'll go counterclockwise around the curve. Each complete counterclockwise turn the vectors make will contribute $+1$, each complete clockwise turn -1. The number these turns define is the *index* of the vector field around the closed curve C. In Fig. 17.13 we see (left to right) three vector fields $\vec{F}_1 = \langle x,y\rangle$, $\vec{F}_2 = \langle -x,-y\rangle$, and $\vec{F}_3 = \langle -x,y\rangle$. In each plot the curve C is the circle $x^2+y^2=1$. Following the vector field vectors around these circles, we see $I_C(\vec{F}_1) = +1$, $I_C(\vec{F}_2) = +1$, and $I_C(\vec{F}_3) = -1$.

Can we compute the index without first plotting the vector field and then relying on our eyes (and memory)? Notice that for the vector field $\vec{F} = \langle P,Q\rangle$, the angle θ the vector field makes with the positive x-axis, as seen in Fig. 17.14, satisfies $\tan(\theta) = Q/P$, and so the angle is given by $\theta = \tan^{-1}(Q/P)$. To measure the number of complete turns the vector makes while traveling around a closed curve C, we integrate $d\theta/dt$ around C. First we find

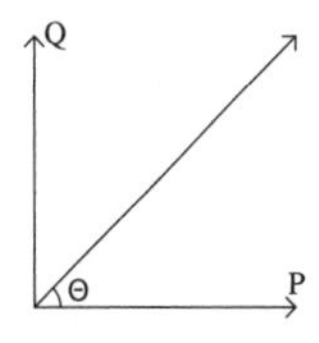

Figure 17.14. Vector components.

$$\begin{aligned}\frac{d\theta}{dt} &= (\tan^{-1}(Q/P))' = \frac{1}{1+(Q/P)^2}\frac{d}{dt}\frac{Q}{P}\\ &= \frac{1}{1+(Q/P)^2}\frac{PQ'-QP'}{P^2} = \frac{PQ'-QP'}{P^2+Q^2}\end{aligned}$$

Now integrating $d\theta/dt$ around C gives the change in θ around C. Because one full counterclockwise turn changes θ by 2π, the number of counterclockwise turns—that is, the index—is given by

$$I_C(\langle P,Q\rangle) = \frac{1}{2\pi}\int_C \frac{d\theta}{dt}\,dt = \frac{1}{2\pi}\int_C \frac{PdQ-QdP}{P^2+Q^2} \tag{17.26}$$

To illustrate the effectiveness of Eq. (17.26), we'll calculate the index of C parameterized by $\vec{r}(t) = \langle\cos(t),\sin(t)\rangle$, $0 \le t \le 2\pi$, for the vector fields $\vec{F}_2 = \langle -x,-y\rangle$ and $\vec{F}_3 = \langle -x,y\rangle$, the second and third vector fields pictured in Fig. 17.13.

$$I_C(\vec{F}_2) = \frac{1}{2\pi}\int_0^{2\pi} \frac{-\cos(t)(-\cos(t)) - (-\sin(t))\sin(t)}{(-\cos(t))^2 + (-\sin(t))^2} = \frac{1}{2\pi}\int_0^{2\pi} dt = 1$$

$$I_C(\vec{F}_3) = \frac{1}{2\pi}\int_0^{2\pi} \frac{-\cos(t)\cos(t) - (\sin(t))\sin(t)}{(\cos(t))^2 + (-\sin(t))^2} = \frac{1}{2\pi}\int_0^{2\pi} -1\, dt = -1$$

For a bit more practice we'll calculate the index around the unit circle C of two more vector fields, $\vec{F}_4 = \langle x^2 - y^2, 2xy\rangle$ (left image of Fig. 17.15) and $\vec{F}_5 = \langle x^3 - 3xy^2, 3x^2y - y^3\rangle$ (the right image).

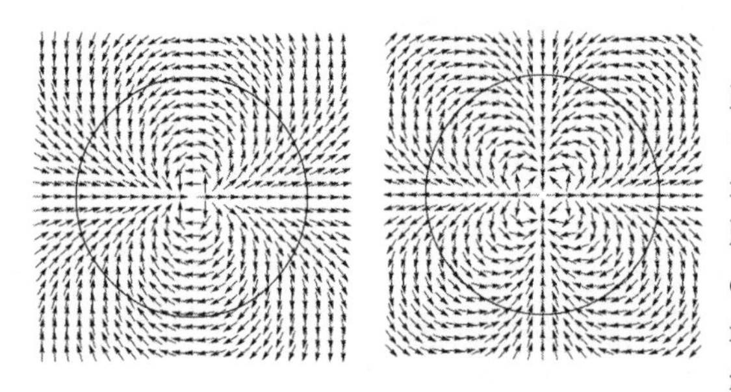

Figure 17.15. Two more vector fields.

These may appear to be peculiar choices of P and Q, but the pattern of coefficients and exponents may look familiar. After we've computed the index of C for both of these vector fields, you may see a relation between the terms of the vector field and the index of the curve. Test your hypothesis in Practice Problem 17.7.1.

For $\vec{F}_4$ we see $P = x^2 - y^2$ and $Q = 2xy$. Substituting $x(t) = \cos(t)$ and $y(t) = \sin(t)$ into the expressions for P and Q, we find

$$P^2 + Q^2 = \cos^4(t) + 2\cos^2(t)\sin^2(t) + \sin^4(t) = (\cos^2(t) + \sin^2(t))^2 = 1$$

so we can ignore the denominator in Eq. (17.26). For these P and Q we see $dP = 2xx' - 2yy'$ and $dQ = 2xy' + 2yx'$. Substituting again for x and y and simplifying, we compute the index

$$\begin{aligned} I_C(\vec{F}_4) &= \frac{1}{2\pi}\int_0^{2\pi} 2(\cos^2(t) - \sin^2(t))^2 + 8(\cos(t)\sin(t))^2\, dt \\ &= \frac{1}{2\pi}\int_0^{2\pi} 2(\cos^4(t) + 2\cos^2(t)\sin^2(t) + \sin^4(t))\, dt \\ &= \frac{1}{2\pi}\int_0^{2\pi} 2(\cos^2(t) + \sin^2(t))^2\, dt = 2 \end{aligned}$$

For $\vec{F}_5$ we see $P = x^3 - 3xy^2$ and $Q = 3x^2y - y^3$. Substituting $x(t) = \cos(t)$ and $y(t) = \sin(t)$ into the expressions for P and Q, we find $P^2 + Q^2 =$

$$\cos^6(t) + 3\cos^4(t)\sin^2(t) + 3\cos^2(t)\sin^4(t) + \sin^6(t) = (\cos^2(t) + \sin^2(t))^3 = 1$$

so again we can ignore the denominator in Eq. (17.26). For these P and Q we see $dP = 3x^2x' - 3y^2x' - 6xyy'$ and $dQ = 6xyx' + 3x^2y' - 3y^2y'$. Substituting $x = \cos(t)$ and $y = \sin(t)$, we find $2\pi I_C(\vec{F}_5) =$

$$\int_0^{2\pi} \left(\cos^3(t) - 3\cos(t)\sin^2(t)\right)\left(-6\cos(t)\sin^2(t) + 3\cos^3(t) - 3\sin^2(t)\cos(t)\right) - \left(3\cos^2(t)\sin(t) - \sin^3(t)\right)\left(-3\cos^2(t)\sin(t) + 3\sin^3(t) - 6\cos^2(t)\sin(t)\right) dt$$

$$= \int_0^{2\pi} 3\left(\cos^6(t) + 3\cos^4(t)\sin^2(t) + 3\cos^2(t)\sin^4(t) + \sin^6(t)\right) dt$$

$$= \int_0^{2\pi} 3\left(\cos^2(t) + \sin^2(t)\right)^3 dt = 6\pi$$

and so $I_C(\vec{F}_5) = 3$.

Up until now we've ignored one potential source of trouble when calculating the index $I_C(\langle P, Q\rangle)$: what happens if a fixed point of the vector field belongs to the curve? Because at a fixed point $P = Q = 0$ and the vector field vector is $\vec{0}$, so the angle θ between the vector and the positive x-axis makes no sense. We can't compute the index for any curve that passes through a fixed point. So long as P and Q are continuously differentiable functions, fixed points are the only problems. This simple observation has some useful consequences.

Despite the generality of these consequences, their proofs are surprisingly simple, might seem too simple. You'll see no calculations, no calculus, not even any algebra. Just a couple of pictures and some simple thinking. Yet these results hold for a vast collection of vector fields. These are examples of topological arguments. So is the proof of the Poincaré-Bendixson theorem (in Sect. A.10), though that proof is quite a bit more complicated than the arguments we'll see next.

Proposition 17.7.1. Suppose C_1 and C_2 are simple closed curves with C_2 lying in the region bounded by C_1. Suppose the vector field $\vec{F}$ is defined and has no fixed points within the region R between C_1 and C_2 as shown in Fig. 17.16. Then $I_{C_1}(\vec{F}) = I_{C_2}(\vec{F})$.

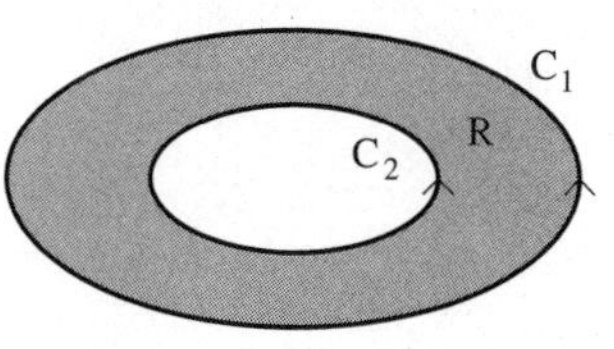

Figure 17.16. The region between two curves.

Proof. So long as $\vec{F}$ is defined and has no fixed points on the region R, the path C_1 can be continuously deformed into the path C_2 and the index is defined for all these intermediate paths. These deformations can be made in very small steps, so the index changes continuously between C_1

and C_2. But—and here's the point—the index can take on only integer values, and the only continuous integer-valued functions are constant. To see this we don't need the technical definition of continuity, just the familiar notion that the graph of a continuous function has no jumps. If an integer-valued function takes more than one value, then there is a big jump where it changes value. So $I_{C_1}(\vec{F}) = I_{C_2}(\vec{F})$. □

That was easy enough, but maybe the utility of Prop. 17.7.1 isn't so obvious. Here's one reason why we need it.

Proposition 17.7.2. If the region bounded by the closed curve C contains no fixed point of the vector field $\vec{F}$, then $I_C(\vec{F}) = 0$.

Proof. Because the region R bounded by C contains no fixed points, we can apply Prop. 17.7.1 to see that $I_C(\vec{F}) = I_{C'}(\vec{F})$ for every simple closed curve C' in the region R, even a very tiny curve C', a curve so small that the continuous vector field $\vec{F}$ is approximately constant around C', and consequently cannot make a complete turn. So $I_{C'}(\vec{F}) = 0$ and by Prop. 17.7.1, $I_C(\vec{F}) = 0$. □

Now we can define the index of a fixed point (x_*, y_*) of a vector field $\vec{F}$. For any simple closed curve C that encloses only one fixed point, (x_*, y_*), we take

$$I_{(x_*, y_*)}(\vec{F}) = I_C(\vec{F})$$

By Prop. 17.7.1, this definition does not depend on the choice of curve C, so long as it satisfies the stated conditions of the curve.

For one class of simple closed curves we can always compute the index.

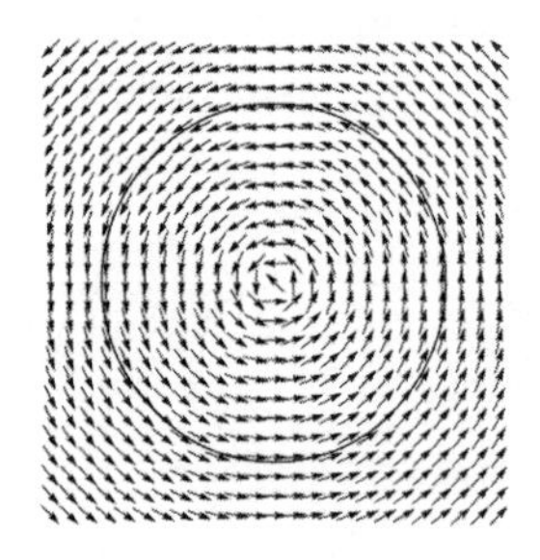

Figure 17.17. Index along a trajectory.

Proposition 17.7.3. If C is a closed trajectory of a vector field $\vec{F}$, then $I_C(\vec{F}) = +1$.

Proof. If the closed curve C is a trajectory of $\vec{F}$, then the tangent vectors of C are vectors of $\vec{F}$. As we go once counterclockwise around C, the tangent vectors of C make one complete counterclockwise circuit. See Fig. 17.17. Consequently, the index $I_C(\vec{F}) = +1$. □

By looking at the vector fields in Fig. 17.13, we see that $I_p(\vec{F}) = -1$ if p is a saddle point and $I_p(\vec{F}) = +1$ if p is an asymptotically stable node or an unstable node. In addition, $I_p(\vec{F}) = +1$ if p is a center (this is contained in Prop. 17.7.3) and also if p is an asymptotically stable or an unstable spiral; you can draw the appropriate pictures for those cases. Notice that unlike eigenvalues, index is not sensitive to stability.

What if a closed trajectory contains more than one fixed point? Just add the indices.

Proposition 17.7.4. If a simple closed curve C encloses several fixed points of a vector field $\vec{F}$, then $I_C(\vec{F})$ is the sum of the indices of these fixed points.

Proof. The proof is pretty simple, building on the extension in Sect. 17.5 of Green's theorem to the situation where the region bounded by the closed curve C isn't simply-connected.

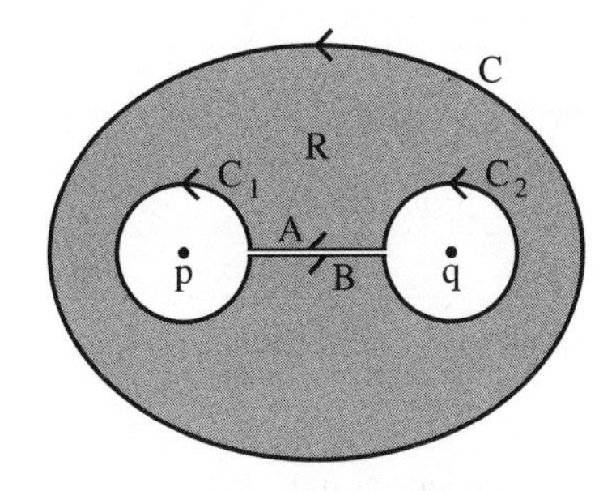

Figure 17.18. A curve enclosing several fixed points

So suppose C is a simple closed curve and the region bounded by C contains two fixed points p and q of the vector field $\vec{F}$. Draw two small circles, C_1 centered on p and C_2 centered on q. Connect C_1 and C_2 by a very narrow path. Call the upper edge of the path A going from C_2 to C_1, and call the lower edge B going from C_1 to C_2. Then as we see in Fig. 17.18, the region R between C and the path consisting of (most of) C_1, B, (most of) C_2, and A contains no fixed points and so by Prop. 17.7.1 both paths bounding R have the same index. Now as the thickness of the very narrow path shrinks to 0, the line integrals along A and B cancel and we're left with the line integrals around C_1 and C_2, that is, the sum of the indices of p and q. This extends in the obvious way if C bounds any (finite) number of fixed points. □

Combining Props. 17.7.3 and 17.7.4, we see

Corollary 17.7.1. Every closed trajectory C of $\vec{F}$ must enclose fixed points whose indices sum to $+1$. In particular, every closed trajectory must enclose at least one fixed point, and if only one, that fixed point cannot be a saddle point.

Before looking at some ways we can apply indices, we'll mention a beautiful generalization. So far, we've studied vector fields defined on the plane or on some subset of the plane. But other geometries are possible, and in some circumstances, required. Vector fields can be defined on a sphere, on a cylinder, on a torus (the crust of a bagel if you're from the north, the glaze on a donut if you're from the south, the inner tube of a tire if you're old enough to remember when tires had inner tubes), on a torus with two holes, with three holes, and so on. Associated with each surface is a number, the *Euler characteristic* (same Euler as in Euler's formula), that depends on only the topology of the surface, that is, on the number of holes in the surface. Table 17.1 shows some examples.

surface	Euler characteristic
sphere	2
torus	0
two-holed torus	−2
three-holed torus	−4
four-holed torus	−6
...	...

Table 17.1. Some Euler characteristics.

There is an amazing relationship between the topology of the surface, as expressed by the Euler characteristic, and the sum of the indices of all the fixed points of a vector field on the surface. This is the *Poincaré-Hopf theorem.* The sum of the indices of the fixed points of a vector field on a surface equals the Euler characteristic of the surface.

The theorem has some hypotheses—the vector field must be continuously differentiable, the surface must be compact and orientable—that we'll mention so you won't try to apply the theorem in inappropriate situations. But we won't explore the hypotheses: Google can show you what they mean.

We mention this theorem for two reasons. First, it is one of the best examples of what topology does so well. It relates one general facet, the topology of the surface, to another, the sum of the indices of the fixed points of every (continuously differentiable) vector field that can be defined on the surface. Why should these be related? The subtle and challenging proof depends not on details of the infinitely many geometric realizations of each of the infinitely many surfaces, or on the form of the infinitely many vector fields that can be defined on each of these realizations. The proof is based on finding which details we can ignore and how to formulate those that are important. It shows how a person could be smitten with topology, as I was in graduate school. Many people spend their intellectual lives with topology exclusively. That's where I started, but as you can see, I've drifted rather far.

The other reason is the authors of the theorem. Poincaré we've met. Two well-known mathematicians are named Hopf. Eberhard Hopf was an analyst who worked in ergodic theory and differential equations. Among other things, he studied how fixed points can change (bifurcate) into limit cycles with the variation of a parameter of a differential equation. He isn't the Hopf of the Poincaré-Hopf theorem. That Hopf is Heinz Hopf, a powerful topologist who contributed much to the development of algebraic topology. One of his fifty Ph.D. students was Hans Samelson, one of whose students is Terry Lawson, one of whose students is the author of this book. The first three people mentioned in this family tree are brilliant mathematicians, then there's your far-from-brilliant author. Should this lineage shake our belief in evolution? Not at all: most mutations are unfavorable (or maybe neutral).

Back to applications. Cor. 17.7.1 tells us something very general about the possible locations of closed trajectories. For example, if a vector field has only two fixed points, and both have index $+1$, then a closed trajectory might enclose one fixed point or the other, but not both, and not neither. This depends on no detail of the vector field other than the indices of the fixed points. We'll see an example in the practice problems.

Now, we aren't guaranteed to find closed trajectories around either fixed point. Recall that the Poincaré-Bendixson theorem gives conditions that guarantee a vector field has a closed trajectory. By contrast, Index theory is like Bendixson's criterion: it gives conditions that preclude closed trajectories, or if closed trajectories are possible, it restricts regions of the plane those trajectories might inhabit.

Practice Problems

17.7.1. We have seen that the vector fields $\vec{F}_1 = \langle x, y \rangle$, $\vec{F}_4 = \langle x^2 - y^2, 2xy \rangle$, and $\vec{F}_5 = \langle x^3 - 3xy^2, 3x^2y - y^3 \rangle$ have index 1, 2, and 3 around C, the circle $x^2 + y^2 = 1$, oriented counterclockwise. Find a vector field with index 4 around C. Check your vector field by evaluating the line integral of Eq. (17.26). By finding a pattern for these examples, for every $n \geq 1$ find a vector field with index n around C. Hint: think of $z = x + iy$.

17.7.2. Compute the index of the vector field $\vec{F} = \langle x^2 - y^2, xy \rangle$ at the fixed point $(0,0)$.

17.7.3. Show the two competing species model

$$x' = (1 - x - 5y)x \quad \text{and} \quad y' = (1 - 2x - 3y)y$$

does not have a closed trajectory. First find the fixed points and compute their indices, which you can do by calculating their eigenvalues. Remember that because x and y represent populations, $x \geq 0$ and $y \geq 0$.

Practice Problem Solutions

17.7.1. The components of $\vec{F}_4$ are terms of $(x-y)^2$. This is a start, ignoring how we would figure out which terms go with P and which with Q. Are the components of $\vec{F}_5$ made of terms of $(x-y)^3$? Not quite. The signs are off: $(x-y)^3 = x^3 - 3x^2y + 3xy^2 - y^3$. In $\vec{F}_5$ the term $3xy^2$ is negative, while it's positive in $(x-y)^3$. The sign of $3x^2y$ is reversed, too.

Here's what works. The origin has index n for the vector field $\vec{F} = \langle P, Q\rangle$ with P the real part of $(x+iy)^n$ and Q the imaginary part of $(x+iy)^n$. Think of the polar representation of $(x+iy)^n$ on the unit circle for a hint at why this works.

For $(x+iy)^4 = (x^4 - 6x^2y^2 + y^4) + i(4x^3y - 4xy^3)$ we have

$$P = x^4 - 6x^2y^2 + y^4 \quad \text{so} \quad dP = 4x^3x' - 12xy^2x' - 12x^2yy' + 4y^3y'$$
$$Q = 4x^3y - 4xy^3 \quad \text{so} \quad dQ = 12x^2yx' + 4x^3y' - 4y^3x' - 12xy^2y'$$

Now substitute $x = \cos(t)$ and $y = \sin(t)$, so $x' = -\sin(t)$ and $y' = \cos(t)$, then evaluate the index by the line integral Eq. (17.26).

First, with these substitutions, $P^2 + Q^2 =$

$$\cos^8(t) + 4\cos^6(t)\sin^2(t) + 6\cos^4(t)\sin^4(t) + 4\cos^2(t)\sin^6(t) + \sin^8(t)$$
$$= (\cos^2(t) + \sin^2(t))^4 = 1$$

so we can ignore the denominator of the integrand in Eq. (17.26).

Next, with these substitutions for x, y, x', and y', we see $PdQ =$

$$4\cos^8(t) - 48\cos^6(t)\sin^2(t) + 152\cos^4(t)\sin^4(t) - 48\cos^2(t)\sin^6(t) + 4\sin^8(t)$$

and

$$QdP = -64\cos^6(t)\sin^2(t) + 128\cos^4(t)\sin^4(t) - 64\cos^2(t)\sin^6(t)$$

After simplifying we find $PdQ - QdP =$

$$4(\cos^8(t) + 4\cos^6(t)\sin^2(t) + 6\cos^4(t)\sin^4(t) + 4\cos^2(t)\sin^6(t) + \sin^8(t)$$
$$= 4(\cos^2(t) + \sin^2(t))^4 = 4$$

After all this,

$$I_C(\langle P, Q\rangle) = \frac{1}{2\pi}\int_0^{2\pi} \frac{PdQ - QdP}{P^2 + Q^2}\, dt = \frac{1}{2\pi}\int_0^{2\pi} 4\, dt = 4$$

That this works out so neatly is very satisfying, but probably not so much that you'd like to try index = 5 in this fashion.

17.7.2. The x-nullcline is $x^2 - y^2 = 0$, that is, $y = \pm x$, and the y-nullcline is $xy = 0$, that is, $x = 0$ and $y = 0$. The nullclines intersect only at the origin, so this is the only fixed point. Consequently, we can compute the index of the origin by integrating around any closed curve containing the origin. We'll take C to be the unit circle, parameterized by $\vec{r}(t) = \langle \cos(t), \sin(t) \rangle$ for $0 \leq t \leq 2\pi$. Then $P = x^2 - y^2 = \cos^2(t) - \sin^2(t)$, $Q = xy = \cos(t)\sin(t)$, $dP = -4\cos(t)\sin(t)$, and $dQ = -\sin^2(t) + \cos^2(t)$. Then the index $I_{(0,0)}(\vec{F})$ can be computed by the integral

$$\begin{aligned}
&\frac{1}{2\pi}\int_C \frac{PdQ - QdP}{P^2 + Q^2} \\
&= \frac{1}{2\pi}\int_0^{2\pi} \frac{(\cos^2(t) - \sin^2(t))^2 - \cos(t)\sin(t)(-4\cos(t)\sin(t))}{(\cos^2(t) - \sin^2(t))^2 + (\cos(t)\sin(t))^2}\, dt \\
&= \frac{1}{2\pi}\int_0^{2\pi} \frac{1}{\cos^4(t) - \cos^2(t)\sin^2(t) + \sin^4(t)}\, dt = 2
\end{aligned}$$

The last integral can be challenging if done by hand. Don't hesitate to call on a computer algebra system.

17.7.3. The x-nullcline is $x = 0$, $y = -(1/5)x + 1/5$ and the y-nullcline is $y = 0$ and $y = -(2/3)x + 1/3$. The fixed points are the intersections of the nullclines: $(0,0)$, $(0,1/3)$, $(1,0)$, and $(2/7,1/7)$. The derivative matrix is

$$\begin{bmatrix} 1 - 2x - 5y & -5x \\ -2y & 1 - 2x - 6y \end{bmatrix}$$

The eigenvalues are $\lambda = 1, 1$ for the fixed point $(0,0)$, $\lambda = -1, -2/3$ for $(0,1/3)$, $\lambda = -1, -1$ for $(1,0)$, and $\lambda = -1, 2/7$ for $(2/7,1/7)$. Consequently, $(0,0)$ is an unstable node, $(0,1/3)$ and $(1,0)$ are asymptotically stable nodes, and $(2/7,1/7)$ is a saddle point. Then these fixed points have indices $+1, +1, +1$, and -1, respectively.

From Cor. 17.7.1 we see that a closed trajectory could enclose any one of the first three fixed points, or the fourth together with any two of the first three. So from index theory should we deduce that this two competing species model can have several closed trajectories? Not at all: the first three fixed points lie on the boundary of the first quadrant, so no trajectory can enclose any of these three. Because the fourth fixed point has index -1, no closed trajectory can enclose only that fixed point. So index theory says that this model has no closed trajectories.

Exercises

17.7.1. For each of these vector fields the origin is a fixed point, either the only fixed point of the vector field, or the only fixed point within the circle C of radius 1. Find the index of the origin by evaluating the line integral of Eq. (17.26).

If you don't have a computer algebra system, evaluating these integrals can be challenging. If you know the results are integer-valued, numerical integration can suffice to find the integer. If neither of these is available, sketch enough of the vectors around C to find the index visually. One of these is enough for anyone to do.

(a) $P = y - x$, $Q = x - y^2$ (b) $P = x + y$, $Q = x^2 + y$
(c) $P = xy$, $Q = x + y^2$ (d) $P = x$, $Q = x - y^2$
(e) $P = y$, $Q = x - y^2$ (f) $P = x - y^2$, $Q = y - x^2$

17.7.2. In these problems the vector fields are defined throughout the plane. The locations and indices of all the fixed points are given. Plot the locations of the fixed points and sketch every type of possible closed trajectory. Here two trajectories have the same type if one can be deformed continuously into the other without crossing a fixed point. For example, if the only fixed points are $(-2,0)$ and $(2,0)$, then the circles $(x+2)^2 + y^2 = 1$ and $(x-2)^2 + y^2 = 1$ have different types, while $(x+2)^2 + y^2 = 1$ and $(x+1)^2 + y^2 = 4$ have the same type.

(a) $(1,1), I = +1$, $(1,0), I = -1$, $(0,0), I = -1$
(b) $(1,1), I = +1$, $(1,0), I = -1$, $(0,0), I = +1$
(c) $(1,1), I = +2$, $(1,0), I = -1$, $(0,0), I = +1$
(d) $(1,1), I = +2$, $(1,0), I = -1$, $(0,0), I = -1$
(e) $(1,1), I = +2$, $(1,0), I = -2$, $(0,0), I = +1$

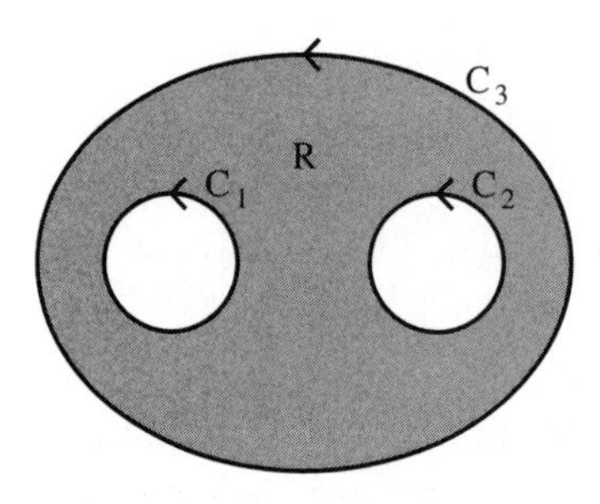

Figure 17.19. Region for Exercise 17.7.3.

17.7.3. Suppose a vector field has exactly three closed trajectories, C_1, C_2, and C_3, arranged as shown in Fig. 17.19. Must the vector field have a saddle point in the region R outside of C_1 and C_2, and inside C_3? Give reasons to support your answer.

17.7.4. Consider the modified two-species population model

$$x' = y(1 + y - x^2) \quad \text{and} \quad y' = x(y - 2)$$

Recall that for population models we require $x \geq 0$ and $y \geq 0$. Show there is no closed trajectory in this system. Comment on the peculiarities of this model: what factors increase the x population, what factors decrease the x population? Same questions for the y population.

17.7.5. Must every index +1 fixed point of a vector field be enclosed by a closed trajectory? Say why if you think the answer is "Yes"; give a counterexample if you think the answer is "No."

17.8 SURFACE INTEGRALS

A curve is 1-dimensional, that is, any point on the curve can be determined by a single variable, for example, the distance along the curve from some starting point. We can parameterize a curve in 3-dimensional space as

$$\vec{r}(t) = x(t)\vec{i} + y(t)\vec{j} + z(t)\vec{k} \quad \text{for all } t \text{ in an interval } I. \qquad (17.27)$$

Similarly, a surface is 2-dimensional: any point on the surface can be determined by two variables. Think of how longitude and latitude determine a location on the globe. In general, we'll use u and v to denote the position on the surface, and so we can parameterize a surface as

$$\vec{r}(u,v) = x(u,v)\vec{i} + y(u,v)\vec{j} + z(u,v)\vec{k} \qquad (17.28)$$

for all (u,v) in a domain D. Here are three examples.

Example 17.8.1. *Parameterization of a sphere.* A sphere of radius R and center $(0,0,0)$ can be parameterized by

$$\vec{r}(\theta,\varphi) = R\cos(\theta)\sin(\varphi)\vec{i} + R\sin(\theta)\sin(\varphi)\vec{j} + R\cos(\varphi)\vec{k}$$

for $0 \le \theta \le 2\pi$, $0 \le \varphi \le \pi$. It's easy to verify that for all θ and φ, $\|\vec{r}(\theta,\varphi)\| = R$, that is, every point $\vec{r}(\theta,\varphi)$ lies on the sphere.

To show that we have parameterized the whole sphere, look at the coordinates. Rather than u and v, here we denote the coordinates by θ, the longitude, and φ, the *co-latitude*. The co-latitude is just the latitude, but with $\varphi = 0$ for the north pole, $\varphi = \pi/2$ for the equator, and $\varphi = \pi$ for the south pole. For each value of θ, the curve $\vec{r}(\theta,\varphi)$ traces out a semicircle on the sphere, from the north pole to the south pole as φ goes from 0 to π. Then as θ goes from 0 to 2π, the semicircle sweeps once around the sphere. □

Example 17.8.2. *Parameterization of a function graph.* Another example is the graph of a function $z = f(x,y)$. Here we'll replace the variables u and v with x and y, because those are the natural variables for this graph. Then this surface can be parameterized by

$$\vec{r}(x,y) = x\vec{i} + y\vec{j} + f(x,y)\vec{k}$$

for (x,y) in a domain D. □

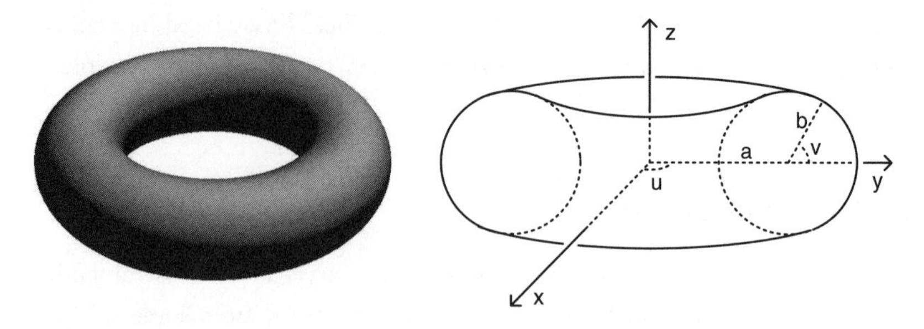

Figure 17.20. Left: a sketch of a torus. Right: how a and b figure into the structure of the torus.

Example 17.8.3. *Parameterization of a torus.* Here we'll keep the variables u and v. The torus can be parameterized this way:

$$\vec{r}(u,v) = (a + b\cos(v))\cos(u)\vec{i} + (a + b\cos(v))\sin(u)\vec{j} + b\sin(v)\vec{k}$$

for $0 \le u \le 2\pi$, $0 \le v \le 2\pi$, and for $a > b$. A schematic picture is presented in Fig. 17.20. Let's see how this parameterization generates a torus.

First note that $\vec{r}(u,0)$ is the circle C_0 in the xy-plane with center $(0,0,0)$ and radius $a+b$. For any fixed v_0, $\vec{r}(u,v_0)$ is the circle C_{v_0} in the plane $z = b\sin(v_0)$ with center $(0,0,b\sin(v_0))$ and radius $a + b\cos(v_0)$. So as v_0 increases from 0 to π, the curves $\vec{r}(u,v_0)$ start as the circle C_0, rise and shrink to the circle $C_{\pi/2}$, fall and shrink further to C_π, continue to fall and now grow to the circle $C_{3\pi/2}$, and finally rise and continue to grow until they return to the circle $C_{2\pi} = C_0$. In this way we see the parameterization covers the torus with horizontal circles.

How do you think the curves $\vec{r}(u_0,v)$ cover the torus? □

At every point $\vec{r}(t_0)$ on a curve $\vec{r}(t)$ in 2 or 3 (or more) dimensions, the curve has a tangent line

$$\vec{L}(t) = \vec{r}(t_0) + t \cdot \vec{r}\,'(t_0) \tag{17.29}$$

The tangent plane at the surface $\vec{r}(u,v)$ is spanned by the vectors

$$\vec{r}_u = \frac{\partial x}{\partial u}\vec{i} + \frac{\partial y}{\partial u}\vec{j} + \frac{\partial z}{\partial u}\vec{k} \quad \text{and} \quad \vec{r}_v = \frac{\partial x}{\partial v}\vec{i} + \frac{\partial y}{\partial v}\vec{j} + \frac{\partial z}{\partial v}\vec{k}$$

The tangent plane is determined by a point $\vec{r}(u_0,v_0)$ on the plane and two independent tangent vectors, the analogue for surfaces of the observation for curves that the tangent line is determined by a point on the line and a tangent vector.

For a surface in 3-dimensional space a tangent plane can be determined in another way, by a point on the plane and a vector perpendicular to the plane.

Vectors perpendicular to the plane are called *normal vectors*. (This is established terminology, despite other uses of the word "normal": normal distribution, normalization factor, and a century ago teacher training colleges were called normal schools.) Here's how to see this. Suppose (x_0, y_0, z_0) is a point on the plane, and $\langle a, b, c\rangle$ is a vector normal to the plane. Now a point (x, y, z) in 3-dimensional space belongs to the plane if and only if the vector $\langle x - x_0, y - y_0, z - z_0\rangle$ lies in the plane, that is, if and only if the vector $\langle x - x_0, y - y_0, z - z_0\rangle$ is perpendicular to the normal vector $\langle a, b, c\rangle$. By the dot product property of Eq. (13.52), these (non-zero) vectors are perpendicular if and only if

$$0 = \langle a, b, c\rangle \cdot \langle x - x_0, y - y_0, z - z_0\rangle = ax + by + cz - (ax_0 + by_0 + cz_0) \quad (17.30)$$

Before we continue, we'll make two observations about this equation. First, the point-slope form of the equation of a line in the xy-plane is

$$y - y_0 = m(x - x_0), \quad \text{that is,} \quad 0 = \langle m, -1\rangle \cdot \langle x - x_0, y - y_0\rangle \quad (17.31)$$

The line equation usually isn't written this way, but this form emphasizes how the plane equation (17.30) is a natural extension of the line equation.

The second observation is about degrees of freedom. Though some care is needed to implement this interpretation for the dimensions of Sect. 6.4, in Euclidean spaces the dimension is the number of degrees of freedom of that space. On a line (1-dimensional) we can move left or right; our position is specified by one number. The line has one degree of freedom. In a plane (2-dimensional), we can move left or right and up or down; our position is specified by two numbers, each adjustable independently of the other. The plane has two degrees of freedom. And so on. To determine a line in a plane, we must reduce the number of degrees of freedom from 2 to 1. The condition imposed by a single scalar equation (17.31) does this. To determine a plane in 3-dimensional space, we must reduce the number of degrees of freedom from 3 to 2. The condition imposed by a single scalar equation (17.30) does this. To determine a line in 3-dimensional space, we must reduce the number of degrees of freedom from 3 to 1. This should require two scalar equations. How can we find these equations from, say, Eq. (17.29)? First, rewrite this vector equation as

$$\langle x, y, z\rangle = \langle x_0, y_0, z_0\rangle + t\langle a, b, c\rangle \quad (17.32)$$

This vector equation gives three scalar equations, but unlike the scalar equations of Eqs. (17.30) and (17.31), each of the three scalar equations from Eq. (17.32) involve the variable t. Eliminating t gives two scalar equations

$$\frac{x - x_0}{a} = \frac{y - y_0}{b} \quad \text{and} \quad \frac{y - y_0}{b} = \frac{z - z_0}{c}$$

just what we need to reduce the degrees of freedom from 3 to 1.

Counting degrees of freedom usually is easy, but sometimes it can give real simplifications. This is another trick to add to your box of tricks.

Finally, we need to construct the normal vector $\vec{n}$ from the tangent vectors $\vec{r}_u$ and $\vec{r}_v$. This we achieve by the vector cross product. For vectors $\vec{a} = \langle a_1, a_2, a_3 \rangle$ and $\vec{b} = \langle b_1, b_2, b_3 \rangle$, the *cross product* is

$$\vec{a} \times \vec{b} = \det \begin{bmatrix} \vec{i} & \vec{j} & \vec{k} \\ a_1 & a_2 & a_3 \\ b_1 & b_2 & b_3 \end{bmatrix} \tag{17.33}$$

A straightforward calculation using Eq. (17.33) shows

$$(\vec{a} \times \vec{b}) \cdot \vec{a} = 0 \quad \text{and} \quad (\vec{a} \times \vec{b}) \cdot \vec{b} = 0 \tag{17.34}$$

That is, $\vec{a} \times \vec{b}$ is perpendicular to the plane spanned by $\vec{a}$ and $\vec{b}$. This is why the cross product is useful for building a normal vector from tangent vectors.

The direction of $\vec{a} \times \vec{b}$ is given by the right-hand rule. Point the fingers of the right hand in the direction of $\vec{a}$ and swing your fingers through the smaller angle between $\vec{a}$ and $\vec{b}$. Then your thumb points in the direction of $\vec{a} \times \vec{b}$. Be sure to use your right hand. If you're right-handed, when you calculate $\|\vec{a} \times \vec{b}\|$ with the pencil in your right hand, and compute the direction with your left hand, you will not find your freshman physics teacher especially sympathetic. Certainly, I didn't.

Another straightforward, but tedious, calculation shows this relation between the cross product and the dot product:

$$\|\vec{a} \times \vec{b}\|^2 = \|\vec{a}\|^2 \|\vec{b}\|^2 - (\vec{a} \cdot \vec{b})^2 \tag{17.35}$$

Together with Eq. (13.52) this gives a formula familiar from basic physics:

$$\|\vec{a} \times \vec{b}\| = \|\vec{a}\| \, \|\vec{b}\| \sin(\theta) \tag{17.36}$$

Now you may be curious about why we didn't mention cross products for 2-dimensional vectors. After all, the dot product can be defined for vectors in all dimensions. What would we require of a general cross product? At least these two properties: $\vec{a} \times \vec{b}$ is perpendicular to both $\vec{a}$ and $\vec{b}$ (Eq. (17.34)), and the length satisfies Eq. (17.35). By a very deep result from algebraic topology (see page 80 of [165]), a vector product satisfying these two conditions can be defined only in 3-dimensional space and in 7-dimensional space. This should *not* seem obvious.

Surface area

Given a surface S parameterized by $\vec{r}(u,v)$ for $(u,v) \in D$, at the point $\vec{r}(u,v)$ the surface has tangent vectors $\vec{r}_u$ and $\vec{r}_v$, and so $\vec{r}_u \times \vec{r}_v$ is a normal vector. From Eq. (17.36) we see that $\|\vec{r}_u \times \vec{r}_v\|$ is the area of the parallelogram spanned by these tangent vectors. Approximate the surface by these little parallelograms and take the appropriate limits. Then the surface area of S is

$$\text{area}(S) = \iint_D \|\vec{r}_u \times \vec{r}_v\| \, dA \tag{17.37}$$

Example 17.8.4. *Area of a sphere.* For the sphere of radius R we have

$$\vec{r}_\theta = -R\sin(\theta)\sin(\varphi)\vec{i} + R\cos(\theta)\sin(\varphi)\vec{j}$$
$$\vec{r}_\varphi = R\cos(\theta)\cos(\varphi)\vec{i} + R\sin(\theta)\cos(\varphi)\vec{j} - R\sin(\varphi)\vec{k}$$

and after some simplification, $\|\vec{r}_\theta \times \vec{r}_\varphi\| = R^2\sqrt{\sin^2(\varphi)} = R^2\sin(\varphi)$. This last equality holds because $0 \le \varphi \le \pi$ implies $\sin(\varphi) \ge 0$, which implies $\sqrt{\sin^2(\varphi)} = \sin(\varphi)$. Then the surface area of the sphere is

$$\text{area(sphere)} = \int_0^{2\pi} \int_0^{\pi} R^2 \sin(\varphi) \, d\varphi \, d\theta = 4\pi R^2$$

This formula was derived by Archimedes, without the use of double integrals, of course. □

Surface integral of a scalar function

Recall that to evaluate the line integral of a scalar function f along a curve C parameterized by $\vec{r}(t)$ for $a \le t \le b$, we evaluate the function on the curve, $f(\vec{r}(t))$, multiply by the arclength correction $\|\vec{r}\,'(t)\|$, and integrate

$$\int_C f \, ds = \int_a^b f(\vec{r}(t))\|\vec{r}\,'(t)\| \, dt$$

The surface integral of a scalar function is defined analogously, with the surface area correction $\|\vec{r}_u \times \vec{r}_v\|$ in place of the arclength correction $\|\vec{r}\,'(t)\|$. That is, the integral of a scalar function $f(x,y,z)$ over a surface S parameterized by $\vec{r}(u,v)$ for (u,v) in a region D is defined by

$$\iint_S f \, dS = \iint_D f(\vec{r}(u,v))\|\vec{r}_u \times \vec{r}_v\| \, dA \tag{17.38}$$

Surface integral of a vector field

Now recall that to evaluate the line integral of a vector field $\vec{F}$ along a curve C parameterized by $\vec{r}(t)$ for $a \le t \le b$, we integrate the component of $\vec{F}(\vec{r}(t))$ in the direction of $\vec{r}\,'(t)$. Then

$$\int_C \vec{F} \cdot d\vec{r} = \int_a^b \left(\vec{F}(\vec{r}(t)) \cdot \frac{\vec{r}\,'(t)}{\|\vec{r}\,'(t)\|} \right) \|\vec{r}\,'(t)\| \, dt = \int_a^b \vec{F}(\vec{r}(t)) \cdot \vec{r}\,'(t) \, dt$$

where the bracketed factors in the second integral are the component of $\vec{F}(\vec{r}(t))$ in the $\vec{r}'(t)$ direction.

The surface integral of a vector field $\vec{F}$ across a surface S is the surface integral of the component of $\vec{F}$ in the direction normal to the surface. We interpret the surface integral as the *flux* of the vector field across the surface.

Each point of the surface has two normal directions, and the sign of the surface integral depends on the choice of normal directions. This choice is part of the input information of evaluating the integral, and is specified by $\vec{r}_u \times \vec{r}_v$. Note that $\vec{r}_v \times \vec{r}_u = -(\vec{r}_u \times \vec{r}_v)$. Then the integral of a vector field $\vec{F}$ over a surface S is defined by

$$\iint_S \vec{F} \cdot d\vec{S} = \iint_S \vec{F} \cdot \frac{\vec{r}_u \times \vec{r}_v}{\|\vec{r}_u \times \vec{r}_v\|}\, dS$$
$$= \iint_D \left(\vec{F} \cdot \frac{\vec{r}_u \times \vec{r}_v}{\|\vec{r}_u \times \vec{r}_v\|} \right) \|\vec{r}_u \times \vec{r}_v\|\, dA = \iint_D \vec{F} \cdot (\vec{r}_u \times \vec{r}_v)\, dA \qquad (17.39)$$

The first equality is the observation that the component of $\vec{F}$ in the direction $\vec{r}_u \times \vec{r}_v$ is the dot product of $\vec{F}$ with the unit vector in the direction of $\vec{r}_u \times \vec{r}_v$. The second equality is just the conversion of the surface integral of this scalar function into a double integral by the surface area correction factor $\vec{r}_u \times \vec{r}_v$.

Example 17.8.5. *Integrals over a saddle.* Suppose S is the saddle parameterized by $\vec{r}(x,y) = x^2 - y^2$ for $0 \leq x \leq 1$ and $0 \leq y \leq 1$. Then the tangent vectors are $\vec{r}_x = \langle 1, 0, 2x \rangle$ and $\vec{r}_y = \langle 0, 1, -2y \rangle$, and the normal vector is

$$\vec{r}_x \times \vec{r}_y = \det \begin{bmatrix} \vec{i} & \vec{j} & \vec{k} \\ 1 & 0 & 2x \\ 0 & 1 & 2y \end{bmatrix} = -2x\vec{i} - 2y\vec{j} + \vec{k}$$

so $\|\vec{r}_x \times \vec{r}_y\| = \sqrt{4x^2 + 4y^2 + 1}$. Because the coefficient of $\vec{k}$ is positive, this is the upward-pointing normal vector.

(a) For $f(x,y) = xy$ the surface integral is

$$\iint_S f dA = \int_0^1 \int_0^1 xy\sqrt{4x^2 + 4y^2 + 1}\, dx\, dy = \frac{61}{60} - \frac{5\sqrt{5}}{24}$$

where for the dx integral we substitute $w = 4x^2 + 4y^2 + 1$ and for the dy integral we substitute $w = 4y^2 + 5$ and $w = 4y^2 + 1$.

(b) For $\vec{F}(x,y) = \langle xy, x^2, y^3 \rangle$ the surface integral is

$$\iint_S \vec{F} \cdot d\vec{S} = \int_0^1 \int_0^1 \langle xy, x^2, y^3 \rangle \cdot \langle -2x, -2y, 1 \rangle\, dx\, dy$$
$$= \int_0^1 \int_0^1 -4x^2y + y^3\, dx\, dy = \int_0^1 -\frac{4}{3} - y^3\, dy = -\frac{13}{12}$$

Often the surface intergal of a vector field is easier to evaluate than is the surface integral of a scalar function. □

If line and surface integrals of scalar and vector functions seem a bit confusing, we'll point out that they have an instructive parallel structure. Suppose C is a curve parameterized by $\vec{r}(t)$, $a \leq t \leq b$, S is a surface parameterized by $\vec{r}(u, v)$ for $(u, v) \in D$, f is a scalar function, and $\vec{F}$ is a vector function. Then

$$\int_C f ds = \int_a^b f(\vec{r}(t)) \|\vec{r}'(t)\| \, dt$$

$$\int_C \vec{F} \cdot d\vec{r} = \int_a^b \vec{F}(\vec{r}(t)) \, dt$$

$$\iint_S f \, dS = \iint_D f(\vec{r}(u, v)) \|\vec{r}_u \times \vec{r}_v\| \, dA$$

$$\iint_S \vec{F} \cdot d\vec{S} = \iint_D \vec{F}(\vec{r}(u, v)) \cdot (\vec{r}_u \times \vec{r}_v) \, dA$$

That is, the role of the tangent vector $\vec{r}'(t)$ for line integrals is taken by the normal vector $\vec{r}_u \times \vec{r}_v$ for surface integrals. Is the function scalar or vector? Is the integral along a curve or across a surface? Those questions determine which of these four formulas is appropriate.

Figure 17.21. Möbius band

We need to mention one more point: the choice of normal direction must be made consistently across the entire surface. If we can do this, the surface is called *orientable*. Every non-orientable surface contains a Möbius band. An image is shown in Fig. 17.21, but many people make these in elementary school. Paper, scissors, and tape are all you need. Google can show you how to do this, if you haven't already. As long as we avoid Möbius bands, the normal vector behaves just fine.

Practice Problems

17.8.1. Find the area of the saddle $z = x^2 - y^2$ for $-1 \leq x \leq 1$ and $-1 \leq y \leq 1$.

17.8.2. For the surface $z = x^2 - y$, $0 \leq x \leq 1$ and $0 \leq y \leq 1$, integrate the function $f(x, y) = x + y$.

17.8.3. Integrate the vector field $\vec{F}(x, y, z) = \langle x^2, y^2, z \rangle$ over the surface $z = x + y^2$ for $0 \leq x \leq 1$ and $0 \leq y \leq 1$, with upward-pointing normal.

Practice Problem Solutions

17.8.1. For this surface $\vec{r}_x = \langle 1, 0, 2x \rangle$ and $\vec{r}_y = \langle 0, 1, -2y \rangle$, so $\|\vec{r}_x \times \vec{r}_y\| = \|\langle -2x, 2y, 1 \rangle\| = \sqrt{4x^2 + 4y^2 + 1}$. Then by Eq. (17.37) the surface area of the saddle S is

$$\text{area(S)} = \int_{-1}^{1} \int_{-1}^{1} \sqrt{4x^2 + 4y^2 + 1}\, dx\, dy \approx 7.446$$

The dx integral can be evaluated by an application of rule 16 from the integral table in Appendix C, but then the dy integral is quite challenging. Mathematica can evaluate this integral exactly, but this answer is not particularly enlightening.

17.8.2. For this surface, $\vec{r}_x = \langle 1, 0, 2x \rangle$ and $\vec{r}_y = \langle 0, 1, -1 \rangle$, so $\|\vec{r}_x \times \vec{r}_y\| = \|\langle -2x, 1, 1 \rangle\| = \sqrt{4x^2 + 2}$. Then

$$\iint_S f\, dS = \int_0^1 \int_0^1 (x + y)\sqrt{4x^2 + 2}\, dx\, dy$$
$$= \frac{1}{12}\left(6^{3/2} - 2^{3/2}\right) + \frac{1}{2}\sqrt{\frac{3}{2}} + \frac{1}{4}\ln\left(1 + \sqrt{\frac{3}{2}}\right) - \frac{1}{8}\ln\left(\frac{1}{2}\right)$$

where we've used rule 16 from the integral table.

17.8.3. We'll apply the definition of Eq. (17.39). For this surface $\vec{r}_x = \langle 1, 0, 1 \rangle$ and $\vec{r}_y = \langle 0, 1, -2y \rangle$, so the normal vector is $\vec{r}_x \times \vec{r}_y = \langle -1, 2y, 1 \rangle$. Because the $\vec{k}$ component is positive, this is the upward-pointing normal. Then the surface integral of this vector field is

$$\iint_S \vec{F} \cdot d\vec{S} = \int_0^1 \int_0^1 \langle x^2, y^2, x + y^2 \rangle \cdot \langle -1, 2y, 1 \rangle\, dx\, dy$$
$$= \int_0^1 \int_0^1 (-x^2 + 2y^3 + x + y^2)\, dx\, dy = \int_0^1 \left(2y^3 + y^2 + \frac{1}{6}\right) dy = 1$$

Exercises

17.8.1. Find the area of the helicoid $\vec{r}(u, v) = \langle u\cos(v), u\sin(v), v \rangle$, for $0 \le u \le 1$ and $0 \le v \le 2\pi$.

17.8.2. (a) Find the surface area of the torus of Example 17.8.3.
(b) Find the area of the part of the torus given by $\pi/2 \le v \le 3\pi/2$.

17.8.3. Integrate $f(x, y, z) = x\sec(z)$ over the helicoid of Exercise 17.8.1.

17.8.4. Integrate $f(x, y, z) = x^2 - y + z$ over the torus of Example 17.8.3.

17.8.5. Integrate $f(x,y,z) = x - y$ over the graph of $z = x^{3/2} - y^{3/2}$ for the surface S defined by $0 \le x \le 1$, $0 \le y \le 1$.

17.8.6. Integrate $\vec{F}(x,y,z) = \langle x,y,z \rangle$ over the helicoid of Exercise 17.8.1. Use the upward-pointing normal vector.

17.8.7. Integrate $\vec{F}(x,y,z) = \langle x,y,1 \rangle$ over the surface defined by $z = x^2 + y^2$ for $0 \le z \le 4$, with the downward-pointing normal.

17.8.8. Integrate $\vec{F}(x,y,z) = \langle x^2, y^2, z \rangle$ over the surface defined by $z = \sin(x)$ for $0 \le x \le \pi$, $0 \le y \le 1$, with the upward-pointing normal.

17.8.9. Integrate $\vec{F}(x,y,z) = \langle y,z,x \rangle$ over the surface defined by $z = \sin(x) + \sin(y)$ for $0 \le x \le \pi$, $0 \le y \le \pi$, with the upward-pointing normal.

17.8.10. Integrate $\vec{F}(x,y,z) = \langle e^y, e^x, 1 \rangle$ over the semicylinder $\vec{r}(u,v) = \langle \cos(u), \sin(u), v \rangle$ for $0 \le u \le \pi$, $0 \le v \le 1$, with normal vector pointing away from the z-axis.

17.9 STOKES' THEOREM

Stokes' theorem is Green's theorem generalized to the setting where the region no longer lies in a plane. The double integral is replaced by a surface integral, the topic of Sect. 17.8. The line integral of a vector field along a curve in 3 dimensions is the obvious generalization of the line integral of a vector field along a curve in the plane. In fact, we can copy the definition of Eq. (17.13) exactly, so long as we understand that $\vec{r}(t)$ is a point in 3 dimensions and $\vec{r}\,'(t)$ is a vector in 3 dimensions. All that remains is to generalize the integrand $\partial Q/\partial x - \partial P/\partial y$ of the double integral in Green's theorem. Here's how to do that.

First, in 3 dimensions write a vector field $\vec{F} = P\vec{i} + Q\vec{j} + R\vec{k}$ and define the vector field $\text{curl}(\vec{F}) = \nabla \times \vec{F}$. That is,

$$\text{curl}(\vec{F}) = \left(\frac{\partial R}{\partial y} - \frac{\partial Q}{\partial z}\right)\vec{i} - \left(\frac{\partial R}{\partial x} - \frac{\partial P}{\partial z}\right)\vec{j} + \left(\frac{\partial Q}{\partial x} - \frac{\partial P}{\partial y}\right)\vec{k} \qquad (17.40)$$

Now suppose S is a surface in 3 dimensions and that S has boundary curve C. (Think of S as the upper hemisphere and C as the equator.) The orientation of a curve is the direction we follow the curve. Recall the curve C is oriented positively if when a person walks in that direction along the curve with their head in the direction of the normal vector $\vec{n}$ of the surface, the surface is on the left. Positive orientation is a relation between the choice of surface normal direction and the direction of traversing the curve. Now we can state Stokes' theorem.

Theorem 17.9.1. Suppose S is a surface with boundary curve C, oriented positively with respect to the choice of normal vectors of S, and the components of $\vec{F}$ have continuous partial derivatives. Then

$$\oint_C \vec{F} \cdot d\vec{r} = \iint_S \text{curl}(\vec{F}) \cdot d\vec{S} \tag{17.41}$$

Proof. We'll prove this in the special case that S is the graph of a function $z = f(x,y)$ for $(x,y) \in D$, with upward-pointing normal. Suppose f has continuous second partial derivatives. The proof is a bit messy but is straightforward: convert the surface integral over S to a double integral over D, convert the line integral over ∂S to a line integral over ∂D, and apply Green's theorem.

The surface S is a graph, parameterized by $\vec{r}(x,y) = (x,y,f(x,y))$ for $(x,y) \in D$, so the surface has tangent vectors $\vec{r}_x = \langle 1,0,\partial f/\partial x\rangle$ and $\vec{r}_y = \langle 0,1,\partial f/\partial y\rangle$ and normal vector

$$\vec{r}_x \times \vec{r}_y = \det \begin{bmatrix} \vec{i} & \vec{j} & \vec{k} \\ 1 & 0 & \partial f/\partial x \\ 0 & 1 & \partial f/\partial y \end{bmatrix} = -\frac{\partial f}{\partial x}\vec{i} - \frac{\partial f}{\partial y}\vec{j} + \vec{k}$$

This is the upward-pointing normal vector because the coefficient of $\vec{k}$ is positive.

First we'll compute the surface integral. $\iint_S \text{curl}(\vec{F}) \cdot d\vec{S} =$

$$\iint_D \left(\left(\frac{\partial R}{\partial y} - \frac{\partial Q}{\partial z} \right)\vec{i} - \left(\frac{\partial R}{\partial x} - \frac{\partial P}{\partial z} \right)\vec{j} + \left(\frac{\partial Q}{\partial x} - \frac{\partial P}{\partial y} \right)\vec{k} \right) \cdot \left(-\frac{\partial f}{\partial x}\vec{i} - \frac{\partial f}{\partial y}\vec{j} + \vec{k} \right) dA$$

$$= \iint_D \left(\frac{\partial Q}{\partial z} - \frac{\partial R}{\partial y} \right)\frac{\partial f}{\partial x} + \left(\frac{\partial R}{\partial x} - \frac{\partial P}{\partial z} \right)\frac{\partial f}{\partial y} + \left(\frac{\partial Q}{\partial x} - \frac{\partial P}{\partial y} \right) dA \tag{17.42}$$

The boundary curve is parameterized by $\vec{r}(t) = (x(t),y(t),f(x(t),y(t)))$, where $(x(t),y(t),0)$, $a \le t \le b$, is a parameterization of ∂D. Then with the chain rule we find that the tangent vectors of ∂S are

$$\vec{r}'(t) = \frac{dx}{dt}\vec{i} + \frac{dy}{dt}\vec{j} + \left(\frac{\partial f}{\partial x}\frac{dx}{dt} + \frac{\partial f}{\partial y}\frac{dy}{dt} \right)\vec{k}$$

and the line integral is $\oint_{\partial S} \vec{F} \cdot d\vec{r} =$

$$\oint_{\partial D} \left(P\vec{i} + Q\vec{j} + R\vec{k}\right) \cdot \left(\frac{dx}{dt}\vec{i} + \frac{dy}{dt}\vec{j} + \left(\frac{\partial f}{\partial x}\frac{dx}{dt} + \frac{\partial f}{\partial y}\frac{dy}{dt}\right)\vec{k}\right) dt$$
$$= \oint_{\partial D} \left(P + R\frac{\partial f}{\partial x}\right) dx + \left(Q + R\frac{\partial f}{\partial y}\right) dy$$
$$= \iint_D \frac{\partial}{\partial x}\left(Q + R\frac{\partial f}{\partial y}\right) - \frac{\partial}{\partial y}\left(P + R\frac{\partial f}{\partial x}\right) dA$$

where the last equality follows from Green's theorem. We use the chain rule to compute the partials. For example,

$$\frac{\partial}{\partial x}Q(x,y,f(x,y)) = \frac{\partial Q}{\partial x} + \frac{\partial Q}{\partial z}\frac{\partial f}{\partial x}$$

On the left, $\partial Q/\partial x$ means compute $\partial/\partial x$ of $Q(x,y,z)$ after $f(x,y)$ is substituted for z to get a function only of x and y. On the right side, by $\partial Q/\partial x$ we mean $\partial Q(x,y,z)/\partial x$. So, to find how $Q(x,y,z)$ changes with x on the surface $z = f(x,y)$, find how $Q(x,y,z)$ changes with x and add to this how $Q(x,y,z)$ changes with z times how $z = f(x,y)$ changes with x.

With the chain rule and the product rule we find

$$\frac{\partial}{\partial x}\left(R(x,y,f(x,y))\frac{\partial f}{\partial y}\right) = \left(\frac{\partial R}{\partial x} + \frac{\partial R}{\partial z}\frac{\partial f}{\partial x}\right)\frac{\partial f}{\partial y} + R\frac{\partial^2 f}{\partial x \partial y}$$

With similar calculations for the $\partial/\partial y$ terms, remove the terms that subtract out (recall Clairaut's theorem) and we have $\int_{\partial S} \vec{F} \cdot d\vec{r} =$

$$\iint_D \left(\frac{\partial Q}{\partial z} - \frac{\partial R}{\partial y}\right)\frac{\partial f}{\partial x} + \left(\frac{\partial R}{\partial x} - \frac{\partial P}{\partial z}\right)\frac{\partial f}{\partial y} + \left(\frac{\partial Q}{\partial x} - \frac{\partial P}{\partial y}\right) dA \qquad (17.43)$$

Combining Eqs. (17.42) and (17.43), we have Stokes' theorem. □

The proof is a bit long and involves some messy calculations, but it did not need a clever trick or insight. With a reminder to use the chain rule, this could have been assigned as homework. Sometimes the straightforward approach works. Of course, knowing *what* to prove may not be so straightforward.

We'll get to examples soon, but first we'll mention some points that can help in applications.

Corollary 17.9.1. If surfaces S and T have the same boundary curve C, oriented positively with respect to the chosen normal directions of both S and T, then

$$\iint_S \text{curl}(\vec{F}) \cdot d\vec{S} = \iint_T \text{curl}(\vec{F}) \cdot d\vec{S} \qquad (17.44)$$

Proof. This is just two applications of Stokes' theorem:

$$\iint_S \operatorname{curl}(\vec{F}) \cdot d\vec{S} = \oint_{\partial S} \vec{F} \cdot d\vec{r} = \oint_{\partial T} \vec{F} \cdot d\vec{r} = \iint_T \operatorname{curl}(\vec{F}) \cdot d\vec{S}$$

where the second equality occurs because $\partial S = C = \partial T$ and the positive orientations agree. □

This is *surface independence* of the surface integral of a curl, analogous to path independence of the line integral of a gradient. In the same way that path independence allows us to replace the line integral of a gradient over a complicated curve with the line integral of the same gradient over a simpler curve, so long as the curves have the same boundary points, surface independence allows us to replace the surface integral of a curl over a complicated surface with the surface integral of the same curl over a simpler surface, so long as the surfaces have the same boundary curves. Sometimes this can be very useful.

In order to use Stokes' theorem to evaluate a surface integral $\iint_S \vec{G} \cdot d\vec{S}$, the vector field $\vec{G}$ must be a curl, $\vec{G} = \operatorname{curl}(\vec{F})$, and we must be able to find the vector field $\vec{F}$.

In order to use Stokes' theorem to evaluate a line integral $\int_C \vec{F} \cdot d\vec{r}$, we must compute $\operatorname{curl}(\vec{F})$ (this is easy), and we must find a surface S with boundary curve C. By surface independence, every surface S with (positively oriented) boundary curve C gives the same integral, so some thought about the choice of surface can simplify the integral.

We can apply Stokes' theorem to finish our analysis of how we can tell if a vector field $\vec{F}$ is conservative, that is, if $\vec{F} = \nabla f$ for some f. First, we'll make a preliminary observation about curl and conservative vector fields.

Corollary 17.9.2. If the potential function f of a conservative vector field $\vec{F}$ has continuous second partials, then $\operatorname{curl}(\vec{F}) = \vec{0}$. If the domain of $\vec{F}$ is simply-connected and $\operatorname{curl}(\vec{F}) = \vec{0}$, then $\vec{F}$ is conservative.

Proof. Suppose $\vec{F}$ is conservative, and $\vec{F} = \nabla f$ where f has continuous second partial derivatives, then $\operatorname{curl}(\vec{F}) = \operatorname{curl}(\nabla f) =$

$$\left(\frac{\partial^2 f}{\partial y \partial z} - \frac{\partial^2 f}{\partial z \partial y}\right)\vec{i} - \left(\frac{\partial^2 f}{\partial x \partial z} - \frac{\partial^2 f}{\partial z \partial x}\right)\vec{j} + \left(\frac{\partial^2 f}{\partial x \partial y} - \frac{\partial^2 f}{\partial y \partial x}\right)\vec{k} = \vec{0}$$

The second equality is an application of Clairaut's theorem.

Now suppose that $\operatorname{curl}(\vec{F}) = \vec{0}$ and the domain D (in $\mathbb{R}^3$) of $\vec{F}$ is simply-connected. Then any closed curve C in D bounds a surface S in D. We apply Stokes' theorem

$$\oint_C \vec{F} \cdot d\vec{r} = \iint_S \text{curl}(\vec{F}) \cdot d\vec{S} = \iint_S \vec{0} \cdot d\vec{S} = 0$$

Then by Cor. 17.2.4, $\int_C \vec{F} \cdot d\vec{r}$ is path independent for all paths C in the domain D. By Cor. 17.2.2 we see that $\vec{F}$ is conservative. □

Examples now, finally.

Example 17.9.1. *Stokes' theorem and line integrals.* The ellipse C is the intersection of the cylinder $x^2 + y^2 = 1$ and the plane $x + z = 2$, oriented counterclockwise when viewed from above. To integrate the vector field $\vec{F} = \langle y^2, x, z^3 \rangle$ around C, parameterize C by $\vec{r}(t) = \langle \cos(t), \sin(t), 2 - \cos(t) \rangle$ for $0 \le t \le 2\pi$. Then $\vec{r}\,'(t) = \langle -\sin(t), \cos(t), \sin(t) \rangle$ and $\vec{F}(\vec{r}(t)) = \langle \sin^2(t), \cos(t), (2 - \cos(t))^3 \rangle$. The line integral is

$$\oint_C \vec{F}(\vec{r}(t)) \cdot d\vec{r} = \int_0^{2\pi} -\sin^3(t) + \cos^2(t) + (2 - \cos(t))^3 \sin(t)\, dt$$

We can evaluate this integral, but it's a bit messy.

Stokes' theorem gives an easier approach, if we pick a simple surface. We'll take S to be the part of the plane $x + z = 2$ bounded by the cylinder $x^2 + y^2 = 1$, so $\partial S = C$. With upward-pointing normals, the positive orientation of C agrees with the counterclockwise orientation of the problem. The computation $\text{curl}(\vec{F}) = (1 - 2y)\vec{k}$ is straightforward. The normal vector of S is the normal vector of the plane, that is, $\vec{i} + \vec{k}$. Then Stokes' theorem gives

$$\oint_C \vec{F} \cdot d\vec{r} = \iint_S \text{curl}(\vec{F}) \cdot d\vec{S} = \iint_D (1 - 2y)\vec{k} \cdot (\vec{i} + \vec{k})\, dA$$

$$= \int_0^{2\pi} \int_0^1 (1 + 2r\sin(\theta))r\, dr\, d\theta = \int_0^{2\pi} \left(\frac{1}{2} + \frac{2}{3}\sin(\theta) \right) d\theta = \pi$$

Direct evaluation of the line integral gives the same answer. □

Example 17.9.2. *Surface independence.* For $\vec{F} = \langle -y^3 + e^{xz}, x^3, e^{xyz} \rangle$, integrate $\text{curl}(\vec{F})$ over the surface S defined by $x^2 + y^2 + (z-2)^2 = 4$ and $z \ge 2 - \sqrt{3}$, with outward-pointing normal vectors.

The first step of the direct calculation is to find $\text{curl}(\vec{F})$. The result is discouraging.

$$\text{curl}(\vec{F}) = \langle xze^{xyz}, -yze^{xyz} + xe^{xz}, 3x^2 + 3y^2 \rangle$$

Do you want to integrate this over S? I don't.

What about an application of Stokes' theorem to convert this surface integral into a line integral along $C = \partial S$. This curve is the circle $x^2 + y^2 = 1$, $z = z_0 = 2 - \sqrt{3}$, parameterized by

$$\vec{r}(t) = \langle \cos(t), \sin(t), 2 - \sqrt{3} \rangle \quad \text{so} \quad \vec{r}\,'(t) = \langle -\sin(t), \cos(t), 0 \rangle$$

Then Stokes' theorem gives

$$\iint_S \text{curl}(\vec{F}) \cdot d\vec{S} = \oint_C \vec{F}(\vec{r}(t)) \cdot d\vec{r}$$

$$= \int_0^{2\pi} \langle -\sin^3(t) + e^{z_0 \cos(t)}, \cos^3(t), e^{\cos(t)\sin(t) z_0} \rangle \cdot \langle -\sin(t), \cos(t), 0 \rangle \, dt$$

This can be evaluated, but it's a bit complicated. Can we do better?

Let's try surface independence, Cor. 17.9.1. The choice of the surface T is clear: the disk $x^2 + y^2 \leq 1, z = z_0$ with normal vector $\vec{k}$. Then

$$\iint_S \text{curl}(\vec{F}) \cdot d\vec{S} = \iint_T \text{curl}(\vec{F}) \cdot d\vec{T}$$

$$= \iint_T \langle xz_0 e^{xyz_0}, -yz_0 e^{xyz_0} + xe^{xz_0}, 3x^2 + 3y^2 \rangle \cdot \langle 0, 0, 1 \rangle \, dA$$

$$= \iint_T (3x^2 + 3y^2) \, dA = \int_0^{2\pi} \int_0^1 3r^2 r \, dr \, d\theta = \frac{3\pi}{2}$$

This works because the dot product with the normal vector of T kills the messy components of $\text{curl}(\vec{F})$. □

Practice Problems

17.9.1. For the vector field $\vec{F} = \langle x^2, x + y^2 + z, xyz \rangle$ integrate $\text{curl}(\vec{F})$ over the surface $S = \{(x, y, z) : x^2 + y^2 + z^2 = 1, z \geq 0\}$.

17.9.2. Integrate the vector field $\vec{F} = \langle x^3, x + y, x + z^3 \rangle$ around the ellipse C parameterized by $\vec{r}(t) = \langle 3\cos(t), 2\sin(t), 2\sin(t) \rangle$ for $0 \leq t \leq 2\pi$, oriented counterclockwise when viewed from above.

Practice Problem Solutions

17.9.1. Although $\text{curl}(\vec{F}) = \langle xz - 1, -yz, 1 \rangle$ is not complicated, we are not enthusiastic about integrating this over the upper hemisphere, so we'll apply Stokes' theorem. The boundary of S is the unit circle C in the xy-plane. Positive orientation is counterclockwise, so $\vec{r}(t) = \langle \cos(t), \sin t, 0 \rangle$, $0 \leq t \leq 2\pi$, is a parameterization of C. Then by Stokes' theorem,

$$\iint_S \text{curl}(\vec{F}) \cdot d\vec{S} = \oint_C \vec{F}(\vec{r}(t)) \cdot d\vec{r}$$
$$= \int_0^{2\pi} \langle \cos^2(t), \cos(t) + \sin^2(t) + 0, 0 \rangle \cdot \langle -\sin(t), \cos(t), 0 \rangle \; dt$$
$$= \int_0^{2\pi} -\cos^2(t)\sin(t) + \cos^2(t) + \cos(t)\sin(t) \; dt = \pi$$

17.9.2. To apply Stokes' theorem we compute $\text{curl}(\vec{F}) = \langle 0, -1, 1 \rangle$. Take S to be the part of the $y = z$ plane inside the ellipse C, so S is parameterized by $\vec{r}(x,y) = \langle x, y, y \rangle$ for $(x,y) \in D$, the region bounded by the ellipse $x^2/9 + y^2/4 = 1$. A picture can guide us to the normal vector, but to be sure, we'll calculate it: $\vec{r}_x = \langle 1, 0, 0 \rangle$ and $\vec{r}_y = \langle 0, 1, 1 \rangle$, so the normal vector is $\vec{r}_x \times \vec{r}_y = \langle 0, -1, 1 \rangle$. Then

$$\oint_C \vec{F}(\vec{r}(t)) \cdot d\vec{r} = \iint_S \text{curl}(\vec{F}) \cdot d\vec{S} = \iint_D \langle 0, -1, 1 \rangle \cdot \langle 0, -1, 1 \rangle \; dA$$

That is, the line integral is twice the area of D. Recall an ellipse with semi-major axis a and semi-minor axis b has area πab. Then the line integral is $2 \cdot (\pi \cdot 3 \cdot 2) = 12\pi$.

Exercises

17.9.1. If S is a flat surface in the xy-plane, show $\vec{r}_x \times \vec{r}_y = \vec{k}$ and that Stokes' theorem reduces to Green's theorem.

17.9.2. Suppose S is the paraboloid defined by $z - 1 - (x^2 + y^2)$ for $x^2 + y^2 \leq 1$, with upward-pointing normals. For the vector field $\vec{F} = \langle y, -x, ze^{xy} \rangle$, evaluate $\iint_S \text{curl}(\vec{F}) \cdot d\vec{S}$.

17.9.3. Evaluate $\oint_C \vec{F} \cdot d\vec{r}$ where $\vec{F} = \langle x^2, 4xy^3, xy^2 \rangle$ and C is the boundary of $S = \{(x,y,0) : 0 \leq x \leq 1, 0 \leq y \leq 3\}$, oriented counterclockwise.

17.9.4. Evaluate $\oint_C \vec{F} \cdot d\vec{r}$ where $\vec{F} = \langle \sin(e^{xy}), y^2 + z^2, x^2 + z^2 \rangle$ and C is the boundary of $S = \{(\cos(\theta), \sin(\theta), z) : 0 \leq \theta \leq \pi/2, 0 \leq z \leq 1\}$, oriented counterclockwise viewed far from the z-axis, with x and y positive.

17.9.5. Evaluate $\oint_C \vec{F} \cdot d\vec{r}$ where $\vec{F} = \langle e^{x^2}, xy, z \rangle$ and C is the boundary of $S = \{(x, y, 1 - y) : 0 \leq x \leq 1, 0 \leq y \leq 1\}$, oriented counterclockwise viewed far from the origin with positive x, y, and z.

17.9.6. Evaluate $\oint_C \vec{F}\cdot d\vec{r}$ where $\vec{F} = \langle e^{x^2}, \sin(y^2), -x+\sin(z^2)\rangle$ and C is the boundary of the half cylinder $S = \{(\cos(\theta), y, \sin(\theta)) : 0 \le \theta \le \pi, 0 \le y \le 1\}$, oriented counterclockwise viewed far above the y-axis.

17.9.7. Evaluate $\oint_C \vec{F}\cdot d\vec{r}$ where $\vec{F} = \langle e^{x^2}, \sin(y^2), y+\sin(z^2)\rangle$ and C is the boundary of the quarter cylinder $S = \{(\cos(\theta), y, \sin(\theta)) : 0 \le \theta \le \pi/2, 0 \le y \le 1\}$, oriented counterclockwise viewed far above the y-axis.

17.9.8. Evaluate $\oint_C \vec{F}\cdot d\vec{r}$ where $\vec{F} = \langle e^{x^3}, x^3+e^{y^3}, e^{z^3}\rangle$ and C is the boundary of the surface $S = \{(x, y, 1-(x^2+y^2)) : 0 \le x^2+y^2 \le 1\}$, oriented counterclockwise viewed from above. Hint: Cor. 17.9.1.

17.9.9. Evaluate $\int_C \vec{F}\cdot d\vec{r}$ where $\vec{F} = \langle x, x+e^{y^2}, y+e^{z^2}\rangle$ and the path C consists of the line segment from $(1,0,0)$ to $(1,1,1)$, followed by the segment from $(1,1,1)$ to $(0,1,1)$, followed by the segment from $(0,1,1)$ to $(0,0,0)$. Hint: close the path by adding the segment from $(0,0,0)$ to $(1,0,0)$.

17.9.10. Evaluate $\int_C \vec{F}\cdot d\vec{r}$ where $\vec{F} = \langle x^2, x^3+e^{y^2}, y+e^{z^2}\rangle$ and the path C is parameterized by $\vec{t}(t) = \langle \cos(t), \sin(t), \sin(t)\rangle$ for $0 \le t \le \pi$, oriented in the direction of increasing t. Hint: close the path by adding the segment from $(-1,0,0)$ to $(1,0,0)$.

17.10 TRIPLE INTEGRALS

Triple integrals are just like double integrals, but with three iterations instead of two. The limits can be a bit more complicated. For example, we may have numerical limits on x, $a \le x \le b$; y can be bounded by functions of x, $f_1(x) \le y \le f_2(x)$; and z can be bounded by surfaces, $g_1(x,y) \le z \le g_2(x,y)$. An example is all we need, for now.

Example 17.10.1. *Cartesian coordinates.* Integrate the scalar function $f(x,y,z) = x+y-z$ over the solid region R defined by $0 \le x \le 1$, $x^2 \le y \le x$, and $0 \le z \le x+y$. The z integral is bounded by functions of x and y, so the z integral must be the innermost. The y integral is bounded by functions of x, so it must be next. The x integral is bounded by constants, so it must be the outermost integral. Then

$$\iiint_R f\,dV = \int_0^1 \int_{x^2}^{x} \int_0^{x+y} x+y-z\,dz\,dy\,dx$$
$$= \int_0^1 \int_{x^2}^{x} \frac{x^2}{2} + xy + \frac{y^2}{2}\,dy\,dx = \int_0^1 \frac{7x^3}{6} - \frac{x^4}{2} - \frac{x^5}{2} - \frac{x^6}{6}\,dx = \frac{71}{840}$$

Perfectly straightforward, so far. □

Complications come when the region is not defined in such an explicit fashion. Suppose the region is defined as the space between two surfaces, say $g_1(x,y) \leq z \leq g_2(x,y)$. So far, so good, but some cleverness may be required to determine the limits of x and of y.

Before we consider this, we'll mention two special coordinate systems that simplify the evaluation of some triple integrals. Remember that double integrals over regions with circular symmetry are much simpler in polar coordinates. In 3 dimensions common symmetries are cylindrical (a section of artery or vein, the axon of a neuron, parts of some bones) and spherical (eyes, lymph nodes, lymphocytes, many bacteria), so we'll introduce cylindrical and spherical coordinates.

Cylindrical coordinates are easy: just polar in the xy-plane plus z. In Cartesian coordinates, $dV = dx\ dy\ dz$; in cylindrical $dV = r\ dr\ d\theta\ dz$. Another way to see this is to parameterize space by cylindrical coordinates: $\vec{r}(r,\theta,z) = \langle r\cos(\theta), r\sin(\theta), z\rangle$. Then

$$\vec{r}_r = \langle\cos(\theta),\sin(\theta),0\rangle, \quad \vec{r}_\theta = \langle -r\sin(\theta), r\cos(\theta), 0\rangle, \quad \vec{r}_z = \langle 0,0,1\rangle$$

For vectors $\vec{a}$, $\vec{b}$, and $\vec{c}$ in 3 dimensions, $\vec{a}\times\vec{b}$ is perpendicular to the parallelogram spanned by $\vec{a}$ and $\vec{b}$ and the length of the cross product is the area of the parallelogram. Then $(\vec{a}\times\vec{b})\cdot\vec{c}$ is the projection of $\vec{c}$ in the direction of $\vec{a}\times\vec{b}$ and so $|(\vec{a}\times\vec{b})\cdot\vec{c}|$ is the volume spanned by these three vectors. This is called the *scalar triple product*. So for cylindrical coordinates,

$$dV = |((\vec{r}_r\ dr)\times(\vec{r}_\theta\ d\theta))\cdot(\vec{r}_z\ dz)| = |(\vec{r}_r\times\vec{r}_\theta)\cdot\vec{r}_z|\ dr\ d\theta\ dz = r\ dr\ d\theta\ dz$$

Spherical coordinates build on the parameterization of the sphere of Example 17.8.1:

$$\vec{r}(\rho,\theta,\varphi) = \langle\rho\cos(\theta)\sin(\varphi), \rho\sin(\theta)\sin(\varphi), \rho\cos(\varphi)\rangle$$

We'll use the scalar triple product to compute

$$dV = |((\vec{r}_\theta\ d\theta)\times(\vec{r}_\varphi\ d\varphi))\cdot(\vec{R}_\rho\ d\rho)| = \rho^2\sin(\varphi)\ d\theta\ d\varphi\ d\rho$$

where we've used the range of the angles, $0 \leq \theta \leq 2\pi$ and $0 \leq \varphi \leq \pi$, so $\sqrt{\sin^2(\theta)} = \sin(\varphi)$.

This is all the background we need. Time for examples.

Example 17.10.2. *Cylindrical coordinates.* Take R to be the region in the first octant ($x \geq 0$, $y \geq 0$, and $z \geq 0$) and under the cone $z = 2 - \sqrt{x^2+y^2}$. To integrate $f(x,y,z) = z\sqrt{x^2+y^2}$ over R in Cartesian coordinates would be messy. In cylindrical coordinates $r = \sqrt{x^2-y^2}$, so the equation of the cone is $z = 2-r$. This is pretty simple, so we'll try the integral in cylindrical coordinates. The first octant is described by $0 \leq \theta \leq \pi/2$ and $z \geq 0$. The cone intersects the xy-plane at $r = 2$, so the region R can be described by

$$0 \leq r \leq 2,\ 0 \leq \theta \leq \pi/2,\ 0 \leq z \leq 2 - r$$

Because the z bounds depend on r, the z integral must be evaluated before the r integral. Because the r and θ bounds are constants, the order of these integrals is immaterial. So

$$\iiint_R f\, dV = \int_0^{\pi/2} \int_0^2 \int_0^{2-r} z\, r\, r\, dz\, dr\, d\theta$$
$$= \int_0^{\pi/2} \int_0^2 \left(2r^2 - 2r^3 + \frac{r^4}{2}\right) dr\, d\theta = \int_0^{\pi/2} \frac{8}{15}\, d\theta = \frac{4\pi}{15}$$

Simple, so long as we can find antiderivatives. □

Example 17.10.3. *Spherical coordinates.* Take R to be the region inside the sphere $x^2 + y^2 + z^2 = 4$ and above the plane $z = 1$, and take the function f to be $f(x, y, z) = z$. Because part of the boundary is a sphere, we'll use spherical coordinates. Then the region inside the sphere is defined by $\rho \leq 2$. To find a spherical coordinate expression for the plane, recall that $z = \rho\cos(\varphi)$, so $z = 1$ becomes $\rho = \sec(\varphi)$. The region R is symmetric around the z-axis, so $0 \leq \theta \leq 2\pi$. To find the limits for φ, note $z = 1$ and $\rho = 2$ give $\cos(\varphi) = 1/2$. This gives the range $0 \leq \varphi \leq \pi/3$. With these bounds, the integral is

$$\iiint_R f\, dV = \int_0^{2\pi} \int_0^{\pi/3} \int_{\sec(\varphi)}^2 z\, \rho^2 \sin(\varphi)\, d\rho\, d\varphi\, d\theta$$
$$= \int_0^{2\pi} \int_0^{\pi/3} \int_{\sec(\varphi)}^2 \big(\rho\cos(\varphi)\big)\, \rho^2 \sin(\varphi)\, d\rho\, d\varphi\, d\theta$$
$$= \int_0^{2\pi} \int_0^{\pi/3} 4\cos(\varphi)\sin(\varphi) - \frac{1}{4}\frac{\sin(\varphi)}{\cos^3(\varphi)}\, d\varphi\, d\theta = \int_0^{2\pi} \frac{9}{8}\, d\theta = \frac{9\pi}{4}$$

Cylindrical coordinates work here, too. The upper hemisphere is $z = \sqrt{4 - r^2}$. The sphere intersects the plane at $r = \sqrt{3}$. Complete the calculation for practice. □

Cylindrical and spherical coordinates were not invented by mathematicians to annoy students. In appropriate settings, the right coordinates can simplify calculations.

Practice Problems

17.10.1. Integrate $f(x, y, z) = z$ over the region R inside the cylinder $(x - 1)^2 + y^2 = 1$ and between $z = 0$ and $z = 1$.

17.10.2. Integrate $f(x,y,z)=z$ over the region R inside the cone $z=\sqrt{x^2+y^2}$ and the sphere $x^2+y^2+z^2=1$.

Practice Problem Solutions

17.10.1. The axis of the cylinder is not the z-axis, so we must work a bit to find a cylindrical coordinate expression for R. Expanding $(x-1)^2+y^2=1$ and using $x^2+y^2=r^2$ gives $r=2\cos(\theta)$. The circle lies only in quadrants I and IV, and the circle is tangent to the z-axis at the origin, so $-\pi/2\le\theta\le\pi/2$. The integral becomes

$$\iiint_R z\,dV=\int_{-\pi/2}^{\pi/2}\int_0^1\int_0^{2\cos(\theta)} z\,r\,dr\,dz\,d\theta$$
$$=\int_{-\pi/2}^{\pi/2}\int_0^1 2z\cos^2(\theta)\,dz\,d\theta=\int_{-\pi/2}^{\pi/2}4\cos^2(\theta)\,d\theta=2\pi$$

17.10.2. The spherical expression for this sphere is $\rho=1$. For the cone, write $z=\sqrt{x^2+y^2}$ in spherical coordinates to obtain $\rho\cos(\varphi)=\rho\sin(\varphi)$; that is, $\varphi=\pi/4$. The spherical representation of the region R is $0\le\rho\le1$, $0\le\theta\le2\pi$, and $0\le\varphi\le\pi/4$. In spherical coordinates, this cone is a rectangular block. Because all the limits of integration are constants, the order of integration is unrestricted. The integral is

$$\iiint_R f\,dV=\int_0^{2\pi}\int_0^{\pi/4}\int_0^1\big(\rho\cos(\varphi)\big)\,\rho^2\sin(\varphi)\,d\rho\,d\varphi\,d\theta$$
$$=\int_0^{2\pi}\int_0^{\pi/4}\frac{1}{4}\cos(\varphi)\sin(\varphi)\,d\varphi\,d\theta=\int_0^{2\pi}\frac{1}{16}\,d\theta=\frac{\pi}{8}$$

Exercises

17.10.1. Evaluate these integrals.

(a) $\displaystyle\int_0^1\int_0^x\int_0^y z+y\,dz\,dy\,dx$ (b) $\displaystyle\int_0^1\int_0^1\int_y^1 e^{x^2}\,dx\,dy\,dz$

17.10.2. Evaluate these integrals.

(a) $\displaystyle\int_0^{\pi}\int_0^z\int_0^z \cos x\,dx\,dy\,dz$ (b) $\displaystyle\int_0^1\int_0^1\int_x^1 \cos(y^2)\,dy\,dx\,dz$

17.10.3. Integrate $f(x,y,z)=z$ over the region R inside the sphere $x^2+y^2+z^2=1$ and between the cones $z=\sqrt{3}\sqrt{x^2+y^2}$ and $z=(1/\sqrt{3})\sqrt{x^2+y^2}$.

17.10.4. Integrate $f(x,y,z) = e^{(x^2+y^2+z^2)^{3/2}}$ over the region R bounded by $x^2 + y^2 + z^2 = 4$.

17.10.5. Integrate $f(x,y,z) = (x^2+y^2+z^2)^{3/2}$ over the region R bounded by $x^2+y^2+z^2 = 1$.

17.10.6. Integrate $f(x,y,z) = ze^{x^2+y^2}$ over the region R bounded by $x^2+y^2 \leq 1$, $z = 0$ and $z = 2$.

17.10.7. Integrate $f(x,y,z) = e^{x^2+y^2}$ over the region R bounded by $x^2+y^2 \leq 1$, $z = 0$ and $z = 2+x$.

17.10.8. Integrate $f(x,y,z) = (x^2+y^2+z^2)^{-1/2}$ over the region R inside $x^2 + y^2 + z^1 = 1$ and above $z = 1/2$.

17.10.9. Integrate $f(x,y,z) = x$ over the region R outside $(x-1)^2+y^2+z^2 = 1$, inside $x^2+y^2+z^2 = 4$, and with $x \geq 0$.

17.10.10. Integrate $f(x,y,z) = x$ over the region R inside the cylinder $(x-1)^2 + y^2 = 1$, above the plane $z = 0$, and below the plane $z = 3-x$.

17.11 GAUSS' THEOREM

So far we've encountered two derivatives related to vectors: the gradient ∇f is a vector derivative of a scalar function, and $\text{curl}(\vec{F})$ is the vector derivative of a vector function. In first semester calculus you learned the scalar derivative of a scalar function. All that remains is the scalar derivative of a vector function. That's the *divergence* of $\vec{F} = P\vec{i} + Q\vec{j} + R\vec{k}$:

$$\text{div}(\vec{F}) = \frac{\partial P}{\partial x} + \frac{\partial Q}{\partial y} + \frac{\partial R}{\partial z} \tag{17.45}$$

Gauss' theorem relates the surface integral of the vector field $\vec{F}$ over the surface $S = \partial E$ to the integral of the divergence of $\vec{F}$ over the region E. Here's the statement.

Theorem 17.11.1. Suppose E is a region in 3-dimensional space with boundary a surface S and normal vectors pointing out of E. If $\text{div}(\vec{F})$ is defined throughout E, then

$$\iint_S \vec{F} \cdot d\vec{S} = \iiint_E \text{div}(\vec{F})\, dV \tag{17.46}$$

Proof. Write Eq. (17.46) as

$$\iint_S (P\vec{i} + Q\vec{j} + R\vec{k}) \cdot d\vec{S} = \iiint_E \left(\frac{\partial P}{\partial x} + \frac{\partial Q}{\partial y} + \frac{\partial R}{\partial z} \right) dV$$

To prove this we'll show

$$\iint_S P\vec{i} \cdot d\vec{S} = \iiint_E \frac{\partial P}{\partial x}\, dV, \quad \iint_S Q\vec{j} \cdot d\vec{S} = \iiint_E \frac{\partial Q}{\partial y}\, dV,$$

$$\text{and} \quad \iint_S R\vec{k} \cdot d\vec{S} = \iiint_E \frac{\partial R}{\partial z}\, dV$$

then add these results. We'll just show the last of these; the arguments for the other two are similar. Also, we'll assume the region E has a shape that simplifies the calculation. If the boundary of E is smooth, this always can be done by chopping E into small enough pieces.

Suppose E is bounded between two surfaces $S_1 = \{(x, y, h_1(x, y)) : (x, y) \in D\}$ and $S_3 = \{(x, y, h_2(x, y)) : (x, y) \in D\}$ and the surface between, S_2, has all normal vectors perpendicular to $\vec{k}$. See Fig. 17.22.

First we'll compute the triple integral

$$\iiint_E \frac{\partial R}{\partial z}\, dV = \iint_D \int_{h_1(x,y)}^{h_2(x,y)} \frac{\partial R}{\partial z}\, dz\, dA$$

$$= \iint_D R(x, y, h_2(x, y)) - R(x, y, h_1(x, y))\, dA$$

Now for the surface integral. Because $S = S_1 \cup S_2 \cup S_3$ and the normal vectors of S_2 are perpendicular to $\vec{k}$,

$$\iint_S R\vec{k} \cdot d\vec{S} = \iint_{S_1 \cup S_2 \cup S_3} R\vec{k} \cdot d\vec{S}$$

$$= \iint_{S_1} R\vec{k} \cdot d\vec{S} + \iint_{S_3} R\vec{k} \cdot d\vec{S}$$

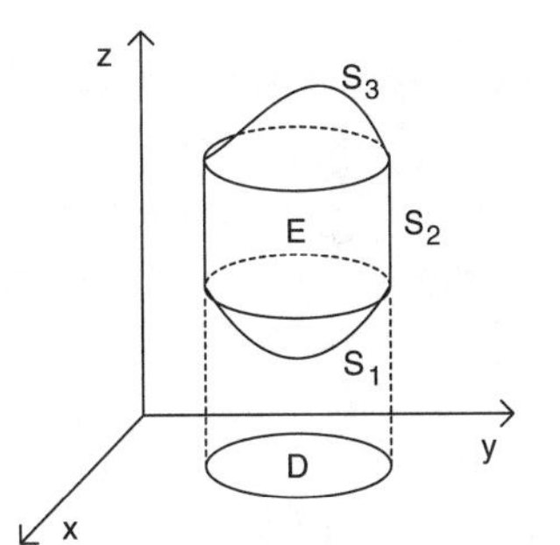

Figure 17.22. A form of E.

where S_1 has downward-pointing normals and S_3 has upward-pointing normals. Because S_1 and S_3 are graphs parameterized by $\langle x, y, h_1(x, y) \rangle$ and $\langle x, y, h_2(x, y) \rangle$ for $(x, y) \in D$, the appropriate normal vectors are $\langle \partial h_1/\partial x, \partial h_1/\partial y, -1 \rangle$ and $\langle -\partial h_2/\partial x, -\partial h_2/\partial y, 1 \rangle$. Then

$$\iint_{S_1} R\vec{k} \cdot d\vec{S} = \iint_D R(x, y, h_1(x, y))(-1)\, dA$$

$$\iint_{S_3} R\vec{k} \cdot d\vec{S} = \iint_D R(x, y, h_2(x, y))(+1)\, dA$$

That is,

$$\iint_S R\vec{k}\cdot d\vec{S} = \iint_D R(x,y,h_2(x,y))\, dA - \iint_D R(x,y,h_1(x,y))\, dA$$

and we see

$$\iiint_E \frac{\partial R}{\partial z}\, dV = \iint_S R\vec{k}\cdot d\vec{S}$$

The other two equalities are proved similarly. □

Some applications are straightforward, some others less so.

Example 17.11.1. *Gauss' theorem on a cube.* Suppose S is the surface, the six square faces, of the unit cube $E = \{(x,y,z) : 0 \leq x \leq 1, 0 \leq y \leq 1, 0 \leq z \leq 1\}$. To evaluate $\iint_S \vec{F}\cdot d\vec{S}$ where $\vec{F}(x,y,z) = (x^2 + y^3)\vec{i} + \sin(x^2)\vec{j} + xyz\vec{k}$, six surface integrals don't sound appealing, so we'll use Gauss' theorem to convert this to a single triple integral.

$$\begin{aligned}\iint_S \vec{F}\cdot d\vec{S} &= \iiint_E \text{div}(\vec{F})\, dV = \int_0^1\int_0^1\int_0^1 (2x + xy)\, dx\, dy\, dz \\ &= \int_0^1 \left(1 + \frac{y}{2}\right) dy\, dz = \int_0^1 \frac{5}{4}\, dz = \frac{5}{4}\end{aligned}$$

We'd still be at work on the individual surface integrals if we'd taken the direct approach. □

Example 17.11.2. *Completing the surface.* With E and $\vec{F}$ as in Example 17.11.1 and S' the five faces of the cube, excluding the top ($z = 1$) face, to evaluate $\iint_{S'} \vec{F}\cdot d\vec{S}$ we can't apply Gauss' theorem directly. So add the top face S'', apply Gauss' theorem, then subtract the surface integral over S''.

$$\begin{aligned}&\iint_{S'} \vec{F}\cdot d\vec{S} = \iint_S \vec{F}\cdot d\vec{S} - \iint_{S''} \vec{F}\cdot d\vec{S} \\ &= \iiint_E \text{div}(\vec{F})\, dV - \iint_{S''} \vec{F}\cdot d\vec{S} = \frac{5}{4} - \int_0^1\int_0^1 \vec{F}(x,y,1)\cdot\vec{k}\, dx\, dy \\ &= \frac{5}{4} - \int_0^1\int_0^1 xy\, dx\, dy = \frac{5}{4} - \frac{1}{4} = 1\end{aligned}$$

This is the surface integral version of closing the curve and applying Green's theorem that we saw in Example 17.5.1. □

You might think that we have four integral theorems involving vectors: FTLI, Green's theorem, Stokes' theorem, and Gauss' theorem. But in Sect. 17.9 we saw that Green's theorem is Stokes' theorem when the surface lies in the xy-plane. So we have three vector integral theorems. Or do we?

Recall that ∂X denotes the boundary of the region X.

- For a solid 3-dimensional region E, ∂E is the surface separating E from the rest of space.
- For a surface S with boundary, ∂S is the curve bounding the surface.
- For a curve C, ∂C is the two endpoints of this curve.

Then the three main integral theorems of vector calculus are

$$\iiint_E \text{div}(\vec{F})\, dV = \iint_{\partial E} \vec{F} \cdot d\vec{S}$$

$$\iint_S \text{curl}(\vec{F}) \cdot d\vec{S} = \oint_{\partial S} \vec{F} \cdot d\vec{r}$$

$$\int_C \nabla f \cdot d\vec{r} = f|_{\partial C}$$

So we see that the integral of the derivative of a function over a region equals the integral of the function over the boundary of the region. These three theorems are expressions, in different dimensions, of a single ur-theorem.

The derivative of a function is a mirror image of the boundary of a region. Another interesting bit of geometry is that a boundary has no boundary. For example, the boundary of the northern hemisphere is the equator, which itself has no boundary. What is the analogue for derivatives of this result for boundaries? We've seen a bit of this already ($\text{curl}(\nabla f) = \vec{0}$) and we'll see another bit in Exercise 17.11.1. It is part of a deep and beautiful connection between geometry and calculus, but about this issue we now have said enough.

Practice Problems

17.11.1. Suppose E is the region inside the cylinder $x^2 + y^2 = 1$, above the plane $z = 0$, and below the surface $z = 3 - x^2 - y^2$. For the vector field $\vec{F} = xy^2\vec{i} + yx^2\vec{j} + z^2\vec{k}$ evaluate $\iint_{\partial E} \vec{F} \cdot d\vec{S}$.

17.11.2. Suppose S is the surface $x^2 + y^2 + z^2 = 1, z \geq 0$ with upward-pointing normal vectors, and $\vec{F}$ is the vector field $\langle x^3, y^3, z + x^2 + y^2 \rangle$. Evaluate the surface integral $\iint_S \vec{F} \cdot d\vec{S}$.

Practice Problem Solutions

17.11.1. The surface ∂E has three pieces: the disc $x^2 + y^2 \leq 1, z = 0$, the cylindrical side $x^2 + y^2 = 1, 0 \leq z \leq 2$, and the cap $x^2 + y^2 \leq 1, z = 3 - x^2 - y^2$.

Rather than evaluate three surface integrals, we'll apply Gauss' theorem. The cylindrical symmetry of the region E suggests cylindrical coordinates, so $\iint_{\partial E} \vec{F} \cdot d\vec{S} =$

$$\iiint_E \operatorname{div}(\vec{F})\, dV = \int_0^{2\pi} \int_0^1 \int_0^{3-r^2} (y^2 + x^2 + 2z) r\, dz\, dr\, d\theta$$
$$= \int_0^{2\pi} \int_0^1 \int_0^{3-r^2} (r^2 + 2z) r\, dz\, dr\, d\theta = \int_0^{2\pi} \int_0^1 (9r - 3r^3)\, dr\, d\theta$$
$$= \int_0^{2\pi} \frac{15}{4}\, d\theta = \frac{15\pi}{2}$$

17.11.2. If $\vec{F} = \operatorname{curl}(\vec{G})$, then we could apply Stokes' theorem, if we could find $\vec{G}$. But how to do this isn't so clear. Direct evaluation of the surface integral is possible, but integrating $\vec{F}$ over the disc $D = \{(x,y,0) : x^2 + y^2 \leq 1\}$ would be easier, so let's close the surface S by adding D and then apply Gauss' theorem. Take $E = \{(x,y,z) : x^2 + y^2 + z^2 \leq 1, z \geq 0\}$ so $\partial E = S \cup D$. Spherical coordinates are the natural choice for the triple integral. For the surface integral over D we'll use polar coordinates. Remember that Gauss' theorem requires the normal vectors to ∂E point out of E, so the normal vector on D points down.

$$\iint_S \vec{F} \cdot d\vec{S} + \iint_D \vec{F} \cdot d\vec{S} = \iiint_E \operatorname{div}(\vec{F})\, dV = \iiint_E (3x^2 + 3y^2 + 1)\, dV$$
$$= \int_0^{2\pi} \int_0^1 \int_0^{\pi/2} (\rho^2 \sin^2(\varphi)) \rho^2 \sin(\varphi)\, d\varphi\, d\rho\, d\theta$$
$$= \int_0^{2\pi} \int_0^1 \frac{2\rho^4}{3}\, d\rho\, d\theta = \int_0^{2\pi} \frac{2}{15}\, d\theta = \frac{4\pi}{15}$$

Now for the surface integral over D.

$$\iint_D \vec{F} \cdot d\vec{S} = \iint_D \langle x^3, y^3, x^2 + y^2 \rangle \cdot \langle 0, 0, -1 \rangle\, dA = \int_0^{2\pi} \int_0^1 -r^2\, r\, dr\, d\theta$$
$$= \int_0^{2\pi} \frac{-1}{4}\, d\theta = -\frac{\pi}{2}$$

Finally,

$$\iint_S \vec{F} \cdot d\vec{S} = \iiint_E \operatorname{div}(\vec{F})\, dV - \iint_D \vec{F} \cdot d\vec{S} = \frac{4\pi}{15} + \frac{\pi}{2} = \frac{23\pi}{30}$$

Exercises

17.11.1. For any vector field $\vec{F} = P\vec{i} + Q\vec{j} + R\vec{k}$, show that if P, Q, and R have continuous second partial derivatives, $\operatorname{div}(\operatorname{curl}(\vec{F})) = 0$.

17.11.2. The three vector derivatives—grad, curl, and div—have nine possible combinations; we've seen curl(grad) $= \vec{0}$ and div(curl) $= 0$. Which combinations are impossible, which are always 0, and which may be something else? Assume every function has continuous second partial derivatives.

17.11.3. For the vector field $\vec{F} = \langle xy, yz, xz \rangle$ compute $\iint_S \vec{F} \cdot \vec{S}$ where $S = \partial E$ for $E = \{(x,y,z) : x \geq 0, y \geq 0, z \geq 0, x+y+z \leq 1\}$, with outward-pointing normals.

17.11.4. For the vector field $\vec{F} = \langle xy^2, xyz, x+y+z \rangle$ compute $\iint_S \vec{F} \cdot d\vec{S}$, where S is the boundary of the unit cube $E = \{(x,y,z) : 0 \leq x \leq 1, 0 \leq y \leq 1, 0 \leq z \leq 1\}$ without the $z = 0$ face, and with outward-pointing normals.

17.11.5. Evaluate $\iint_S \langle xy^2, yx^2, z^2 \rangle \cdot d\vec{S}$ where $S = \partial E$ with normals pointing out of E and $E = \{(x,y,z) : x^2 + y^2 \leq 1, 0 \leq z \leq x\}$.

17.11.6. Evaluate $\iint_S \langle xy^2, yx^2, xyz \rangle \cdot d\vec{S}$ where $S = \partial E$ with normals pointing out of E and

$$E = \{(x,y,z) : 1 \leq x^2 + y^2 \leq 4, 0 \leq z \leq 1, 0 \leq x, 0 \leq y\}.$$

17.11.7. For any region E, show $\iint_{\partial E} \langle x, y, z \rangle \cdot d\vec{S} = 3\text{volume}(E)$, where the normals of ∂E point out of E. This may remind you of the planimeter.

17.11.8. Suppose E is the region between the spheres $x^2 + y^2 + z^2 = 1$ and $x^2 + y^2 + x^2 = 4$ and in the first octant ($x \geq 0, y \geq 0, z \geq 0$). Evaluate $\iint_{\partial E} \langle x^3, y^3, z(x^2 + y^2) \rangle \cdot d\vec{S}$, where the normals of ∂E point out of E.

17.11.9. For the region E and the vector field $\vec{F}$ of Exercise 17.11.8, evaluate $\iint_S \vec{F} \cdot d\vec{S}$ where S consists of the two spherical portions of ∂E, with normal vectors pointing out of E.

17.11.10. For the vector field $\vec{F} = \langle x^3, y^3, z^3 \rangle$ evaluate $\iint_{\partial E} \vec{F} \cdot d\vec{S}$ where $E = \{(x,y,z) : x^2 + y^2 + z^2 \leq 4, z \leq 1\}$, with normals pointing out of E.

17.12 DIFFUSION

We haven't seen any biological examples for a while, so now we'll present a big one: the diffusion equation. We think the cleanest introduction to the diffusion equation is through its formulation as the heat equation, a model of how heat diffuses through a medium. Denote by

$T(x,y,z,t)$ the temperature at time t and position (x,y,z),
ρ the energy density per unit volume
$\vec{J}$ the energy flux

First two simple observations. For some positive constants a and b,

$$\rho = aT \qquad \text{and} \qquad \vec{J} = -b\nabla T \tag{17.47}$$

That is, the energy density is proportional to the temperature: higher energy per unit volume means higher temperature, though the equation says a bit more than that. It says *how* the temperature increases with the energy density. The second equation states that a higher temperature gradient means a faster flow of energy, in the direction opposite the temperature gradient. Heat flows from warmer regions to cooler regions.

The total energy content of a region E is $\iiint_E \rho \, dV$. In this setting the law of conservation of energy implies that the only way the total energy of E changes is by flowing across ∂E. That is,

$$\frac{d}{dt}\iiint_E \rho \, dV = -\iint_{\partial E} \vec{J} \, d\vec{S} \tag{17.48}$$

where the minus sign is there because the surface integral uses the outward-pointing normal. Move both terms to the same side of the equation and apply Gauss' theorem to the surface integral.

$$\frac{d}{dt}\iiint_E \rho \, dV + \iiint_E \operatorname{div}(\vec{J}) \, dV = 0$$

$$\iiint_E \frac{\partial \rho}{\partial t} \, dV + \iiint_E \operatorname{div}(\vec{J}) \, dV = 0$$

$$\iiint_E \left(\frac{\partial \rho}{\partial t} + \operatorname{div}(\vec{J}) \right) dV = 0 \tag{17.49}$$

The derivative d/dt in the first equation turns into $\partial/\partial t$ in the second equation because $\iiint_E \rho \, dV$ depends only on t, while ρ depends on x, y, z, and t.

Now Eq. (17.49) holds for all solid regions E, so

$$\frac{\partial \rho}{\partial t} + \operatorname{div}(\vec{J}) = 0 \tag{17.50}$$

Here's why. Suppose $(\partial\rho/\partial t + \operatorname{div}(\vec{J}))(x_0, y_0, z_0) > 0$ at some point $(x_0, y_0, z_0) \in E$. Continuity implies that $(\partial\rho/\partial t + \operatorname{div}(\vec{J}))(x,y,z) > 0$ for all $(x,y,z) \in E$ for some small enough solid region E. So for that E, the integral of Eq. (17.49) would be positive.

By the second equation of Eq. (17.47),

$$\begin{aligned}\operatorname{div}(\vec{J}) &= \operatorname{div}(-b\nabla T) = -b \operatorname{div}\left(\frac{\partial T}{\partial x}\vec{i} + \frac{\partial T}{\partial y}\vec{j} + \frac{\partial T}{\partial z}\vec{k}\right) \\ &= -b\left(\frac{\partial^2 T}{\partial x^2} + \frac{\partial^2 T}{\partial y^2} + \frac{\partial^2 T}{\partial z^2}\right) = -b\,\nabla^2 T\end{aligned}$$

where $\nabla^2 T$ is called the *Laplacian* of T.

Finally, by the first equation of Eq. (17.47) we see

$$\frac{\partial \rho}{\partial t} = \frac{\partial (a\,T)}{\partial t} = a\,\frac{\partial T}{\partial t}$$

Substitution of these expressions for $\partial\rho/\partial t$ and for $\operatorname{div}(\vec{J})$ into Eq. (17.50) we obtain the *heat equation*

$$\frac{\partial T}{\partial t} = -\frac{b}{a}\nabla^2 T \tag{17.51}$$

If we change temperature to concentration and adjust the constants, the heat equation becomes the diffusion equation. Change concentration to the wave function ψ and put Planck's constant and an i among the constants, and then we have Schrödinger's equation. Adjust the constants again and change $\partial/\partial t$ to $\partial^2/\partial t^2$ and we have the wave equation. Such apparent similarities between so many dynamical processes may seem mysterious, and maybe it is, but think about this. If we allow only first or second derivatives, there are not many possible equations.

In situations other than heat diffusion (call the variable Z) we can add a source term $f(x,y,z,t)$ to the heat equation (17.51). With adjustment of the constants this gives

$$\frac{\partial Z}{\partial t} = \nabla^2 Z + f$$

For example, modified to allow variation of predator $u(x,y,t)$ and prey $v(x,y,t)$ with location (x,y), the Lotka-Volterra system is

$$\frac{\partial u}{\partial t} = a\,u(1-v) + b\left(\frac{\partial^2 u}{\partial x^2} + \frac{\partial^2 u}{\partial y^2}\right) \quad \frac{\partial v}{\partial t} = c\,v(u-1) + b\left(\frac{\partial^2 u}{\partial x^2} + \frac{\partial^2 u}{\partial y^2}\right)$$

Our previous models include the assumption that the populations are distributed uniformly throughout the space. While this is a reasonable first-order approximation, it certainly misses some of the details of how species interact.

Epidemiological models are more realistic if they include diffusion. Genetic drift, speciation, transport of ions across transmembrane channels, the development of leukemias ...so many models in biology are more accurate if they involve both space and time. The differential equations of this section are called partial differential equations because they involve partial derivatives. Treating them in any serious fashion is another book or ten. But not this book.

Chapter 18 A glimpse of systems biology

We've seen that reductionism is the notion that we can explain a complex system by pulling it apart, finding out how the pieces work, and glueing them back together. This works well for some things, mechanical clocks, for example. My grandfather was a watchmaker. Occasionally he gave me old clocks to disassemble. The dance of the gears, the twirl of the balance wheel, how one gear convinced another to turn at just the right rate: magic to a six-year-old, but magic I could understand. I have appreciated reductionism for over sixty years.

But reductionism doesn't work for every system. In a network of many interacting elements, often self-regulation is achieved through feedback loops and nonlinear responses (the magnitude of the response is not proportional to the magnitude of the stimulus). Here reductionism is manifestly a wrong approach. We are not a collection of organs working independently, organs are not a collection of cells working independently, and let's not forget that DNA does not synthesize proteins or copy bits of itself in a simple, straightforward way.

We do need to know how the individual pieces work, but how a network of pieces work together must be studied as a collective process. For that matter, the ways that many of the pieces work cannot be determined by studying the pieces in isolation. Systems biology is a study of biological networks, of how living bits talk with one another.

Some analyses of biological networks are presented in [187] and [240]. Synchronization of weakly-coupled networks was an early focus of applications of systems theory to biology, but the study has grown to model the complex dynamical interactions of the elements of metabolic, transcription, and gene regulatory networks, among many others.

Though the application of systems theory to biology began in the 1960s, substantial developments resulted from the convergence of three factors: the availability of fast computers with large memories, the massive data sets produced by functional genomics, and the application of dynamical systems and network theories to problems in biology. Since the 2000s, many universities and medical schools have opened systems biology institutes. Now we find conferences, conference proceedings, courses, and texts [3, 9, 29, 365] on systems biology. Here we'll see a tiny bit of this field.

Networks can be represented by directed graphs whose vertices are components of the network. An edge $A \to B$ represents a network process that transforms A directly to B. The qualifier "directly" is important. If every transformation of A to B must go through the component C, then the graph has no edge $A \to B$, but rather has $A \to C \to B$. Even for networks with many states, the graph edges can be found by through correlations detected with high-throughput instruments. For example, a mass spectrometer can make simultaneous measurements of the concentrations of tens of thousands of proteins.

To make testable predictions, the graph must be converted into a system of differential equations, and to do this we must have the transition rates. Except for the simplest networks, deducing these from first principles is difficult, maybe impossible. While direct measurement can be achieved for some networks—for example, neural nets by use of patch clamps—typically this is impossible with current technology. The *encyclopedia approach* of simulations, the systematic exploration of parameter space, can give insight for small networks, but the computational time grows exponentially with the number of vertices of the graph.

Another approach for transition rate estimation is to compare measured values $x_1, \dots, x_N$ of the system with simulated values $y_1, \dots, y_N$ obtained by solving the system of differential equations with given values of the transition rates $r_1, \dots, r_M$. Then adjust the transition rates to minimize $(x_1 - y_1)^2 + \cdots (x_N - y_N)^2$, that is, to make the simulated behavior as close as possible to the measured behavior. Each of the y_i is determined by the transition rates, so we can write

$$(x_1 - y_1)^2 + \cdots (x_N - y_N)^2 = f(r_1, \dots, r_M)$$

Several strategies can be used to minimize f:

- *Gradient descent*. As described in Sect. 15.3, for the current rates $r_1, \dots, r_M$, change the rates in the direction $-\nabla f$. This guarantees the quickest approach to a minimum, but this search may become trapped at a local minimum.
- *Genetic algorithms*. Start with a random collection of transition rate sets $(r_1, \dots, r_M)$ and for each compute $f(r_1, \dots, r_M)$. Delete the sets with high values of f because with these parameters the simulation doesn't agree well with measurements. Reproduce those with the lowest values of f with crossover and occasional mutation. That is, if $(r_1, \dots, r_M)$ and $(s_1, \dots, s_M)$ have low f values, keep them in the population and add the pair $(r_1, \dots, r_k, s_{k+1}, \dots, s_M)$ and $(s_1, \dots, s_k, r_{k+1}, \dots, r_M)$ and occasionally change (mutate) one of the values. Then repeat.
- *Simulated annealing*. To the gradient descent method, add a slowly decreasing amount of noise (heat) to the system. When a local minimum is reached, return to the initial transition rates and run gradient descent again with a different noise. Continue for a while and take the lowest of the local minima.

Simulated annealing has a cute cartoon description. Suppose you want to find the lowest point of an old, beat-up cast iron bathtub. Drop a marble in the tub and it will come to rest at a local minimum. This is gradient descent. Now start again and give the bathtub to a gorilla. As the gorilla shakes the bathtub, the marble bounces around, and eventually comes to rest in a local minimum when the gorilla gets tired. As the gorilla tires, the magnitude of shaking decreases: this is the "slowly decreasing amount of noise." Get a new gorilla and start the process again. Keep at it until you run out of gorillas.

These are only a few of the minimization strategies that can speed the search through transition rate space.

To simplify comparisons of modeling approaches, some cross-platform languages have been developed. One is SBML, Systems Biology Markup Language. On the website [324] we find "Computational modeling of biological systems is no longer a fringe activity—it's a *requirement* for us to make sense of our vast and ever-expanding quantities of data." Sounds reasonable, don't you think?

Much of the analysis in systems biology is based on the mathematics of network theory, a field which is itself still developing. Good references for network theory are [32, 97, 257].

This is an immense, and immensely interesting, field. Here we'll get only a tiny sample. After a first sketch of an engineered network with two elements, we'll consider in a bit of detail two gene transcription networks in *E. coli*, then a brief mention of metabolic networks and neural networks and conclude with some thoughts on why evolution should have produced networks that we can understand.

18.1 GENETIC TOGGLE SWITCH

We'll start with a very simple example from [107], a gene regulatory network that consists of two repressors, proteins that bind to the promoter of a gene and thus inhibit the expression of that gene. In this network, each repressor inhibits the synthesis of the other repressor. In [129] Timothy Gardner, Charles Cantor, and James Collins realize this model of a genetic toggle switch in *E. coli*. From the dynamics viewpoint, under some circumstances this is a *bistable system*: it has two stable fixed points. (This might remind you of final state sensitivity of the Duffing system of Sect. 13.8.) Under sufficient chemical or thermal perturbation, the system flips between these fixed points.

Let x denote the concentration of repressor 1 and y the concentration of repressor 2. Then a simple model for the dynamics of x and y, derived from the rate equation of gene expression, is

$$\frac{dx}{dt} = -x + \frac{a_1}{1+y^{b_1}} \qquad \frac{dy}{dt} = -y + \frac{a_2}{1+x^{b_2}} \tag{18.1}$$

Here the $-x$ and $-y$ terms reflect the dilution effect of each repressor on its own growth rate. The other terms represent the growth rates: a_i is the synthesis rate of repressor i in the absence of the other repressor, and the denominators model the negative effect of one repressor population on the growth rate of the other. The exponent b_i (called the *Hill coefficient* in enzyme kinetics) is called the cooperativity, because if, for example, $b_i = 2$, then pairs of one repressor impede the growth of the other.

The x-nullcline is the curve $x = a_1/(1+y^{b_1})$, and the y-nullcline is the curve $y = a_2/(1+x^{b_2})$. In Fig. 18.1 we plot these graphs for $a_1 = a_2 = 1, 2, 3$ and $b_1 = b_2 = 2$. The x-nullcline has a vertical asymptote along the y-axis; the y-nullcline has a horizontal asymptote along the x-axis.

To simplify the arithmetic, we'll take $a_1 = a_2 = a$ and $b_1 = b_2 = b$. Then for all positive a and b, the system (18.1) has one fixed point (x_0, x_0) on the diagonal $y = x$. Unless $b = 1$, an exact solution of $x = a/(1+x^b)$ is at the very least messy, so we'll argue indirectly.

Define $f(x) = -x + a/(1+x^b)$. Then

$$f(0) = a > 0,\ f(a) = -\frac{a^{b+1}}{1+a^b} < 0, \text{ and } f'(x) = -1 - \frac{abx^{b-1}}{(1+x^b)^2} < 0$$

and so there is a unique solution of $f(x)=0$, that is, a unique fixed point (x_0, x_0) of the system (18.1) on the diagonal.

In Fig. 18.1 we plot both nullclines for $b=2$. For $a \leq 2$ the system has only one fixed point, (x_0, x_0). For $a > 2$, the system has three fixed points. How general is this result?

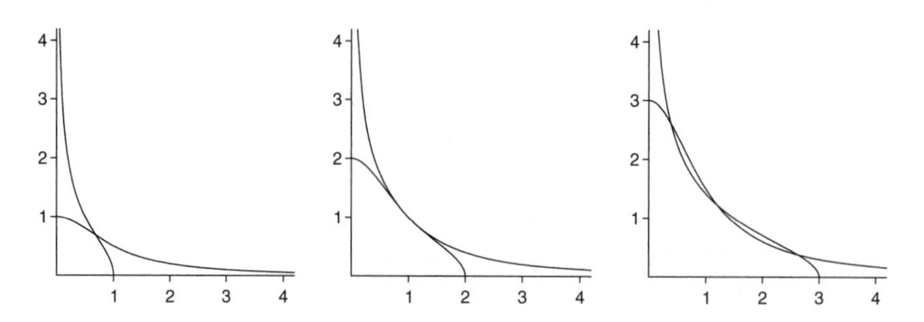

Figure 18.1. The x- and y-nullclines for $a_1 = a_2 = 1, 2, 3$ and $b_1 = b_2 = 2$.

In the left graph of Fig. 18.1 we see that at (x_0, x_0), the tangent of the x-nullcline has a larger magnitude slope than the tangent of the y-nullcline; in the right graph we see that the relative magnitudes of the slopes are reversed; in the middle graph the slopes of the tangents are equal. So at the bifurcation of a single fixed point (x_0, x_0) into three, at (x_0, x_0) the x- and y-nullclines have tangents with the same slopes. We'll still take $a_1 = a_2 = a$ and $b_1 = b_2 = b$. For the y-nullcline, $y = a/(1+x^b)$ from the second equation of (18.1), and so

$$\frac{dy}{dx} = \frac{d}{dx}\left(\frac{a}{1+x^b}\right) = -\frac{abx^{b-1}}{(1+x^b)^2} \tag{18.2}$$

For the x-nullcline,

$$\frac{dy}{dx} = \frac{1}{dx/dy} = -\frac{(1+y^b)^2}{aby^{b-1}}$$

where the first equality is the inverse function theorem: if for $y = f(x)$, $f'(x_0) \neq 0$, then for all x near enough to x_0, the function f has an inverse function g, that is, $x = g(y)$ for all y near $y_0 = f(x_0)$. Moreover, $df/dx = 1/(dg/dy)$. That is, $dy/dx = 1/(dx/dy)$.

Because $y = x$ on the diagonal, at (x_0, x_0) the x-nullcline has

$$\frac{dy}{dx} = -\frac{(1+x^b)^2}{abx^{b-1}} \tag{18.3}$$

Using the x-nullcline equation, at (x_0, x_0) we have

$$x_0 = \frac{a}{1+x_0^b}, \quad \text{or} \quad 1+x_0^b = \frac{a}{x_0}, \quad \text{or} \quad x_0^{b+1} = a - x_0 \tag{18.4}$$

Substitute the second in the right-hand sides of Eqs. (18.2) and (18.3) and equate these right-hand sides. (This is the tangents with the same slope condition.) Then a bit of algebra and the third equality of (18.4) give

$$x_0 = \frac{a(b-1)}{b} \tag{18.5}$$

Finally, insert this into $x_0 = a/(1+x_0^b)$ to obtain

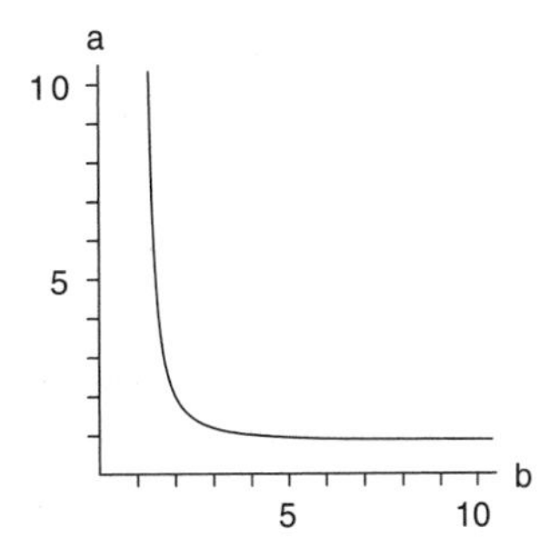

Figure 18.2. Eq. (18.6) graph.

$$a = b(b-1)^{-(1+b)/b} \tag{18.6}$$

The graph of Eq. (18.6) is shown in Fig. 18.2. As we see from the three graphs of Fig. 18.1, for parameters (b,a) below the curve of Fig. 18.2 the system has only one fixed point, (x_0,x_0). For (b,a) above the curve, the system has three fixed points, (x_0,x_0) and two off the diagonal.

An increase of a, the rate of repressor synthesis, increases the distance between the two non-diagonal fixed points, which increases the difficulty of thermal or chemical perturbations to induce jumps between these fixed points because the distance the system moves is proportional to the magnitude of the perturbation. The greater the distance between the fixed points, the larger the perturbation to cause a jump between the fixed points.

To test the stability of the fixed points, we use the derivative matrix

$$D\vec{F}(x,y) = \begin{bmatrix} -1 & \dfrac{-aby^{b-1}}{(1+y^b)^2} \\ \dfrac{-abx^{b-1}}{(1+x^b)^2} & -1 \end{bmatrix}, \quad D\vec{F}(x_0,x_0) = \begin{bmatrix} -1 & -\dfrac{bx_0^{b-1}}{a} \\ -\dfrac{bx_0^{b-1}}{a} & -1 \end{bmatrix}$$

where we have used the second equality of (18.3) for (x_0,x_0). This matrix has eigenvalues

$$\lambda_{\pm} = -1 \pm \frac{bx_0^{b+1}}{a}$$

Because a, b, and x_0 are positive, λ_- is negative; the sign of λ_+ depends on a and b.

Below the curve of Fig. 18.2, $a < b(b-1)^{-(1+b)/b}$ and the diagonal fixed point (x_0,x_0) has $x_0 = a(b-1)/b$ (from Eq. (18.6)). From these, a bit of algebra shows that $\lambda_+ < 0$ and so (x_0,x_0) is an asymptotically stable node for (b,a) below the curve. A similar argument shows that for (b,a) above the curve, $\lambda_+ > 0$ and (x_0,x_0) is a saddle point.

General expressions for the off-diagonal fixed points are ...complicated. We'll work through one example: $b = 2$. The off-diagonal fixed points are

$$(x_1, y_1) = \left(\frac{a + \sqrt{a^2 - 4}}{2}, \frac{a - \sqrt{a^2 - 4}}{2} \right) \text{ and } (x_2, y_2) = (y_1, x_1) \qquad (18.7)$$

Now if you're concerned about the $a^2 - 4$ inside the square root, note that from Fig. 18.2 for $b = 2$, in order to be in the region with off-diagonal fixed points, we must have $a > 2$. The eigenvalues of both $D\vec{F}(x_1, y_1)$ and $D\vec{F}(x_2, y_2)$ are

$$\lambda_{\pm} = \frac{\pm 2 - a}{a}$$

Because $a > 2$, both eigenvalues are negative and both fixed points are asymptotically stable nodes. This is precisely the set-up for a system toggling between stable fixed points. We'll see some simulations in the practice problems.

But first, we'll menton that [129] is not only a report of a mathematical model: Gardner, Cantor, and Collins constructed genetic toggle switches with repressors inserted by plasmids into *E. coli* plasmids. The experimental technique is a clever application of molecular cloning. While early genetic engineering controlled behaviors by the modification of proteins and other regulatory elements, here the authors manipulate regulatory network architecture. This was a step in the construction of programmable gene circuits to control cell behaviors.

Practice Problems

18.1.1. For Eq. (18.1) with $a_1 = a_2 = a = 2.5$ and $b_1 = b_2 = b = 2$,
(a) find the off-diagonal fixed points (x_1, y_1) and (x_2, y_2) of Eq. (18.7).
(b) Use different-frequency sines to simulate noise: add $c\sin(t/10)$ to the dx/dt equation and $c\sin(t/15)$ to the dy/dt equation. With $c = 0.05$, $x(0) = 0.5$, and $y(0) = 1.5$ plot $x(t)$ and $y(t)$ for $0 \leq t \leq 300$. Mathematica codes are in Sect. B.22.
(c) Repeat (b) with $x(0) = 1.5$ and $y(0) = 0.5$.

18.1.2. Repeat Practice Problem 18.1.1 (b) with $c = 0.08$. Interpret the graphs in terms of the toggle switch dynamics.

Practice Problem Solutions

18.1.1. (a) From Eq. (18.7) we see that the off-diagonal fixed points are $(x_1, y_1) = (2.0, 0.5)$ and $(x_2, y_2) = (0.5, 2.0)$.

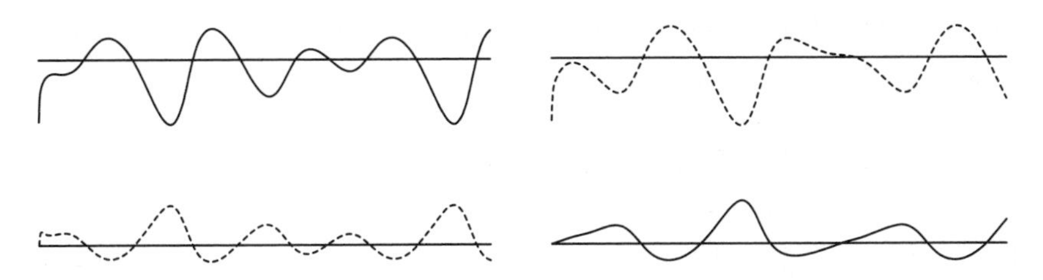

Figure 18.3. Plots of $x(t)$ (dashed) and $y(t)$ (solid) for Practice Problem 18.1.1 (b), left, and (c), right.

(b) The plots of $x(t)$ (dashed) and $y(t)$ (solid) are in the first image of Fig. 18.3. The lower horizontal line is at height y_1, the upper at x_1. Note that $x(t)$ oscillates around $y_1 = x_2$ and $y(t)$ around $x_1 = y_2$; the perturbations force this trajectory to oscillate around the fixed point (x_2, y_2).
(c) The plots, in the second image of Fig. 18.3, show this trajectory oscillates around the fixed point (x_1, y_1).

18.1.2. In Fig. 18.4 we see plots of $x(t)$ (dashed) and $y(t)$ (solid) when the perturbation strength has increased to $c = 0.08$. When the dashed curve is near the upper line and the solid curve is near the lower, the system oscillates around (x_1, y_1); when the dashed curve is near the lower line and the solid curve is near the upper, the system oscillates around (x_2, y_2).

Note that we never see both curves near the upper line or both curves near the lower line. The system jumps between oscillating around one off-diagonal fixed point and oscillating around the other. Longer-time plots might reveal an interesting distribution of durations between jumps.

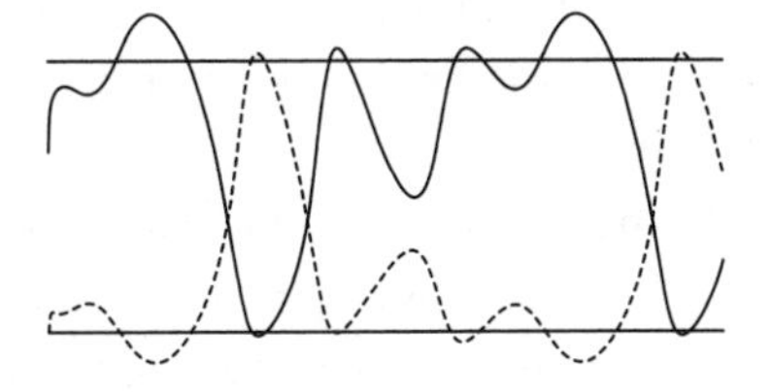

Figure 18.4. Plots of $x(t)$ and $y(t)$ for Practice Problem 18.1.2.

Exercises

18.1.1. In the genetic toggle switch model (18.1) take $a_1 = a_2 = a$ and $b_1 = b_2 = b$. Start from parameters $a = b = 2$. Plot the off-diagonal fixed points when
(a) a is held constant and b is increased from 3 to 10 in steps of 1, and
(b) b is returned to 2 and a is increased from 3 to 10 in steps of 1.
(c) Do these parameter increases uniformly increase the distance between the

off-diagonal fixed points, and so uniformly increase the size perturbation needed to switch between the fixed points?

18.1.2. (a) Plot the off-diagonal fixed points when $a_1 = 2$, $b_1 = b_2 = 2$, and a_2 increases from 3 to 10 in steps of 1.
(b) Plot the off-diagonal fixed points when $a_1 = a_2 = 2$, $b_1 = 2$, and b_2 increases from 3 to 10 in steps of 1.
(c) Do these parameter increases uniformly increase the distance between the off-diagonal fixed points, and so uniformly increase the size perturbation needed to switch between the fixed points?

18.1.3. For $a_1 = a_2 = 2.5$, $b_1 = b_2 = 2$, $x(0) = 1.5$, and $y(0) = 0.5$, add $c\sin(t/10)$ to the dx/dt equation and $c\sin(t/15)$ to the dy/dt equation. To the nearest 0.005 find the smallest value of c for which you observe toggling between the fixed points.

18.1.4. In the configuration of Exercise 18.1.3, change $t/10$ to $t/(3\pi)$ and super-perturb the system by setting $c = 1.0$. Do you still observe toggling between fixed points?

18.1.5. For $a_1 = a_2 = 4$ and $b_1 = b_2 = 2$, add $c\sin(t/(7\sqrt{2}))$ to the dx/dt equation and $c\sin(t/(5\sqrt{3}))$ to the dy/dt equation. Take the initial values $x(0) = 1.5$ and $y(0) = 0.5$, take $c = 0.4$, and take $0 \le t \le 1000$.
(a) Plot the $x(t)$ and $y(t)$ curves. Note that the periods when the system toggles between the fixed points is interrupted by periods of oscillation about a single fixed point.
(b) Increase a_1 and a_2 to 5 and plot the $x(t)$ and $y(t)$ curves.
(c) Increase a_1 and a_2 to 6 and plot the $x(t)$ and $y(t)$ curves.
(d) Comment on how the length of the periods of oscillation about a single fixed point varies with the magnitude of $a_1 = a_2$.

18.2 TRANSCRIPTION NETWORKS

A *transcription network* is an arrangement of genes that regulate the production of proteins in response to conditions inside and outside the cell. A *transcription factor* is a protein that has two states, *active* and *inactive*. Environmental conditions can cause a transcription factor to switch rapidly between these states. An active transcription factor binds to DNA to regulate the rate at which the target gene is transcribed. In gene transcription, RNA polymerase (RNAp) binds to a site in the gene promoter, a regulatory region of the DNA that precedes the gene and controls the production rate of mRNA that corresponds to the gene coding sequence. The mRNA is translated into protein, called the *gene product*.

A transcription network can be encoded as a graph. The vertices of the graph represent genes; the edges, $X \to Y$, represent the regulation of the gene Y by the protein product of gene X. This regulation can increase or decrease the transcription rate when the product of X binds to the promoter of Y. If increase, then X is called an *activator* of Y; if decrease, a repressor.

The strength of this influence of the active transcription factor X^* on the target transcription rate is called the *input function* f and has several models. The *step function* is

$$f(X^*) = \begin{cases} \beta & \text{if } X^* > K \\ 0 & \text{if } X^* \leq K \end{cases} \quad \text{and} \quad f(X^*) = \begin{cases} \beta & \text{if } X^* \leq K \\ 0 & \text{if } X^* > K \end{cases} \tag{18.8}$$

The first holds if X^* is an activator, the second if X^* is an inhibitor. The constant K is the *activation threshold*; the constant β is the *maximum expression rate* of Y.

The other common model is the *Hill function* defined by

$$f(X^*) = \frac{\beta (X^*)^n}{K^n + (X^*)^n} \quad \text{and} \quad f(X^*) = \frac{\beta}{1 + (X^*/K)^n} \tag{18.9}$$

for activator (left) and inhibitor (right). The exponent n is called the *Hill exponent*. (Some authors call n the Hill coefficient—see Eq. (18.1)—but clearly n is not a coefficient.) In Fig. 18.5 we see plots of the Hill functions (the starts low and ends high) for $n = 5$, $n = 10$, and $n = 50$, left to right. As n increases, the Hill functions approach the step functions.

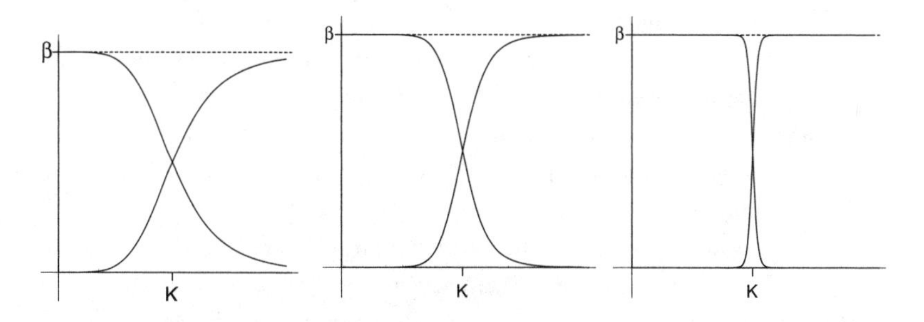

Figure 18.5. Graphs of Hill functions for $n = 5$, $n = 10$, and $n = 50$.

Mutation can alter a transcription network in several ways.

- A graph edge can be eliminated by the alteration of a promoter letter, if that alteration prevents the binding of a transcription factor.

- A graph edge can be added by a promoter letter mutation that enables transcription factor binding, or by appropriate duplication or relocation of a piece of the genome.
- Even if the graph topology remains the same, mutation of promoter letters or relocation of the binding site can alter β or K, and so change the strength of the transcription factor influence.

These realizations give a way to identify functionally important graph patterns, called *network motifs*. A pattern is a network motif if it occurs significantly more often than it does in a suitable random graph. What's a suitable random graph for a real transcription network? At the very least, the random graph should have the same number of vertices and of edges as the real network. A bit more subtle is for each vertex of the random graph to have the same *in-degree* (number of edges pointing to the vertex) and the same *out-degree* (number of edges pointing from the vertex) as the corresponding vertex of the real network. To be sure, there are other factors, but we'll stick with these.

Before we do some calculations, we'll mention a few properties of real transcription networks. Appendix C of Uri Alon's book [9] is a good reference.

- Transcription networks are sparse. A graph with V vertices has at most V^2 edges: for each vertex an edge can lead to each of the V vertices, including itself. A graph is *sparse* if $E \ll V^2$, where E is the number of edges. For typical transcription networks, $E < 0.001V^2$. Because mutation can break graph edges, only edges that enhance survival are preserved through selection. If an important edge is broken, the cell is less likely to survive long enough to reproduce.
- Most transcription networks have a few vertices, called *global regulators*, with much higher out-degrees than the network average out-degree, E/V. (Do you see why the network average in-degree also is E/V?) The transcription factors of these vertices represent important environmental conditions—glucose starvation is an example—that regulate a large number of genes to generate possible responses.
- Over some range, the distribution of network out-degrees satisfies a power law. Several mechanisms have been proposed, including self-organized criticality [20, 21, 23] and optimized tolerance [52].
- For most microorganism transcription networks, the in-degrees cluster around the average. The reason is space: a typical promoter region is only long enough to include a few binding sites. For more complex organisms, development requires sophisticated computation, which can be done through vertices with higher in-degrees. This is achieved by the effect of transcription factors on parts of the DNA some distance from the binding site.

Now, for a transcription network with V vertices and E edges, what can we say about a random graph with V vertices and E edges? For that matter, how can we construct such a random graph? Each edge must connect two vertices, so is determined by a pair of vertices. There are V^2 pairs of vertices, so we can select the edges uniformly randomly from among these. Each pair of vertices has probability E/V^2 of being connected by an edge. Properties of these graphs were studied by Paul Erdös and Alfréd Rényi [108, 109].

Suppose a transcription network T with V vertices and E edges contains a subgraph t with v vertices and e edges. What is the expected number of times $\mathbb{E}(N(t))$ that a random network with V vertices and E edges also contains the subgraph t? First, we must select v vertices from among the V. This can be done in

$$\binom{V}{v} = \frac{V!}{v!(V-v)!}$$

ways. For each pair of these v vertices that corresponds to vertices of t connected by an edge, we want to find an edge in T. But we've seen that the probability of finding an edge between any pair of vertices in T is E/V^2. Then we have $\mathbb{E}(N(t)) =$

$$\frac{1}{S}\frac{V!}{v!(V-v)!}\left(\frac{E}{V^2}\right)^e = \frac{1}{S}\frac{V(V-1)\cdots(V-v+1)}{v(v-1)\cdots 1}\left(\frac{E}{V^2}\right)^e \tag{18.10}$$

where the factor S counts the symmetry of the subgraph g. By this we mean the number of ways we can relabel the vertices and produce an equivalent subgraph. Fig. 18.6 illustrates how to calculate S.

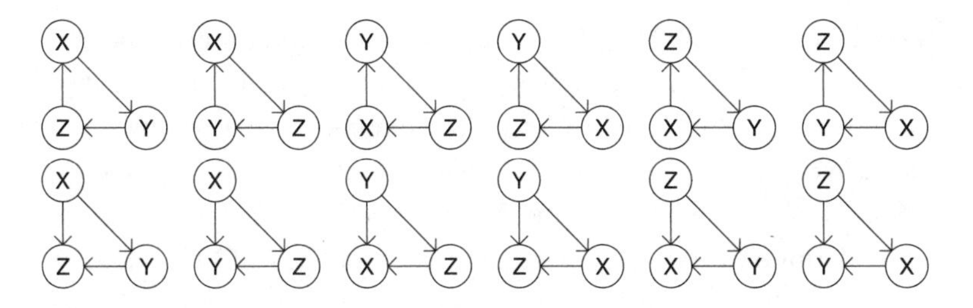

Figure 18.6. Variations on a cycle (top) and an FFL (bottom).

In the top row we see a subgraph with 3 vertices and 3 edges, arranged in a cycle. The first image shows the transcription pattern $X \to Y$, $Y \to Z$, and $Z \to X$ of the subgraph. We'll keep the arrangement of vertices and edges, and permute the labels X, Y, and Z. A graph with 3 vertices has 6 arrangements of the vertex labels. These are the 6 graphs of the top row of Fig. 18.6. The second is not equivalent to the first, because the first has $Y \to Z$ while the second has $Z \to Y$. The third and sixth are equivalent to the first; the second, fourth, and fifth are not. So for this cyclic arrangement we have $S = 3$.

In the second row we see a subgraph called a *feed-forward loop*, or *FFL*. Not one of the five vertex permutations in this row is equivalent to the first, so $S = 1$ for the FFL.

The *E. coli* transcription network of Figure 3.2 of [9] has $V = 420$ and $E = 520$. Then by Eq. (18.10) the expected number of FFLs in a random graph with these V and E values is 0.3. In the *E. coli* transcription network we count 42 FFLs. This number is so much larger than the expected number in the random graphs that we can call the FFL a network motif of the *E. coli* transcription network. In a moment we'll see why this might be the case.

First, though, we'll point out that Alon gets a different expected value. On page 45 of [9] we find 1.7 instead of 0.3. Alon simplifies the calculation of (18.10) by replacing $V(V-1)\cdots(V-v+1)$ with V^v. For $v \ll V$ this is close because $V-1 \approx V$, ..., $V-v+1 \approx V$. But this simplification misses the $v!$ in the denominator, and this factor accounts for the difference. Now Alon was looking for a power law relation between the expected value and the network parameters, and for that purpose the approximation is appropriate.

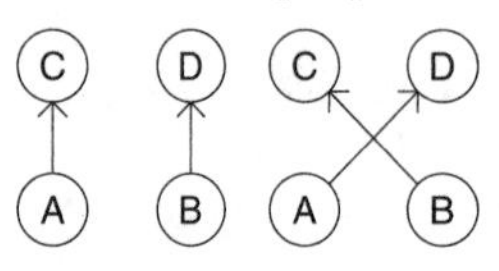

Figure 18.7. Switch incoming arrows at C and D, or outgoing arrows at A and B.

Suppose we calculate the expected value not among all random graphs, but among those for which the in-degree and the out-degree of each vertex matches the in-degree and the out-degree of the corresponding vertex of the real transcription network. How can we construct such a graph? Start with the real transcription network and switch incoming arrows at two randomly selected vertices, or switch outgoing arrows at two randomly selected vertices. Fig. 18.7 is an example. Do this a lot of times and we get a degree-preserving random network. These networks, with the V and E of the *E. coli* transcription network, have an expected number 7 of FFLs (page 45 of [9]), still far short of the actual number, 42. So even with this more stringent notion of random graph, we'll say that FFLs are network motifs.

Now we'll see why the FFL is a network motif. Each arrow $X \to Y$, $Y \to Z$, and $X \to Z$ can be an activator or a repressor, giving 8 varieties of FFL, shown in Table 18.1. These varieties fall into two categories: *coherent* if both paths $X \to Z$ and $X \to Y \to Z$ activate the transcription of Z or both repress the transcription of Z; and *incoherent* if these paths have the opposite effect on the transcription of Z. Remember that both repression and activation occur when gene products bind with the promoter region of Z. We'll see situations that call for coherent FFLs and other situations that benefit from incoherent FFLs.

For example, if X activates the transcription of both Y and Z, and Y represses the transcription of Z, then the path $X \to Y \to Z$ represses Z because X activates Y and Y represses Z. That is, the direct path $X \to Z$ and the indirect path $X \to Y \to Z$ have opposite effects on Z.

$X \to Y$	$Y \to Z$	$X \to Z$		$X \to Y$	$Y \to Z$	$X \to Z$	
A	A	A	C	A	R	A	I
R	A	R	C	R	R	R	I
A	R	R	C	A	A	R	I
R	R	A	C	R	A	A	I

Table 18.1. The 8 varieties of FFL. Here A, R, C, and I denote activator, repressor, coherent, and incoherent, respectively.

While all 8 are observed in the *E. coli* transcription network, the two varieties represented by the top row of Table 18.1 occur far more often than the other six.

The last detail to discuss is how Z combines the signals from X and Y. The most common options are AND and OR: both X and Y must be activated in order to transcribe Z, or Z is transcribed if either X or Y is activated. We'll work through the dynamics of the top coherent FFL of Table 18.1 and describe those of the top incoherent FFL.

We'll study the dynamics of the top coherent FFL with the signals $X \to Z$ and $Y \to Z$ combined by AND. That is, both X and Y must be activated in order to transcribe Z. Suppose the F_Y signal is present, so the Y transcription factor is active, and that the signal F_X begins. The rapid appearance of X^* binds with the promoter of Y and with the promoter of Z. The protein Y generated by the action of X^* itself is activated by F_Y. But the AND means that the presence of a high concentration of X^* is not sufficient to start the transcription of Z. The Y^* concentration must rise above its activation threshold. (This delay is illustrated in Fig. 18.10.) The time required for the Y^* concentration to pass this threshold means that this coherent FFL with an AND combiner at Z has a response delay, and that means it can filter out short bursts of F_X. We'll see that this behavior is used in the transcription network of *E. coli*. To see how evolution has fine-tuned some of the system parameters, we'll need to study this system in a bit of detail.

We'll use the interaction $X \to Y$ as a model. When the signal F_X appears, X rapidly activates to X^*. As soon as the concentration of X^* rises above the activation threshold K_{XY} (because we'll have several activation thresholds, we'll label each with the genes at the start and end of network edge $X \to Y$), the protein Y is produced at a rate β_Y and degrades at a per capita rate α_Y. That is, the Y concentration dynamics are governed by

$$\frac{dY}{dt} = \begin{cases} -\alpha_Y Y & \text{if } X^* < K_{XY} \\ \beta_Y - \alpha_Y Y & \text{if } X^* \geq K_{XY} \end{cases} \tag{18.11}$$

When $X^* < K_{XY}$, Y undergoes exponential decay, $Y(t) = Y(0)e^{-\alpha_Y t}$. When $X^* \geq K_{XY}$ the dynamics are more complicated, but we can apply the method of Sect. 10.9.

$$\begin{aligned} Y &= a_0 + a_1 t + a_2 t^2 + a_3 t^3 + a_4 t^4 + \cdots \\ Y' &= a_1 + a_2 2t + a_3 3t^2 + a_4 4t^3 + \cdots \\ \beta_Y - \alpha_Y Y &= \beta_Y - \alpha_Y \left(a_0 + a_1 t + a_2 t^2 + a_3 t^3 + a_4 t^4 + \cdots\right) \end{aligned}$$

As usual, $a_0 = Y(0)$. Matching coefficients of like powers of t in Y' and $\beta_Y - \alpha_Y Y$ we find that for $n \geq 1$

$$a_n = (-1)^{n-1} \frac{\alpha^{n-1}}{n!} \beta_Y + (-1)^n \frac{\alpha_Y^n}{n!} Y(0) \tag{18.12}$$

With a bit of algebra this gives

$$Y(t) = \frac{\beta_Y}{\alpha_Y} + \left(Y(0) - \frac{\beta_Y}{\alpha_Y}\right) e^{-\alpha_Y t} \tag{18.13}$$

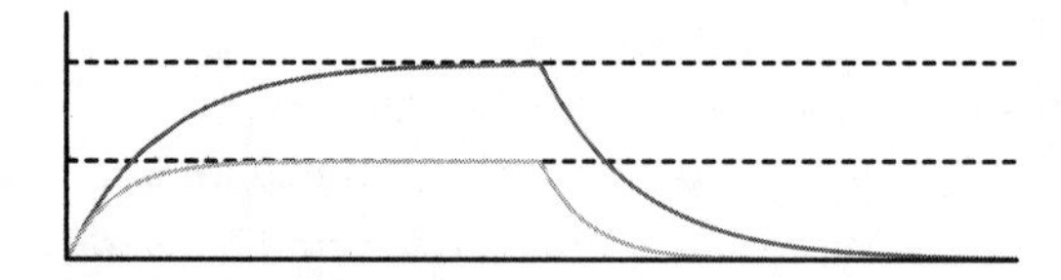

Figure 18.8. Plots of Y, same α_Y, different β_Y.

For example, in Fig. 18.8 we plot $Y(t)$ for $1 = \alpha_Y = \beta_Y$ (darker) and $1 = \alpha_Y = 2\beta_Y$ (lighter). The second line of Eq. (18.11) defines dY/dt for the left half of the time range where $X^* \geq K_{XY}$; the first line of Eq. (18.11) defines dY/dt for the right half where $X^* < K_{XY}$. The dashed lines are at the asymptotic values β_Y/α_Y.

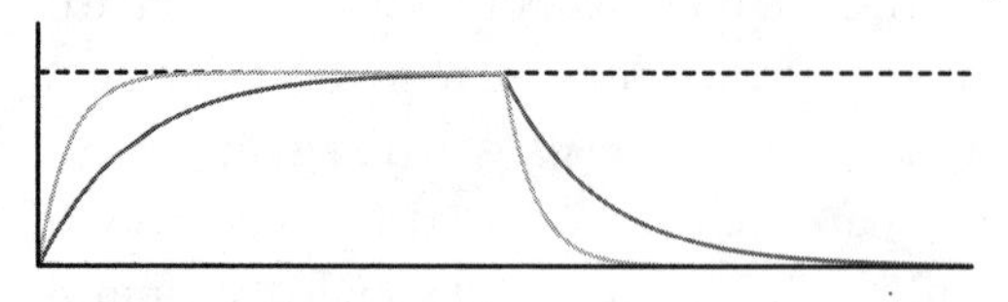

Figure 18.9. More plots of Y.

One behavior suggested by Fig. 18.8 is that higher β_Y causes faster initial growth of Y. We show this more clearly in Fig. 18.9. The darker curve is a plot of $Y(t)$ for $1 = \alpha_Y = \beta_Y$, the lighter curve for $3 = \alpha_Y = \beta_Y$. Note that not only does $Y(t)$ increase more rapidly when $X^*(t)$ rises above K_{XY}, but also it falls off more rapidly when $X^*(t)$ drops below K_{XY}.

Now we'll write the differential equation for the dynamics of Z determined by this FFL. Say K_{XZ} and K_{YZ} are the activation thresholds for X^* and Y^* to initiate transcription of Z.

$$\frac{dZ}{dt} = \begin{cases} -\alpha_Z z & \text{if } X^* < K_{XZ} \text{ or } Y^* < K_{YZ} \\ \beta_Z - \alpha_Z Z & \text{if } X^* \geq K_{XZ} \text{ and } Y^* \geq K_{YZ} \end{cases} \quad (18.14)$$

For example, in Fig. 18.10 we see plots of X^* (darkest, a step function), Y^* (lightest), and Z (medium). The graph of $X^*(t)$ is above K_{XY} and K_{XZ} (dashed line) for $0 \leq t \leq 5$ and below both thresholds for $t > 5$. Note that Y^* grows all the time X^* lies above K_{XY}, and declines as soon as X^* drops below K_{XY}. On the other hand, Z begins to increase only when $X^* \geq K_{XZ}$ and $Y^* \geq K_{YZ}$, so the growth of Z waits until Y^* reaches K_{YZ}. That is, adding the path $X \to Y \to Z$ and combining this with the path $X \to Z$ by an AND builds a delay into the transcription of Z, and the duration of this delay can be adjusted by evolution.

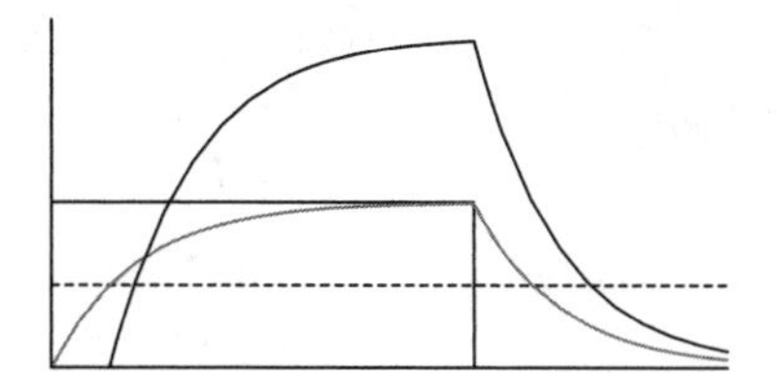

Figure 18.10. Plots of X, Y, and Z.

Another consequence of the AND combiner for Z transcription is that Z begins to decline as soon as X^* drops below K_{XZ}, even though Y^* remains above K_{YZ} for a while. The growth of Z has a delay; the decline of Z is immediate.

The growth delay of Z has an interesting corollary: this FFL ignores, or filters out, short bursts of X^*. In Fig. 18.11 we see a pulse of X^* too short to allow Y^* to reach K_{YZ}. Then the X^* signal is turned off and Y^* declines, but not to 0. At the second, longer X^* pulse the time for Y^* to rise to K_{YZ} is shorter than in Fig. 18.10 where Y^* starts from 0.

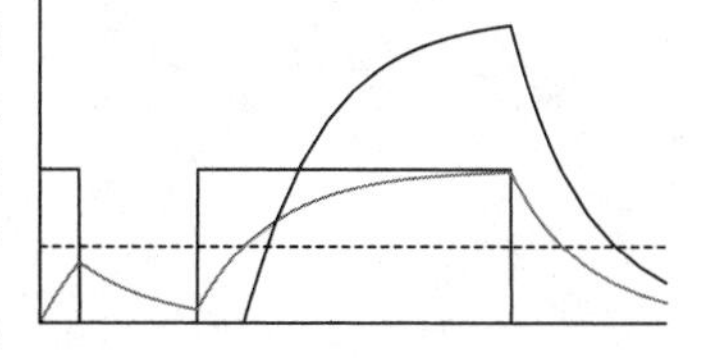

Figure 18.11. A This FFL is a short burst filter.

This FFL appears in many transcription networks, including that of *E. coli*. Here is one example, from Sect. 4.6.5 of [9]. The best energy and carbon source is glucose, but in the absence of glucose, the sugar arabinose can be used. The lack of glucose is signaled by the molecule cAMP. The transcription activator CRP (X in our graph) responds to cAMP. That is, CRP is activated when a lack of glucose is detected. The second transcription activator araC (Y in our graph) responds to the presence of arabinose. Coupled by an AND, these act on the promoter of the arabinose degradation genes (Z in our graph). Experiments with *E. coli* showed a delay of about 20 minutes between CRP expression and arabinose degradation expression. This is about the same duration as transient low glucose signals that can accompany changes in growth conditions. Evolution has fine-tuned this FFL delay to prevent the transition to arabinose metabolism unless there is a real glucose shortage.

The first coherent FFL with an OR combiner generates a delay when the X signal is turned off, but no delay when the X signal is turned on. This FFL is involved in the production of the motor that spins the *E. coli* flagellum.

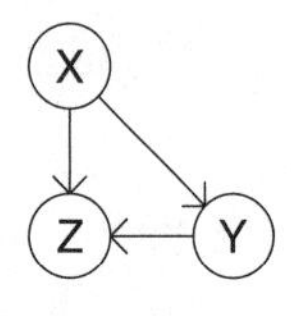

Figure 18.12. An incoherent FFL.

Finally, we'll discuss the first incoherent FFL: $X \to Y$ and $X \to Z$ are activators, $Y \to Z$ is a repressor, and at Z the signals are combined with an AND, shown in Fig. 18.12. Suppose $X^* \geq K_{XY}$ and $X^* \geq K_{XZ}$. Then Y^* grows according to

$$\frac{dY^*}{dt} = \beta_Y - \alpha_Y Y^*$$

and Z according to

$$\frac{dZ}{dt} = \begin{cases} \beta_Z - \alpha_Z Z & \text{when } Y^* < K_{YZ} \\ \beta'_Z - \alpha_Z Z & \text{when } Y^* \geq K_{YZ} \end{cases} \tag{18.15}$$

That is, while $Y^* < K_{YZ}$, only the activator X^* binds to the Z promotor, so Z grows at the rate β_Z and decomposes at the per capita rate α_Z. Once Y^* reaches the threshold K_{YZ}, the repressor Y^* binds to the Z promoter and the combined effect of the bound activator and repressor is to lower the Z growth rate to $\beta'_Z < \beta_Z$.

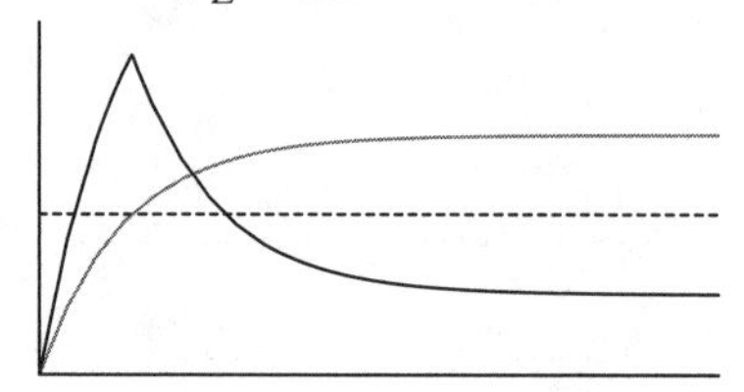

Figure 18.13. Plots of Z (darker), Y^* (lighter), and K_{YZ} (dashed).

The plot of Fig. 18.13 shows the graphs of Z (the darker curve) and Y^* (the lighter curve), along with the dashed line at K_{YZ}. So long as $Y^* < K_{YZ}$, the dynamics of Z are governed by the first case of Eq. (18.15) and we see rapid growth. As soon as $Y^* \geq K_{YZ}$ the effect of the repressor is seen and the Z curve converges to the fixed point of the second case, β'_Z/α_Z. Remember the simplest way to find this fixed point is to set the right-hand side of the differential equation to 0 and solve for Z.

Just focusing on the graph of Z we see one behavior of this incoherent FFL: it can generate a pulse in Z. The larger the ratio β'_Z/β_Z, the steeper the pulse. See Exercise 18.2.6.

In Fig. 18.14 we illustrate another feature of this incoherent FFL: it speeds up the expression of Z. To compare this with a simple $Y_0 \to Z_0$ network (subscripts to distinguish this network from the previous one), we must adjust the parameters β_{Z_0} and α_{Z_0} of the dZ_0/dt equation so that the asymptotic value β_{Z_0}/α_{Z_0} equals the asymptotic value β_Z/α_Z, because we imagine that this asymptotic value has been fine-tuned by evolution.

Because both curves converge to this asymptotic value but never reach it, we can't talk about the time to reach that value. So instead, we'll note the time each curve takes to attain half the asymptotic value. This is the time that the curves cross the dashed line in Fig. 18.14. Certainly, the incoherent FFL crosses this line much sooner than does the simple network.

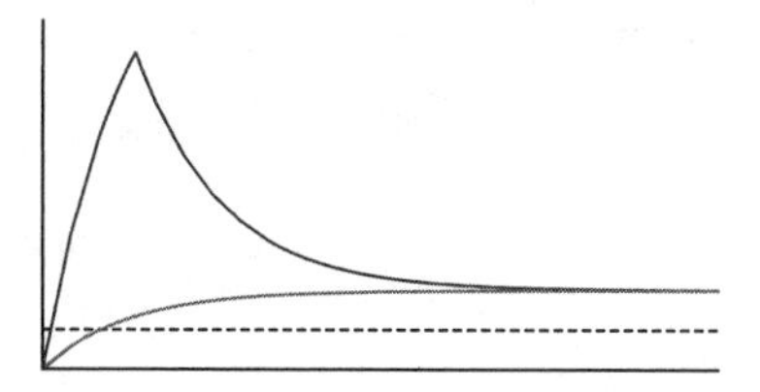

Figure 18.14. Plots of Z (darker), Z_0 (lighter), and half the asymptotic value (dashed).

Now we could make the simple network cross this line more quickly by increasing β_{Z_0}, but to maintain the same asymptotic value, we'd have to increase α_{Z_0} by the same factor. While there may be some uses for a protein that is produced very quickly and decays very quickly, this would consume a lot of resources. The FFL is a more sensible, and more often used, solution.

This FFL also occurs in the *E. coli* transcription network, in the system to utilize the sugar galactose. As in the arabinose network, the activator CRP plays the role of vertex X and responds to a glucose shortage signaled by cAMP. The Y vertex is the repressor GalS; the Z vertex is galactose utilization genes. In the absence of galactose, the GalS repressor keeps the expression of the galactose utilization genes at a low level. The presence of galactose inactivates the GalS repressor, so the absence of glucose and the presence of galactose rapidly (this is the speed-up effect of this FFL) starts the galactose utilization network.

The brilliance of this, the sheer *aptness* of evolution to discover subtle patterns, takes my breath away. Every. Single. Time.

Practice Problems

18.2.1. (a) For the differential equation $dY/dt = \beta_Y - \alpha_Y Y$ with $Y(0) = 0$, find the time T for which $Y(T) = K_{YZ}$.
(b) Interpret the expression for T if $K_{YZ} > \beta_Y/\alpha_Y$.

18.2.2. For the coherent FFL system (18.11) and (18.14) with $\alpha_Y = 1$, $\beta_Y = 1$, $\alpha_Z = 1$, $\beta_Z = 2$, $K_{XY} = 0.5$, $K_{XZ} = 1.0$, and $K_{YZ} = 0.75$,
(a) sketch the solution curves Y^* and Z for $X^*(t) = 1.25$ for $0 \leq t \leq 10$.
(b) Sketch these curves for $X^*(t) = 1.25$ for $0 \leq t \leq 5$ and $X^*(t) = 0.55$ for $5 \leq t \leq 10$.

(c) Sketch these curves for $X^*(t) = 1.25$ for $0 \le t \le 5$ and $X^*(t) = 0.45$ for $5 \le t \le 10$.

(d) Describe the differences between the graphs of (a), (b), and (c).

Practice Problem Solutions

18.2.1. (a) Use Eq. (18.13) with $Y(T) = K_{YZ}$ and $Y(0) = 0$:

$$Y(t) = \frac{\beta_Y}{\alpha_Y} + \left(Y(0) - \frac{\beta_Y}{\alpha_Y}\right)e^{-\alpha_Y t} \quad \text{and so} \quad K_{YZ} = \frac{\beta_Y}{\alpha_Y}\left(1 - e^{-\alpha_Y T}\right)$$

We can solve for T:

$$T = -\frac{1}{\alpha_Y}\ln\left(1 - \frac{K_{YZ}}{\beta_Y/\alpha_Y}\right)$$

(b) If $K_{YZ} > \beta_Y/\alpha_Y$, then the argument of the logarithm is negative, which would be a problem. But $Y(t)$ converges monotonically to β_Y/α_Y. So $K_{YZ} > \beta_Y/\alpha_Y$ implies $Y(t)$ never reaches K_{YZ}.

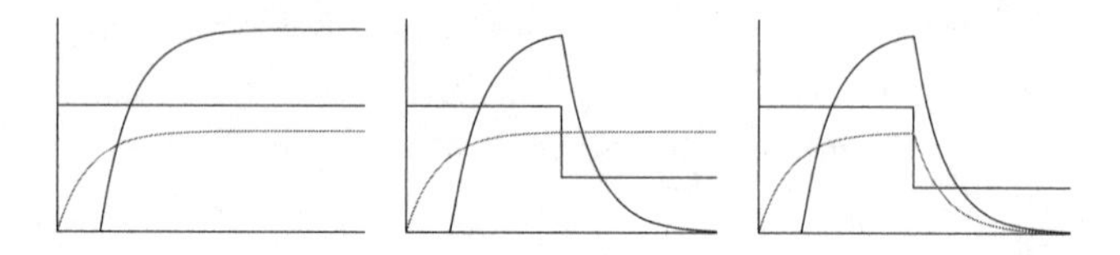

Figure 18.15. The graphs for Practice Problem 18.2.2 (a), (b), and (c).

18.2.2. In Fig. 18.15 we see the plots for the systems in (a), (b), and (c): Y^* is the lighter curve, Z is the darker curve, and X^* is made of straight lines. Code to generate figure (b) is in Sect. B.23.

(d) In (a) $X^*(t) > \max\{K_{XY}, K_{XZ}\}$ for all t. The production of Y^* begins right away and continues monotonically to its asymptotic value $\beta_Y/\alpha_Y = 1$. The production of Z begins as soon as $Y^* \ge K_{YZ} = 0.75$ and continues monotonically to its asymptotic value $\beta_Z/\alpha_Z = 2$.

In (b) $X^*(t) > \max\{K_{XY}, K_{XZ}\}$ for $0 \le t \le 5$. The production of Y^* begins right away and continues monotonically until $t = 5$. The production of Z begins as soon as $Y^* \ge K_{YZ} = 0.75$ and continues monotonically until $t = 5$. For $t > 5$, $X^*(t) > K_{XY}$ and so Y^* continues to increase monotonically to its asymptotic value $\beta_Y/\alpha_Y = 1$. On the other hand, for $t > 5, X^*(t) < K_{XZ}$ and so Z decreases monotonically toward $Z = 0$.

In (c) $X^*(t) > \max\{K_{XY}, K_{XZ}\}$ for $0 \le t \le 5$. The production of Y^* begins right away and continues monotonically until $t = 5$. The production of Z begins as soon as $Y^* \ge K_{YZ} = 0.75$ and continues monotonically until $t = 5$. For $t > 5$, $X^*(t) < K_{XY}$ and $X^*(t) < K_{XZ}$ so both Y^* and Z decrease monotonically toward $Y^* = 0$ and $Z = 0$.

Exercises

18.2.1. Explain how the FFLs of the second and third entries of the second column of Table 18.1 are incoherent.

18.2.2. For the differential equation $dY/dt = \beta_Y - \alpha_Y Y$,
(a) verify that the terms of the series solution are given by Eq. (18.12),
(b) verify that this series defines the solution of Eq. (18.13), and
(c) check that this Y is a solution of the differential equation.

18.2.3. (a) For the system (18.11) and (18.14), suppose that $X^* \geq K_{XZ}$ and $Y^* \geq K_{YZ}$ for a long time. Find the asymptotic value of Z.
(b) Suppose X^* drops to 0 at $t = T_1$, when both Y^* and Z have reached their asymptotic values. Find the time T_2 at which Y^* drops to K_{YZ}.
(c) Find the time at which Z drops below K_{YZ}.

18.2.4. Here we'll elaborate on the example of Fig. 18.11. Take $\alpha_Y = \beta_Y = \alpha_Z = 1$, $\beta_Z = 2$, and $K_{XY} = K_{YZ} = 0.5$. Suppose

$$X^*(t) = \begin{cases} 1 & \text{for } 0 \leq t \leq 0.5 \text{ and for } 0.5 + \delta \leq t \leq 1 + \delta \\ 0 & \text{for } 0.5 < t < 0.5 + \delta \text{ and for } 1 + \delta < t \end{cases}$$

To the nearest hundredth, find the smallest δ for which Z is not activated.

18.2.5. Continue with the parameters of Exercise 18.2.4 and change the AND combiner at Z to an OR.
(a) Take $\delta = 1$ and plot $Y(t)$ and $Z(t)$ for $0 \leq t \leq 5$. Explain why this Z graph differs from that of Exercise 18.2.4.
(b) What changes if $\delta = 0.5$?

18.2.6. For the system (18.11) and (18.15) with $\alpha_Y = \beta_Y = \alpha_Z = \beta_Z = 1$, $K_{XY} = K_{YZ} = 0.5$, and $X^*(t) = 1$ for all t, plot the Z curves for $\beta'_Z = 0.25, 0.1$, and 0.01. Describe the effect of the ratio β_Z/β'_Z on the steepness of the initial decay of the Z curve. Can the asymptotic values of these curves account for the different steepnesses?

18.3 SOME OTHER NETWORKS

In Sect. 15.6.2 we mentioned gene regulatory networks and said a bit about how the network topology and parameters are deduced. Here we'll discuss a few other examples.

Sensory transcription networks, for example, those described in Sect. 18.2, detect and respond to environmental circumstances. In addition to the FFL, these networks exhibit three other kinds of motif: an *autoregulator* (a gene that regulates its own transcription), a *single-input module (SIM)* (a single transcription factor is the sole regulator of several genes), and a *dense overlapping regulon (DOR)* (several transcript factors regulate several genes). An SIM can transcribe genes one-by-one in a specified order, and so produce proteins in the order they are needed. A DOR allows for more complex responses: a collection of environmental factors regulate the transcription of a collection of genes. Note that DORs have only two levels, input genes that link directly to output genes. Cascades, where the output of one DOR acts as input for another DOR, are very rare in sensory transcription networks, probably because of the long time scale implied by the sequencing of a cascade. On the other hand, developmental networks, which work over much longer times, do exhibit multi-layer cascades.

Sensory transcription networks must respond quickly and reversibly (environmental conditions can return to previous states) and so must process information quickly. In contrast, *developmental transcription networks*, active when a cell differentiates into another type of cell, work on longer time scales, generally along an irreversible sequence of stages.

Within a cell we find also networks for protein-protein interactions, metabolism, and signal transduction. These networks are superimposed but not necessarily disjoint: some motifs function in several networks. The different kinds of networks have their own collections of motifs, though some motifs are shared by several network types. Alon's book [9] analyzes many more examples.

On a larger size scale, the most prominent network in the body is the neuronal net. Between animals (and plants [291]) we have networks of predators and prey, resource distribution, the spread of pathogens. First we assumed the populations were spread homogeneously and so we considered only total populations, not how or where these populations interacted. Then in Sect. 17.12 we saw how to model variations in space as well as in time. But even these models have some degree of homogeneity: changes are determined only by local concentrations. Interaction was determined by place and time alone. A better approximation is to consider networks of interactions [371]. If we can estimate who interacts with whom, then the dynamics equations would take place on that network.

Metabolic networks are collections of biochemical reactions between metabolites and enzymes. First discovered was the pathway *glycolysis*, a sequence of ten reactions converting glucose to pyruvate, and occuring in almost all organisms. Among other metabolic networks are the Krebs cycle, oxidative phosphorylation, and the pentose phosphate pathway. Some steps along these networks can be regulated; it is because of these variable rates that metabolic networks contribute to homeostasis. The collection of pathways within a cell, or an organ, or an organism, constitute the metabolic network of that cell, organ, or organism.

One reason to study metabolic networks is that in the 1920s Nobel laureate Otto Warburg observed that malignant tumors use far more glucose than regular cells, and that this glucose is converted to lactate by fermentation, a process called *aerobic glycolysis* [374, 375], or the *Warburg effect*. Otto Warburg [373] hypothesized that aerobic glycolysis results from dysfunctional mitochondria. This is the basis for the *Warburg hypothesis,* that cancer results from malfunctioning mitochondria. Subsequent discovery of oncogenes and tumor suppressor genes led some to believe that the Warburg effect is a consequence, rather than a cause, of mutations in these genes. Other work [328, 112] suggests that the Warburg effect is an essential ingredient in tumor growth. This is an active area of research; [208] is a good summary.

The Warburg effect is the basis for PET (positron emission tomography) scans. The higher uptake rate by tumors of radioactively tagged glucose analogs is used to diagnose and monitor the treatment of some cancers.

We'll end with the most complicated network known in biology, the neural net of the human brain. How complicated? In Sect. 13.6 we saw that the average human brain has about 86 billion neurons, and on average each is connected to a few thousand other neurons. There are a hundred trillion or more connections in this network. How did this network become you? How does it continue to learn, to make inferences, discover new interests; then begin to forget, to lose words, then names, then you? I have no memory of the first part of this journey, but I watched the growth of the minds of my sister, brother, and nephew. Watched with delight, but could not figure out how this happened. Also without much understanding I've watched beloved family make the last part of this journey, but now that I've begun this myself, I hope to take better notes.

Though some of the concepts of neural nets were proposed in the late 19th century, major steps were the threshold logic models of Warren McCullouch and Walter Pitts [231], the neural plasticity of Donald Hebb [160], the perceptron of Frank Rosenblatt [313, 236], and the back propagation algorithm of Paul Werbos [381, 382]. Now we'll give an example of the last.

In Fig. 18.16 we see an example of a very simple neural net with three layers. The input layer consists of two neurons, I_1 and I_2, that receive signals from other neurons. The hidden layer consists of two neurons, H_1 and H_2, that receive signals from the input layer. The output layer consists of a single neuron, O, that receives signals from the hidden layer and passes its output on to other neurons. Typically each layer has many more neurons, and there can be several hidden layers.

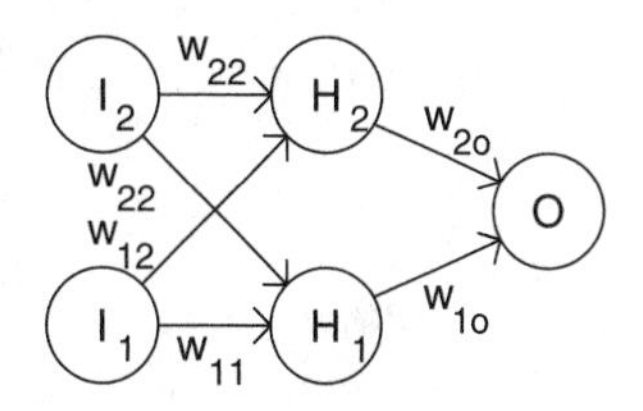

Figure 18.16. A neural net.

Each neuron sums the inputs it receives and if the sum is above a certain level, the *threshold* of the neuron, it sends a signal to the neurons connected to it. For this example, we'll take the threshold to be 0. That is, if the summed inputs are ≤ 0, the neuron sends no signal; if the summed inputs are positive, the neuron sends the signal 1.

In a neural net, the connections have different weights, the numbers w_{ij} and w_{io} in the figure, that represent the strengths of the connections between the neurons. The neural net learns by adjusting these weights. For artificial neural nets this is achieved by training sets, collections of inputs and the desired outputs. In the next two paragraphs we'll sketch the general architecture. Then we'll give a concrete example.

For a given set of inputs, the net computes an output. If this output differs from the training set output, the difference is used to adjust the weights by back propagation. Then we run the same inputs through the net and back propagate the difference to adjust the weights. We repeat until the net computes an output that is close enough. Error tolerances are part of the training data. Now we move to the next set of training data and repeat the adjustments until the output is close enough. After we've gone through the entire collection of training sets, we return to the first training set. Subsequent weight adjustments likely have degraded the net's performance on this training set, so we retrain it: run through the sets again and again, until the net gets all calculated outputs close enough to the target outputs. When this is finished, the net is trained and is ready to respond to new inputs.

This seems to be more or less how we learn, too. And it gives an interesting window on the process. When we learn an unfamiliar topic, first we learn a few of the main points. Then we fill in the gaps with the nearest things we already know, neural weights we have not adjusted yet. To gain a firmer grasp, we must unlearn how these gaps are filled, propagate errors back to adjust weights further. "Unlearning is a part of learning" was a mantra I discovered as a kid. Now I understand, at least as metaphor, why this is true.

So, how does back propagation work? Suppose the initial weights of the Fig. 18.16 net are

$$w_{11} = 2.1,\ w_{12} = 1.2,\ w_{21} = -1.4,\ w_{22} = 2.5,\ w_{1o} = 1.4,\ w_{2o} = -1.6$$

and the training set is $I_1 = 1$, $I_2 = 0$, and $O = 1$.

The first step is to compute the net's output with this input. The hidden layer inputs are

$$I_1 w_{11} + I_2 w_{21} = 1 \cdot 2.1 + 0 \cdot (-1.4) = 2.1 \qquad \text{so } H_1 = 1 \tag{18.16}$$

$$I_1 w_{12} + I_2 w_{22} = 1 \cdot 1.2 + 0 \cdot 2.5 = 1.2 \qquad \text{so } H_2 = 1 \tag{18.17}$$

and this generates the output

$$H_1 w_{1o} + H_2 w_{2o} = 1 \cdot 1.4 + 1 \cdot (-1.6) = -0.2 \qquad \text{so } O = 0 \qquad (18.18)$$

This isn't the target output of the training set, so we need to adjust the weights. The output error e_o is the difference between the target output and the computed output, that is, $e_o = 1 - 0 = 1$. We use the weights w_{1o} and w_{2o} to propagate this error back to the hidden level

$$e_{h1} = e_o \cdot w_{1o} = 1.4 \qquad \text{and} \qquad e_{h2} = e_o \cdot w_{2o} = -1.6$$

If the network has more hidden layers, these errors would propagate back to each hidden layer. Obviously, the errors cannot propagate to the input layer.

These errors determine the changes δw in the weights

$$\delta w_{1o} = e_o \cdot H_1 = 1 \qquad \delta w_{2o} = e_o \cdot H_2 = 1$$
$$\delta w_{11} = e_{h1} \cdot I_1 = 1.4 \qquad \delta w_{21} = e_{h1} \cdot I_2 = 0$$
$$\delta w_{12} = e_{h2} \cdot I_1 = -1.6 \qquad \delta w_{22} = e_{h2} \cdot I_2 = 0$$

Then the adjusted weights $\hat{w} = w + \delta w$ are

$$\hat{w}_{1o} = w_{1o} + \delta w_{1o} = 2.4 \qquad \hat{w}_{2o} = w_{2o} + \delta w_{2o} = -0.6$$
$$\hat{w}_{11} = w_{11} + \delta w_{11} = 3.5 \qquad \hat{w}_{21} = w_{21} + \delta w_{21} = -1.4$$
$$\hat{w}_{12} = w_{12} + \delta w_{12} = -0.4 \qquad \hat{w}_{22} = w_{22} + \delta w_{22} = 2.5$$

With these weights and the original input in Eq. (18.16) the adjusted net computes $H_1 = 1$ (the input of H_1 is 3.5) and $H_2 = 0$. Then with these values for the hidden neurons and the adjusted weights, Eq. (18.18) gives $O = 1$. This matches the training set.

Typically many training sets are required, and each one can disturb the weights of all the previous sets. Both carbon and silicon neural nets learn by repetition.

(Another approach to neural net computation is *neural differential equations*, where discrete hidden layers are replaced by specifying derivatives. This is a continuous model incorporating some of the ideas of neural net architecture. A good introduction is [57].)

Still, how could this process, even with many billions of neurons and many trillions of connections, build your mind? This we still do not know, but one interesting approach is Nobel laureate Gerald Edelman's neural Darwinism [98, 99, 100, 101, 102, 103]. Here many groups of neurons compete to perform tasks. The more successful are reinforced, the less successful find other things to do.

All of this—every poem, piece of art, book, act of kindness, act of selfishness, every child who has looked at the night sky and imagined falling into the deep dark space between the stars—all of this comes from a lump of living matter about the size of your two fists side by side. How can you not think about this? If network theory and systems biology can help even a tiny bit, that's reason to look at them.

18.4 WHY SHOULD THIS WORK?

Biological systems, even a single cell, are immensely complicated. Some scientists believe fine attention to detail is the only reasonable way to model a system. As a caution, we mention Jorge Luis Borges's story "On exactitude in science" [47]. In that story, Borges describes a country where cartography was the national obsession, so much that eventually the inhabitants produced a map that corresponded, point for point, with the country. Later generations lost interest in the map. Weather tore it to pieces and animals made nests from the bits. More detail is not necessarily so useful.

In his essay "A calculus of purpose" [201], Arthur Lander expresses the view that the complexity of biological networks may complicate the search for underlying laws. But there may be no underlying general laws. Recall the Feynman quotation in Sect. 6.3: there may be no ultimate laws of nature, but just layer after layer. While this may be true, we adopt the more optimistic view of Ellner and Guckenheimer on page 130 of [107]: "We would say that biological phenomena are generated by evolution, which uses large complex networks for some tasks ...Natural selection has apparently preserved—or repeatedly discovered—successful solutions to the basic tasks that cells face." And so there is reason to expect (sometimes) simple dynamical rules because "designs produced by evolution exhibit ...modularity and recurring design elements ...."

Certainly, biology is complicated. But the unending patience of evolution can discover simple solutions and use them in apparently dissimilar situations. Systems biology is a route to recognize these solutions.

Evolution directs not only how the bits of our bodies grow, but also how the bits work together. Evolution occurs also in an abstract world of dynamical networks. And evolution may occur in dimensions still unseen.

Chapter 19 What's next?

The earlier chapters have introduced a collection of mathematical tools that help us to understand biological systems and in some cases to design approaches to fix problems in these systems. This collection may seem to be quite large, but in fact it is just the tip of the iceberg. And unlike physical icebergs, this one is growing, rapidly. In this last chapter we'll sketch three examples of other directions. The first two (Sects. 19.1 and 19.2) are well established and already have shown impressive successes. The third (Sect. 19.3) is at an earlier stage of development, but is an approach that I think shows some promise. Of course, I may be mistaken about this.

Likely the most interesting developments are fuzzy thoughts, puzzles on the edge of consciousness of researchers immersed in lab work or of coders staring at screens. In fast-moving fields, every book becomes an account of history before the first copy comes off the press. Nevertheless, in this chapter I present my guesses at some approaches that come next.

19.1 EVOLUTIONARY MEDICINE

"Should doctors and medical researchers think about evolution?" This question is the first sentence of *Evolution in Health and Disease* [344] by Stephen Stearns and Jacob Koella. The evolution of antibiotic-resistant bacteria is an obvious, but not the only, reason for the answer "Yes." Twenty years ago, a colleague enthusiastic in his hypochondria thought every sniffle or sneeze should be treated by antibiotics. He kept changing physicians until he found one who would write an unending collection of antibiotic prescriptions. Unless these were placebos, always a possibility I suppose, this physician should have known better. Every person with a prescription pad needs to be aware of the evolution of antibiotic-resistant bacteria.

Important historical examples are summarized in *Why We Get Sick: The New Science of Darwinian Medicine* [255] by George Williams and Randolph Nesse, a popular book based on their paper [386]. Although the completion of the human genome project was almost a decade in the future when this book was published, it contains fascinating information and is a delight to read. If you haven't thought about the evolutionary aspects of disease and its treatment, this is a fine place to start. In addition to many interesting examples, Nesse and Williams include apt literary examples and themselves appear as characters in the narrative. More recent information is in [254, 344, 345], also well-written volumes. On pages 53–54 of [255] we find the following examples. In 1941 all staphylococcal bacteria were susceptible to penicillin. Within three years some strains had evolved the ability to produce enzymes that break down penicillin, and by the early 1990s some penicillin resistance was exhibited by 95% of staphylococcal strains. Other antibiotics show similar fates: although they are effective initially, eventually mutation produces a variation of the pathogen that ignores the antibiotic. Descendants of this strain survive while the antibiotic clears the ancestral strain from the population. Natural selection switches the pathogen population to the antibiotic-resistant strain.

This problem is even worse than it at first appears: Nobel laureate Joshua Lederberg pointed out that antibiotic resistance can jump from one bacteria species to another by genetic transduction [402]. Public health will benefit when medical researchers can anticipate the evolutionary response of pathogens to antibiotics, but right now all of us can be responsible in our use of antibiotics: complete the course of treatment even if you feel better part way through, and please don't try to convince your physician that you need an antibiotic. That's her or his call. Soberingly, more than this may be needed to reduce the growth of antibiotic resistant bacteria [149].

Alone, this would be reason enough, but there are so very many more. The diversity of human viruses—Dengue, Ebola, hantavirus, hepatitis B, HIV, influenza, SARS-CoV, and now COVID-19—can be best understood through the lens of evolution. For example, 58% of human pathogens are *zoonotic* [305, 390], human diseases that arose in animals.

The notion of evolutionary medicine was introduced by Williams and Nesse in the early 1990s, and by now there has been some success with incorporating evolutionary thinking into the medical curriculum; see [13, 345] and the on-line journal *The Evolution & Medicine Review* [110]. Evolutionary perspectives in medicine and public health are presented in [253, 346, 347, 380], and [191] surveys recent applications of high-throughput (next-generation) DNA sequencing, described in Sect. 19.2, to study the complicated ways that infectious diseases affect the evolution of human genomes.

Still, the effort to get physicians and medical researchers to think seriously about evolutionary medicine has not been a simple process. One problem is this question: if natural selection drives evolution to maximize fitness, why is there any disease at all? Some accidental injury can't be avoided, but shouldn't evolution have removed all illness? We'll see that this question is based on a confused notion of fitness. Here are some examples of how evolution can help us understand disease.

1. As mentioned in our discussion of fitness landscapes in Sect. 15.3, in natural selection the term "fitness" refers to reproductive success and nothing else. Not freedom from disease, not happiness, not longevity, just the number of offspring. (So from the point of view of natural selection, I am unfit because I have no kids.) Consequently, evolution cannot select against any trait that is favorable up through the reproductive years, regardless of its effect, including potentially fatal diseases, later in life. J. B. S. Haldane [154] seems to be the first person to have made this observation, in 1942. For example, high levels of insulin growth factor-1 favor early growth and so are selected because of fitness advantage, even though in later life they are associated with increased risk of prostate cancer and breast cancer [141]. Hemochromatosis causes excess iron absorption. After age forty, this can cause organ failure and death; but earlier it may protect against iron deficiency anemia [4].

2. Pathogens have an advantage in the evolutionary arms race [191] against our natural and manufactured defenses. Bacteria divide on average about once an hour, so a single human generation corresponds to about 150,000 bacterial generations. Consequently, pathogens can respond much more quickly to immune system evolution or the introduction of new drugs than the immune system or pharmacology researchers can respond to the pathogen's adaptation. For example, the average time between the introduction of a new antibiotic and the appearance of bacteria resistant to that antibiotic is about two years.

Biologists call this evolutionary arms race the *Red Queen principle*, named after the conversation between the Red Queen and Alice in Lewis Carroll's *Through the Looking-Glass*.

3. Evolution depends on history, on selective pressures, positive and negative, experienced by the predecessors of the current state. This has produced some suboptimal adaptations. For example (pgs. 127–129 of [255]), in the eyes of vertebrates the blood vessels and nerves lie in front of the retina, causing a blind spot where the nerves exit the eye, and also many smaller blind spots that are shadows of the blood vessels. Our brains compensate for these blind spots by keeping the eye in constant motion so nothing in the field of vision is obscured for long, and the brain assembles a coherent image from these views. On the other hand, in squid eyes the nerves and blood vessels lie on the outside of the retina, so they cause no blind spots. In addition, the nerves anchor the squid retina to the back of its eyeball, good protection against a detached retina. (After I lost the sight in one eye through unsuccessful surgeries to repair a detached retina, I became envious of squids, or at least of their eyes.) Evolution is constrained by its history. Vertebrate eyes and squid eyes evolved independently and along different paths.

Moreover, every person has a (perhaps only slightly) different life history evolution [343], and so responds differently to diseases and to drugs. These differences don't matter much when we take aspirin for headaches, but they can be very important in the selection of clinical trial participants and in the use of targeted therapies.

4. Some genes present trade-offs between benefits and risks. Here are three examples.

- Sickle cell anemia occurs mostly in parts of Africa where malaria is present. Heterozygosity for the sickle cell gene gives considerable protection from malaria. Homozygosity with both sickle cell alleles results in the disease. Victims usually die before they reproduce, so selection acts to remove this state. Homozygosity with both alleles normal grants no protection against malaria, so selection also works to clear this state in environments where malaria is common. See [4, 255, 344, 345].

- One copy of the cystic fibrosis gene offers some protection against typhoid [255, 397]; both alleles cause cystic fibrosis. With this disease the average lifetime is a bit under forty years, and although the patient's life will be difficult, selection pressure to remove the patient's genome will be weak.

- About 14% of European Caucasians have a deletion in the CCR5 gene that gives some high level of resistance to HIV, but also increases the risk of dying from a West Nile virus infection [139, 141].

5. Some ailments result from different time scales for biological and social evolution. For example, many of our metabolic processes evolved when our ancestors were hunter-gatherers in Africa. High-fat foods were rare and every bit of nutrition precious, so preference for these foods is a left over from those earlier times. The few thousand years that have passed since then have been enough to put hamburger joints on every other street corner (fast social evolution), but not long enough to remove desire for these foods (slow biological evolution). A good deal of obesity, diabetes, and hypertension can be attributed to these mismatched time scales. See Chapter 11 of [255].

6. Some circumstances thought to be harmful effects of diseases and consequently in need of alleviation are adaptive defenses against these diseases.

- Fever is a tool in the fight against infection. Raising the body temperature a few degrees helps our defense against pathogens, and if the fever persists for only a day or two, does little damage other than using more of the body's resources. Treatment of a high fever can be helpful; treatment of a low fever can slow recovery.
- Invading bacteria need iron to reproduce, so we have evolved several mechanisms to keep iron from pathogens. Patients with infections often present low iron levels, but sometimes the standard treatment of iron supplements accelerates the infection.

  The evolutionary roots of fever and iron deficiency are important in the effort to understand when to treat and when to leave it be. See pgs. 27–31 of [255].

Our vulnerability to cancer has an evolutionary basis, a loss of regulation of cell division. See [233], [344], Chapter 6 of [345], Chapters 21 and 22 of [344], Chapter 12 of [258], and Chapter 12 of [255], for example. Cancer is not a single disease, so the existence of a single cure is unlikely. Molecular and genetic approaches have brought relief—sometimes only temporarily, but relief nonetheless—but so much more remains to be done. An evolutionary approach to the mechanisms of cancer already has been helpful and shows great promise for future developments. We'll give one example, treatment of metastatic prostate cancer.

In [400] Jingsong Zhang and coworkers apply evolutionary dynamics to the treatment of metastatic prostate cancer (mPC). The first line of treatment is androgen deprivation therapy (ADT), but nearly all men with mPC eventually develop metastatic castrate-resistant prostate cancer (mCRPC). ADT resistance often is a consequence of increased expression of CYP17A1, an enzyme that is key to androgen synthesis. The drug abiraterone acetate is a CYP17A1 inhibitor. It lowers prostate specific antigen (PSA) levels and increases mean survival times. To understand the several therapy protocols, we must note that tumor cells have three competing phenotypes:

- T_+ cells require exogenous androgen,
- TP cells express CYP17A1 and produce testosterone, and
- T_- cells are androgen independent and resistant to abiraterone.

More resources are required by T_- cells, so without treatment T_+ and TP cells have a selective advantage in the limited-resource microenvironments of tumors.

Once mCRPC develops, a conventional therapy protocol is to administer abiraterone at its maximum tolerated dose (MTD). But this eliminates the T_+ and TP populations, removing the competitive disadvantage of T_- cells, which reproduce to a lethal level, so this therapy doesn't work for very long.

The intermittent, or metronomic, protocol is to administer abiraterone at MTD for an eight month period, then stop treatment. Resume treatment when PSA rises above 4ng/ml; stop when PSA drops below 4, then repeat. The reason for this strategy is that when treatment stops the TP population recovers rapidly and its selection advantage will keep the T_- population low. However, between successive treatment pulses, the TP population recovers a bit less and eventually the T_- population takes off, (very) bad news for the patient.

The [400] protocol is based on a more careful attention to evolutionary dynamics. The authors' idea is to preserve enough of the therapy-sensitive T_+ and TP populations to keep low the selectively disadvantaged T_- population. To do this, stop abiraterone treatment when the PSA drops to half of its pre-treatment level. Simulation shows this improves survival time by a factor of about 2, which may not sound like much unless you are the patient. Would you be content to live only another two years, or would you prefer to live another four? Think of this as your actual life and an additional two years become a very big deal. Even with this approach, we observe a slow increase in the T_- population between each treatment period, but perhaps this method can be refined.

In the pilot clinical trial, 10 of the 11 patients maintained stable oscillations of tumor burden with a median time to progression (TTP) of at least 27 months (probably it's longer; this number was imposed by the time the paper went to press) and a total drug use under half that of the conventional therapy. With conventional therapy, the median TTP is about 16.5 months. Certainly, this evolutionary approach is promising.

Because it recalls the competing species model of Eq. (7.7), we'll state the model of [400] for the growth of the tumor cell populations. Denote by x_1, x_2, and x_3 the populations of T_+, TP, and T_- cells. The model is

$$\frac{dx_i}{dt} = r_i x_i \left(1 - \frac{a_{i1}x_1 + a_{i2}x_2 + a_{i3}x_3}{K_i}\right)$$

The growth rates r_i are measured by the doubling times of appropriate cell lines. The carrying capacities K_i are estimated by rough models of the processes involved. For example, assume the TP cells are the source of testosterone for the T_+ cells, say that without therapy $K_1 = 1.5x_2$—that is, each TP cell supports the growth of 1.5 T_+ cells—and say that with therapy $K_1 = 0.5x_2$. The competition factors a_{ij} are a bit more involved and are studied in [395]. This matrix $[a_{ij}]$ is a construction from evolutionary game theory, a type of model in which fitness depends on the relative frequencies of the various phenotypes. We gave a brief sketch in Sect. 15.6.1. Refer to Chapter 3 of [258] and to [331] for more detailed expositions of game theory in evolution.

Aging is a consequence of our interaction with the clock and calendar, but aging is not a disease. Senescence, the deterioration of body and mind that occurs with age, may be considered an evolutionary disease [385]. The maximum life spans recorded [198] of some wild species are shown in Table 19.1.

mayfly	1 d	cat	36 y
queen termite	50 y	human	122 y
lobster	170 y	bowhead whale	211 y
bristlecone pine	1000s y	jellyfish	immortal

Table 19.1. Lifetimes of some species.

Over the last few hundred years, while the average length of a human life has steadily increased, the maximum life span has not. Why is this? Why are we programmed to die? If the goal of selection is to maximize individual fitness, individual reproductive success, then programmed death makes no sense. But longer life accompanied by ever-increasing senescence does not give more years to reproduce and raise children. So the better question is why haven't we evolved to have a much longer robust life. At the moment I write this sentence, I am 68 and three years ago gave up teaching, the only thing I didn't do terribly, because I could no longer do it well enough. Why can't I have another century to teach fractal geometry and biomath, to pass on the importance of curiosity and of kindness, to tell stories about cats? How many nights have my tears dampened the pillow, have I swallowed howls of frustration, because my ability to do the only thing I have loved to do has dissolved? Maybe the cognitive aspects of senescence are a gift: when my mind goes, I won't so much notice the departure of my body.

But again, selection favors reproductive success, not health or happiness. An early attempt at an evolutionary explanation for senescence held that old creatures are removed in order to make room for new. But this argument is based on the assumption that selection works to improve the fitness of the population, while really it works to improve the fitness of individuals. After all, the genome belongs to the individual, not to the population.

A more convincing explanation is based on point 1 near the start of this section. Some aspect of our genes, something that increases fitness during the reproductive years, brings down the curtain at our century mark or earlier. But what would that aspect, or combination of aspects, be? A good question, don't you think?

In Chapter 23 of [344] we read, "If humans did not age and could preserve their pristine physiological peak when mortality rates are lowest (typically near the age of puberty), then life expectancy in the United States in the year 2000 would have been about 5,000 years." Putting aside the wonderful scientific questions, if senescence is an evolutionary disease, do we really want to find a complete cure? Imagine the current world with Julius Caesar a middle-aged man of 2,000. Think of what this implies.

19.2 TRANSLATIONAL BIOINFORMATICS

We'll begin this section with a brief sketch of the mechanics of DNA sequencing. Because DNA is so important to contemporary medicine, every physician should be a bit familiar with how the data are obtained.

The role of DNA in genetics was established in 1944 with experiments by Oswald Avery, Colin MacLeod, and Maclyn McCarty [14]. Based on Rosalind Franklin's X-ray crystallographic studies of DNA, in 1953 James Watson and Francis Crick proposed the double helix model of DNA molecules, though they did not credit Franklin's work [221]. The helices are composed of four nucleotides, cytosine (C), adenine (A), thymine (T), and guanine (G). Each nucleotide of one helix is paired with the complementary nucleotide, C with G and A with T, so whatever information is encoded in one helix also is encoded in the other.

In 1977 Frederick Sanger and his coworkers [320] sequenced the first full genome, that of the bacteriophage φX174. For sequencing first proteins and then DNA, Sanger was awarded two Nobel Prizes. Though later refined and improved, Sanger's *chain-termination method* was the basis for sequencing the human genome [202, 363]. Since then, many researchers have developed next-generation sequencing (NGS) techniques that have resulted in great economies in both time and cost. And this journey is just beginning: even more advanced third-generation approaches have been designed, the genomes of (small) populations sequenced, evolutionary pathways mapped more clearly, and therapies targeted to specific mutations. All this depends on a deep understanding of molecular biology, and also on the math, statistics, and computer science to use the immense data sets generated by sequencing.

Here we'll sketch the main points of polymerase chain reactions (PCR), a method to make many copies of (to amplify) a DNA sequence, and Sanger's chain-termination method. Also, we'll describe how NGS is an extension of Sanger's method. All these use some common concepts listed below. First, in Sect. A.22 we mention that DNA strands have direction. One end of a nucleotide is called the $3'$-end; the other is called the $5'$-end. In the double helix, the two strands are arranged in the opposite directions.

1. *DNA polymerase* is an enzyme that copies all or part of a DNA sequence by adding nucleotides one by one to the $3'$-end of a DNA strand. Often *Taq* polymerase is used in PCR and other processes because it maintains its integrity during the high-temperature part of the process. (*Taq* polymerase was isolated from the *Thermus aquaticus* bacterium that lives in the hot springs of Yellowstone Park.) Polymerase can synthesize DNA in one direction only, from the $3'$ end toward the $5'$ end.
2. A *primer* is a short single strand of DNA that marks the starting point of DNA synthesis. The primer must occur at only one location in the entire DNA sequence. With a method called BLAST (basic local alignment search tool)[10], which we described a bit in Sect. 15.5, the entire genome can be searched for (unwanted) occurrences of the primer sequence at other locations. If the primer sequence appears at more than one location, another is tested.
3. The *DNA template* is the strand to be sequenced, and can be anything from a fragment of a gene to the whole genome.

Sketched in Fig. 19.1, the *polymerase chain reaction* a method to amplify, that is, make millions or billions of copies of, a region of DNA, works in the following way. Primers are identified for each end of the DNA template. A tube is filled with DNA template, primers, nucleotides to build the copies, and DNA polymerase. Then PCR has four steps:

i. Heat the sample to about 96°C to *denature*, that is, separate, the DNA helix into two single strands. These single strands contain the DNA templates to be copied.
ii. *Anneal*, that is, slowly cool the sample to about 60°C so the primer can bind to the ends of the template.

iii. Heat the sample to 72°C, the temperature at which *Taq* polymerase is most active. The polymerase *extends* the primers, adding nucleotides to the single strands by complementary base-pairing, duplicating the template.
iv. Repeat steps i, ii, and iii, between 25 and 35 times.

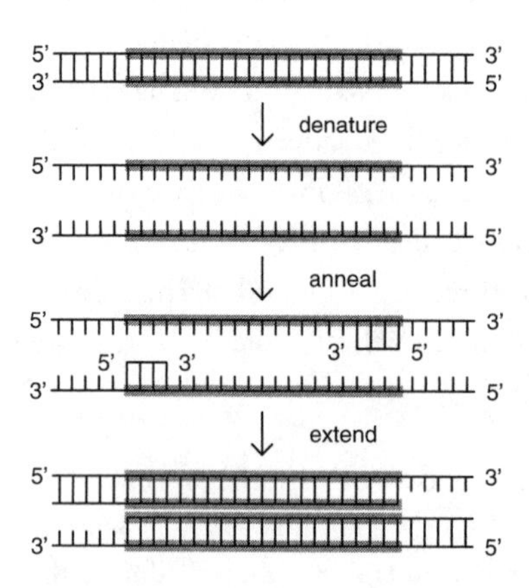

Figure 19.1. PCR steps. The template region is shaded gray.

Because the sample is prepared with an adequate supply of primer, *Taq* polymerase, and nucleotides, the number of copies of the template approximately doubles with each iteration. Each copy of the template, including those that have just been produced, can be duplicated with each step. So 25 to 35 iterates will produce between 30 million and 30 billion copies of the template from each sample copy with which the process began. Depending on the length of the template, PCR takes between 2 and 4 hours.

Sanger's chain-terminating method has an additional ingredient.

4. *Chain-terminating* versions of the four nucleotides. Call these dC, dA, dT, and dG. The standard nucleotides have an hydroxyl group that acts as a hook to which the next nucleotide attaches. In chain-terminating nucleotides, this hydroxyl group is replaced with a hydrogen. This destroys the possibility of extending the chain beyond this nucleotide, hence the name "chain-terminating nucleotide." In order to read the sequence, these nucleotides are marked with dyes: blue for dC, green for dA, red for dT, and purple for dG.

In Sanger's method, the DNA template, the fragment to be sequenced, is mixed with primer, polymerase, DNA nucleotides, and much smaller quantities of the four dye-labeled, chain-terminating nucleotides. The process is similar to PCR:

i. Heat the mixture to denature the template DNA.
ii. Anneal the sample so primer can bind to a single-strand template.
iii. Heat the sample so the polymerase can synthesize new DNA starting from the primer.
iv. Continue until by chance polymerase adds a chain-terminating nucleotide, which stops the growth of this chain.
v. Repeat steps i, ii, iii, and iv.

Eventually, at every position of the template after the primer, a chain-terminating nucleotide will have been added to some chain. So the sample will contain chains of different lengths, from the primer length to the whole template length, and the last nucleotide of each fragment is marked with a dye that corresponds to that nucleotide. This idea is so simple, and so brilliant.

The last step is to sort these fragments by length and read the color of the terminal dyes. One approach is to run the sample through a long, thin tube filled with a gel. The longer the fragment, the longer the transit time through the gel. This method is called *gel electrophoresis*. Then the end of the tube is illuminated with a laser and the color of the dye is recorded.

A cartoon example: suppose the template is *AATCGCCGTAT* and the primer is *AATC*. Of course we don't need to sequence the primer, because we already know it. Sanger's method will produce these fragments:

AATCG *AATCGC* *AATCGCC* *AATCGCCG*
AATCGCCGT *AATCGCCGTA* *AATCGCCGTAT*

The underlined letters represent chain-terminating nucleotides. The shortest fragment is the first through the tube, which is signaled by the first observed fluorescence being purple. Next comes blue, because the next-to-the-shortest fragment reaches the end of the tube next, and so on.

Sequencing all 3 billion base pairs of the human genome is a bit more involved, because Sanger's approach will work for templates of length up to about 900. But with cleverness and patience, and the work of hundreds of scientists, this is how the first human genome was sequenced.

Next generation sequencing (NGS) is an umbrella term for a variety of techniques, but all are massively parallel. The processes are carried out on such a small scale that many, many reactions can be performed on a single chip. Very roughly, NGS consists of simultaneously running millions of sequencings. Common techniques include pyrosequencing, ion semiconductor sequencing, and nanoball ligation.

Pyrosequencing was developed by Pål Nyren, Bertil Pettersson, and Mathias Uhlen [264]. After denaturing, single strands of DNA are immobilized on solid beads. The four nucleotides are added one at a time. DNA polymerase attaches the nucleotide complementary to the next strand nucleotide that is unattached to its complement. Luciferase, the enzyme that turned the fireflies of my childhood into constellations close to the ground and first induced vertigo when perspective flipped and a stroll among fireflies became a ramble through the deep dark space between the stars, tags the polymerase reaction when the complementary nucleotide is attached. Jonathan Rothberg and many coworkers

[223] showed that pyrosequencing can be performed in parallel using DNA microarrays. That is, pyrosequencing can be a basis for NGS.

Ion semiconductor sequencing was developed by Nicole Rusk [315] and coworkers. Here a microwell containing a single strand of the DNA template is filled with each of the four nucleotides, one at a time. When DNA polymerase adds the complementary nucleotide, a H^+ ion is released and detected by an ion-sensitive field-effect transistor (ISFET). Parallelism is achieved with an array of many microwells and detectors.

Sequencing by *DNA nanoballs* was developed by Radoje Drmanac [96] and coworkers, and by Gregory Porreca [288]. The DNA to be sequenced first is broken into fragments of a few hundred base pairs. The fragments are amplified by PCR and then the ends of each fragment are glued together to form a circle. By a method called *rolling circle replication* the circular piece is replicated and then untied into a single strand that contains several copies of the original fragment. This strand self-assembles into a *nanoball*, a tight clump of DNA. These nanoballs are charged and repel one another, minimizing tangling between the strands. The nanoballs are attached to a charged silicon wafer. Fluorescent-tagged nucleotides are added and high-resolution images of the color of each nanoball indicate the base attached and its location.

Many other approaches also are used. The point of brief sketches of these few is to emphasize that some problems admit several approaches. Conversations with people having different perspectives can reveal unexpected avenues.

The human genome was sequenced first in 2001. The project took a decade and cost 3 billion dollars. The papers [202, 363] that reported the result each had hundreds of coauthors. Technology developed so rapidly that fifteen years later a next-generation system could sequence a genome in a couple of days for a cost of about a thousand dollars.

By now so many genomes have been sequenced that we have amassed an enormous collection of data. New computational, statistical, and analytic tools have been developed to map this genomic data onto disease phenotypes. Along with big data science, bioinformatics has blossomed. Translational bioinformatics is the application of these techniques to translate genomic data into treatment and prevention strategies.

In [327] Tony Shen and coworkers survey some clinical applications of NGS. For example, genetic markers (*BRCA1*, *BRCA2*, *HER2*, among others) in breast cancer patients enable improved diagnostics and targeted therapies. Rather than general purpose chemotherapies, poisons that harm the whole body and that kill the tumor before it can kill the patient because the tumor grows more quickly, targeted therapies attack specific aspects of cancer cells distinct from all or most other cells. Because cancer is caused by somatic mutations (mutations that occur after conception), it is a disease of the genome. Consequently, with bioinformatics we can process the mountains of data, discover nuanced differences and similarities. An especially interesting development along these lines is the pediatric cancer genome project at the St. Jude Children's Research Hospital [350].

The Cancer Genome Atlas [51] is an assembly of a mass of sequencing data. Analysis has produced interesting results. For instance, in [360] the authors report the results of their large-scale study of the relations between tumors and evolutionary history. Genes expressed in tumor cells were dated by phylostratigraphy, to identify the most evolutionarily distant species that carries the gene. All seven types of cancer studied exhibited activation of genes that appeared in our unicellular ancestors and repression of genes from our multicellular ancestors. They identified twelve genes that mediate the interaction of unicellular and multicellular genes, and suggested these as targets for therapies. This is just one example. The atlas website [50] presents many more.

Still more examples are described in [214] and in the extensive translational bioinformatics resource [285].

Finally, we mention that sequencing and analyzing the entire genome of pathogens can detect drug sensitivities and identify the source of outbreaks. For example, whole genome analysis traced a methicillin resistant staphylococcus aureus (MRSA) outbreak in a neonatal intensive care clinic to single staff member [59].

These are promising developments at the intersection of molecular genetics, statistics, computer science, and math.

19.3 TOPOLOGICAL METHODS

We'll end with a way to visualize patterns generated by symbolic dynamics, a notion we introduced in Sect. 2.4. The geometric technique is called *iterated function systems* (IFS), developed by John Hutchinson [180] as a method to generate fractals, and popularized by Michael Barnsley [30, 31] as a way to compress images. This has become an immense field in math and computer science, a very simple way to generate some exceedingly complex images.

Here's how IFS work. Suppose $T_1, \dots, T_n$ are contractions of the plane. A function T is a *contraction* if it shrinks distances; for example, the distance between the points (x_1, y_1) and (x_2, y_2) is greater than the distance between the points $T(x_1, y_1)$ and $T(x_2, y_2)$. The central theorem of IFS is that there is a unique subset A of the plane for which

$$T_1(A) \cup \cdots T_n(A) = A \tag{19.1}$$

(Technically, we need to add the modifier "compact" to "subset." For subsets of the plane, compact means closed and bounded. Here bounded means the subset is contained inside some large circle, and closed means the complement is open. That is, for any point p not in the set, a small enough disc centered at p lies entirely outside the set.)

Another important result is that if we start with any point (x_0, y_0) and apply the functions one at a time in random order, each function applied to the point just generated,

$$(x_1, y_1) = T_{i_1}(x_0, y_0),\ (x_2, y_2) = T_{i_2}(x_1, y_1) = T_{i_2} T_{i_1}(x_0, y_0),\ \ldots$$

where the subscripts $i_1, i_2, i_3, \ldots$ are are selected randomly from $\{1, \ldots, n\}$, then the sequence of points (x_0, y_0), (x_1, y_1), $(x_2, y_2), \ldots$ fill up the set A. (Once again, there is a technical detail: the set of limit points of the sequence of the (x_i, y_i) equals A.) This method of generating fractals is called the random IFS algorithm.

But we won't use IFS to generate fractals, at least not in the usual way. Rather, we'll always use these four functions:

$$T_1(x,y) = \left(\frac{x}{2}, \frac{y}{2}\right) \qquad T_2(x,y) = \left(\frac{x}{2}, \frac{y}{2}\right) + \left(\frac{1}{2}, 0\right) \tag{19.2}$$

$$T_3(x,y) = \left(\frac{x}{2}, \frac{y}{2}\right) + \left(0, \frac{1}{2}\right) \qquad T_4(x,y) = \left(\frac{x}{2}, \frac{y}{2}\right) + \left(\frac{1}{2}, \frac{1}{2}\right)$$

These do not constitute a very interesting IFS: for these functions the shape A of Eq. (19.1) is the filled-in unit square $A = \{(x,y) : 0 \leq x \leq 1, 0 \leq y \leq 1\}$.

3	4
1	2

33	34	43	44
31	32	41	42
13	14	23	24
11	12	21	22

Figure 19.2. Addresses.

With these functions we can divide the square A into subsquares. For example, $T_1(A) = \{(x,y) : 0 \leq x \leq 1/2, 0 \leq y \leq 1/2\}$, so we can assign address 1 to the lower left subsquare. The left image of Fig. 19.2 shows the four length-1 addresses. The right image shows the sixteen length-2 addresses. For example, address 14 is assigned to the region determined by the composition $T_1(T_4(A))$: we know $T_4(A)$ is the upper right subsquare and $T_1(A)$ is the lower left, so $T_1(T_4(A))$ is the upper right corner of the lower left subsquare. Clearly, this can be continued for any length address we wish.

Also, note that most of the edges of these subsquares do not have unique addresses. For a simple example, the point $(0,0)$ has address $111\cdots = 1^{\infty}$ and the point $(1,0)$ has address 2^{∞}, so the point $(1/2,0)$ has two addresses: $2(1^{\infty})$ and $1(2^{\infty})$.

The random IFS algorithm guarantees that if we apply the T_i one at a time in random order, the points come as close as we like to every point of the unit square. But what if we apply the functions in some order that is not random? For example, take the DNA sequence for amylase. Read the sequence in order and apply T_1 when we encounter C, T_2 for A, T_3 for T, and T_4 for G. We call this image a *driven IFS*. This produces the first image of Fig. 19.3, the IFS *driven* by the amylase sequence. Then the address of a length-n square that contains a driven IFS point corresponds to the last n bases read to plot that point.

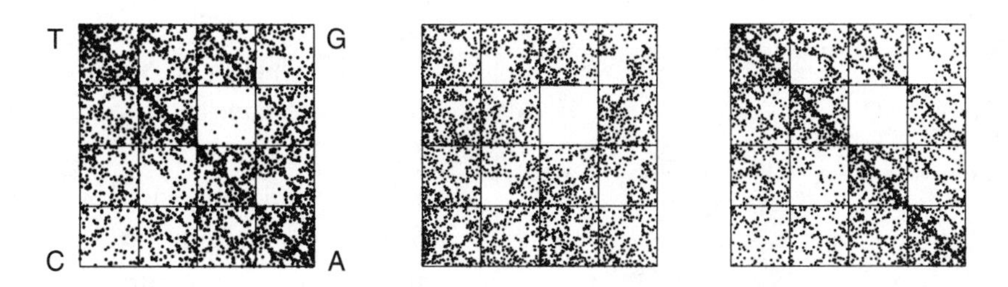

Figure 19.3. Left: IFS driven by the amylase DNA sequence. Middle and right: Surrogates of the amylase-driven IFS.

One of the most obvious features of the amylase-driven IFS is that address 41 is almost empty. This is because in the amylase sequence, only rarely does G immediately follow C. And because 41 is almost empty, few points lie in every address containing 41. These include 141, 241, 341, and 441. The middle plot of Fig. 19.3 contains the same number of points, 3,957, as the left image; the functions (19.2) are applied uniformly randomly, except that T_4 never immediately follows T_1. With this single condition, the middle plot looks reasonably close to the left plot. The most obvious difference is that the middle plot misses the strong diagonal between the lower right (A) and upper left (T) corners. A source of this problem is that the middle plot has about the same number of points in each of the four address length 1 subsquares, while the amylase sequence has 589 Cs, 1,305 As, 1,389 Ts, and 674 Gs. The right plot is produced by never applying T_4 immediately after T_1 and the transformations are applied with probability $p_1 = 589/3957 \approx 0.149$, T_2 with $p_2 \approx 0.330$, T_3 with $p_3 \approx 0.351$, and T_4 with $p_4 \approx 0.170$. The right plot comes close to matching the strong diagonal of the amylase plot.

Close, but not quite right. A lesson from this technique is that some of the most obvious features of the driven IFS plot result from very simple patterns in the driving sequence. We can filter out those features and focus on the remaining differences. This is the potential strength of driven IFS: it is a quick visual way to identify some important and easily explained patterns, and consequently to identify more subtle patterns whose source may be closer to the dynamics of the biological system.

The use of driven IFS to study DNA patterns was pioneered by H. Joel Jeffrey [182, 183]. Jeffrey saw driven IFS as a way to recognize forbidden or unlikely strings through the empty or nearly empty subsquares whose addresses are those strings. In particular, complicated empty regions that are not the result of a single forbidden string are easy to see in the driven IFS but would be difficult to detect by standard statistical methods.

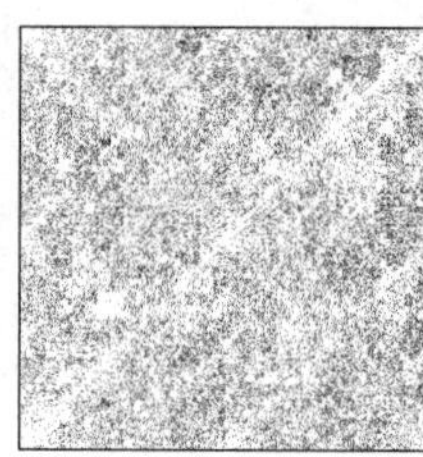

Figure 19.4. Bacteriophage-driven IFS.

While applications of driven IFS to analyze nucleotide sequences in DNA and amino acid sequences in proteins began slowly, work has grown steadily and has seen a few surprising developments. For example, in Fig. 19.2 we see that the natural squares that correspond to specific length-n addresses have side length $1/2^n$. The square with lower left corner $(1/4, 1/4)$ and side length $1/2^4$ has address 1411, for instance. In [8] we see that division of the unit square into subsquares of side length not of the form $1/2^n$ can reveal information about sequence redundancy. Observe that any horizontal or vertical line in the unit square is the limit of the image of a side of the square under sequences of transformations. For example, consider the repeating infinite sequence of transformations $(T_1(T_1(T_1(T_2))))^n$ for $n = 1, 2, 3, \ldots$ If we start at any point $(1, y)$, summing the geometric sequences we see these converge to $(1/15, 0)$. Replacing T_1 by T_3 and T_2 by T_4 the corresponding sequence converges to $(1/15, 1)$. Appropriately mixing T_1 and T_3, and appropriately mixing T_2 and T_4, gives a series that converges to $(1/15, y)$ for any y. We can recognize the edges of any square with rational corners by finite or limits of infinite sequences of transformations. But subtle issues are involved with these limits, and this is where sequence redundancy can be seen. Jonas Almeida's review [7] is a good survey of the first quarter-century of driven IFS study of DNA and of proteins.

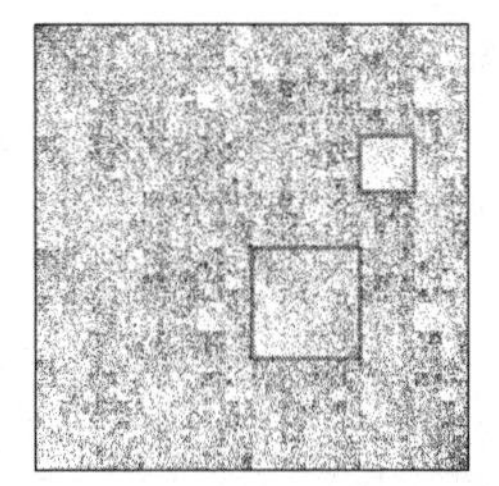
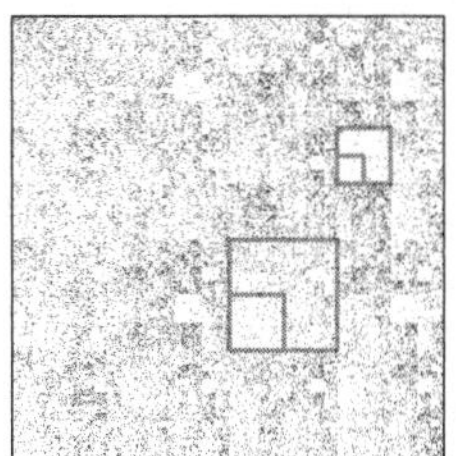

Figure 19.5. Bacteriophage addresses.

From GenBank we can find the complete genomes of all sorts of things. For example, the left image of Fig. 19.4 is the driven IFS for Enterobacteria phage lambda (GenBank: J02459.1), and the right image is the driven IFS for bacteriophage T7 (GenBank: V01146.1). The length of the first is 48,502; the length of the second is 39,937. The sources of these genomes are GenBank sites [133].

On the left of Fig. 19.5 we see the left image of Fig. 19.4 with boxes outlining the subsquares with addresses 23 and 423. The square 423 corresponds to the string T, A, G. Note the orders. The left-most address symbol corresponds to the last transformation applied, and so too the right-most entry of the string.

The subsquare with address 423 is less well filled than its neighbors. Every point in this subsquare results from applying T_4 to points in the subsquare with address 23, the larger subsquare outlined in the left plot of Fig. 19.5. The larger square has far more points than the smaller, so most of the points in 23 appear to go to 123, 223, or 323.

The right image of Fig. 19.5 is the IFS driven by the first half of the genome producing the left IFS. We see far fewer points in address 432 than in the left plot, so the composition $T_4(T_3(T_2))$ is more likely in the second half of the genome than in the first half. The driven IFS suggests what to study; then we can surgically isolate the portion of the genome that contains the subsequence and plot the IFS driven by that portion. The isolation process can be automated by noting the genome locations where each occurrence of the subsequence starts. The target portion of the genome lies between the start of the first occurrence of the subsequence and the subsequence length plus the start of the last occurrence.

To understand the most obvious feature of the right image of Fig. 19.4 first we must discuss IFS and lines. From Eq. (19.1) we see that the IFS $\{T_1, T_2\}$ generates the line between $(0,0)$ and $(1,0)$. Similarly, the other five IFS $\{T_i, T_j\}$ generate the other three sides and the two diagonals of the square. In particular, the IFS $\{T_1, T_4\}$ generates the diagonal between $(0,0)$ and $(1,1)$. Then in the right plot of Fig. 19.4 the almost empty diagonal between $(0,0)$ and $(1,1)$ means that long strings that consist of just C and G are rare, because from any point in the unit square, a long string of C and G generates a sequence of points that converge quickly to this diagonal. Note that C and G occur often, because the square with address 1 and the square with address 4 have many points. The light diagonal means most occurrences of C and of G are followed by A or T.

In general, departures from uniform fill of the unit square tell us something about the sequence driving the IFS. The trick is to learn to read the story.

Can this method be applied to a sequence $x_1, \dots, x_n$ of numerical measurements? Certainly, if we first divide the range $[\min\{x_i\}, \max\{x_i\}]$ into four bins, one for each T_i, with numbers B_1, B_2, and B_3, the *bin boundaries*. To drive the IFS, for each data point x_i apply

$$T_1 \text{ if } x_i < B_1 \qquad T_2 \text{ if } B_1 \le x_i < B_2$$
$$T_3 \text{ if } B_2 \le x_i < B_3 \qquad T_4 \text{ if } B_3 \le x_i$$

Without special information, we have three clear choices for bin boundaries. For these, $R = \max\{x_i\} - \min\{x_i\}$. Then we have:
Equal size. The boundaries are

$$B_1 = \min\{x_i\} + R/4, \quad B_2 = B_1 + R/4, \quad \text{and} \quad B_3 = B_2 + R/4$$

Equal weight. The B_i are selected so that about the same number of data points lie in each of the four bins.
Median centered. Take B_2 to be the median of the sequence (the middle value when the sequence is sorted in increasing order), $B_1 = B_2 - \delta \cdot R$, and $B_3 = B_2 + \delta \cdot R$. Here δ is an adjustable parameter that lies between 0 and 1. So median centered bins are a family of bin choices.

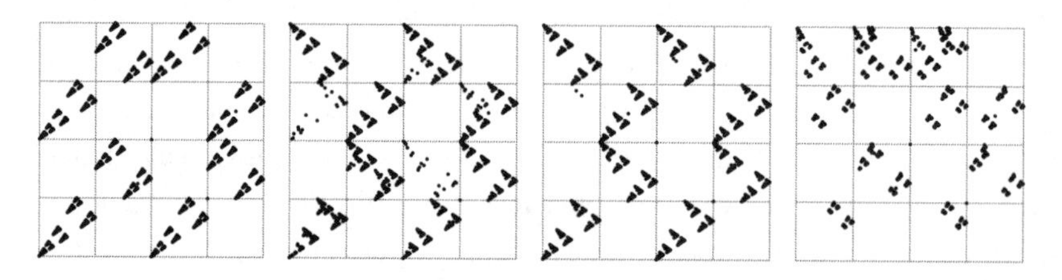

Figure 19.6. IFS generated by logistic and tent map iterates.

A natural way to generate data sequences is to iterate a function, $x_{i+1} = f(x_i)$, say, where f is a logistic map or a tent map. In Fig. 19.6 we see IFS plots for (left to right)

the $r = 4$ logistic map with equal size bins,
the $r = 4$ logistic map with equal weight bins,
the $r = 2$ tent map with equal size bins, and
the $r = 3.95$ logistic map with equal size bins.

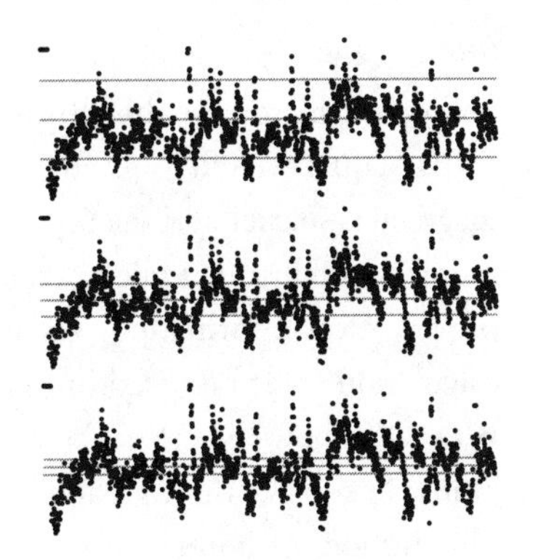

Figure 19.7. Cardiac data with different bin bondaries.

The similarities of the second and third plots are not so hard to understand with the notion of a Markov partition (pgs. 380–383 of [120] for example), but this is too far afield for our analysis now. In some sense, the first and third plots are very simple: all the empty addresses contain an empty length-2 address; the fourth image is much more complex. Driven IFS give a way to quantify the complexity of symbolic dynamics patterns.

In Fig. 19.7 we see a sequence of successive intervals between heartbeats. Because these measurements are ordered by time, this is a time series. In the top we see equal size bins; in the middle, equal weight bins; and in the bottom image median centered bins with $\delta = 0.05$.

None of these points depart very far from the others, but if some values are much higher or much lower than the others, then with equal size bins, the top bin or the bottom bin might have very few points; this condition pushes most of the points into three or maybe only two bins, reducing the utility of the driven IFS. In such instances, equal weight bins give a more instructive plot.

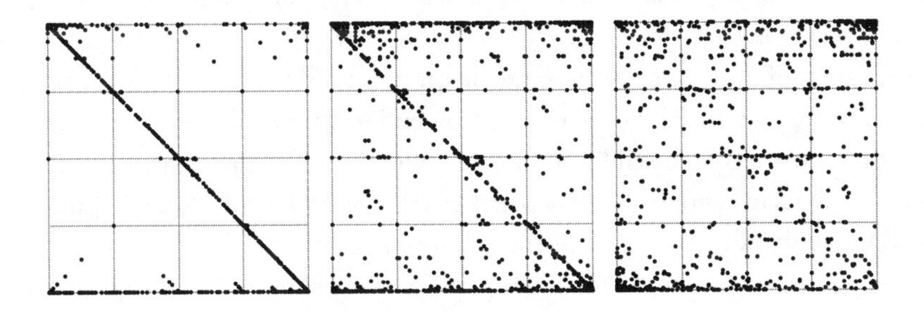

Figure 19.8. Equal size, equal weight, and median centered driven IFS.

In Fig. 19.8 we plot the IFS driven by this cardiac data, with equal size (left), equal weight (middle), and median centered (right) bins. For equal size bins, if most changes are relatively small, points usually lie on the backward Z we see in that plot. That the diagonal is the most completely filled part of the plot is a consequence of the fact that most of the data points lie in bins 2 and 3. The faint echoes of the diagonal in square 4 are the images of points on the diagonal under one or several applications of T_4. Other than the distribution of the points along the top of the square, this driven IFS has little else to reveal.

For equal weight bins, the driven IFS still shows a backward Z, but the diagonal is sparser and the top is more filled in than in the left plot. Because this plot is generally more filled in than the equal size plot, the equal weight plot may provide more information. The plot will guide our investigation, but any serious calculation must be based on the number of points in each subsquare. In Sect. B.24 we'll show code to generate these driven IFS plots and count the number of points in squares through address length 3. Here's a sample of what we can do. The number of points with address 32 is $N(32) = 89$. Every point in address 32 must go to one of address 132, 232, 332, or 432. We count $N(132) = 7$, $N(232) = 32$, $N(332) = 38$, and $N(432) = 12$. From this we deduce that

$$P(132|32) = \frac{7}{89} \approx 0.079 \qquad P(232|32) \approx 0.360$$

$$P(332|32) \approx 0.427 \qquad P(432|32) \approx 0.135$$

So, for example, when an interbeat interval slightly below the median is followed by an interval slightly above the median, the next interval is more likely to be slightly above the median again.

In the median centered plot on the right of Fig. 19.8 one apparent feature is the line of points along the top edge of square 23 or along the bottom edge of square 41. Can we tell which from the plot? These points come from the bottom edge of square 1 by an application of T_4, or from the top edge of square 3 by an application of T_2. Comparison of the density of the fills suggests the former. That is, a string of intervals below the median, ending in several more than $\delta \cdot R$ below the median, is followed by an interval more than $\delta \cdot R$ above the median.

Because median centered bins are parameterized by δ, how the density of the plots varies with δ can give more nuanced information.

If we perform this analysis for other patients, will we find similar patterns? Do patients with circulatory or cardiac ailments present different patterns from healthy patients? So far as I know, this has not been investigated.

We'll mention one more application of IFS driven by cardiac time series, an approach developed by Noelle Driver in her 2014 applied math senior thesis [95] at Yale. The idea is to find how much of the behavior of a time series can be explained by simple processes. Because we'll use networks of logistic maps and we may want to assign different r-values to different maps, we'll write $L_r(x) = r \cdot x \cdot (1 - x)$; that is, we've added the subscript r to the function name. For example, a network of three coupled logistic maps is

$$
\begin{aligned}
x_{n+1} &= c_{11} \cdot L_{r_1}(x_n) + c_{12} \cdot L_{r_2}(y_n) + c_{13} \cdot L_{r_3}(z_n) \\
y_{n+1} &= c_{21} \cdot L_{r_1}(x_n) + c_{22} \cdot L_{r_2}(y_n) + c_{23} \cdot L_{r_3}(z_n) \\
z_{n+1} &= c_{31} \cdot L_{r_1}(x_n) + c_{32} \cdot L_{r_2}(y_n) + c_{33} \cdot L_{r_3}(z_n)
\end{aligned}
\tag{19.3}
$$

To guarantee that all the variables stay in the range $[0,1]$, we put

$$c_{11} + c_{12} + c_{13} = 1, \; c_{21} + c_{22} + c_{23} = 1, \text{ and } c_{31} + c_{32} + c_{33} = 1$$

We'll build a dictionary of coupled maps and compare the driven IFS of a time series with those of the dictionary entries. To compare time series, we'll use differences of address occupancies. For each dictionary entry, we first adjust the data bin boundaries so the length-1 address occupancies match, as closely as possible, the length-1 address occupancies of the dictionary entry. Then we denote by X_{ij} the number of data-driven IFS entries with address ij, and by D_{ij} the number of dictionary entries with this address. The 2-*address correlation* is

$$\kappa_2(X,D) = \frac{N - \frac{1}{2}\sum_{i,j=1}^{4} |X_{ij} - D_{ij}|}{N}$$

We employ the heuristic that we ignore any address with fewer than five points because we want to describe the bulk behavior of the system and not rare transitions that may be artifacts of the bin boundary placement. The correlations $\kappa_3(X,D)$, $\kappa_4(X,D)$, and $\kappa_5(X,D)$ are defined similarly.

For a given time series X, the dictionary comparison tool calculates $\kappa_2(X,D)$ through $\kappa_5(X,D)$ for each dictionary entry D. The best matches are determined by these address correlations.

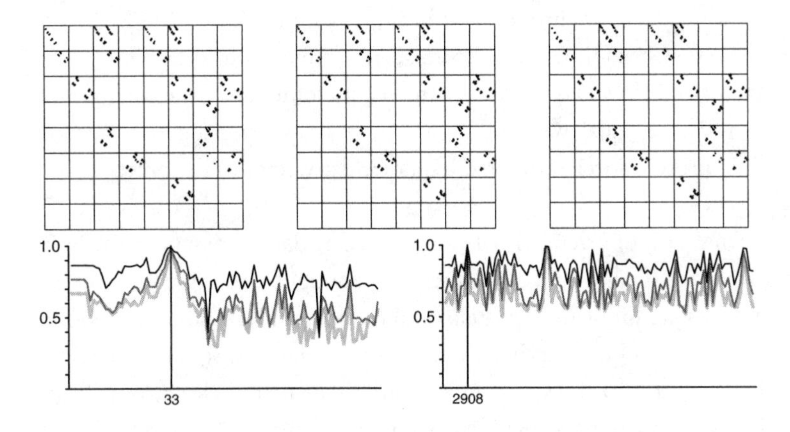

Figure 19.9. Top left: the test-driven IFS. Top middle and left: two matches from the dictionary. Bottom: graphs of κ_2, κ_3, and κ_4 for the portions of the dictionary giving these matches.

As a test, take the time series generated by the system (19.3) with

$$r_1 = 3.754 \quad c_{11} = 0.3 \quad c_{12} = 0.4 \quad c_{13} = 0.3$$
$$r_2 = 3.985 \quad c_{21} = 0.4 \quad c_{22} = 0.4 \quad c_{23} = 0.2$$
$$r_3 = 3.971 \quad c_{31} = 0.3 \quad c_{32} = 0.2 \quad c_{33} = 0.5$$

The top left image of Fig. 19.9 is the driven IFS for the average values, $(x_n + y_n + z_n)/3$. None of the dictionary entries are generated by this test system because all the dictionary entries use simple symmetric coupling determined by a single parameter c: $c_{11} = c_{22} = c_{33} = 1 - c$ and all other $c_{ij} = c/2$. The dictionary comparison tool found several close matches, for example, a single $r = 3.9$ logistic map (the middle top picture of Fig. 19.9), and two coupled logistic maps with $r_1 = 3.94$, $r_2 = 3.86$, and $c = 0.5$ (the right top picture).

To illustrate how these were selected, the bottom graphs of Fig. 19.9 plot κ_2, κ_3, and κ_4, shown with the black, dark gray, and light gray curves, for dictionary entries 1 through 100 and 2901 through 3000. The top middle picture was generated by dictionary entry 33, the top right by dictionary entry 2908.

That the test series driven IFS matches that of a single logistic map suggests that the test logistic maps have synchronized. And the dictionary entry 2908 pair of logistic maps have synchronized to the same $r = 3.9$ logistic map. One interpretation is that the driven IFS statistics stratify the dictionary parameter space. If this coarse-graining of behavior holds more generally, we have some reason to hope that this method will be effective in the study of necessarily noisy experimental data.

The first test against cardiac data revealed a good match, with $\kappa \approx 0.73$, for four uncoupled chaotic maps, suggesting that the heart cells act in four autonomous groups, all synchronized within their own group.

While these first steps show some promise, much more work remains.

Our final example involves sequences ordered by something other than time. In Fig. 19.10 we see equal weight IFS driven by mass spectrograph data of amniotic fluid protein species, ordered by molecular weight of the proteins, for sixteen patients.

The patterns fall into several visual categories. The data were sent to me by a researcher from the Yale School of Medicine. He sought a correlation between the distribution of amniotic fluid proteins and the onset of cervical cancer. Sadly, our correspondence did not survive several server migrations. It may be backed up on one of about 400 discs in a box on the floor of my closet, but likely I'll never know. And I don't know if the study was completed or moved off in a different direction. With no analysis to present, I admit that I included this example because the plots exhibit such distinct structures, and a mystery for the conclusion has some appeal.

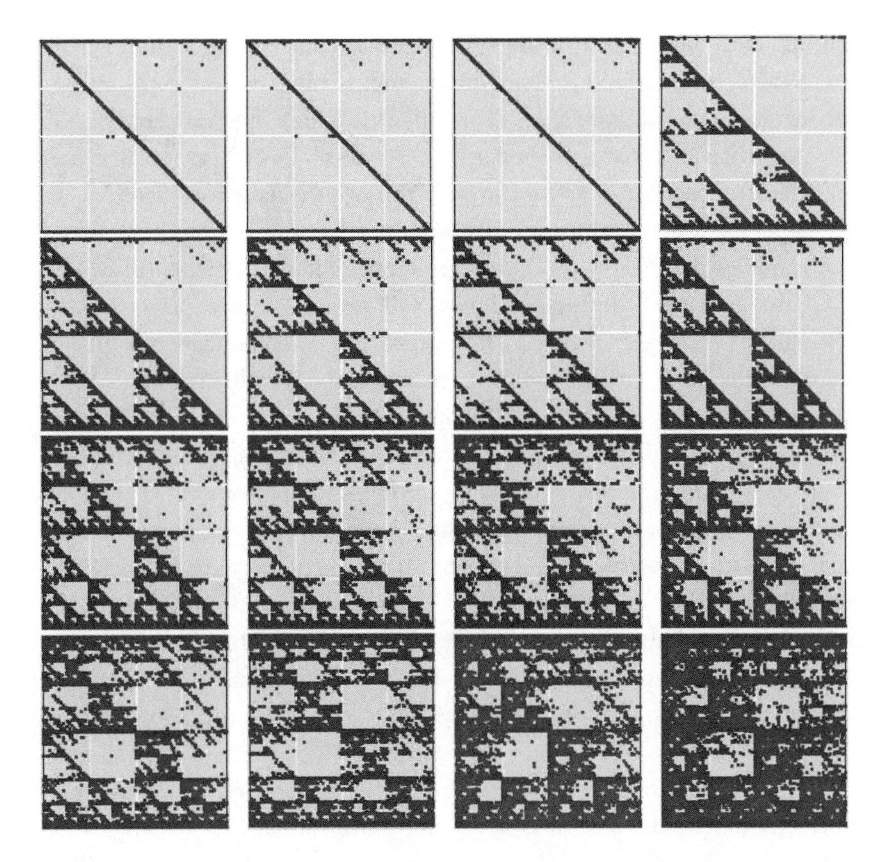

Figure 19.10. Equal weight IFS driven by molecular weights.

Driven IFS is a way to see patterns of symbol strings, that is, a way to visualize symbolic dynamics. This method is fairly young and has not yet been tested thoroughly.

19.4 NO, REALLY, WHAT'S NEXT?

The true answer is that I don't have a clue. But I've got a guess, a hope. I'll end by describing that guess.

So far, my life has had only ordinary problems for someone my age: death and illness of some beloved family members have been the main heartbreaks. But I'm still here, spending my days thinking about geometry, talking with my wife, and taking care of cats. So maybe it isn't such a surprise that I'm an optimist about many things. What I'm most optimistic about, what makes me smile as my thoughts drift up into the night sky after a long evening's work with a cat or two curled up beside my laptop, is the power of curiosity. My first memories are of wondering about the sky. I hope my last thought will be, "Oh, that's interesting. I wonder how it works." Doesn't so much matter if it's a new geometrical construction, or if through the haze of Alzheimer's I'm contemplating a button on my shirt. Wondering, wanting to know how and why, is one of the most, I think the very most, profound experiencse we can have. This is what will save us, if anything will.

My guess is that we are seeing the first steps of a period of brilliant new growth in both biology and mathematics. These fields will speak with one another more often, and more deeply, than before. Biology is asking beautiful questions, many informed by our understanding of genetics and of networks, that mathematics cannot answer. My tribe doesn't do a lot of things well. Most of us aren't good company in casual conversation, many have no idea why a hammer has a claw, some think separating two eggs means to place the eggs on different shelves in the refrigerator. But we do have an intense sense of curiosity; we love puzzles, and biology is posing some puzzles of a depth we've not seen from it before. A lot of biologists and mathematicians know this, and are working on the boundary between biology, math, computer science, and statistics. I expect some halting, complicated progress, then an elegant reformulation of question and solution, new math and new biology. Brilliant applications of these techniques to other, apparently unrelated, or perhaps newly developed, fields. New paradigms for clinicians.

My guess is revolutions in math and biology, whole new worlds, unseen by us now, familiar to those who come next.

I'm old, so I won't see these revolutions. But likely you will. While in general I don't envy you, coming of age in a politically uncertain world running unchecked into climatological disaster orchestrated by greed and stupidity, in this regard I do envy you: you'll get to explore these next worlds uncovered by our curiosity about biology and math. Keep your eyes and mind open. Wonders unimagined await.

Appendix A Technical Notes

Here we'll see some details that are too complicated, or would have taken us too far afield, to include in the main text. Each section of the notes references the sections of the book in which the appendix section'it

A.12 SOME MORE LINEAR ALGEBRA

Here we'll organize and extend a few notions that we've used and present a tiny bit of the underlying theory to support the linear algebra used in Sects. 13.1, 13.9, 14.1, 14.2, and 14.4.

A.12.1 MATRIX EXPONENTIATION

Here we'll see why the series

$$e^B = I + B + \frac{1}{2!}B^2 + \frac{1}{3!}B^3 + \cdots$$

converges for an $n \times n$ matrix $B = [B_{ij}]$. Let $b = \max\{|B_{ij}|\}$. Then

$$|(B^2)_{ij}| \leq \left|\sum_{k=1}^{n} B_{ik}B_{kj}\right| \leq \sum_{k=1}^{n}\left|B_{ik}B_{kj}\right| = \sum_{k=1}^{n}\left|B_{ik}\right|\left|B_{kj}\right| \leq \sum_{k=1}^{n} b^2 = nb^2$$

$$|(B^3)_{ij}| \leq \left|\sum_{k,m=1}^{n} B_{ik}B_{km}B_{mj}\right| \leq \sum_{k,m=1}^{n} b^3 = n^2 b^3$$

and in general $|(B^q)_{ij}| \le n^{q-1}b^q$. Now

$$\left|\left(e^B\right)_{ij}\right| \le \left|\sum_{q=0}^{\infty}\frac{(B^q)_{ij}}{q!}\right| \le \sum_{q=0}^{\infty}\frac{|(B^q)_{ij}|}{q!} \le \sum_{q=0}^{\infty}\frac{n^{q-1}b^q}{q!} \le \sum_{q=0}^{\infty}\frac{(nb)^q}{q!} = e^{nb}$$

That is, the series that defines the (i,j) term of $|e^B|$ is bounded above by a series that converges, specifically, the series that converges to e^{nb}. Then the Comparison Test implies the series that defines each term of e^B converges absolutely.

A.12.2 LINEAR DIFFERENTIAL EQUATIONS, AGAIN

Here we'll apply the notion of matrix exponentiation from Sect. 16.2 to solve the linear differential equations $\vec{v}\,' = M\vec{v}$ of Chapter 8. First, we establish the matrix version of $(e^{at})' = ae^{at}$.

Lemma A.12.1. For an $n \times n$ matrix M we have $\left(e^{Mt}\right)' = Me^{Mt}$.

Proof. This is a straightforward calculation based on the definition of the derivative and the series definition of e^{Mt}.

$$\left(e^{Mt}\right)' = \lim_{h\to 0}\frac{e^{M(h+t)} - e^{Mt}}{h} = \lim_{h\to 0}\frac{e^{Mh}e^{Mt} - e^{Mt}}{h} = \lim_{h\to 0}\left(\frac{e^{Mh} - I}{h}\right)e^{Mt}$$

$$= \lim_{h\to 0}\left(\frac{(I + Mh + M^2h^2/2 + M^3h^3/3! + \cdots) - I}{h}\right)e^{Mt}$$

$$= \lim_{h\to 0}\left(M + \frac{M^2h}{2} + \frac{M^3h^2}{3!} + \cdots\right)e^{Mt} = Me^{Mt}$$

The only place in this calculation where some care is required is the claim $e^{Mh+Mt} = e^{Mh}e^{Mt}$. This is a consequence of

$$e^{A+B} = e^Ae^B \quad \text{if } AB = BA \tag{A.1}$$

Here's why this is true. By the binomial expansion and the condition $AB = BA$,

$$(A+B)^n = \sum_{i=0}^{n}\binom{n}{i}A^iB^{n-i} = \sum_{i=0}^{n}\frac{n!}{i!(n-i)!}A^iB^{n-i}$$

Then

$$e^{A+B} = \sum_{n=0}^{\infty}\frac{(A+B)^n}{n!} = \sum_{n=0}^{\infty}\sum_{i=0}^{n}\frac{((n!)/(i!(n-i)!))A^iB^{n-i}}{n!}$$

$$= \sum_{n=0}^{\infty}\sum_{i=0}^{n}\frac{A^i}{i!}\frac{B^{n-i}}{(n-i)!} = \sum_{i=0}^{\infty}\frac{A^i}{i!}\sum_{j=0}^{\infty}\frac{B^j}{j!} = e^Ae^B$$

To see the penultimate equality, note that the first first few terms of the left side of that equality are

$$\underbrace{\frac{A^0}{0!}\frac{B^0}{0!}}_{n=0}+\underbrace{\left(\frac{A^0}{0!}\frac{B^1}{1!}+\frac{A^1}{1!}\frac{B^0}{0!}\right)}_{n=1}+\underbrace{\left(\frac{A^0}{0!}\frac{B^2}{2!}+\frac{A^1}{1!}\frac{B^1}{1!}+\frac{A^2}{2!}\frac{B^0}{0!}\right)}_{n=2}+\cdots$$

and the first few of the right side of that equality are

$$\begin{aligned}&\left(\frac{A^0}{0!}+\frac{A^1}{1!}+\frac{A^2}{2!}+\cdots\right)\left(\frac{B^0}{0!}+\frac{B^1}{1!}+\frac{B^2}{2!}+\cdots\right)\\&=\frac{A^0}{0!}\left(\frac{B^0}{0!}+\frac{B^1}{1!}+\frac{B^2}{2!}+\cdots\right)+\frac{A^1}{1!}\left(\frac{B^0}{0!}+\frac{B^1}{1!}+\frac{B^2}{2!}+\cdots\right)\\&+\frac{A^2}{2!}\left(\frac{B^0}{0!}+\frac{B^1}{1!}+\frac{B^2}{2!}+\cdots\right)+\cdots\\&=\frac{A^0}{0!}\frac{B^0}{0!}+\left(\frac{A^0}{0!}\frac{B^1}{1!}+\frac{A^1}{1!}\frac{B^0}{0!}\right)+\left(\frac{A^0}{0!}\frac{B^2}{2!}+\frac{A^1}{1!}\frac{B^1}{1!}+\frac{A^2}{2!}\frac{B^0}{0!}\right)+\cdots\end{aligned}$$

With this we see that if $AB=BA$ then $e^{A+B}=e^Ae^B$. □

Now we can solve the differential equation $\vec{v}\,'=M\vec{v}$.

Theorem A.12.1. The differential equation $\vec{v}\,'=M\vec{v}$ with the condition $\vec{v}(0)=\vec{w}$ has a solution $\vec{v}(t)=e^{Mt}\vec{w}$, and this solution is unique.

Proof. By Lemma A.12.1,

$$(\vec{v}(t))'=\left(e^{Mt}\vec{w}\right)'=\left(e^{Mt}\right)'\vec{w}=Me^{Mt}\vec{w}=M\vec{v}(t)$$

and $\vec{v}(0)=\vec{w}$. That is, $\vec{v}(t)=e^{Mt}\vec{w}$ is a solution of $\vec{v}\,'=M\vec{v}$. To see this is the only solution, suppose $\vec{u}(t)$ is another solution, that is, $\vec{u}\,'=M\vec{u}$ and $\vec{u}(0)=\vec{w}$. Take $\vec{z}(t)=e^{-Mt}\vec{u}(t)$. Then by the product rule $(\vec{z})\,'=$

$$\left(e^{-Mt}\right)'\vec{u}+e^{-Mt}\left(\vec{u}\right)'=-Me^{-Mt}\vec{u}+e^{-Mt}M\vec{u}=\left(-M+M\right)e^{-Mt}\vec{u}=\vec{0}$$

where the penultimate equality uses $e^{-Mt}M=Me^{-Mt}$, easy to verify with the series expansion of e^{-Mt}. Consequently, $\vec{z}(t)$ is constant. Because $\vec{z}(0)=e^{-M0}\vec{u}(0)=I\vec{w}=\vec{w}$, we see $\vec{w}=e^{-Mt}\vec{u}(t)$. Now take $B=-M$ in A.1 to see $e^{-M}=\left(e^M\right)^{-1}$. Consequently, $\vec{u}(t)=e^{Mt}\vec{w}$ and the solution is unique. □

Let's compare this with case 2 of Sect. 8.5. There

$$M=\begin{bmatrix}1&0\\0&2\end{bmatrix},\quad\text{so}\quad e^{Mt}=\begin{bmatrix}e^t&0\\0&e^{2t}\end{bmatrix}$$

and by Theorem A.12.1 the solution is

$$\vec{v}(t) = \begin{bmatrix} e^t & 0 \\ 0 & e^{2t} \end{bmatrix} \begin{bmatrix} x(0) \\ y(0) \end{bmatrix} = \begin{bmatrix} x(0)e^t \\ y(0)e^{2t} \end{bmatrix} = x(0)e^t \begin{bmatrix} 1 \\ 0 \end{bmatrix} + y(0)e^{2t} \begin{bmatrix} 0 \\ 1 \end{bmatrix}$$

We see that the solution of Theorem A.12.1 is equivalent to the solution, based on eigenvalues and eigenvectors of M, of Sect. 8.5.

Why do we bother with this approach when we already have solved these problems? The matrix exponentiation method simplifies these solutions.

If $M = \text{diag}(\lambda_1, \dots, \lambda_n)$, then $e^{Mt} = \text{diag}(e^{\lambda_1 t}, \dots, e^{\lambda_n t})$. Now by Theorem A.7.1 if the eigenvectors of M form a basis $\vec{e}_1, \dots, \vec{e}_n$ and $W = [\vec{e}_1, \dots, \vec{e}_n]$, then

$$W^{-1}MW = \text{diag}(\lambda_1, \dots, \lambda_n)$$

and so

$$e^{Mt} = We^{\text{diag}(\lambda_1, \dots, \lambda_n)t}W^{-1} = W\text{diag}(e^{\lambda_1 t}, \dots, e^{\lambda_n t})W^{-1}$$

We'll extend this idea to complex eigenvalues and to repeated eigenvalues with geometric multiplicity less than their algebraic multiplicity. Proofs are given in [166]. See also [167] and [281]. First, an example.

Example A.12.1. Solve $\vec{v}\,' = M\vec{v}$ where $M = \begin{bmatrix} 1 & 2 \\ 1 & 0 \end{bmatrix}$ and $\vec{v}(0) = \begin{bmatrix} x(0) \\ y(0) \end{bmatrix}$.

The eigenvalues of M are $\lambda_1 = 2$ and $\lambda_2 = -1$ and the eigenvectors of M are $\vec{w}_1 = \begin{bmatrix} 1 \\ 1/2 \end{bmatrix}$ and $\vec{w}_2 = \begin{bmatrix} 1 \\ -1 \end{bmatrix}$. Then

$$W = \begin{bmatrix} 1 & 1 \\ 1/2 & -1 \end{bmatrix} \text{ and } W^{-1} = \begin{bmatrix} 2/3 & 2/3 \\ 1/3 & -2/3 \end{bmatrix} \text{ so } W^{-1}MW = \begin{bmatrix} 2 & 0 \\ 0 & -1 \end{bmatrix}$$

The general solution of the differential equation is

$$\begin{aligned} \vec{v}(t) = e^{Mt}\vec{v}(0) &= W \begin{bmatrix} e^{2t} & 0 \\ 0 & e^{-t} \end{bmatrix} W^{-1} \begin{bmatrix} x(0) \\ y(0) \end{bmatrix} \\ &= \begin{bmatrix} (e^{-t} + 2e^{2t})/3 & (-2e^{-t} + 2e^{2t})/3 \\ (-e^{-t} + e^{2t})/3 & (2e^{-t} + e^{2t})/3 \end{bmatrix} \begin{bmatrix} x(0) \\ y(0) \end{bmatrix} \end{aligned}$$

Substitution of the values $x(0)$ and $y(0)$ gives the particular solution. □

Suppose M has complex eigenvalues $\lambda_j = a_j \pm ib_j$ with eigenvectors $\vec{w}_j = \vec{u}_j \pm i\vec{v}_j$. With the matrix $W = [\vec{v}_1, \vec{u}_1, \dots, \vec{v}_n, \vec{u}_n]$,

$$W^{-1}MW = \text{diag}\left(\begin{bmatrix} a_j & -b_j \\ b_j & a_j \end{bmatrix} \right)$$

and so

$$e^{Mt} = W\text{diag}\left(e^{a_j t}\begin{bmatrix}\cos(b_j t) & -\sin(b_j t)\\ \sin(b_j t) & \cos(b_j t)\end{bmatrix}\right)W^{-1}$$

Example A.12.2. Solve $\vec{v}\,' = M\vec{v}$ where $M = \begin{bmatrix}1 & -1\\ 4 & 1\end{bmatrix}$ and $\vec{v}(0) = \begin{bmatrix}x(0)\\ y(0)\end{bmatrix}$.

The eigenvalues of M are $\lambda_1 = 1+2i$ and $\lambda_2 = 1-2i$; an eigenvector of λ_1 is $\vec{w}_1 = \vec{u}_1 + i\vec{v}_1 = \begin{bmatrix}1\\ 0\end{bmatrix} + i\begin{bmatrix}0\\ -2\end{bmatrix}$. Then

$$W = \begin{bmatrix}0 & 1\\ -2 & 0\end{bmatrix} \text{ and } W^{-1} = \begin{bmatrix}0 & -1/2\\ 1 & 0\end{bmatrix} \text{ so } W^{-1}MW = \begin{bmatrix}1 & -2\\ 2 & 1\end{bmatrix}$$

then $a = 1$ and $b = 2$. The general solution is

$$\begin{aligned}\vec{v}(t) = e^{Mt}\vec{v}(0) &= W\begin{bmatrix}e^t\cos(2t) & -e^t\sin(2t)\\ e^t\sin(2t) & e^t\cos(2t)\end{bmatrix}W^{-1}\begin{bmatrix}x(0)\\ y(0)\end{bmatrix}\\ &= \begin{bmatrix}e^t\cos(2t) & -(e^t/2)\sin(2t)\\ 2e^t\sin(2t) & e^t\cos(2t)\end{bmatrix}\begin{bmatrix}x(0)\\ y(0)\end{bmatrix}\end{aligned}$$

The $\cos(2t)$ and $\sin(2t)$ factors with different coefficients mean elliptical motion; the e^t factor means trajectories spiral away from the origin. □

Now suppose an $n \times n$ matrix M has eigenvalues $\lambda_1, \dots, \lambda_n$, where in this list the number of times each eigenvalue appears equals its algebraic multiplicity. If for each eigenvalue the geometric multiplicity equals the algebraic multiplicity, then the matrix W whose columns are the eigenvectors is invertible and we can apply the approach of Example A.12.1. If for some eigenvalues the algebraic multiplicity exceeds the geometric multiplicity, we must try something different.

For this we need a few preliminary notions. A matrix N is *nilpotent* of order k if k is the smallest integer with $N^k = 0$. If the algebraic multiplicity of an eigenvalue is m, then for $k = 1, 2, \dots, m$ a *generalized eigenvector* of M is a non-zero solution $\vec{v}$ of $(M - \lambda I)^k\vec{v} = \vec{0}$. Note that $k = 1$ gives the familiar, non-generalized, eigenvectors.

In Appendix III of [166] Morris Hirsch and Stephen Smale prove this result. There is a basis of $\mathbb{R}^n$ consisting of generalized eigenvectors $\vec{w}_1, \dots, \vec{w}_n$, and for any such basis the matrix $W = [\vec{w}_1, \dots, \vec{w}_n]$ is invertible. Then the matrix $S = W\text{diag}(\lambda_1, \dots, \lambda_n)W^{-1}$ is diagonalizable. We write the equation this way to show how to find S once we know W. Then $N = M - S$ is nilpotent of order k and the solution of $\vec{v}\,' = M\vec{v}$ is

$$\vec{v}(t) = W\mathrm{diag}(e^{\lambda_1 t}, \dots, e^{\lambda_n t})W^{-1}\left(I + Nt + \cdots + \frac{N^{k-1}t^{k-1}}{(k-1)!}\right)\vec{v}(0)$$

Is this really any simpler than the approach we used in Sect. 8.5? An example might help you decide.

Example A.12.3. For $M = \begin{bmatrix} 1 & -1 & 0 \\ 1 & 3 & 0 \\ 2 & 1 & -1 \end{bmatrix}$ and $\vec{v}(0) = \vec{v}_0 = \begin{bmatrix} x(0) \\ y(0) \\ z(0) \end{bmatrix}$ solve the differential equation $\vec{v}\,' = M\vec{v}$.

The eigenvalues of M are $\lambda_1 = -1$ and $\lambda_2 = \lambda_3 = 2$. An eigenvector of λ_1 is $\vec{w}_1 = \begin{bmatrix} 0 \\ 0 \\ 1 \end{bmatrix}$; all eigenvectors of λ_2 are multiples of $\vec{w}_2 = \begin{bmatrix} 1 \\ -1 \\ 1/3 \end{bmatrix}$. That is, λ_2 has algebraic multiplicity 2 and geometric multiplicity 1, and so the eigenvectors of M do not constitute a basis for $\mathbb{R}^3$. We must find a generalized eigenvector $\vec{w}_3$ of λ_2 so $\{\vec{w}_1, \vec{w}_2, \vec{w}_3\}$ are linearly independent.

Because the algebraic multiplicity of λ_2 is 2, the generalized eigenvector $\vec{w}_3$ is a nonzero solution of $(M - 2I)^2\vec{w}_3 = \vec{0}$. This gives the single relation $-7x - 4y + 9z = 0$. Take $x = 1$ and $y = 0$, for example. This shows $\vec{w}_3$ independent of $\vec{w}_1$ and $\vec{w}_2$. Then

$$W = \begin{bmatrix} 0 & 1 & 1 \\ 0 & -1 & 0 \\ 1 & 1/3 & 7/9 \end{bmatrix} \text{ so } W^{-1} = \begin{bmatrix} -7/9 & -4/9 & 1 \\ 0 & -1 & 0 \\ 1 & 1 & 0 \end{bmatrix}$$

Then $S =$

$$W\begin{bmatrix} -1 & 0 & 0 \\ 0 & 2 & 0 \\ 0 & 0 & 2 \end{bmatrix}W^{-1} = \begin{bmatrix} 2 & 0 & 0 \\ 0 & 2 & 0 \\ 7/3 & 4/3 & -1 \end{bmatrix} \text{ and } N = \begin{bmatrix} -1 & -1 & 0 \\ 1 & 1 & 0 \\ -1/3 & -1/3 & 0 \end{bmatrix}$$

where $N = M - S$ is a nilpotent matrix of order 2. Finally, the general solution is

$$\vec{v}(t) = W\begin{bmatrix} e^{-t} & 0 & 0 \\ 0 & e^{2t} & 0 \\ 0 & 0 & e^{2t} \end{bmatrix}W^{-1}(I + Nt)\vec{v}(0)$$

We'd write this out, except it's pretty messy. Nevertheless, we (more precisely, Mathematica) checked that this $\vec{v}(t)$ satisfies $\vec{v}\,'(t) = A\vec{v}(t)$ and $\vec{v}(0) = \vec{v}_0$. □

For higher-dimensional systems we can find pairs of complex eigenvalues as well as real eigenvalues whose geometric multiplicities are less than their algebraic multiplicities. In this circumstance the matrix can be converted into a bunch of diagonal blocks that are similar to those we have seen in the examples. One way to do this is called the Jordan canonical form. Consult [166] or [281], or Google, for some details.

A.12.3 DECOMPOSING A MATRIX

Now we'll derive Eq. (13.50)

$$M = \lambda_u \vec{e}_u \vec{f}_u + \lambda_s \vec{e}_s \vec{f}_s \tag{A.2}$$

that we used in Sect. 13.9. Recall that $\lambda_u > 0$ is the unstable eigenvalue, $\lambda_s < 0$ is the stable eigenvalue, and $\vec{e}_u$ and $\vec{e}_s$ are unstable and stable (column) eigenvectors. Because the eigenvalues λ_u and λ_s are not equal, by Prop. A.7.2 we know the eigenvectors are linearly independent and consequently the matrix M is diagonalizable by Theorem A.7.1.

Also recall the matrices

$$W = \begin{bmatrix} \vec{e}_u & \vec{e}_s \end{bmatrix} = \begin{bmatrix} e_{u1} & e_{s1} \\ e_{u2} & e_{s2} \end{bmatrix} \quad \text{and} \quad W^{-1} \begin{bmatrix} \vec{f}_u \\ \vec{f}_s \end{bmatrix} = \begin{bmatrix} f_{u1} & f_{u2} \\ f_{s1} & f_{s2} \end{bmatrix} \tag{A.3}$$

Next,

$$\lambda_u \vec{e}_u \vec{f}_u = \lambda_u \begin{bmatrix} e_{u1} \\ e_{u2} \end{bmatrix} \begin{bmatrix} f_{u1} & f_{u2} \end{bmatrix} = \begin{bmatrix} \lambda_u e_{u1} f_{u1} & \lambda_u e_{u1} f_{u2} \\ \lambda_u e_{u2} f_{u1} & \lambda_u e_{u2} f_{u2} \end{bmatrix}$$

$$\lambda_s \vec{e}_s \vec{f}_s = \lambda_s \begin{bmatrix} e_{s1} \\ e_{s2} \end{bmatrix} \begin{bmatrix} f_{s1} & f_{s2} \end{bmatrix} = \begin{bmatrix} \lambda_s e_{s1} f_{s1} & \lambda_s e_{s1} f_{s2} \\ \lambda_s e_{s2} f_{s1} & \lambda_s e_{s2} f_{s2} \end{bmatrix}$$

Combining these,

$$\lambda_u \vec{e}_u \vec{f}_u + \lambda_s \vec{e}_s \vec{f}_s = \begin{bmatrix} \lambda_u e_{u1} f_{u1} + \lambda_s e_{s1} f_{s1} & \lambda_u e_{u1} f_{u2} + \lambda_s e_{s1} f_{s2} \\ \lambda_u e_{u2} f_{u1} + \lambda_s e_{s2} f_{s1} & \lambda_u e_{u2} f_{u2} + \lambda_s e_{s2} f_{s2} \end{bmatrix} \tag{A.4}$$

Now Eq. (A.17) can be rewritten as

$$M = W \operatorname{diag}(\lambda_u, \lambda_s) W^{-1} \tag{A.5}$$

Substituting in the expressions for W and W^{-1} from Eq. (A.3), we see that the right-hand side of Eq. (A.4) equals the right-hand side of Eq. (A.5). That is, we have shown that Eq. (13.50) is true.

A.12.4 EIGENVECTORS OF SYMMETRIC MATRICES

Here we'll derive a result about parallel eigenvectors used in Sect. 13.9. Suppose M is a symmetric 2×2 matrix with distinct eigenvalues λ_1 and λ_2 and corresponding right eigenvectors $\vec{e}_1$ and $\vec{e}_2$,

$$M\vec{e}_1 = \lambda_1\vec{e}_1 \qquad M\vec{e}_2 = \lambda_2\vec{e}_2$$

Let U be the matrix formed from the column vectors $\vec{e}_1$ and $\vec{e}_2$, and let $\vec{f}_1$ and $\vec{f}_2$ be the row vectors of U^{-1}:

$$U = \begin{bmatrix} \vec{e}_1 & \vec{e}_2 \end{bmatrix} \qquad U^{-1} = \begin{bmatrix} \vec{f}_1 \\ \vec{f}_2 \end{bmatrix}$$

We'll show that $\vec{e}_1$ is parallel to $\vec{f}_1$. Now $\vec{e}_1$ is a column vector and $\vec{f}_1$ is a row vector, so when we say they are parallel, we mean that there is a constant k for which

$$\vec{e}_{1,1} = k\vec{f}_{1,1} \quad \text{and} \quad \vec{e}_{2,1} = k\vec{f}_{2,1}$$

There are clever ways to do this, but it's important to remember that sometimes plain old calculation will suffice. This is one of those times. Write the symmetric matrix M as $M = \begin{bmatrix} a & b \\ b & c \end{bmatrix}$. The eigenvalues are

$$\lambda_1 = \frac{a+c+\sqrt{(a-c)^2+4b^2}}{2} \quad \text{and} \quad \lambda_2 = \frac{a+c-\sqrt{(a-c)^2+4b^2}}{2}$$

The eigenvectors are

$$\vec{e}_1 = \begin{bmatrix} \dfrac{a-c+\sqrt{(a-c)^2+4b^2}}{2b} \\ 1 \end{bmatrix} \quad \text{and} \quad \vec{e}_2 = \begin{bmatrix} \dfrac{a-c-\sqrt{(a-c)^2+4b^2}}{2b} \\ 1 \end{bmatrix}$$

Then

$$U^{-1} = \begin{bmatrix} \dfrac{b}{\sqrt{(a-c)^2+4b^2}} & \dfrac{-a+c+\sqrt{(a-c)^2+4b^2}}{2\sqrt{(a-c)^2+4b^2}} \\ \dfrac{-b}{\sqrt{(a-c)^2+4b^2}} & \dfrac{a-c+\sqrt{(a-c)^2+4b^2}}{2\sqrt{(a-c)^2+4b^2}} \end{bmatrix}$$

The first row is $\vec{f}_1$,

$$\vec{f}_1 = \begin{bmatrix} \dfrac{b}{\sqrt{(a-c)^2+4b^2}} & \dfrac{-a+c+\sqrt{(a-c)^2+4b^2}}{2\sqrt{(a-c)^2+4b^2}} \end{bmatrix}$$

Multiply both entries of $\vec{f_1}$ by the reciprocal of $\vec{f}_{1,2}$. With a little algebra we see that

$$\frac{2\sqrt{(a-c)^2+4b^2}}{-a+c+\sqrt{(a-c)^2+4b^2}}\vec{f}_{1,1}$$
$$=\frac{2\sqrt{(a-c)^2+4b^2}}{-a+c+\sqrt{(a-c)^2+4b^2}}\frac{b}{\sqrt{(a-c)^2+4b^2}}$$
$$=\frac{a-c+\sqrt{(a-c)^2+4b^2}}{2b}=\vec{e}_{1,1}$$

and

$$\frac{2\sqrt{(a-c)^2+4b^2}}{-a+c+\sqrt{(a-c)^2+4b^2}}\vec{f}_{1,2}$$
$$=\frac{2\sqrt{(a-c)^2+4b^2}}{-a+c+\sqrt{(a-c)^2+4b^2}}\frac{-a+c+\sqrt{(a-c)^2+4b^2}}{2\sqrt{(a-c)^2+4b^2}}=1=\vec{e}_{1,2}$$

A similar argument shows that $\vec{f_2}$ is parallel to $\vec{e}_2$.

A.12.5 RETURN MAP EIGENVECTORS

Now for use in the OGY method of Sect. 13.9 we'll consider the special case of the eigenvectors of the return map

$$\begin{bmatrix} x_{n+2} \\ x_{n+1} \end{bmatrix} = \begin{bmatrix} a & b \\ c & d \end{bmatrix} \begin{bmatrix} x_{n+1} \\ x_n \end{bmatrix}$$

Look at the second equation $x_{n+1} = cx_{n+1} + dx_n$. The most obvious parameters are $c = 1$ and $d = 0$, which give the eigenvalues and eigenvectors

$$\lambda_\pm = \frac{a \pm \sqrt{a^2+4b}}{2} \quad \text{and} \quad \vec{e}_\pm = \begin{bmatrix} \dfrac{a \pm \sqrt{a^2+4b}}{2} \\ 1 \end{bmatrix}$$

Note that the eigenvalues are real if $a^2 > -4b$; if $b > 0$ then regardless of the sign of a, λ_- is negative and λ_+ is positive. If $b < 0$ and the eigenvalues are real, then the eigenvalues have the same sign. In order to apply the OGY control method, we'll suppose a and b take values making the origin a saddle point.

The matrices W and W^{-1} are

$$W = \begin{bmatrix} \vec{e}_+ & \vec{e}_- \end{bmatrix} = \begin{bmatrix} \dfrac{a-\sqrt{a^2+4b}}{2} & \dfrac{a-\sqrt{a^2+4b}}{2} \\ 1 & 1 \end{bmatrix}$$

and

$$W^{-1} = \begin{bmatrix} \vec{f}_+ \\ \vec{f}_- \end{bmatrix} = \begin{bmatrix} \dfrac{1}{\sqrt{a^2+4b}} & -\dfrac{a}{2\sqrt{a^2+4b}} + \dfrac{1}{2} \\ -\dfrac{1}{\sqrt{a^2+4b}} & \dfrac{a}{2\sqrt{a^2+4b}} + \dfrac{1}{2} \end{bmatrix}$$

The condition that $\vec{e}_+$ is parallel to $\vec{f}_+$, viewed as a column vector, is that there is a constant k that gives $\vec{e}_{+,1} = k\vec{f}_{+,1}$ and $\vec{e}_{+,2} = k\vec{f}_{+,2}$. Substituting in the values of these vector components and solving for k, we find that $b = 1$. Because $c = 1$ for the return map system, we see that as in the general 2×2 matrix, $\vec{e}_+$ is parallel to $\vec{f}_+$, and so we use $\vec{f}_u$, rather than the more obvious choice $\vec{e}_u$, in the OGY formula, only in the case of a symmetric matrix. The special form of the return map does not give any additional justification to this choice.

A.13 SOME MARKOV CHAIN PROPERTIES

Now we'll derive a few properties of Markov chains, which we used in Sects. 14.1, 14.2, 14.3, 14.4, 16.1, 16.3, 16.4, and 16.5. The book [195] by John Kemeny and J. Laurie Snell and Chapter 15 of [357] by Henk Tijms are excellent sources.

From Sect. 14.1 recall we partition the states into two classes, the transient T, and the ergodic E. Once the state leaves the transient class, it can never re-enter T; once the state enters the ergodic class, it can never exit E.

For example, consider the Markov chain defined by the matrix M_1,

$$M_1 = \begin{bmatrix} 0.5 & 0 & 0 & 0 & 0 \\ 0 & 0.5 & 0.25 & 0.25 & 0 \\ 0.25 & 0 & 0.25 & 0 & 0.5 \\ 0 & 0.25 & 0.5 & 0.75 & 0 \\ 0.25 & 0 & 0 & 0 & 0.5 \end{bmatrix}$$

We see that $T = \{1,3,5\}$ and $E = \{2,4\}$. If this isn't clear, draw the transition graph for M_1.

To simplify our study of the long-term behavior of Markov chains, the first step is to group together the states of the ergodic class, and group together the states of the transient class. So we'll number the rows and columns this way: 2, 4, 1, 3, 5: ergodic, then transient. This gives the matrix M_2,

$$M_2 = \begin{bmatrix} 0.5 & 0.25 & 0 & 0.25 & 0 \\ 0.5 & 0.75 & 0 & 0.5 & 0 \\ 0 & 0 & 0.5 & 0 & 0 \\ 0 & 0 & 0.25 & 0.25 & 0.5 \\ 0 & 0 & 0.25 & 0 & 0.5 \end{bmatrix}$$

For example, because of the reordering of the states, row 1 of M_2 is row 2 of M_1 and column 2 of M_2 is column 4 of M_1. This gives

$$(M_2)_{12} = (M_1)_{24} = 0.25$$

The matrix M_2 can be split into blocks:

$$M_2 = \begin{bmatrix} S & R \\ 0 & Q \end{bmatrix} \text{ where } S = \begin{bmatrix} 0.5 & 0.25 \\ 0.5 & 0.75 \end{bmatrix}, \; R = \begin{bmatrix} 0 & 0.25 & 0 \\ 0 & 0.5 & 0 \end{bmatrix},$$

$$Q = \begin{bmatrix} 0.5 & 0 & 0 \\ 0.25 & 0.25 & 0.5 \\ 0.25 & 0 & 0.5 \end{bmatrix}, \text{ and } 0 = \begin{bmatrix} 0 & 0 \\ 0 & 0 \\ 0 & 0 \end{bmatrix}$$

The entries of the matrix S are the probabilities of transition among states of the ergodic class, Q the transitions among the transient states, and R the transitions from transient to ergodic. That there are no transitions from ergodic to transient states is the reason for the lower left 0 matrix of M_2.

We're interested in how long on average it takes for the system to exit the transient class and enter the ergodic class. Consequently, we're interested in what occupies the lower right corner of M_2^n. It's easy to verify that

$$\begin{bmatrix} S & R \\ 0 & Q \end{bmatrix}^2 = \begin{bmatrix} S^2 & SR+RQ \\ 0 & Q^2 \end{bmatrix}, \begin{bmatrix} S & R \\ 0 & Q \end{bmatrix}^3 = \begin{bmatrix} S^3 & S^2R+SRQ+RQ^2 \\ 0 & Q^3 \end{bmatrix}$$

and so on. In particular, the lower right corner of M_2^n is Q^n.

Recall that for every Markov chain transition matrix M, the entry $(M^n)_{ij}$ is the probability of being in state i n steps after being in state j. We'll show that $Q^n \to 0$ as $n \to \infty$ and conclude from this that eventually the transient states empty out.

Proposition A.13.1. If Q is the transition matrix for the transient states of a Markov chain, then $\lim_{n\to\infty} Q^n = 0$.

Proof. For each transient state t_i there are paths through the transition graph to ergodic states. Say the shortest of these paths has length m_i. Then if we start from t_i there is some positive probability of reaching an ergodic state in m_i iterates. Let p_i be the probability of *not* reaching an ergodic state in m_i iterates when starting from t_i, so $p_i < 1$. Now take m max$\{m_i\}$ and p max$\{p_i\}$. Then starting from every transient state, the probability of not reaching an ergodic state in m iterates is $< p$, the probability of not reaching an ergodic state in $2m$ iterates is $< p^2$, and so on. Because $p < 1$, as $k \to \infty$, $p^{km} \to 0$. That is, the probability of staying in the transient class goes to 0 and so $\lim_{n\to\infty} Q^n = 0$. □

The main application of this proposition is to construct the fundamental matrix of a Markov chain that has ergodic states.

Proposition A.13.2. If $\lim_{n\to\infty} Q^n = 0$, then $I - Q$ is invertible, and

$$(I-Q)^{-1} = I + Q + Q^2 + Q^3 + \cdots \tag{A.6}$$

Proof. First note that when the two factors on the left side of

$$(I-Q)(I+Q+Q^2+\cdots+Q^{n-1}) = I - Q^n \tag{A.7}$$

are multiplied, almost all the terms subtract out, leaving only the two terms on the right side. As $n \to \infty$ we know $I - Q^n \to I$. Because the determinant $\det(I) = 1$, for large enough n $\det(I - Q^n) \neq 0$. Now $\det(AB) = \det(A)\det(B)$, so by Eq. (A.7)

$$\det(I-Q)\det(I+Q+Q^2+\cdots+Q^{n-1}) = \det(I-Q^n) \neq 0$$

and we see that $\det(I - Q) \neq 0$. A square matrix is invertible if and only if its determinant is non-zero, so we see $(I - Q)^{-1}$ exists. Multiply both sides of Eq. (A.7) on the left by $(I - Q)^{-1}$ to obtain

$$I + Q + Q^2 + \cdots + Q^{n-1} = (I-Q)^{-1}(I-Q^n)$$

Let $n \to \infty$ and Eq. (A.6) follows. □

Write $N = (I - Q)^{-1}$. Our interest in the matrix N comes from the interpretation given in the next proposition.

Proposition A.13.3. The matrix entry N_{ij} is the expected number of times the Markov chain will be in the transient state i, given that its initial (transient) state was j.

Proof. Suppose t_i and t_j are transient states and define a random variable X by

$$X(k) = \begin{cases} 1 & \text{starting at } t_j\text{, the system is in } t_i \text{ at step } k \\ 0 & \text{starting at } t_j\text{, the system is not in } t_i \text{ at step } k \end{cases}$$

Then

$$P(X(k)=1) = (Q^k)_{ij} \quad \text{and} \quad P(X(k)=0) = 1 - (Q^k)_{ij}$$

and consequently by Eq. (11.14),

$$\mathbb{E}(X(k)) = 0 \cdot P(X(k)=0) + 1 \cdot P(X(k)=1) = (Q^k)_{ij}$$

From this we see that the expected number of times in the first n steps that, starting from t_j, the chain visits t_i is

$$\begin{aligned}\mathbb{E}(X(0)+X(1)+\cdots+X(n)) &= \mathbb{E}(X(0))+\mathbb{E}(X(1))+\cdots+\mathbb{E}(X(n))\\ &= (Q^0)_{ij}+(Q^1)_{ij}+\cdots+(Q^n)_{ij}\end{aligned}$$

where $Q^0 = I$ because if the system starts in transient state i, after 0 time steps, the system remains in state t_i. Take the $n \to \infty$ limit and we see that the expected number of times the Markov chain will be in state i, given that its initial state was j, is

$$\begin{aligned}\mathbb{E}(X(0)+X(1)+\cdots) &= (Q^0)_{ij}+(Q^1)_{ij}+\cdots\\ &= (Q^0+Q^1+Q^2+\cdots)_{ij}\\ &= ((I+Q)^{-1})_{ij} && \text{by Eq. (A.6)}\\ &= N_{ij}\end{aligned}$$

This is what we wanted to prove. □

Now suppose we start in the transient state t_i. We'd like to compute the expected number of steps until the chain enters the ergodic class. This is straightforward using the matrix N.

Proposition A.13.4. The expected number of steps before a Markov chain enters the absorbing class when starting from a transient initial state t_i is the ith entry of $\vec{1}N$, where $\vec{1}$ is the row vector with all entries 1.

Proof. The calculation when N is a 2×2 matrix suffices to illustrate the ideas.

$$\begin{bmatrix}1 & 1\end{bmatrix}\begin{bmatrix}N_{11} & N_{12}\\ N_{21} & N_{22}\end{bmatrix} = \begin{bmatrix}N_{11}+N_{21} & N_{12}+N_{22}\end{bmatrix}$$

The first entry is $N_{11}+N_{21}$, the expected number of steps the system spends in state t_1 when starting from state t_1, plus the expected number of steps the system spends in state t_2 when starting from state t_1. Because t_1 and t_2 are all the transient states, once the system departs from these, it enters the ergodic class. So $N_{11}+N_{21}$ is the expected number of steps in the transient class from the initial state t_1. The process is similar for initial state t_2. When the system exits the transient class, it enters the ergodic class. □

Eventually, iterates from the transient class land in the ergodic class. The ideas we've developed can be used to find the probability that a transient class j eventually lands in an ergodic class i. Here's how.

Proposition A.13.5. For any transient class j and ergodic class i, the probability of landing in i when starting from j is B_{ij} where

$$B = [B_{ij}] = RN \tag{A.8}$$

and where $N = (I - Q)^{-1}$ and R is the matrix giving the transition probabilities from transient states to ergodic states.

Proof. To get from j to i we could go directly, with probability R_{ij}, or for any transient state k we can go from i to k in n steps with probability $(Q^n)_{kj}$, then from k to i with probability R_{ik}. Summing over k and n we have

$$\begin{aligned} B_{ij} &= \sum_n \sum_k R_{ik}(Q^n)_{kj} = \sum_k \sum_n R_{ik}(Q^n)_{kj} \\ &= \sum_k R_{ik} \sum_n (Q^n)_{kj} = \sum_k R_{ik} N_{kj} = RN \end{aligned}$$

where the third equality follows because R_{ik} does not depend on n, and the fourth by Eq. (A.6) and $N = (I - Q)^{-1}$. □

Recall from Sect. 14.2 that a matrix M is primitive if for some n, every entry of M^n is positive. Now we'll show that the calculation of Example 14.1.1 is not a fluke.

Theorem A.13.1. Suppose the $q \times q$ transition matrix M of a Markov chain is primitive. Then $\lim_{n\to\infty} M^n = W$, where all the columns of W are equal to a vector $\vec{w}_\infty$, and all the entries of $\vec{w}_\infty$ are positive. Moreover, $\vec{w}_\infty$ is the unit eigenvector of the dominant eigenvalue of M.

Proof. We'll show that M^n converges to a matrix W with constant rows $\begin{bmatrix} c & c & \cdots & c \end{bmatrix}$; consequently, all the columns of W are equal. Now suppose $\vec{z}$ is a row vector with all entries 0, except for a single 1 in location i. Then $\vec{z}W$ is the ith row of W, so it's enough to show that for any row vector $\vec{w}$, $\lim_{n\to\infty} \vec{w}M^n$ is a constant row vector.

Because M is primitive, for some k, all entries of M^k are positive. We'll consider the limit $\lim_{j\to\infty} M^{jk}$, then show this suffices. Suppose δ is the smallest entry of M^k, a_0 and b_0 the smallest and largest entries of $\vec{w}$, and a_k and b_k the smallest and largest entries of $\vec{w}M^k$. We'll show

$$b_k - a_k \leq (1 - 2\delta)(b_0 - a_0)$$

Because $q \geq 2$ (a 1-state Markov chain is not interesting), δ is the smallest entry in any column of M^k, and the entries of each column sum to 1, we see $0 < \delta < 1/2$

and so $0 < 1 - 2\delta < 1$. That is, the difference between the largest and smallest elements of the row vectors decrease under repeated application of M^k.

Here's why this is true. The largest possible value of $\vec{w}M^k$ would occur if the largest value b_0 of $\vec{w}$ occurs in $q-1$ entries, the smallest value a_0 occurs in one entry, and in $\vec{w}M^k$ this smallest entry has the smallest possible weight, δa_0. Consequently,

$$b_k \leq (1-\delta)b_0 + \delta a_0$$

Similarly, the smallest possible value of $\vec{w}M^k$ would occur if the smallest value a_0 of $\vec{w}$ occurs in $q-1$ entries, the largest value b_0 occurs in one entry, and in $\vec{w}M^k$ this largest entry has the smallest possible weight, δb_0. Consequently,

$$a_k \geq (1-\delta)a_0 + \delta b_0$$

Combining these bounds on b_k and a_k we find

$$b_k - a_k \leq (1-\delta)b_0 + \delta a_0 - ((1-\delta)a_0 + \delta b_0) = (1-2\delta)(b_0 - a_0)$$

Iterating this argument, we find

$$b_{jk} - a_{jk} \leq (1-2\delta)^j (b_0 - a_0)$$

and so as $j \to \infty$, the difference between the maximum (b_{jk}) and minimum (a_{jk}) entries of $\vec{w}M^{jk}$ goes to 0. That is, $\lim_{j\to\infty} M^{jk} = W$ and all the columns of W equal the same vector $\vec{w}_\infty$.

The reason $\lim_{j\to\infty} M^{jk} = W$ implies that $\lim_{n\to\infty} M^n = W$ is for all i, $b_{i+1} - a_{i+1} \leq b_i - a_i$. This follows from another application of the convex combination argument in the proof of Prop. 14.2.2.

Next, $\vec{w}_\infty$ is an eigenvector of W with eigenvalue $\lambda = 1$. Illustrating this in the 3×3 case suffices.

$$\begin{bmatrix} a & a & a \\ b & b & b \\ c & c & c \end{bmatrix} \begin{bmatrix} a \\ b \\ c \end{bmatrix} = \begin{bmatrix} a(a+b+c) \\ b(a+b+c) \\ c(a+b+c) \end{bmatrix} = \begin{bmatrix} a \\ b \\ c \end{bmatrix}$$

The last equality follows because the sum of the entries of each column of W, $a+b+c$ in this example, is 1.

In fact, $\vec{w}_\infty$ is an eigenvector with $\lambda = 1$ of the original matrix M. To see this, first we observe that

$$MW = M \lim_{n\to\infty} M^n = \lim_{n\to\infty} MM^n = \lim_{n\to\infty} M^{n+1} = W \qquad \text{(A.9)}$$

Then

$$\vec{w}_\infty = W\vec{w}_\infty, \quad \text{so} \quad M\vec{w}_\infty = MW\vec{w}_\infty = W\vec{w}_\infty = \vec{w}_\infty$$

The third equality follows by Eq. (A.9) and the fourth because $\vec{w}_{\text{inf}}$ is an eigenvector of W. By Prop. 14.2.2, every eigenvalue λ of M satisfies $|\lambda| \leq 1$. Consequently, $\lambda = 1$ is the dominant eigenvalue of M and by the Perron-Frobenius theorem all the entries of its eigenvectors are positive. □

We'll end with two results about dynamics in the ergodic class: the mean time to return to an ergodic state, and the mean time to go from one ergodic state to another. So suppose M is the $r \times r$ transition matrix for ergodic states. We'll use this important result, Theorem A.13.2. See Sect. 4.2 of [195] for a proof.

Recall a square stochastic matrix M is primitive if for some k, every element of M^k is positive. Let $\vec{v}$ be the dominant unit eigenvector of M, $M\vec{v} = \vec{v}$. The point of this theorem is to show that the entries v_i of $\vec{v}$ represent the fraction of time the Markov chain spends in state i.

Theorem A.13.2. Denote by $T_i^{(n)}$ the fraction of n iterates a primitive Markov chain spends in state i. Then for every $\epsilon > 0$,

$$\lim_{n\to\infty} P(|T_i^{(n)} - v_i| > \epsilon) = 0$$

Now given an irreducible Markov chain, a Markov chain whose transition matrix M is irreducible, suppose we want to find the average time to first visit a state i, given we started in state j. This is called the *mean first passage time* from j to i. We'll denote it by μ_{ij}. In fact, this is an easy application of Prop. A.13.4: change the matrix M to M' by setting $m_{ii} = 1$ and $m_{ki} = 0$ for all $k \neq i$. Until state i is encountered, M and M' determine the same dynamics, so the mean time to absorption given by Prop. A.13.4 is the mean first passage time to state i. Let's see an example.

Example A.13.1. *Mean first passage time.* For the irreducible Markov chain with transition matrix M, to find the mean first passage time to state 2 first form the matrix M' by setting $m_{22} = 1$, and $m_{12} = m_{32} = 0$. Then to identify the matrix Q, form M'' by interchanging states 1 and 2.

$$M = \begin{bmatrix} 0.3 & 0.7 & 0.2 \\ 0.2 & 0.1 & 0.5 \\ 0.5 & 0.2 & 0.3 \end{bmatrix} \quad M' = \begin{bmatrix} 0.3 & 0 & 0.2 \\ 0.2 & 1 & 0.5 \\ 0.5 & 0 & 0.3 \end{bmatrix} \quad M'' = \begin{bmatrix} 1 & 0.2 & 0.5 \\ 0 & 0.3 & 0.2 \\ 0 & 0.5 & 0.3 \end{bmatrix}$$

Then

$$Q = \begin{bmatrix} 0.3 & 0.2 \\ 0.5 & 0.3 \end{bmatrix} \text{ and } N = (I - Q)^{-1} = \begin{bmatrix} 1.7949 & 0.5128 \\ 1.2821 & 1.7949 \end{bmatrix}$$

and the mean first passage time is

$$\begin{bmatrix}\mu_{21} & \mu_{23}\end{bmatrix} = \vec{1}N = \begin{bmatrix}1 & 1\end{bmatrix}\begin{bmatrix}1.7949 & 0.5128 \\ 1.2821 & 1.7949\end{bmatrix} = \begin{bmatrix}3.077 & 2.3077\end{bmatrix}$$

That is, a bit over 3 iterates to reach state 2 from state 1, and a bit over 2 iterates to reach state 2 from state 3. □

Another natural question for irreducible Markov chains is mean number of steps to return to a state i. This is called the *mean recurrence time* of state i, denoted by ρ_i. We'll derive some relations involving μ_{ij} and ρ_i, and use these to find an elegant expression for ρ_i.

First, suppose $i \neq j$. Then we could get from i to j in 1 step, with probability m_{ij}. Or in 1 step we could go to some state $k \neq i$, with probability m_{kj} and from there take a mean number μ_{ik} of steps to reach state i. Consequently, these paths contribute $\mu_{ik} + 1$ steps from j to i. This gives the first equation below,

$$\begin{aligned}\mu_{ij} &= 1 \cdot m_{ij} + \sum_{k \neq i}(\mu_{ik} + 1)m_{kj} = \sum_{k} m_{kj} + \sum_{k \neq i} \mu_{ik} m_{kj} \\ &= 1 + \sum_{k \neq i} \mu_{ik} m_{kj} \qquad \text{(A.10)}\end{aligned}$$

where the second equality follows by gathering all the m_{kj} terms, and the third because each column of M sums to 1.

Now to find the mean recurrence time ρ_i, first note that we must leave state i before we can return to it. The possibility that we stay in state i adds 1 step with probability m_{11}. Then for $k \neq i$, with probability m_{ki} we leave state i for state k. From state k on average we need μ_{ik} steps to return to state i, therefore $\mu_{ik} + 1$ steps to go from i to k and back to i. This gives the first equation below.

$$\begin{aligned}\rho_i &= m_{ii} + \sum_{k \neq i}(\mu_{ik} + 1)m_{ki} \\ &= \sum_{k} m_{ki} + \sum_{k \neq i} \mu_{ik} m_{ki} = 1 + \sum_{k \neq i} \mu_{ik} m_{ki} \qquad \text{(A.11)}\end{aligned}$$

Now let $\begin{bmatrix}1\end{bmatrix}$ denote the $n \times n$ matrix with all entries 1, A the matrix with entries $A_{ij} = \mu_{ij}$ for $i \neq j$ and $A_{ii} = 0$, and B the matrix with $B_{ii} = \rho_i$ and $B_{ij} = 0$ for $i \neq j$. Then applying Eqs. (A.10) and (A.11) we see

$$A = AM + \begin{bmatrix}1\end{bmatrix} - B \qquad \text{(A.12)}$$

To see why this is true, it's enough to take $n=3$ and consider the $(1,1)$ and $(1,2)$ entries. The $(1,1)$ entry of Eq. (A.12) is

$$\begin{aligned} 0 &= \mu_{12}m_{21} + \mu_{13}m_{31} + 1 - \rho_1 \\ &= \mu_{12}m_{21} + \mu_{13}m_{31} + 1 - (1 + \mu_{12}m_{21} + \mu_{13}m_{31}) = 0 \end{aligned}$$

where we've applied Eq. (A.11) for the penultimate equality. The $(1,2)$ entry is

$$\mu_{12} = \mu_{12}m_{22} + \mu_{13}m_{32} + 1$$

which is a consequence of Eq. (A.10).

Finally, we're ready for an interpretation of the mean recurrence time.

Proposition A.13.6. Suppose $\vec{v}$ is the right eigenvector of $\lambda = 1$, the dominant eigenvalue of an irreducible Markov chain. Then $\rho_i = 1/v_i$.

Proof. Recall $M\vec{v} = \vec{v} = I\vec{v}$ so $(I-M)\vec{v} = \vec{0}$. Next, Eq. (A.12) can be rewritten as

$$A - AM = \begin{bmatrix} 1 \end{bmatrix} - B$$

Multiplying on the right by $\vec{v}$,

$$\begin{aligned} A(I-M)\vec{v} &= (\begin{bmatrix} 1 \end{bmatrix} - B)\vec{v} \\ \vec{0} &= \begin{bmatrix} 1 \end{bmatrix}\vec{v} - B\vec{v} \qquad \text{because } (I-M)\vec{v} = \vec{0} \end{aligned}$$

Now

$$\begin{bmatrix} 1 \end{bmatrix}\vec{v} = \begin{bmatrix} 1 & 1 & \cdots & 1 \\ 1 & 1 & \cdots & 1 \\ & & \cdots & \\ 1 & 1 & \cdots & 1 \end{bmatrix} \begin{bmatrix} v_1 \\ v_2 \\ \cdots \\ v_n \end{bmatrix} = \begin{bmatrix} v_1 + v_2 + \cdots + v_n \\ v_1 + v_2 + \cdots + v_n \\ \cdots \\ v_1 + v_2 + \cdots + v_n \end{bmatrix} = \begin{bmatrix} 1 \\ 1 \\ \cdots \\ 1 \end{bmatrix}$$

The last equality follows because $\vec{v}$ is a probability vector. Consequently, $\begin{bmatrix} 1 \end{bmatrix}\vec{v} = B\vec{v}$ gives

$$\begin{bmatrix} 1 \\ 1 \\ \cdots \\ 1 \end{bmatrix} = \begin{bmatrix} \rho_1 & 0 & \cdots & 0 \\ 0 & \rho_2 & \cdots & 0 \\ & & \cdots & \\ 0 & 0 & \cdots & \rho_n \end{bmatrix} \begin{bmatrix} v_1 \\ v_2 \\ \cdots \\ v_n \end{bmatrix} = \begin{bmatrix} \rho_1 v_1 \\ \rho_2 v_2 \\ \cdots \\ \rho_n v_n \end{bmatrix}$$

This gives $\rho_i = 1/v_i$, as intended. □

A.14 CELL MEMBRANE CHANNELS

In this section we'll give a bit of background for Sects. 13.6, 16.3, and 16.4.

We'll begin with cell membranes. In an environment rich in water, fat molecules are a natural way to separate the inside of a cell from the outside. In

1925 Evert Gorter and François Grendel [143] proposed that the cell membrane is a lipid bilayer. Each layer consists of a hydrophilic phosphate head and hydrophobic tail made of two fatty acid chains. The layers are arranged tail to tail, so on both sides the phosphate head separates the membrane from its exterior. The membrane has low permeability to most ions and to water-soluble (hydrophilic) molecules, so the membrane is very good at separating the inside of the cell from its environment.

Also, the membrane anchors the cell's cytoskeleton, the structure that maintains the cell's shape. *Ion pumps* are transmembrane proteins that use energy to pump particular ion species across the cell membrane, against the concentration gradient, in order to maintain an electrical potential difference across the membrane. Some other transmembrane channels are passive: appropriate ions flow across the cell membrane propelled by electrochemical gradients.

We're mostly interested in two aspects of the cell membrane:

- *transmembrane channels* are proteins that span the membrane, and
- *cell surface receptors* transport information and play a central role in immune response.

We discuss cell surface receptors in Sect. A.15. In this section we'll present some of the properties of transmembrane channels and say a bit about the remarkable experiments that gave these observations. Chapter 3 of [107] is a good reference.

These are the properties we'll use.

1. Each channel allows only one ion species, Na^+, K^+, or Ca^{2+}, to flow through it. In contrast to ion pumps, transmembrane channels allow ions to flow only in the direction of the concentration gradient.
2. Each channel admits only a small number of conformations, physical placements of channel molecules, that influence the channel conductance and the flow rate of ions through the channel.
3. At a constant transmembrane potential, the ionic current through a channel can take only a small number of values, and the transition between these current values is abrupt.
4. Switching channel conformations occurs rapidly and randomly. The transition probabilities are determined by the transmembrane potential (*voltage dependence*, *K* channels for example) or by certain molecules bound to the membrane (*ligand dependence*, *Na* and *Ca* channels, for example).

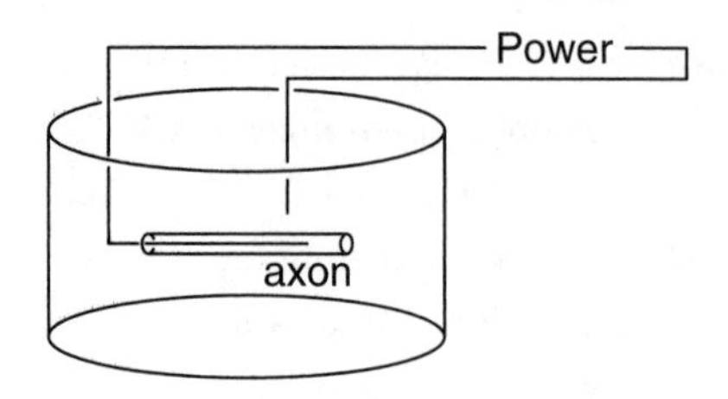

Figure A.1. A voltage clamp.

How can we know these things? Two techniques, the voltage clamp and the patch clamp, enabled the experimental basis for many of these discoveries.

The voltage clamp technique was developed in the late 1940s by Kenneth Cole and George Marmount. Because the voltage clamp requires inserting an electrode into a cell, a cell with a large cylindrical profile was necessary. The giant axon of the squid *Loligo* was ideal. This is the cell type and electrode configuration that Hodgkin and Huxley used.

An electrode is inserted along the axis of the axon; the cell is immersed in a saline solution in which the reference electrode is placed. The name "voltage clamp" comes from Cole's observation that with appropriate negative feedback the membrane potential can be held constant, that is, clamped, throughout the experiment. With this technique, Hodgkin and Huxley discovered the functional form of the ionic conductances $g_{Na}(V)$ and $g_K(V)$ in Eq. (13.36), and also the coefficients in the dn/dt, dm/dt, and dh/dt equations, (13.33), (13.34), and (13.35).

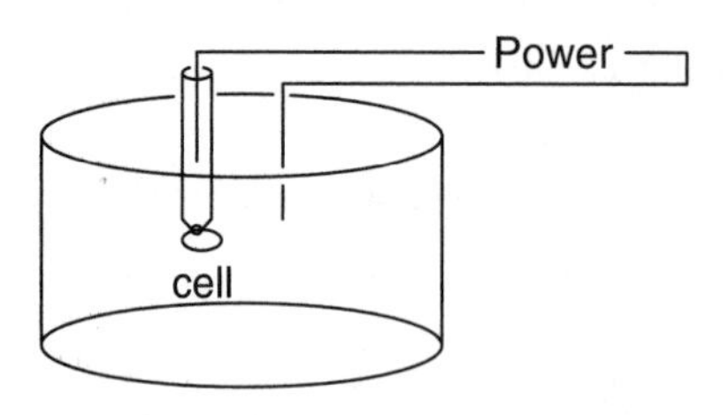

Figure A.2. A patch clamp.

This voltage clamp configuration provided useful data, but only for the aggregate behavior of all the transmembrane channels of the cell. In the late 1970s Erwin Neher and Bert Sakmann developed the *patch clamp*, a technique that permits the measurement of current through a single transmembrane channel. See Fig. A.2. For this method and its applications [249, 317], Neher and Sakmann won the 1991 Nobel Prize for Physiology or Medicine. With a microscope and micromanipulators (fascinating gadgets—look for images on Google) on a vibration-dampening table, a micropipette with tip diameter around a μmeter is placed against a cell membrane and held there with a mild suction. One electrode is in the micropipette, the other in the solution containing the cell. The small size of the pipette tip makes it likely that only one, or at most a few, transmembrane channels are covered by the tip. How to recognize when a single channel is covered by the patch clamp is discussed in [77]. Taking this into account, we can study current flow through a single transmembrane channel while the voltage is constant (voltage clamp, again, though now we measure the current through a single channel). Or the current can be clamped and the change in voltage measured.

These are delicate measurements: about 10^7 ions/sec pass through an open ion channel. This gives a current of a few picoamperes. The channel molecules usually are in configurations that locally minimize energy. Thermal impact of nearby molecules can push the configuration from one energy minimum to another, which explains the randomness in property 4 above. Each local minimum in the energy of the configurations can make the transition to any other local minimum with a probability that depends only on the energies of those configurations and the energy of the peak between those minima. In particular, the transition probability does not depend on the sequence of configurations leading up to the current configuration. This is the justification for using Markov chains to model channel openings and closings.

Mostly we focus on math, but every now and again we should look at why the math is connected to reality. Mandelbrot often reminded people of the story of how Hercules defeated Antaeus by keeping him separated from the earth. Math needs to touch the earth, to recall its roots in experiment, or else face the fate of Antaeus. But when we look at the cleverness that goes into experimental design, we see that this, too, can be a wonderful way to spend a life.

A.15 VIRUSES AND THE IMMUNE SYSTEM

This is background for Sects. 13.4 and 13.5. Viruses are between 20×10^{-9}m and 300×10^{-9}m in length. For comparison, lymphocytes have diameter about 5×10^{-6}m and human hairs have diameters between about 20×10^{-6}m and 180×10^{-6}m. Yes, viruses are tiny. And they are plentiful, about 10^{31} in our biosphere. Lined up one after another, they'd stretch 10^8 light-years, a thousand times the diameter of our galaxy.

Viruses aren't alive, at least by most definitions of life. They can't reproduce themselves, but they can invade other cells, including bacteria, and hijack their reproductive mechanisms to make copies of the virus rather than copies of the cell they've invaded, the *host cell*. Sometimes. Sometimes the virus makes the host cell wait a long time, or wait until specified conditions arise, before copying the virus. Sometimes the virus makes the host cell divide, producing two infected cells. Some viruses make their host cells divide uncontrollably; that is, these viruses cause cancer. A frightening example of this type is hepatitis B, infecting about 300 million people worldwide, and about one-quarter of these could die from liver cancer caused by the virus.

Other viruses are scary, too. Influenza A killed 20 million people in 1919 (when the world's population was about 1.8 billion, compared to 7.5 billion in 2017). According to the WHO, about 70 million people have been infected with HIV, about half of whom have died from AIDS. We see that we have plenty of reasons to understand how viruses work, the better to figure out how to fight them.

Most viruses consist of three parts:

- The *core* is made of DNA or RNA, 10^3 to 10^6 bases, and is responsible for recruiting the host cell's enzymes to build copies of the virus.
- The *capsid* is a protein coat that covers the core.
- The *envelope* is a lipid membrane that covers the capsid. Some viruses don't have an envelope. These are called *naked viruses*.

A virus can't attack just any cell it encounters. Cell surfaces are dotted with receptors, proteins that enable communication between the cell and its environment. An important class of receptors are gates for transmembrane ion channels. We've discussed these in Sects. 16.3 and A.14. Each virus can attach to a specific type of receptor. For example, hepatitis viruses can attach to and infect only liver cells.

Once a virus attaches to a host cell, the virus injects its genetic material into the host. In one of several ways, the viral genetic material recruits the host cell's enzymes. These copy the virus genetic material and also copy viral proteins. Inside the host cell, these virus particles are assembled into a new virus. The completed virus leaves the cell and searches for a new host. The number of virions, infective viruses outside the host cell, released from a cell is the burst size, in the low double digits for some viruses, but in the low thousands for influenza A and for HIV.

This sounds like very bad news, and would be had not evolution equipped animals with a complex immune system to identify and combat viruses, bacteria, fungi, and parasites. Far too many people still die from viruses and other infections, so it's still bad news, but without the immune system, the news would be much worse. Let's sketch some of how the immune system works.

At first (and second) glance, the immune system appears quite complicated. And it is, so we need a guide. In 2019 an Amazon keyword search for "immune system" returned titles of 4000 books. I'm sure many of these are excellent, but for a readable, overall view of the immune system Lauren Sompayrac's *How the Immune System Works* [338] is a wonderful choice. With some study, bafflement at the complexity of the immune system will be replaced by awe at its elegance.

First, the immune system has three parts.

1. The *physical barrier* of the skin ($2m^2$) and the mucous membrane ($400m^2$) that lines the respiratory, digestive, and reproductive systems. Like the airways of the

lungs, the small intestine lining has fractal characteristics [142]. This gives more surface area for nutrient absorption, but we have to wonder if fractality confers immunological advantage.

2. The *innate immune system* (IIS) is found, in one form or another, in all animals. Some aspects of this system appeared half a billion years ago. Agents of the IIS—the complement system, the interferon system, macrophages, neutrophils, natural killer (NK) cells, dendritic cells—respond quickly to familiar pathogens. Often, they are enough to counter the attack. But when they aren't, the immune system has another part.
3. The *adaptive immune system* (AIS) has agents—B cells, helper T cells, killer T cells—that learn to recognize and combat new pathogens. All vertebrates have some form of AIS. Evolution rediscovers, or reuses, successful constructions.

Antigens are chemicals, surface proteins on viruses for example, that the immune system recognizes as belonging to a foreign entity, possibly a pathogen. One way the immune system is alerted to the presence of pathogens is through the *major histocompatibility complex* (MHC) proteins. These come in two classes. Class I MHC proteins are found on most of our cells. If a cell has been infected by a virus, class I MHC molecules carry bits of the virus to the cell surface where they are read by killer T cells which then kill the infected cell. Class II MHC proteins are found on *antigen presenting cells* (APC), macrophages, B cells, and dendritic cells. For example, when a macrophage eats a bacterium, the macrophage's class II MHC molecules carry bits of the bacterium to the macrophage surface where they are read by helper T cells, which then emit cytokines, chemicals that signal the attack.

In a bit more detail, in a cell proteins that have been assembled incorrectly or have broken down are chopped into pieces (*peptides*) by protein complexes called *proteasomes*. Most peptides are further broken down into amino acids for reuse in the cell, but some are carried by transporter proteins (*TAP*1 and *TAP*2) into the endoplasmic reticulum (ER). There they are loaded on MHC I molecules, carried to the cell surface, and bound to the surface. The MHC I molecules give an inventory of bits of all the proteins made in the cell. If some of these are virus proteins, a killer T cell with appropriate receptors will recognize the virus and alert the immune system. To give an up-to-date picture of the cell's interior, peptides displayed by MHC I molecules are replaced frequently.

In APC cells MHC II molecules form inside the ER. To keep peptides in the ER from binding to the MHC II molecule, a protein called the *invariant chain* (IC) binds to it. The MHC II + IC complex leaves the ER in a vesicle called an *endosome*. Invaders are engulfed by the APC cell and held in a vesicle called a *phagosome* where its proteins are broken into peptides. The endosome and phagosome merge, the protein HLA-DM removes IC and attaches a peptide from the invader proteins, and the MHC II + peptide complex is carried to the cell surface and bound there.

In order to recognize unfamiliar pathogens, AIS cells go through a complicated genetic mix-and-match to insure a great variety of their receptors. To prevent AIS agents from attacking the body's cells, careful testing is done before the AIS cells enter the blood stream. Elements of the IIS alert the AIS if an unfamiliar cell is dangerous. Then untrained (called *naive* or *virgin*) AIS cells are trained to recognize these pathogens and a multi-component attack begins. B cells produce antibodies that bind to the pathogen, marking it for attack by agents of the IIS and the AIS.

When the pathogen invasion has been overcome, most of the trained (*experienced*) AIS cells are eliminated, but some remain, retaining the ability to identify this pathogen and respond quickly if it returns. And there are back-ups, redundancies, clever use of multiple pathways in the lymphatic system. But ...even hundreds of millions of years of evolution can't prepare a system that anticipates every problem. Some mistakes occur and also some conflicts: the steps the immune system takes to avoid autoimmune problems oppose how it might fight cancer. Because autoimmune diseases can attack young people while cancer usually attacks older people, often people past their reproductive age, the evolutionary understanding of disease (Sect. 19.1) helps explain this preference for preventing autoimmune diseases over preventing cancers.

Altogether, this is a beautiful picture of the remarkable brilliance of evolution. And I am smitten. Learning some of the dynamics of the immune system has been such a joy that I must share a bit of it with you. Plus, wouldn't you like to know that there is an immunological basis for the instinct of mothers to kiss their babies?

To begin, let's look at the IIS in a bit more detail. The *complement system* (CS) is a collection of about 20 proteins, mostly produced in the liver and present in high concentrations in the blood and tissues. The most abundant complement molecule is $C3$. The CS has three *activation pathways*: the *classical* (based on antibodies so discussed after we've talked about antibodies), the *alternative*, and the *lectin*. The CS is always ready to go, and it works very rapidly.

In the alternative pathway, $C3$ spontaneously breaks into two pieces, $C3a$ and $C3b$. The $C3b$ fragment binds to surface molecules of many pathogens if it is close enough to the pathogen when the $C3$ is broken apart. But it must be really close, because unbound $C3b$ is neutralized in about 60μsec. If $C3b$

does bind, it and other complement proteins build a *membrane attack complex* (MAC) that punches a hole in the pathogen wall, killing the invader. Why doesn't the complement kill us? The surface of our cells contains enzymes (*MCP*) and proteins (*DAF*) that deactivate *C3b* before it can build an MAC.

The lectin pathway exploits mannose, a carbohydrate on the surfaces of many pathogens, and uses *mannose-binding lectin* (MBL), a protein that binds to mannose on pathogen surfaces, but not to carbohydrates on healthy human cells. When MBL binds to mannose, it also binds to a protein *MASP* in the blood. This protein clips *C3*, and because it is very close to the pathogen, the *C3b* piece binds to the pathogen and builds an MAC that kills the pathogen.

Complement acts in other ways, too. A *C3b* molecule bound to a pathogen can be further clipped to a fragment called *iC3b*. Macrophages have receptors that can bind to *iC3b*, and the macrophage eats a pathogen that is bound to it. We say the *iC3b* acts to *opsonize* the pathogen, that is, prepare the pathogen to be eaten. Also, fragments of complement molecules (including *C3a*—nothing is wasted) signal other IIS agents to join the fight.

Macrophages (from the Latin for "big eater," where "big" refers to both the magnitude of the macrophage's appetite and to the physical size of the cell) are another type of IIS agent. They reside throughout the body, mostly in the tissues below the physical barrier. In the absence of a threat, macrophages clean up the debris from the million or so cells that die in our bodies every second. When cytokines alert a macrophage to an attack, the macrophage upregulates the expression of class II MHC molecules. Nearby pathogens are engulfed in phagosomes that merge with *lysosomes*, vesicles that contain enzymes to destroy pathogens. As mentioned earlier, fragments of the destroyed pathogen are transported to the macrophage surface by class II MHC molecules, where they are read by helper T cells. When presented with both cytokines and pathogens, macrophages stop proliferating. They grow large enough to engulf a single-cell parasite (remember our second interpretation of "macro"), kill determinedly, and secrete cytokines and *tumor necrosis factor* (TNF) that can kill tumor cells and virus-infected cells. And macrophages can produce complement proteins when they have been depleted from a region. Mostly, though, think of macrophages as sentinels of the immune system.

Macrophages detect invaders with *pattern-recognition receptors* (PRR), for example, *toll-like* receptors, named because of their similarity to the protein encoded by the toll gene of *Drosophila*. PRR are general-purpose: some detect patterns that occur on many pathogens, while others detect molecules released when body cells are killed. This second feature is crucial. Receptors respond to *damage-associated molecular patterns*, so the IIS will raise the alarm even when they do not recognize the pathogen. The PRRs that are pathogen-associated have evolved to key to features of the pathogen—for example, molecules essential to construct the cell membrane—that are not easily replaced by mutation. So PRRs have reached a configuration relatively stable in the evolutionary arms race with pathogens.

In contrast to the versatility of macrophages, *neutrophils* do one thing: kill stuff. Like all blood cells, neutrophils are produced in the bone marrow. When they leave the bone marrow, neutrophils circulate in the blood and live for about five days. Once they exit the blood and enter tissue, neutrophils engulf and kill pathogens, secrete cytokines to invite others to the fight, and secrete a toxic brew that kills invaders and also some of the body's cells. One of my students called neutrophils the badasses of the immune system. Limitation of this collateral damage is a reason for the short lifespan of neutrophils. Because they do so much damage when they leave the circulatory system, the IIS has several safeguards against unnecessary exits of neutrophils from capillaries. Three different molecules must be expressed in order for neutrophils to exit, and one of these, *selectin*, takes about six hours to express. That is, the supply of cytokines must be maintained for at least six hours in order for neutrophils to join the fray. If other IIS players handle the invasion by then, neutrophils aren't needed.

Before viruses enter a cell, the IIS can fight them by opsonizing the virus with complement proteins, that is, tagging the virus with complement proteins so macrophages find and digest the virus. But what about after the virus has entered the cell? This, too, the IIS can handle. When a cell's PRR detects a virus attack, the cell secretes *type I interferon*, cytokines that alert neighboring cells to the presence of virus invaders by binding to the type I interferon receptors on the cell surface. This produces proteins in the cell that slow viral reproduction, and also if this neighboring cell is attacked by the virus, the cell dies, killing all viruses that have entered it. But cellular suicide, apoptosis, is triggered only if the cell is invaded by a virus. Otherwise, the cell keeps right at its job. We should not be surprised that evolution has imposed efficiencies in the immune system: kill something only when it needs to be killed.

The *natural killer* (NK) cells constitute another class of actors of the IIS. When they respond to a pathogen, NK cells release cytokines to spread the alarm, and they employ several chemicals that cause the target cell to self-destruct. A different feature of NK cells is their target identification method: two types of receptors, *activation* and *inhibition*. Activation receptors respond to unusual molecules that signal the potential target cell has been infected by a pathogen; inhibition receptors recognize class I MHC molecules on the potential target cell surface. If the activation signal is stronger than the inhibition signal, the NK cell attacks the target cell. Why are NK cells useful? Recall that class I MHC molecules present virus fragments from an infected cell so killer T cells can detect the infection and attack the infected cell. If in the evolutionary arms race a virus found a way to stop the expression of class I MHC molecules on an infected cell, a killer T cell would not attack the infected cell. But an NK cell would. The immune system has a remarkable array of back-ups.

Another class of sentinels of the IIS are the dendritic cells, located throughout the body below the epithelial cells (the physical barrier, part 1 of the immune system). The dendritic cells are able to activate naive T cells. In 2011 Ralph Steinman won the Nobel Prize in Physiology or Medicine "for his discovery of the dendritic cell and its role in adaptive immunity." But Steinman died three days before the prize was announced. The prize committee had not heard of his death, so although Nobel Prizes are not awarded posthumously, here the committee made an exception. Dendritic cells are activated by TNF secreted by neutrophils and macrophages fighting invaders, by chemicals given off by dying cells, and by signals from their PRR. Activated dendritic cells hold their position for about six hours while they collect antigens from the attack. Then they travel to a lymph node (this takes about a day), where their class I and II MHC molecules present peptides from the invaders, and produce the co-stimulatory protein *B7*. Some important steps in the immune response—B cell activation for example—require two signals, crosslinked B cell receptors (BCR) and a second, co-stimulatory signal from a helper T cell, for instance. As naive T cells travel through the lymphatic system, in these lymph nodes they are activated by dendritic cells. Dendritic cells live about a week in a lymph node. If the attack has been overcome by then, no further naive T cells are activated. If the attack persists, newly activated dendritic cells give an updated picture of the state of the fight.

Now we'll turn to the AIS. We'll discuss antibodies, B cells and T cells, how they are activated, and what they do when they are activated. The lymphatic system has an important role in the activation of these cells, as does the geometry of several paths through the circulatory system. And we'll say a bit about how the immune system remembers previous attackers, and how it learns to not attack our own cells.

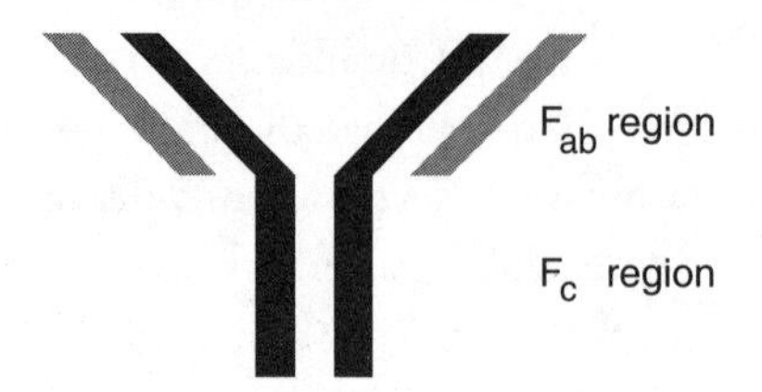

Figure A.3. A monomer antibody. The heavy chains are dark; the light chains are light.

B cells are antibody factories, so we'll begin by describing some of the classes of *antibodies*, protein complexes that can bind to antigens. Three classes, IgD, IgE, and IgG are monomers (Fig. A.3), shaped like a Y and made of two light chains (*Lc*) and two heavy chains (*Hc*). Two other geometries are observed: IgA antibodies are dimers, IgM antibodies are pentamers(Fig. A.4).

When naive B cells are activated, the antibody they initially produce is IgM. IgM opsonizes pathogens for consumption by macrophages and neutrophils, and it mediates the classical activation pathway of the complement system, which we'll discuss soon. The antibody IgA, most abundant in the body, protects the mucous membrane of the physical barrier, and is secreted in a mother's milk to help protect her child while its immune system develops. The dimer structure of IgA gives it some tolerance for stomach acids and digestive enzymes. IgA antibodies prevent invaders in the digestive tract from attaching to intestinal cells and can bind these invaders together into large enough clumps to be eliminated in the obvious way. (About 30% of the solid waste you eliminate consists of bacteria removed by IgA.) The antibody IgG, most common in the blood, opsonizes pathogens, helps NK cells attack invaders, and can cross the placenta. This is another way the mother confers some of her immune system agents to her child while the child's immune system gets started. The antibody IgE defends against parasites. The type of antibody is determined by its F_c region.

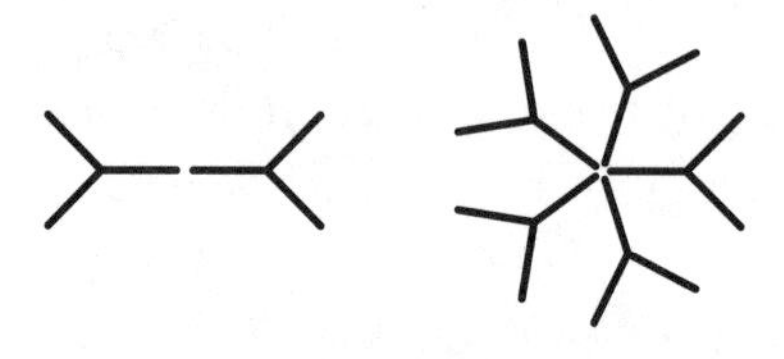

Figure A.4. Dimer and pentamer.

Antibodies also can bind to viruses before they enter a cell. These are called *neutralizing antibodies*. They can prevent viruses from reproducing when they enter a cell. Another attack modality is to bind to the part of the virus that would bind to the cellular receptor. This opsonizes the virus and it is eaten by a macrophage or a neutrophil.

An antigen binds to the F_{ab} region of an antibody; the F_c region of an antibody binds to receptors on immune system agents, macrophages for instance. In order for an antibody to bind to a particular antigen, the F_{ab} region must have a particular shape. The specific antigen to which an antibody binds is called the *cognate antigen*; this binding occurs at an *epitope*, the small region of the antigen to which an antibody binds. To protect against all possible pathogens, an estimated 10^8 different antibodies are needed. Given that the human genome consists of about 3.2×10^9 base pairs, it seems unlikely that each antibody is encoded explicitly in the genome. Rather, evolution has discovered a brilliant way for B cells to produce all these antibodies using only a little information.

Until the late 1970s it was believed that all the body's cells (other than mature red blood cells, which have no nuclei) have the same DNA, modulo copying errors. Then in the late 1970s, Susumu Tonegawa [48, 358, 359] and colleagues showed this is not always the case. In immature B cells the part of chromosome 14 that encodes Hc has 40 kinds of gene segments of a type called V, 25 of type D, 6 of type J, and 10 of type C; the part that encodes Lc has V, J, and C types. As each B cell matures in the bone marrow, it randomly selects one of each type. Combined with *junctional diversity*, the insertion or deletion of a few bases when the gene segments are joined together, this strategy gives rise to 10^8 different types of mature B cells. Each mature B cell can produce only one type of antibody, so this mechanism can provide all the kinds of antibodies we'll need. For this idea, and experiments that verify it, Tonegawa was awarded the 1987 Nobel Prize in Physiology or Medicine.

A little arithmetic reveals two problems. Typically, our blood contains about 3×10^9 mature B cells. Spread over 10^8 different types, we have only about 30 of each type. Only one type can fight a given pathogen, and 30 B cells aren't nearly enough. So when the immune system figures out which is the right type of B cell, it needs to build a lot more of that type. The second problem is that with such a tiny number of B cells of the right type, and the small number of pathogens early in the attack, the principle of mass action implies that encounters, necessary to start the growth of an appropriate defense, are very unlikely. We'll see that the lymph nodes are central to the solution of this problem.

The surface of every mature B cell is decorated with up to 10^5 identical B cell receptors (BCR). A BCR is identical to (a monomer, a single Y, if the antibody is IgA or IgM) the antibody this B cell can produce, except the BCR has an extra protein complex at the base of the F_c region to anchor the BCR to the B cell surface. A complete BCR has two more bits: transmembrane proteins $Ig\alpha$ and $Ig\beta$ that signal the cell nucleus when the F_{ab} region has bound to its cognate antigen.

Before we study how B cells are activated, we'll describe the classical activation pathway for the complement system. Typically, complement proteins in the blood are bound in large groups called $C1$ *complexes*. In the $C1$ complex, complement proteins are bound to an inhibitor that prevents the activation of the cascade that forms an MAC. However, when two $C1$ complexes are nearby, the inhibitors detach, $C3$ is clipped to $C3b$, and this leads to the construction of an MAC. When IgM binds to a pathogen, $C1$ complexes can bind to the F_c region of IgM. Because IgM has five F_c regions, all close together, several $C1$ complexes can be bound in close proximity, release their inhibitors, and build an MAC. This is the classical activation pathway of the complement system.

Now for naive B cell activation. There are two main mechanisms, *T cell dependent* and *T cell independent*. Both require BCR to send an activation signal to the B cell nucleus. A single BCR bound to a single epitope will not signal the nucleus. A signal will be generated if many BCR are bound to many epitopes on an antigen, or to epitopes on many antigens clumped together on a pathogen. We say the BCR are *crosslinked*, although the term "clustered" is a better match for the geometry. This brings together enough of the $Ig\alpha$ and $Ig\beta$ proteins to initiate a signal to the nucleus. A second signaling mechanism involves a *co-receptor*, a complement receptor on the B cell surface. If a complement molecule is bound to an antigen, then the complement system has identified the invader as dangerous and far fewer BCR must be crosslinked to send the signal.

But a signal from crosslinked BCR is not sufficient to activate a B cell. One of two additional steps is needed: a signal from an activated helper T cell (the T cell dependent pathway, obviously), or binding to the B cell's toll-like receptors (the T cell independent pathway). The T cell independent activation can recognize carbohydrate and fat antigens; helper T cells recognize only protein antigens. Together, these two B cell activation mechanisms are sensitive to a vast range of antigens. This second step is necessary to prevent B cells from creating antibodies for our own cells. We'll see that B cell training removes almost all B cells that could initiate an autoimmune reaction. For those that slip through, the second step, requiring a clear indication of an attack, prevents B cells from producing antibodies to our own cells.

Once a naive B cell is activated, the immune system has determined that the antibodies produced by that B cell are needed for the fight. The B cell enters a *proliferation* phase that lasts about a week, the number of cells doubling about every 12 hours. This produces a clone of about 16,000 identical B cells, so the process is called *clonal selection.*

After the proliferation phase, B cell maturation involves three steps. Initially, these B cells produce IgM antibodies, with F_{ab} region the same as that of the BCR of this cell. Recall that an antibody is made of the same F_{ab} and F_c regions as the BCR, except the antibody lacks the bit that binds the BCR to the B cell. As the B cell matures, it can undergo *class switching*. Cytokines present data about the type of antibody needed and the B cell makes appropriate modifications of the F_c region and leaves the F_{ab} region unaltered.

Also, maturation involves *somatic hypermutation*, a process that increases the antibody affinity for its cognate antigen. The standard human cell mutation rate is about one per 10^7 bases per generation; in the part of the B cell chromosome that encodes the antibody F_{ab} region, the mutation rate can be as high as about one per 10^3 bases per generation. This allows much wider exploration of DNA sequence space. Then natural selection amplifies the mutations with higher affinities.

Finally, a B cell can become a *plasma B cell* and produce up to 2000 antibodies per second throughout its several-day lifetime. Or it can become a *memory B cell*. These live much longer and because they remember previous encounters with their cognate antigen can produce a rapid response if this antigen appears again. The mechanism of the choice of plasma B cell or memory B cell is unclear.

These are the main points we need to know about B cells. On to T cells.

The bone marrow produces T cells, but while B cells mature in the bone marrow, T cells mature in the thymus. About 95% of T cells have receptors (TCR) that consist of two antigen recognition proteins, α and β, and a signal complex *CD*3 that consists of four different proteins. As with BCR, the recognition proteins α and β are assembled by randomly selecting gene segments. In addition, immature T cells express two types of co-receptors, *CD*4 and *CD*8. If a cell's TCR has a structure consistent with recognizing class I MHC molecules, as the cell matures it downregulates expression of the *CD*4 co-receptor and becomes a killer T cell. If a cell's TCR recognizes class II MHC molecules, it downregulates *CD*8 and becomes a helper T cell. A ml of normal human blood contains 500 to 1200 helper T cells and 150 to 100 killer T cells.

To become experienced, naive T cells need both crosslinked TCR bound to MHC + peptide, and also co-stimulation, for example, a *B*7 protein on an APC, bound to a *CD*28 receptor on the T cell membrane.

In lymph nodes helper T cells scan dendritic cells for the cognate antigen of the T cell, a signal that this T cell will be able to help fight the pathogen invaders. If a T cell finds such a dendritic cell, its connection with the dendritic cell is stabilized to keep them together for several hours. This allows the helper T cell to activate fully, and the dendritic cell to amplify its expression of MHC and co-stimulus. In addition, this bond increases the life of the dendritic cell. By clonal selection, an activated helper T cell produces about 10^4 daughters in a week.

Killer T cells are thought to be activated in a similar fashion, through contact with a dendritic cell with class I MHC presenting the cognate antigen, and contact with a co-stimulus. Helper T cells secrete growth factors that increase the killer T cell proliferation rate.

We've mentioned that dendritic cells travel from the attack site to the lymph nodes. Through their PRR and cytokine receptors, dendritic cells deduce the type of attacker. Different regions of the body produce cytokine combinations characteristic to the location. The dendritic cells transmit this information to appropriate helper T cells, those whose TCR match the presented antigen. These then secrete a combination of cytokines (a *cytokine profile*) that activate appropriate members of the immune system and tell them the location of the attack. Cytokines secreted by helper T cells have a limited physical range, allowing the body to mount different simultaneous defenses.

Activated killer T cells proliferate, leave the lymph nodes, move to the location of the attack, and start killing. Killer T cells have two primary weapons, *Fas ligand (FasL)* and *perforin–granzyme B*. Both weapons induce apoptosis: cell membrane breached, cell contents (including any pathogen DNA or RNA) encased in vesicles and eaten by macrophages. This is the clean way to kill a cell. The other route, *necrosis* due to injury, spills the cell contents, including viruses, back into the body. To paraphrase a bit of old movie dialogue, apoptosis doesn't want to just kill the cell, it wants to *KILL* the cell.

Now we'll address the issue of how in the early stages of an infection, the small population of antigens can encounter a small population of naive B and T cells with receptors for which those antigens are cognate. The principle of mass action makes this seem unlikely. The answer involves some of the secondary lymphoid organs (SLO): lymph nodes, Peyer's patches, and the spleen. Antigens enter the lymph nodes from the lymph, Peyer's patches from the small intestine, and the spleen from the blood. All three organ types have regions for B cells and regions for T cells, and the B cell regions of all three have lymphoid follicles, aggregates of follicular dendritic cells (FDC) that bind to complement-opsonized antigens and present them to B cells. The FDC hold antigens close together, so then can crosslink BCR. These B cells proliferate, their number doubling every six hours. Those that are co-stimulated by helper T cells undergo somatic hypermutation and class switching. The T cell region also contains dendritic cells that present antigens to T cells. The movements of lymphocytes within these organs is governed by the downregulation and upregulation of receptors for various *chemokines*, chemoattractive cytokines, and by the production of chemokines of limited range. Helper T cells know which B cells to help because the B cells absorb some of the bound antigens and present them on their class II MHC molecules, for helper T cells to read. The B cell and helper T cell each contribute to the maturation of the other.

Naive T cells express an array of adhesion molecules that allow them to circulate throughout the entire collection of SLO. If the cell doesn't find its cognate antigen, it travels through all these organs in about a day. In this fashion, the T cell can survive for about six weeks.

Experienced T cells express different adhesion molecules that depend on where the T cell was activated. Some of these direct the T cell to return to the type of organ where it was activated, and others direct the T cell to exit the blood at the site of the infection, where killer T cells kill infected cells and helper T cells secrete cytokines.

Naive B cells express adhesion molecules that direct them to revisit the SLO, seeking cognate antigens. On the other hand, adhesion molecules direct experienced B cells to stay in the SLO or bone marrow and generate antibodies.

Now, a brief aside: why do mothers kiss their babies? (If you replace "kiss" by "lick," this observation holds for mammals in general.) Babies do not produce their own IgG until about two months after they are born. Mother's IgG can cross the placenta to supply her baby; mother's IgA is secreted in with her milk. These provide the baby with general immunological response, but what about responses to the specific pathogens on the baby? Most enter the baby through its mouth and nose. Mom kisses baby and samples some of those pathogens. Antigens arrive at Mom's SLO where memory B cells generate antibodies for that pathogen. These antibodies enter the mother's milk and so her baby has some adaptive immunological defense before its own adaptive immune system begins to work.

Once the invasion has been handled, the immune response must be turned off and most of the activated cells decommissioned. One aspect is obvious: when the pathogen population drops, so does the antigen supply and consequently so does the activation rate of immune system cells.

But the immune system has evolved additional off-switches. To activate naive T cells, *B7* proteins on APC bind to *CD*28 receptors and act as co-stimulators for the T cells. T cells have another receptor, *CTLA*-4, to which *B7* can bind. This blocks the "activate" signal from any *B7* bound to *CD*28, and the affinity of *B7* for *CTLA*-4 is thousands of times greater than the affinity of *B7* for *CD*28. So why does *B7* ever bind to *CD*28? Because almost all the *CTLA*-4 of a naive T cell lies within the cell, not on the surface. But two days after the T cell is activated, *CTLA*-4 begins to move to the cell surface, where it impedes ligation of *B7* to *CD*28, which makes more difficult the reactivation of these T cells.

Also, after activation of a T cell, expression of the protein *PD*-1 ("programmed death"-1—really) increases on the T cell surface. The ligand of *PD*-1, *PD*-1*L*, is common in tissues under attack, and the effect of binding *PD-1L* to *PD-1* is to reduce the proliferation of activated T cells.

In addition, cell mortality helps to clear out unneeded immune agents. Neutrophils die in a few days, NK cells in a week or so. Because NK cells produce a cytokine (*IFN-γ*) that keeps macrophages attacking, as the NK cell population drops macrophages return to their day job, garbage collection. Dendritic cells survive about a week when they reach the lymph nodes. Plasma B cells expire in about five days. Antibodies are short-lived; the longest is IgG, which lasts about three weeks.

On the other hand, T cells have longer lives because they may need to circulate many times through the SLO in order to find their cognate antigens. One way killer T cells kill is to bind the protein *FasL* on the T cell surface to the target cell protein *Fas*, which induces target cell apoptosis. Before activation and early in its fighting life, killer T cells are insensitive to ligation of their own *Fas* by their own *FasL*. To keep our bodies from filling up with T cells that fought off the previous attack but likely are ineffective against the next, after many re-stimulations T cells lose this insensitivity to their own *Fas* ligation. Apoptosis removes more than 90% of this population.

Some trained AIS cells are retained to insure a rapid response if the pathogen that trained them returns some day. Here's how we think the AIS memory works.

When B cells are activated, three types of B cells are produced.

- *(Short-lived) plasma B cells* are produced in the SLO, travel to the bone marrow or spleen, produce lots of antibodies, and die in a few days.
- *Long-lived plasma B cells* are one type of memory B cell. Also produced in the SLO and also moved to the bone marrow, these produce much smaller amounts of antibodies and survive longer. These can provide long-term immunity to reinfections.
- *Central memory B cells* are the other type of memory B cell. These live in the SLO, gradually proliferate to sustain their population and to replace long-lived plasma B cells that have died, and during another attack by the same pathogen, rapidly produce more short-lived plasma B cells.

For T cells the memory structure is a bit different. The activated T cells that leave the blood and enter the tissues are the *effector T cells*. We mentioned that after the attack is over, about 90% of the effector T cells die by apoptosis.

- *Memory effector T cells* are those effector T cells that remain in the tissues after the attack has ended.
- *Central memory T cells* are activated T cells that do not go to the site of the attack, but stay in the SLO.

If the same pathogen attacks again, the central memory T cells proliferate, and most of the daughters move to the battle site, while the remainder stay in the SLO, ready for yet another attack.

Usually, a second attack is dispatched more quickly than the first. At the first attack only about one-millionth of the B and T cells have receptors that bind to the pathogen's antigens. Because not all of the initial populations of activated B and T cells die after the battle, if a second attack occurs about one-thousandth of the B and T cells have receptors that bind to the pathogen's antigens. The buildup of the population of activated B and T cells is much faster. Also, memory cells can be activated by a lower concentration of antigens than can naive cells, so the immune response to the attack can be triggered earlier.

Of course, the IIS has a memory, that is, genes for receptors that recognize common attackers. The IIS responds quickly to common threats, sometimes without the need to alert the AIS. On the other hand, permanent memory of all possible attackers would consume much of the genome, so evolution outfitted the AIS to remember attackers only after they attack, and still be ready to respond to just about anything, though response a to an unfamiliar attacker will take longer to mount.

By now perhaps you are tired of reading "This is an immense subject and we've touched only a few points," but that statement is more fundamentally accurate about the immune system than the corresponding statement has been about any other topic. More than all the years I've thought about paleontology, the few months I've spent thinking about the immune system have flattened me by the beauty of the intricate networks evolution can discover.

Nevertheless, we have one more question to consider: how do immune system agents know not to attack our healthy cells? That is, why aren't we done in by autoimmune disease? In the thymus T cells learn to recognize antigens presented on MHC molecules (positive selection), and are eliminated if they recognize self antigens (negative selection). No step in this training process is foolproof, but the consequences of autoimmune disease, of attacking your own healthy cells, can be so dire that training to recognize and not attack selfantigens has multiple backups.

- In the thymus any T cell that recognizes abundant self antigens dies by apoptosis.
- T cell traffic patterns keep naive T cells circulating in the SLO where they encounter roughly the same kinds and concentrations of self antigens as in the thymus. Higher levels of self antigens could be in the tissues, but almost all naive T cells are restricted from going there.
- If tissue injury releases into the blood or lymphatic system a self antigen that is rare in the SLO and consequently some T cells that recognize this antigen have

not been eliminated, *natural regulatory T cells* in the lymph nodes suppress this activation.

- Naive T cells that depart from their traffic pattern and enter tissue where they encounter self antigen cognate to their TCR are unlikely to receive co-stimulation. These cells die or are *anergized*, blocked from their immunological tasks.
- If these naive T cells in tissues encounter self antigens in concentrations sufficient to crosslink their receptors and the cell whose MHC molecules present the self antigen also can provide co-stimulation, then the high density of cells presenting self antigens leads to multiple stimulations, which leads to activation-induced cell death.

The first impression brings to mind the dialogue between the Scarecrow and the Cowardly Lion: "What if it was an elephant?" "I'd wrap him up in celephant." "What if it was a brontosaurus?" "I'd show him who was king of the forest." Every time the training to recognize self antigens revealed an incompleteness, evolution found yet another patch. Tens of millions of years of evolution can discover a lot of patches.

What about B cells?

- B cells whose BCR bind with self antigens die in the bone marrow.
- Naive B cells circulate through the SLO. If the concentration in the bone marrow of a self antigen is too low to remove a B cell whose BCR bind to that antigen, then the concentration in the SLO is too low to activate the B cell. Traffic control protects against B cells generating self antibodies.
- Naive B cells that depart from their traffic pattern and enter tissue where they encounter self antigens cognate to their BCR can be anergized if they do not receive a signal from a helper T cell, and helper T cells whose TCR could bind to self antigens likely already have been removed.

Because it alters the BCR, somatic hypermutation complicates the training of B cells to ignore self antigens. If the BCR of a hypermutated B cell binds to a self antigen, the B cell is unlikely to be activated by that self antigen. This is because in the SLO antigens are presented to B cells by FDC, FDC present only opsonized antigens, and usually self antigens aren't opsonized. We see that B cell preparation has several layers before the cells are released into the body.

A final bit of training involves NK cells. Recall that some viruses downregulate the expression of class I MHC in the invaded cell to hide from killer T cells. To counter that step in the evolutionary arms race, NK cells kill cells that do not display class I MHC on their surface. Each NK cell expresses many inhibitory receptors, and if none of these recognizes any of your class I MHC molecules,

NK cells will attack your healthy cells. Probably in the bone marrow, NK cells that don't recognize at least one of your class I MHC molecules are deactivated.

The immune system is vastly more complex than the brief sketch given here. We've presented this overview to help us understand some aspects of the models of Sects. 13.4 and 13.5, and also because everyone should appreciate the remarkable architecture of the immune system. Half a billion years of evolution can discover systems of great subtlety.

A.16 THE NORMAL DENSITY FUNCTION

First of all, we don't have an antiderivative for e^{-x^2}, so we can't evaluate $\int_{-\infty}^{\infty} e^{-(x-\mu)^2/2\sigma^2}dx$ in the usual way. But there's a trick, a very clever trick.

Recall that what we call the variable of integration doesn't matter, so

$$\int_{-\infty}^{\infty} e^{-(x-\mu)^2/2\sigma^2}dx = \int_{-\infty}^{\infty} e^{-(y-\mu)^2/2\sigma^2}dy$$

and we see that

$$\left(\int_{-\infty}^{\infty} e^{-(x-\mu)^2/2\sigma^2}dx\right)^2$$

$$\begin{aligned}
&= \left(\int_{-\infty}^{\infty} e^{-(x-\mu)^2/2\sigma^2}dx\right)\cdot\left(\int_{-\infty}^{\infty} e^{-(y-\mu)^2/2\sigma^2}dy\right) \\
&= \int_{-\infty}^{\infty}\left(\int_{-\infty}^{\infty} e^{-(x-\mu)^2/2\sigma^2}dx\right)e^{-(y-\mu)^2/2\sigma^2}dy \\
&= \int_{-\infty}^{\infty}\int_{-\infty}^{\infty} e^{-(x-\mu)^2/2\sigma^2}e^{-(y-\mu)^2/2\sigma^2}dxdy
\end{aligned}$$

The second equality follows because the *dx* integral is a constant and so factors inside the *dy* integral. The last equality is an application of the observation that y and x are independent variables and so factors that involve y can be pulled inside the *dx* integral.

Next, change variables by $u = (x-\mu)/(\sqrt{2}\sigma)$ and $v = (y-\mu)/(\sqrt{2}\sigma)$. The corresponding differentials are $du = dx/(\sqrt{2}\sigma)$ and $dv = dy/(\sqrt{2}\sigma)$. Then the double integral above becomes

$$\int_{-\infty}^{\infty}\int_{-\infty}^{\infty} e^{-u^2}e^{-v^2}2\sigma^2dudv = 2\sigma^2\int_{-\infty}^{\infty}\int_{-\infty}^{\infty} e^{-(u^2+v^2)}dudv$$

Finally, convert to polar coordinates $u = r\cos(\theta)$ and $v = r\sin(\theta)$. Then $u^2 + v^2 = r^2$ and as we saw in Eq. (17.18), $dudv = rdrd\theta$. In polar coordinates the double integral can be evaluated by simple substitution. Note that the limits $-\infty < u < \infty$, $-\infty < v < \infty$ become $0 \leq \theta \leq 2\pi$, $0 \leq r < \infty$.

$$2\sigma^2 \int_0^{2\pi} \int_0^{\infty} e^{r^2} r dr d\theta = 2\sigma^2 \theta \Big|_0^{2\pi} \int_0^{\infty} e^{r^2} r dr$$

$$= 4\pi\sigma^2 \frac{-e^{-r^2}}{2} \Big|_0^{\infty} = 0 - (-2\pi\sigma^2)$$

Remember, this is the integral of the square of $\int_{-\infty}^{\infty} e^{-(x-\mu)^2/2\sigma^2} dx$, so we have

$$\int_{-\infty}^{\infty} e^{-(x-\mu)^2/2\sigma^2} dx = \sqrt{2\pi}\sigma$$

and consequently the normalization factor is $1/(\sqrt{2\pi}\sigma)$.

We can't evaluate an integral, so we try to evaluate its square. Who would have thought this could work? Some tricks aren't necessarily things we'd think up on our own, but are to be added to our list of tricks.

A.17 A SKETCH OF THE PERRON-FROBENIUS PROOF

Here we'll sketch the proof of part of the Perron-Frobenius theorem, stated in Sect. 14.2 and restated in a moment. The theorem was proved for positive matrices by Oskar Perron [282] and extended to irreducible matrices by Ferdinand Frobenius [125]. Many different proofs of this theorem are available in the literature, Sect. 8.4 of [177], Sect. 4.2 of [211], [220], and Chapter 9 of [349], for example. One elegant proof [41] is based on the CMP, but it requires some sophisticated background work, including the introduction of Hilbert's projective metric. As beautiful as this approach is, we'll take a more direct path.

In the preface of [349], Shlomo Sternberg writes about the Perron-Frobenius theorem, "For some reason that I do not understand, this important theorem has disappeared from the linear algebra curriculum." The first place I recall seeing this theorem is in a graduate physics course, but it has clear applications in biology, from populations of ions crossing cell membranes to populations of individuals in the wide world.

Theorem A.17.1. Suppose M is a primitive matrix. Then
(1) M has a dominant (real) eigenvalue λ_0, that is, $\lambda_0 > |\lambda|$ for all other eigenvalues λ of M,
(2) λ_0 has an eigenvector $\vec{v}_0$ with all elements positive, and
(3) all eigenvectors of λ_0 are multiples of $\vec{v}_0$.

We'll show (1) for positive M and give the idea for how to extend this for primitive M. For this we need a few preliminary results.

Lemma A.17.1. If every entry of a matrix M is multiplied by a constant k, then every eigenvalue of M is multiplied by k.

Proof. Suppose $M\vec{v} = \lambda\vec{v}$. Then $kM\vec{v} =$

$$\begin{bmatrix} km_{11} & \dots & km_{1n} \\ & \dots & \\ km_{n1} & \dots & km_{nn} \end{bmatrix} \begin{bmatrix} v_1 \\ \dots \\ v_n \end{bmatrix} = \begin{bmatrix} k(m_{11}v_1 + \dots + m_{1n}v_n) \\ \dots \\ k(m_{n1}v_1 + \dots + m_{nn}v_n) \end{bmatrix} = (k\lambda)\vec{v}$$

That is, the eigenvalues of kM are k times the eigenvalues of M. Note that if $\vec{v}$ is an eigenvector of λ for M, then $\vec{v}$ also is an eigenvector of $k\lambda$ for kM. □

Lemma A.17.2. If λ is an eigenvalue of a matrix M, then λ^k is an eigenvalue of the matrix M^k, and an eigenvector of λ for M also is an eigenvector of λ^k for M^k.

Proof. Suppose $M\vec{v} = \lambda\vec{v}$. Then

$$(M^2)\vec{v} = M(M\vec{v}) = M(\lambda\vec{v}) = \lambda(M\vec{v}) = \lambda^2\vec{v}$$

The generalization to M^k is straightforward. □

Recall that a complex number $z = x + iy$ has a polar representation $z = re^{i\theta}$, where $r = \sqrt{x^2 + y^2}$ is the modulus of z, and $\theta = \tan^{-1}(y/x)$ is the argument of z. Multiplication is particularly simple in the polar representation: multiply the moduli and add the arguments. See Exercise 7.1.8. The result for moduli is simple algebra; for arguments the key step is the formula for $\tan(\alpha + \beta)$.

Now if the spectral radius $\rho(M) \neq 1$, replace M by $N = (1/\rho(M))M$. By Lemma A.17.1, if λ is an eigenvalue of M, then $\lambda/\rho(M)$ is an eigenvalue of N, and we see that $\rho(N) = 1$. Consequently, at least one eigenvalue λ_0 of N lies on the unit circle $|\lambda| = 1$. For the purpose of contradiction, let's assume $|\lambda_0| = 1$ and $\lambda_0 \neq 1$.

Because M is primitive, for some positive integer p, M^p is a positive matrix. To simplify the argument, we'll assume M is a positive matrix. If it isn't, replace M with M^p in the following steps.

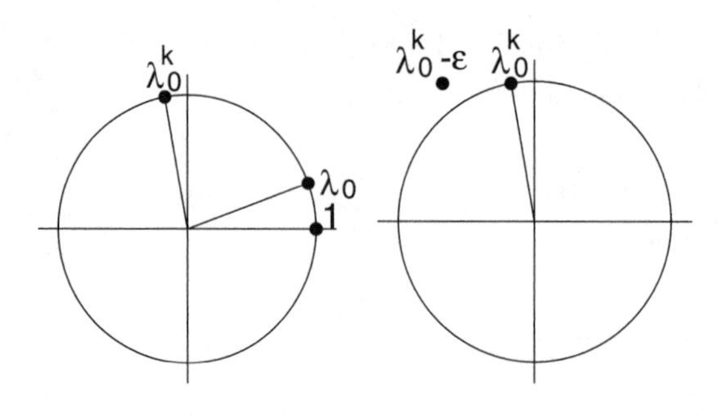

Figure A.5. Left: λ_0^k. Right: $\lambda_0^k - \epsilon$.

By the polar representation of complex multiplication, we see that for large enough k, λ_0^k has negative real part. See the left image of Fig. A.5. Then by Lemma A.17.2, the matrix N^k has an eigenvalue λ_0^k with negative real part.

Because N is a positive matrix, so is N^k. Take ϵ to be half the smallest diagonal entry of N^k. Then $A = N^k - \epsilon I$ also is a positive matrix.

Suppose $\vec{v}$ is an eigenvector of λ for N. Then by Lemma A.17.2, $\vec{v}$ also is an eigenvector of N^k with eigenvalue λ^k, and so

$$A\vec{v} = (N^k - \epsilon I)\vec{v} = N^k\vec{v} - \epsilon I\vec{v} = \lambda^k\vec{v} - \epsilon\vec{v} = \left(\lambda^k - \epsilon\right)\vec{v}$$

That is, $\lambda^k - \epsilon$ is an eigenvalue of A.

Because λ_0^k lies on the unit circle and has negative real part, $\lambda_0^k - \epsilon$ lies outside the unit circle (see the right image of Fig. A.5), and because $\lambda_0^k - \epsilon$ is an eigenvalue of A, we see that $\rho(A) \geq |\lambda_0^k - \epsilon| > 1$. To finish, we need a relation between $\rho(A)$ and $\rho(N)$. For this we use *Gelfand's formula*. For a square matrix M, define a *matrix norm* of M by $\|M\| = \max_i\left(\sum_j |m_{ij}|\right)$. This is a *submultiplicative* norm, that is, $\|CD\| \leq \|C\|\|D\|$. Gelfand's formula is

$$\rho(M) = \lim_{k\to\infty} \|M^k\|^{1/k} \qquad \text{(A.13)}$$

Soon we'll say a little bit about this formula, but for now let's use it to prove our assumption that $\lambda_0 \neq 1$ leads to a contradiction.

Now for all i and j we have $0 < A_{ij} \leq (N^k)_{ij}$, which we'll write as $A \leq N^k$. Then for each positive integer n we have $A^n \leq (N^k)^n$. (If this isn't clear, work out the $n = 2$ case for 2×2 matrices: if $C \leq D$, then $C^2 \leq D^2$.) Now if $C \leq D$, then $\|C\| \leq \|D\|$ (again, check the 2×2 case if this isn't clear), so $\|A^n\| \leq \|(N^k)^n\|$. From this we have $\|A^n\|^{1/n} \leq \|(N^k)^n\|^{1/n}$ and $\lim_{n\to\infty} \|A^n\|^{1/n} \leq \lim_{n\to\infty} \|(N^k)^n\|^{1/n}$. Then apply Gelfand's formula:

$$\rho(A) = \lim_{n\to\infty} \|A^n\|^{1/n} \leq \lim_{n\to\infty} \|(N^k)^n\|^{1/n} = \rho(N^k) = (\rho(N))^k$$

The last equality follows from Lemma A.17.2. We know that $\rho(N) = 1$ and we've seen that $\rho(A) > 1$, which contradicts $\rho(A) \leq \rho(N)$.

That is, the only eigenvalue of N on the unit circle is $\lambda_0 = 1$, and all other eigenvalues λ satisfy $|\lambda| < 1$. For the original matrix M this shows that there is a real eigenvalue $\lambda_0 = \rho(M)$, and all other eigenvalues λ of M satisfy $|\lambda| < \rho(M)$.

Figure A.6. Plot of $\|M^n\|^{1/n}$.

Finally, an illustration of the convergence rate of Gelfand's formula. We'll use the matrix $\begin{bmatrix} 1 & 2 \\ 1 & 3 \end{bmatrix}$. The eigenvalues are $2 \pm \sqrt{3}$, so $\rho(M) = 2 + \sqrt{3}$. In Fig. A.6 we plot $\|M^n\|^{1/n}$ for $n = 1, \ldots 200$. The horizontal gray line had $y = \rho(M)$. For more details, Google is your friend.

The Perron-Frobenius theorem is an important result, central to so much of our analysis of systems modeled by Markov chains or Leslie matrices. In addition, outside of biology the Perron-Frobenius theorem has many applications, one of which you use often, perhaps every day: the Google PageRank is the dominant eigenvector [269] of the Google adjacency matrix.

In physics the result can be extended from matrices to transfer operators on infinite-dimensional spaces. In applications to statistical mechanics, the dominant eigenvalue is related to the free energy per particle of the system. The Perron-Frobenius theorem has an analogue for quantum stochastic maps. And in economics the theorem is central to the proof of the Hawkins-Simon condition for macroscopic stability. One mathematical idea can touch many disciplines.

A.18 THE LESLIE MATRIX CHARACTERISTIC EQUATION

Here we'll derive the characteristic equation (14.19) for $A = 5$. The characteristic equation of the Leslie matrix L is

$$0 = \det(L - \lambda I) = \det \begin{bmatrix} f_0 - \lambda & f_1 & f_2 & f_3 & f_4 & f_5 \\ p_0 & 0 - \lambda & 0 & 0 & 0 & 0 \\ 0 & p_1 & 0 - \lambda & 0 & 0 & 0 \\ 0 & 0 & p_2 & 0 - \lambda & 0 & 0 \\ 0 & 0 & 0 & p_3 & 0 - \lambda & 0 \\ 0 & 0 & 0 & 0 & p_4 & 0 - \lambda \end{bmatrix}$$

Expanding along the top row, this becomes

$$= (f_0 - \lambda) \det \begin{bmatrix} -\lambda & 0 & 0 & 0 & 0 \\ p_1 & -\lambda & 0 & 0 & 0 \\ 0 & p_2 & -\lambda & 0 & 0 \\ 0 & 0 & p_3 & -\lambda & 0 \\ 0 & 0 & 0 & p_4 & -\lambda \end{bmatrix}$$

$$-f_1 \det \begin{bmatrix} p_0 & 0 & 0 & 0 & 0 \\ 0 & -\lambda & 0 & 0 & 0 \\ 0 & p_2 & -\lambda & 0 & 0 \\ 0 & 0 & p_3 & -\lambda & 0 \\ 0 & 0 & 0 & p_4 & -\lambda \end{bmatrix} + f_2 \det \begin{bmatrix} p_0 & -\lambda & 0 & 0 & 0 \\ 0 & p_1 & 0 & 0 & 0 \\ 0 & 0 & -\lambda & 0 & 0 \\ 0 & 0 & p_3 & -\lambda & 0 \\ 0 & 0 & 0 & p_4 & -\lambda \end{bmatrix}$$

$$-f_3 \det \begin{bmatrix} p_0 & -\lambda & 0 & 0 & 0 \\ 0 & p_1 & -\lambda & 0 & 0 \\ 0 & 0 & p_2 & 0 & 0 \\ 0 & 0 & 0 & -\lambda & 0 \\ 0 & 0 & 0 & p_4 & -\lambda \end{bmatrix} + f_4 \det \begin{bmatrix} p_0 & -\lambda & 0 & 0 & 0 \\ 0 & p_1 & -\lambda & 0 & 0 \\ 0 & 0 & p_2 & -\lambda & 0 \\ 0 & 0 & 0 & p_3 & 0 \\ 0 & 0 & 0 & 0 & -\lambda \end{bmatrix}$$

$$-f_5 \det \begin{bmatrix} p_0 & -\lambda & 0 & 0 & 0 \\ 0 & p_1 & -\lambda & 0 & 0 \\ 0 & 0 & p_2 & -\lambda & 0 \\ 0 & 0 & 0 & p_3 & -\lambda \\ 0 & 0 & 0 & 0 & p-4 \end{bmatrix}$$

Expanding these six determinants (details in a moment), we see the characteristic equation becomes

$$\begin{aligned} 0 = (f_0 - \lambda)(-\lambda)^5 - f_1 p_0 (-\lambda)^4 + f_2 p_0 p_1 (-\lambda)^3 - f_3 p_0 p_1 p_2 (-\lambda)^2 \\ + f_4 p_0 p_1 p_2 p_3 (-\lambda) - f_5 p_0 p_1 p_2 p_3 p_4 \end{aligned}$$

Multiplying the first term, distributing the minus signs, and isolating λ^6 on the left-hand side, we obtain the characteristic equation (14.19) for $A = 5$,

$$\lambda^6 = f_0 \lambda^5 + f_1 p_0 \lambda^4 + f_2 p_0 p_1 \lambda^3 + f_3 p_0 p_1 p_2 \lambda^2 + f_4 p_0 p_1 p_2 p_3 \lambda + f_5 p_0 p_1 p_2 p_3 p_4$$

Now, about expanding the determinants of the 5×5 matrices above:

$(f_0 - \lambda)$ **term.** The matrix is lower triangular, so the determinant is the product of the diagonal entries, $(-\lambda)^5$.

$-f_1$ **term.** The matrix is lower triangular, so the determinant is the product of the diagonal entries, $p_0(-\lambda)^4$.

f_2 **term.** This matrix is neither upper triangular nor lower triangular. However, we can expand the determinant down the first column, pulling out a factor of p_0 for the submatrix obtained by deleting the row and the column containing p_0, and pulling out a factor of 0 for all the other submatrices down the first column, so we can ignore them. The submatrix of p_0, is lower triangular and so its determinant is the product of the diagonal entries, $p_1(-\lambda)^3$. So altogether the determinant of the f_2 term is $f_2 p_0 p_1 (-\lambda)^3$.

$-f_3$ **term.** Expand down the first column, pulling out a factor of p_0. The p_0 submatrix is neither upper triangular nor lower triangular, so expand down its first column, pulling out a factor of p_1. The submatrix of p_1 is lower triangular, so its determinant is $p_2(-\lambda)^2$. Altogether then, the determinant of the $-f_3$ term is $p_0p_1p_2(-\lambda)^2$.

f_4 **term.** This matrix is upper triangular, so the determinant is the product of the diagonal entries, $p_0p_1p_2p_3(-\lambda)$.

$-f_5$. **term** This matrix is upper triangular, so the determinant is the product of the diagonal entries, $p_0p_1p_2p_3p_4$.

For larger A the calculation is similar, though longer.

A.19 LIAPUNOV EXPONENTS

Recall that Liapunov exponents measure the divergence of nearby trajectories. In Sect. 2.4 we saw how to compute the Liapunov exponent for iterates of a 1-dimensional (single-variable) function. Now we'll see how to extend this notion to differential equations in three and more dimensions.

The number of Liapunov exponents equals the dimension of the system. Because each Liapunov exponent measures the exponential growth rate in one direction (not necessarily the direction of a coordinate axis), finding the long-term average growth rate can have a complication: growth in the direction of the largest Liapunov exponent can dominate the growth in all other directions. When we must rely on numerical estimates (and this we must do for all but a very few systems), the effect of dominant growth can induce errors in the straightforward calculation of the other exponents. There are methods to handle this problem; we'll introduce one.

For differential equations the core idea is similar, with iterates replaced by trajectory points sampled at equal time intervals. The computation of the largest Liapunov exponent is straightforward, but there is a complication if we want to compute all the Liapunov exponents. Briefly, the problem is that the largest Liapunov exponent eventually overpowers all the others. So every little while we need to re-orthogonalize the vectors corresponding to each Liapunov exponent. Here's the method devised by Giancarlo Benettin and coworkers [36, 37] and independently by Ippei Shimada and Tomomasa Nagashima [329], see also [389]. We'll present this method for a 3-dimensional system because this dimension exhibits enough detail to allow generalization to all dimensions.

Suppose the differential equation is

$$\vec{x}' = \begin{bmatrix} x_1' \\ x_2' \\ x_3' \end{bmatrix} = \vec{F}(x_1, x_2, x_3) = \begin{bmatrix} f_1(x_1, x_2, x_3) \\ f_2(x_1, x_2, x_3) \\ f_3(x_1, x_2, x_3) \end{bmatrix} \tag{A.14}$$

and the tangent vectors at $\vec{x}(t) = \langle x_1(t), x_2(t), x_3(t) \rangle$ evolve according to

$$\begin{bmatrix} v_1' \\ v_2' \\ v_3' \end{bmatrix} = D\vec{F}_{\vec{x}(t)} \begin{bmatrix} v_1 \\ v_2 \\ v_3 \end{bmatrix} = \begin{bmatrix} \dfrac{\partial f_1}{\partial x_1} & \dfrac{\partial f_1}{\partial x_2} & \dfrac{\partial f_1}{\partial x_3} \\ \dfrac{\partial f_2}{\partial x_1} & \dfrac{\partial f_2}{\partial x_2} & \dfrac{\partial f_2}{\partial x_3} \\ \dfrac{\partial f_3}{\partial x_1} & \dfrac{\partial f_3}{\partial x_2} & \dfrac{\partial f_3}{\partial x_3} \end{bmatrix}_{\vec{x}(t)} \begin{bmatrix} v_1 \\ v_2 \\ v_3 \end{bmatrix} \tag{A.15}$$

Given an initial position $\vec{x}(0) = \langle x_1(0), x_2(0), x_3(0) \rangle$, let $\vec{x}(t)$ denote the trajectory that is the solution of Eq. (A.14) with initial position $\vec{x}(0)$. Then given any initial vector $\vec{v}(0)$ at $\vec{x}(0)$, let $\vec{v}(t)$ denote the solution of Eq. (A.15) along the trajectory $\vec{x}(t)$.

Think of the tangent vectors as indicating directions in which the initial point $\vec{x}(0)$ can be perturbed; comparing the trajectory from the initial position $\vec{x}(0)$ and from a small perturbation of $\vec{x}(0)$ is how we'll quantify sensitivity to initial conditions. There's no reason to believe the maximum growth should occur along one of the coordinate axes. Any perturbation is a linear combination of three standard basis vectors $\vec{e}_1 = \vec{i}$, $\vec{e}_2 = \vec{j}$, and $\vec{e}_3 = \vec{k}$.

For almost any initial vector $\vec{v}(0)$, the largest Liapunov exponent, λ_1, is given by

$$\lambda_1 = \lim_{t \to \infty} \frac{1}{t} \ln\left(\frac{\|\vec{v}(t)\|}{\|\vec{v}(0)\|} \right) \tag{A.16}$$

Recall that $\|\vec{v}\|$ denotes the length of the vector $\vec{v}$, so λ_1 measures the highest exponential growth rate of vectors along a trajectory. The existence of this limit, under pretty general conditions, is guaranteed by the multiplicative ergodic theorem of Valery Oseledec [266].

A first approach to compute the other Liapunov exponents is to measure the growth rate of areas spanned by two linearly independent vectors transported along the trajectory by Eq. (A.15), and by the growth rate of volumes spanned by three linearly independent transported vectors.

$$\lambda(2) = \lim_{t\to\infty} \frac{1}{t}\ln\left(\frac{\text{area}(\vec{v}_1(t),\vec{v}_2(t))}{\text{area}(\vec{v}_1(0),\vec{v}_2(0))}\right) \tag{A.17}$$

$$\lambda(3) = \lim_{t\to\infty} \frac{1}{t}\ln\left(\frac{\text{vol}(\vec{v}_1(t),\vec{v}_2(t),\vec{v}_3(t))}{\text{vol}(\vec{v}_1(0),\vec{v}_2(0),\vec{v}_3(0))}\right) \tag{A.18}$$

Then

$$\lambda(2) = \lambda_1 + \lambda_2 \quad \text{and} \quad \lambda(3) = \lambda_1 + \lambda_2 + \lambda_3$$

and so

$$\lambda_2 = \lambda(2) - \lambda_1 \quad \text{and} \quad \lambda_3 = \lambda(3) - \lambda_1 - \lambda_2$$

This comes close to working. The problem is that the transported vectors need not stay linearly independent. We can fix this problem by occasionally adjusting the transported vectors using a process called *Gram-Schmidt orthogonalization* (see Prop. 15.2.7 of [336]), which we'll describe now. (Also, this will explain a line from the film *Hidden Figures* that occurs during the first conversation between Taraji P. Henson's character and Kevin Costner's character.) Here's how Gram-Schmidt works for a 3-dimensional basis $\{\vec{v}_1, \vec{v}_2, \vec{v}_3\}$.

First, recall that in Sect. 13.9 we saw $\|\vec{b}\|\cos(\theta)$ is the length of the component of $\vec{b}$ in the direction of $\vec{a}$. Next, from the definition of dot product (13.52) we see that this length is $(\vec{a}\cdot\vec{b})/\|\vec{a}\|$. To find the component of $\vec{b}$ in the direction of $\vec{a}$, multiply this length by the unit vector in the $\vec{a}$ direction. Then because $\|\vec{a}\|^2 = \vec{a}\cdot\vec{a}$

$$\text{component of } \vec{b} \text{ in the direction of } \vec{a} = \frac{\vec{a}\cdot\vec{b}}{\|\vec{a}\|}\frac{\vec{a}}{\|\vec{a}\|} = \frac{\vec{a}\cdot\vec{b}}{\vec{a}\cdot\vec{a}}\vec{a}$$

Then the Gram-Schmidt process turns the basis $\{\vec{v}_1, \vec{v}_2, \vec{v}_3\}$ into

$$\vec{w}_1 = \vec{v}_1$$

$$\vec{w}_2 = \vec{v}_2 - \frac{\vec{w}_1\cdot\vec{v}_2}{\vec{w}_1\cdot\vec{w}_1}\vec{w}_1$$

$$\vec{w}_3 = \vec{v}_3 - \frac{\vec{w}_1\cdot\vec{v}_3}{\vec{w}_1\cdot\vec{w}_1}\vec{w}_1 - \frac{\vec{w}_2\cdot\vec{v}_3}{\vec{w}_2\cdot\vec{w}_2}\vec{w}_2$$

That is, $\vec{w}_2$ is $\vec{v}_2$ minus the component of $\vec{v}_2$ in the direction of $\vec{w}_1$, so $\vec{w}_2$ is orthogonal to $\vec{w}_1$. Also, $\vec{w}_3$ is $\vec{v}_3$ minus the component of $\vec{v}_3$ in the direction of $\vec{w}_1$ and minus the component of $\vec{v}_3$ in the direction of $\vec{w}_2$. How to generalize this to higher-dimensional spaces is clear.

One final adjustment. In order to keep the vectors from growing too long, convert each of the Gram-Schmidt basis vectors into unit vectors

$$\vec{u}_1 = \frac{\vec{w}_1}{\|\vec{w}_1\|}, \quad \vec{u}_2 = \frac{\vec{w}_2}{\|\vec{w}_2\|}, \quad \text{and} \quad \vec{u}_3 = \frac{\vec{w}_3}{\|\vec{w}_3\|}$$

We call these the *Gram-Schmidt orthonormal basis.*

At given time intervals, replace the evolved vectors $\vec{v}_1(t)$, $\vec{v}_2(t)$, and $\vec{v}_3(t)$ with the Gram-Schmidt orthonormal basis constructed from the evolved vectors. With this adjustment, the Liapunov exponents λ_1, λ_2, and λ_3 can be computed from Eqs. (A.16), (A.17), and (A.18).

We'll mention one application. Suppose for an n-dimensional system we write the Liapunov exponents in non-increasing order, $\lambda_1 \geq \lambda_2 \geq \cdots \geq \lambda_n$. In order for the trajectories to converge to an attractor (a fixed point, a limit cycle, or something more complicated if $n \geq 3$), the sum of the Liapunov exponents must be negative. Take m to be the integer determined by

$$\lambda_1 + \cdots + \lambda_m \geq 0 \quad \text{and} \quad \lambda_1 + \cdots + \lambda_{m+1} < 0$$

Then the *Kaplan-Yorke conjecture* [122, 189] is that the dimension of the attractor is

$$d = m + \frac{\lambda_1 + \cdots + \lambda_m}{|\lambda_{m+1}|} \tag{A.19}$$

Often this equals the box-counting dimension of the attractor.

A.20 STOCHASTIC RESONANCE AND THE DUFFING EQUATION

Stochastic resonance is easier to understand once we have a clear example in mind. The forced Duffing equation (13.43)

$$x'' + ax' + bx + cx^3 = d\cos(\omega t + \varphi)$$

will be our guide. In Sect. 13.8 we saw that the unforced ($d = 0$) Duffing equation has three fixed points, $(0,0)$ and $(\pm\sqrt{-b/c}, 0)$. We'll take $a = 0.05$, $b = -0.5$, and $c = 0.5$. Then the eigenvalues of the derivative matrix at $(0,0)$ are $\lambda \approx 0.6825$ and $\lambda \approx -0.7325$, so the origin is a saddle point. At the fixed points $(\pm 1, 0)$ the eigenvalues of the derivative matrix are $\lambda \approx -0.025 \pm 0.9997i$, so these fixed points are asymptotically stable spirals.

In the forced Duffing equation we'll take $\omega = 0.05$ and $\varphi = 0$. First we'll explore the effect of the forcing amplitude d. The left graph of Fig. A.7 was generated with the forcing amplitude $d = 0.1$. In all these figures, the upper horizontal line occurs at the fixed point $x = 1$, presented on the vertical axis, while the lower occurs at $x = -1$. The gray middle horizontal line is the graph of the added noise, 0 in both plots of Fig. A.7. The middle dark graph is the forcing cosine function. The initial values are $x(0) = 1$ and $x'(0) = 0$. The left plot shows a rapid decay to match the forcing cosine function, oscillating around $x = 1$. The forcing amplitude is not large enough to move the system state from $x = 1$ to $x = -1$.

In contrast, the right plot of Fig. A.7 was generated with a forcing amplitude $d = 0.2$. Here we see that roughly from the max to the min of the forcing signal, $x(t)$ oscillates around $x = 1$, while from the min to the max of the forcing signal, $x(t)$ oscillates around $x = -1$. In this sense, $x(t)$ has been been locked in a periodic oscillation sharing many of the properties of the forcing cosine. The forcing signal and $x(t)$ are in resonance.

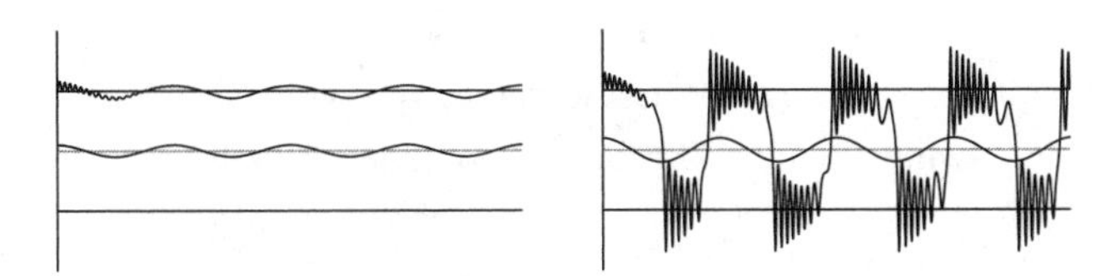

Figure A.7. An effect of the forcing amplitude, $d = 0.1$ (left) and $d = 0.2$ (right).

Now we'll change the forced Duffing equation to

$$x'' + ax' + bx + cx^3 = d\cos(\omega t + \varphi) + \xi(t)$$

where $\xi(t)$ is a random noise term. We'll keep $d = 0.1$, too small to lock $x(t)$ into resonance with the forcing signal. (Recall the left plot of Fig. A.7.) In Fig. A.8 we see two examples for different random signals $\xi(t)$, both uniformly distributed in the range $-0.15 \le \xi(t) \le 0.15$. This is stochastic resonance. The addition of a small amplitude noise is enough to lock $x(t)$ into approximate resonance with the forcing signal, in the sense that $x(t)$ oscillates around $x = 1$ when the forcing signal is near its max, and $x(t)$ oscillates around $x = -1$ when the forcing signal is near its min. Not as cleanly as in the right image of Fig. A.7, because the randomness added to x'' alters x, with some delays and complicated interaction with the forcing signal.

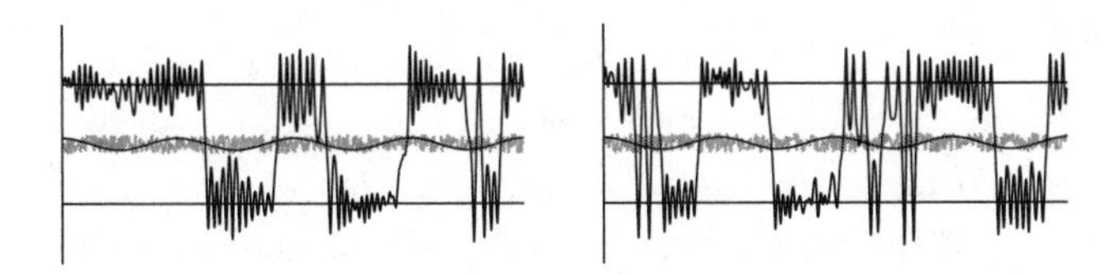

Figure A.8. Two examples of stochastic resonance.

The amplitude of the added noise is chosen to be in what we'd call the "Goldilocks zone," if astronomers searching for exoplanets had not already made this term their own. In the context of stochastic resonance, we mean this:

much lower amplitudes don't move $x(t)$ from a neighborhood of $x = 1$ to a neighborhood of $x = -1$; much higher amplitudes induce $x(t)$ to hop between $x = 1$ and $x = -1$ in a way more or less determined by the swings of the noise term $\xi(t)$.

Fig. A.9 illustrates this. The left plot is generated with $-0.05 \leq \xi(t) \leq 0.05$, the right with $-0.30 \leq \xi(t) \leq 0.30$. Indeed, we see the left plot is a fuzzy cosine graph centered along $x = 1$, while the right jumps randomly between about $x = 1$ and $x = -1$. While in some regions the graph of $x(t)$ swings rapidly between extremes, in others it oscillates about one of these extremes for a while. We expect lot of swings, but not only swings. Is there a simple way to understand these three regions, too cold (no big jumps), just right (stochastic resonance, most jumps tied to sign changes of the forcing function), and too hot (lots of jumps)?

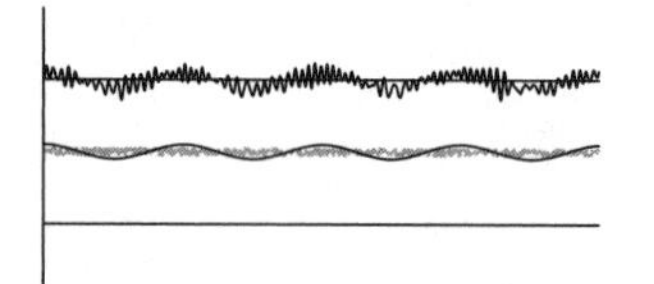

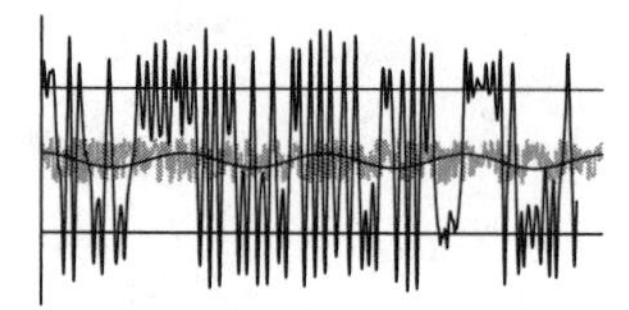

Figure A.9. Random amplitude too low (left) and too high (right).

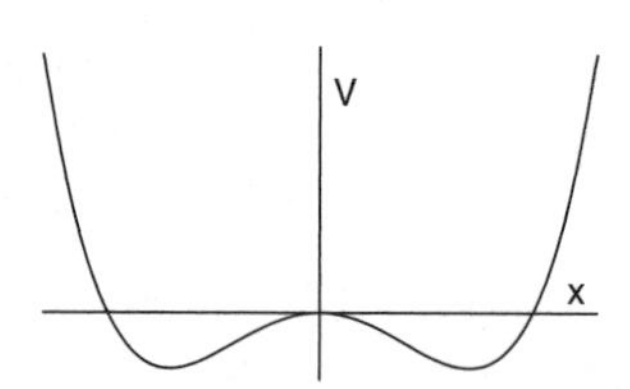

Figure A.10. Two-well potential.

Think of the potential function for the unforced Duffing equation sketched in Fig. 13.27, reproduced in Fig. A.10. We'll give a model now that isn't quite right, but it's pretty close. Imagine the state of the system is described by a marble rolling along this potential curve. As we mentioned in Sect. 13.8, we can view the forcing function $d\cos(\omega t + \varphi) + \xi(t)$ as shaking the potential curve left and right. If $\xi(t) = 0$ and d is small, the marble rolls back and forth in whichever well it started: the forcing amplitude isn't sufficient to push the marble over the barrier separating the two wells. This is the left image of Fig. A.7. Increase d enough and the marble hops between wells in time with the forcing. This is the right image of Fig. A.7.

Reset d to the lower level, and add a small amount of random noise to the shaking. The marble stays in its well, but now its motion reflects the periodic shaking plus noise, a fuzzy cosine. This is the right side of Fig. A.9. If we add a bit more noise, sometimes the noise plus the periodic motion is enough to make the marble hop across the barrier, more or less at the maximum left extent or maximum right extent of the periodic shaking. This is stochastic resonance, illustrated in Fig. A.8. Increase the noise amplitude enough and the loudest noise makes the marble hop between wells. This is the right image of Fig. A.9.

Okay, that's just a cartoon of how stochastic resonance works. We used the Duffing equation for an illustration because its two asymptotically stable fixed points are the wells between which the system oscillates. But of course we can find stochastic resonance in systems more complicated than this.

Many people have studied stochastic resonance. In particular, James Collins and the late Frank Moss have been very active in this field, in its theory and especially in the breadth of applications. We'll mention a few, to help you decide if you want to learn a bit more about this topic. The surveys [243, 383] are good places to start.

In [297, 298], Collins and his coworkers presented experimental evidence that vibrations applied to the soles of the feet below the level of conscious sensory notice can act by stochastic resonance to amplify somatosensory feedback and improve balance. Applications [299] to improved balance control in stroke patients and in patients with diabetic neuropathy are encouraging. Using noise to increase tactile sensitivity through stochastic resonance is explored in [76].

In [157, 262, 263], Collins and Moss and their coworkers showed that in some circumstances (experiments with rat neurons, and analytical methods), $1/f$ noise can be more effective than white noise at enhancing response to a weak signal through stochastic resonance. Also, in addition to the periodic forcing signals of our examples, stochastic resonance can amplify weak aperiodic signals [75, 162]; observations of this effect in the Fitzhugh-Nagumo equations suggest avenues for sensory neurons to detect weak signals. Synchronization in neuron models under stochastic resonance is studied by Collins, Moss, and colleagues in [251, 252, 25].

In addition, Collins, Moss, and others [38, 39, 40, 126, 241, 242, 244] applied stochastic resonance to networks, for example, two-state ion channels, to investigate the role of internal noise, for example, thermal fluctuations in ion gate molecular configurations. Moss and coworkers [333, 332] use stochastic resonance to measure human visual sensitivity in the presence of optical noise, and obtained fMRI data that suggests the visual cortex is involved in computation.

Stochastic resonance is observed in non-humans, too, reinforcing its evolutionary advantage. For example, Moss, Sonya Bahar, and coworkers found stochastic resonance in the caudal photoreceptor and mechanoreceptor systems of the crayfish (and in analog and digital computer simulations of these systems) [17, 18, 19, 93, 94, 215], synchronization in the electrosensory system of the paddlefish [250], and in how juvenile paddlefish locate plankton swarms [124, 145] (plankton movement generates the noise field), in optimal foraging paths

[82]. Non-human instances of stochastic resonance remind us that evolution reuses useful techniques it has discovered.

A.21 PROOF OF LIENARD'S THEOREM

Here we'll sketch the proof of Lienard's criterion, Thm. 9.8.1, for the existence of a limit cycle for a Lienard system (9.27), rewritten here so you won't need to flip back through a lot of pages.

$$x' = y - F(x) \qquad \text{and} \qquad y' = -g(x) \tag{A.20}$$

We'll follow the presentation in Sect. 3.8 of Lawrence Perko's excellent text [281].

Theorem A.21.1. Suppose F and g are odd functions. The Lienard system (A.20) has a single limit cycle, and that limit cycle is stable, if
1. $xg(x) > 0$ for all $x \neq 0$,
2. $F(0) = 0$,
3. $F'(0) < 0$,
4. F has a single positive zero that is at $x = a$, and
5. for all $x \geq a$, $F(x)$ increases monotonically to ∞ as $x \to \infty$.

Proof. First we'll show that the origin is the only fixed point. The x-nullcline is $y = F(x)$ and the y-nullcline is $g(x) = 0$. Because g is an odd function, $g(0) = 0$, while by hypothesis 1, $g(x) \neq 0$ for $x \neq 0$. Consequently, the y-nullcline is $x = 0$. Then by hypothesis 2, the x-nullcline $y = F(x)$ intersects the y-nullcline $x = 0$ at the point $(0, F(0)) = (0,0)$.

Second, for the vector field $\vec{V}(x,y) = \langle y - F(x), -g(x)\rangle$, note that

$$\begin{aligned}\vec{V}(-x,-y) &= \langle -y - F(-x), -g(-x)\rangle = \langle -y + F(x), g(x)\rangle \\ &= -\langle y - F(x), -g(x)\rangle = -\vec{V}(x,y)\end{aligned}$$

where the second equality follows because both F and g are odd. Now the transformation $(x,y) \to (-x,-y)$ is reflection across the y-axis and reflection across the x-axis, and this is equivalent to rotation by $180°$ about the origin. So we've shown that if we rotate a point (x,y) by $180°$ around the origin, the vector field at the rotated point is the $180°$ rotation of the vector field at the original point. This means, for example, that the trajectories on the left side of the y-axis are the rotation of the trajectories on the right side of the y-axis. This will give us a clean way to formulate the existence of a limit cycle.

We'll focus on the right side of the y-axis. Because hypothesis 1 guarantees that $g(x) > 0$ for $x > 0$, we see that $y' < 0$ for all points to the right of the y-axis.

Next, $x' > 0$ for $y > F(x)$ and $x' < 0$ for $y < F(x)$. That is, to the right of the y-axis the vector field points SE at points above $y = F(x)$ and SW at points below $y = F(x)$.

Now suppose a trajectory Γ^+ starts at a point $P_0 = (0, y_0)$ on the y-axis. The trajectory will continue SE until it crosses the x-nullcline $y = F(x)$ vertically (downward) at the point $P_2 = (x_2, y_2)$, then continues SW until it reaches the y-axis at $P_4 = (0, y_4)$. Here's where we use the rotational symmetry observation above. To the left of the y-axis, at points below the x-nullcline the vector field points NW, and at points above the x-nullcline the vector field points NE. Now continue Γ^+ to the left side of the y-axis. Call this trajectory Γ^-. If $y_4 = -y_0$, then rotational symmetry shows that Γ^- is the rotation of Γ^+ and so $\Gamma = \Gamma^+ \cup \Gamma^-$ is a closed trajectory.

To show $y_4 = -y_0$, we'll use the Liapunov function (Sect. 9.4)

$$V(x,y) = \frac{y^2}{2} + G(x) \qquad \text{(A.21)}$$

where $G(x) = \int_0^x g(t)\, dt$. Because $G(0) = 0$ we see that $y_4 = -y_0$ if and only if $V(0, y_0) = V(0, y_4)$, so that's what we'll show.

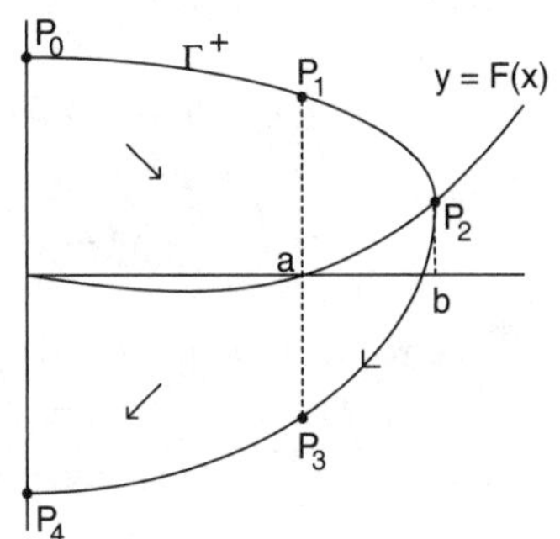

Figure A.11. Lienard geometry

Recall that trajectories can cross only at fixed points. Because the only fixed point is the origin, trajectories cannot cross except at the origin. So if we start another trajectory Γ' from a point $(0, y_0')$, it will lie entirely inside the region bounded by Γ^+ and the y-axis, or entirely outside that region. The right-most point of a trajectory occurs at the maximum value of the x-coordinate; that is, where $x' = 0$. In other words, where the trajectory crosses the x-nullcline. Because trajectories cannot cross except at the origin, to the right of the y-axis two trajectories cannot share their intersection with the x-nullcline and so the x-coordinate of the right-most point of a trajectory is unique to that trajectory. Call this x-coordinate b. Then we have shown that this line integral

$$\varphi(b) = \int_{\Gamma^+} \nabla V \cdot d\vec{r} = V(0, y_4) - V(0, y_0) \qquad \text{(A.22)}$$

defines a function of b. The second equality follows from an application of the fundamental theorem of line integrals, Thm. 17.2.1.

We'll use several reformulations to evaluate this line integral.

$$\int_{\Gamma^+} \nabla V \cdot d\vec{r} = \int_{\Gamma^+} \frac{\partial V}{\partial x}\, dx + \frac{\partial V}{\partial y}\, dy = \int_{\Gamma^+} g(x)\, dx + y\, dy \qquad \text{(A.23)}$$

In the second expression of (A.23), use the differential equation (A.20) to deduce $dx = (y - F(x))\, dt$ and $dy = (-g(x))\, dt$. Take t_0 to be the time when Γ^+ begins at P_0 and t_4 the time when it ends at P_4, we have

$$\int_{\Gamma^+} \frac{\partial V}{\partial x}\, dx + \frac{\partial V}{\partial y}\, dy = \int g(x)(y - F(x))\, dt + y(-g(x))\, dt$$
$$= \int F(x)(-g(x))\, dt = \int_{\Gamma^+} F(x)\, dy \tag{A.24}$$

For the third formulation, use the chain rule and the differential equation (A.20) to deduce

$$\frac{dy}{dx} = \frac{dy/dt}{dx/dt} = \frac{-g(x)}{y - F(x)} \tag{A.25}$$

Then return to Eq. (A.23) and find

$$\int_{\Gamma^+} g(x)\, dx + y\, dy = \int_{\Gamma^+} \left(g(x) + y\frac{dy}{dx} \right) dx$$
$$= \int_{\Gamma^+} \left(g(x) + y\frac{-g(x)}{y - F(x)} \right) dx = \int_{\Gamma^+} \frac{-F(x)g(x)}{y - F(x)}\, dx \tag{A.26}$$

Back to the function φ. The trajectory Γ is closed if and only if $\varphi(b) = 0$. We'll show that $\varphi(b)$ has exactly one 0, at $b = b_0$, and that $b_0 > a$, where we recall that a is defined by $F(a) = 0$.

For $b < a$, all along the x-values of Γ^+ we have $F(x) < 0$. For all $x > 0$, hypothesis 1 gives $g(x) > 0$. Because we follow Γ^+ in the direction of the vector field, we have $dt > 0$ all along Γ^+. Then $dy = -g(x)\, dt < 0$ so $F(x)\, dy > 0$. Combining these pieces and using Eq. (A.24), we find

$$\varphi(b) = \int_{\Gamma^+} \nabla V \cdot d\vec{r} = \int_{\Gamma^+} F(x)\, dy > 0$$

if $b < a$.

For $b \geq a$ we'll show that $\varphi(b)$ is a decreasing function of b. This argument is a bit complicated; the final step is to show that $\varphi(b) \to -\infty$ as $b \to \infty$.

We begin by splitting Γ^+ into three pieces: Γ_1^+ between P_0 and P_1, Γ_2^+ between P_1 and P_3, and Γ_3^+ between P_3 and P_4. And let $\varphi_1(b)$, $\varphi_2(b)$, and $\varphi_3(b)$ denote $\varphi(b)$ restricted to Γ_1^+, Γ_2^+, and Γ_3^+.

Along Γ_1^+ and Γ_3^+ we have $F(x) < 0$ and $g(x) > 0$. Recalling $dx/dt = y - F(x)$, we see $dx/(y - F(x)) = dt > 0$. Then by Eq. (A.26) we have

$$\varphi_1(b) > 0 \quad \text{and} \quad \varphi_3(b) > 0$$

Along Γ_2^+ the only change from the Γ_1^+ and Γ_3^+ calculation is that now $F(x) > 0$ and so we find

$$\varphi_2(b) < 0$$

What happens to $\varphi_1(b)$, $\varphi_2(b)$, and $\varphi_3(b)$ as b increases? Because these trajectories can cross only at the origin, increasing b raises Γ_1^+ and lowers Γ_3^+, but in both cases the x limit, $x = a$, remains unchanged. See the left image of Fig. A.12. Let's consider Γ_1^+. The range of x-values, $0 \leq x \leq a$, is unchanged, while y increases along this raised trajectory. Consequently, the integrand $-F(x)g(x)/(y - F(x))$ of Eq. (A.26) decreases and so $\varphi_1(b)$ decreases as b increases. A similar argument shows that $\varphi_3(b)$ decreases with increasing b.

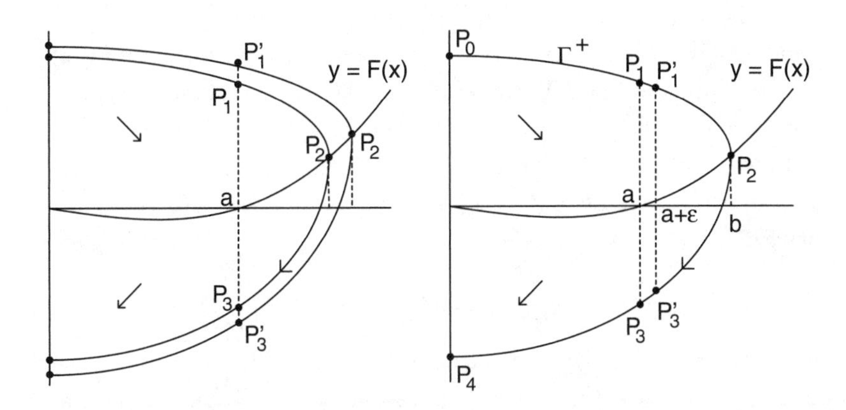

Figure A.12. The effect of increasing b (left). Reducing the middle segment of the trajectory (right).

For Γ_2^+, increasing b moves $P_1 = (a, y_1)$ upward to $P_1' = (a, y_1')$ and moves $P_3 = (a, y_3)$ downward to $P_3' = (a, y_3')$. We'll use Eq. (A.24) to calculate the change in $\varphi_2(b)$. Because $F(x) > 0$ for $x > a$, and because we evaluate the line integral moving downward along this path so $dy < 0$, integrating over this larger range of y-values gives a more negative result. That is, $\varphi_2(b)$ decreases (becomes more negative) as b increases.

Combining the arguments of the last two paragraphs, we see that for $b > a$, $\varphi(b)$ is a decreasing function.

Finally, we'll show that as $b \to \infty$, $\varphi(b) \to -\infty$. Because $\varphi_1(b)$ and $\varphi_3(b)$ decrease with b, it's enough to show that $\varphi_2(b) \to -\infty$. We'll use Eq. (A.24) to evaluate $\varphi_2(b)$. For small $\epsilon > 0$, let $P_1' = (a+\epsilon, y_1 - \epsilon')$ and $P_3' = (a+\epsilon, y_3 + \epsilon'')$ be points on Γ_2^+. See the right side of Fig. A.12. Let $\Gamma_{2'}^+$ be the portion of Γ_2^+ between P_1' and P_3'. Then

$$\varphi_2(b) = \int_{\Gamma_2^+} F(x)\, dy \approx \int_{\Gamma_{2'}^+} F(x)\, dy$$

and the approximation can be made as close as we like by taking ϵ very small. By hypothesis 5, for $x \geq a+\epsilon$, $F(x) \geq F(a+\epsilon) > 0$. Consequently,

$$\int_{\Gamma_{2'}^{+}} F(x)\,dy \geq \int_{\Gamma_{2'}^{+}} F(a+\epsilon)\,dy = F(a+\epsilon)\int_{\Gamma_{2'}^{+}} dy$$
$$= F(a+\epsilon)((y_3+\epsilon'')-(y_1-\epsilon')) \geq F(a+\epsilon)(-(y_1-\epsilon'))$$

The last inequality follows because $y_3+\epsilon'' < 0$. Again by hypothesis 5 we know that $y_1 \to \infty$ as $b \to \infty$ and consequently $\varphi_2(b) \to -\infty$, and so $\varphi(b) \to -\infty$ as $b \to \infty$. Because $\varphi(b) > 0$ for $b < a$, $\varphi(b)$ is monotonically decreasing for $b \geq a$, and because $\varphi(b) \to -\infty$ as $b \to \infty$, we see there is exactly one value of b, $b = b_0$, at which $\varphi(b_0) = 0$. If you remember where we began this argument, this shows that the differential equation (A.20) has exactly one limit cycle.

Finally, we show this limit cycle is stable. Recall that $\varphi(b) > 0$ for $b < b_0$ and $\varphi(b) < 0$ for $b > b_0$. Because $G(0) = 0$, from Eq. (A.21) we see $V(0,y) = y^2/2$, and then Eq. (A.22) gives

$$\text{for } b < b_0 0 < \varphi(b) = \frac{y_4^2}{2} - \frac{y_0^2}{2} \qquad \text{and for } b > b_0 0 > \varphi(b) = \frac{y_4^2}{2} - \frac{y_0^2}{2}$$

That is, for $b < b_0$ we have $y_4^2 > y_0^2$ and so the trajectory spirals out toward the limit cycle. For $b > b_0$ we have $y_4^2 < y_0^2$ and so the trajectory spirals in toward the limit cycle. This shows the limit cycle is stable. □

A.22 A BIT OF MOLECULAR GENETICS

Here we provide a snapshot of some of the biochemical mechanisms whose effects we study in Sects. 13.4 and 13.5, and in Chapters 15, 16, 18, and 19. That is, we present just enough of the mechanics of genetics to provide background for these sections and chapters.

DNA is a double helix sugar-phosphate backbone, the two spirals connected by rungs that are pairs of four bases: adenine (A), cytosine (C), guanine (G), and thymine (T). An A on one spiral always is paired with a T on the other, and a G with a C. We say A and T are *complementary*, as are G and C. The nucleotides A and G are purines (caffeine is another), C and T (and uracil and vitamin B_1) are pyrimidines. The discovery of the structure of DNA is an amazing scientific detective story, recounted in many places. But for a clear portrait of the struggle of a scientific quest, of the impossible privilege of understanding how a bit of the world works, I recommend the novel *The Gold Bug Variations* [290] by Richard Powers.

RNA usually is a single spiral sugar-phosphate backbone to which the bases A, C, G, and U (uracil) are attached. The base U is complementary to A. Arranged in triples, these four bases produce $4^3 = 64$ combinations, called *codons*. These code for the 20 amino acids, so some redundancy arises necessarily. In a moment we'll say what "code for" means. The 20 amino acids, their abbreviations, and their codons are shown in Table A.1.

Why do codons consist of three bases? From the bases A, C, G, and T, length-2 codons could encode only $4^2 = 16$ amino acids. Yet evolution discovered the need for 20, so 3 is the smallest codon length that can direct the construction of 20 amino acids. Some redundancy is a consequence: in all but two cases, at least 2 base combinations code an amino acid.

The order in which the codon is read is important: while GCU gives alanine, UCG gives serine. So codons have a direction. At one end of the nucleotide chain, called the $5'$ end, the nucleic acid chain terminates at the fifth carbon of the sugar ring. At the other end, the $3'$ end, the chain terminates with the hydroxyl group of the third carbon of the sugar ring. With a phosphate group attached to the $5'$-end, the $5'$-end of one nucleotide can bind to the hydroxyl group of the third carbon on the $3'$-end of another. Codons are read from the $5'$ end to the $3'$ end. The same order, from $5'$ to $3'$, is followed when DNA sequences replicate. Evolution reuses, or rediscovers, successful motifs.

Here's a sketch of how DNA synthesizes proteins. An enzyme called RNA polymerase opens part of the DNA strand and constructs a strand of mRNA with bases complementary to those of the DNA strand. The construction of this mRNA strand is called *transcription*.

How does a codon of mRNA bases encode an amino acid? This is a job of tRNA (transfer RNA). Very roughly, one end of a tRNA molecule has a loop, called the *anticodon loop*, which contains the complement of the codon whose amino acid it encodes. The amino acid is attached to the other end of the tRNA molecule. For example, suppose the next codon on the mRNA is UCG. The tRNA with anticodon AGC has serine attached to the other end. This anticodon binds with the UCG and delivers serine to the ribosome that constructs the protein. This is called *translation*.

Nobel laureate Venki Ramakrishnan's book *Gene Machine* [303] is a delightful sketch of the science, and the politics, of the discovery of the structure and function of ribosomes. If you've ever wondered what X-ray crystallography has to do with the structure of biomolecules, this book will show you. And it contains a detailed explanation of how ribosomes do the remarkable things they do, and how they may be leftovers from an earlier RNAworld.

The *central dogma* of molecular biology was proposed first by Francis Crick [79] in the late 1950s. In its original form it stated that information can be transferred between nucleic acids, or from nucleic acid to protein, but not

alanine (ala, A)	GCU, GCC, GCA, GCG
arginine (arg, R)	CGU, CGC, CGA, CGG
asparagine (asn, N)	AAU, AAC
aspartic acid (asp, D)	GAU, GAC
cysteine (cys, C)	UGU, UGC
glutamine (gln, Q)	CAA, CAG
glutamic acid (glu, E)	GAA, GAG
glycine (gly, G)	GGU, GGC, GGA, GGG
histidine (his, H)	CAU, CAC
isoleucine (ile, I)	AUU, AUC, AUA
leucine (leu, L)	UUA, UUG
lysine (lys, K)	AAA, AAG
methionine (met, M), the start codon	AUG
phenylalanine (phe, F)	UUU, UUC
proline (pro, P)	CCU, CCC, CCA, CCG
serine (ser, S)	UCU, UCC, UCA, UCG
threonine (thr, T)	ACU, ACC, ACA, ACG
tryptophan (trp, W)	UGG
tyrosine (tyr, Y)	UAU, UAC
valine (val, V)	GUU, GUC, GCA, GUG
the stop codons	UAA, UAG, UGA

Table A.1. Amino acids and their codons.

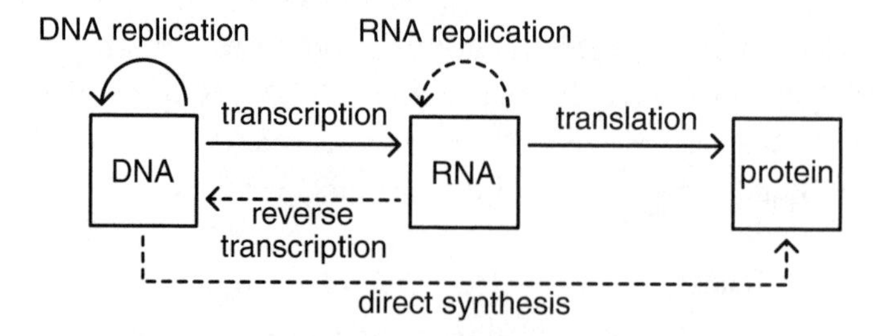

Figure A.13. The central dogma of molecular biology.

between proteins and not from proteins to nucleic acids. The standard pathways in the central dogma are indicated by the solid arrows in Fig. A.13; pathways that can occur under some circumstances are shown by the dashed arrows. The main operations allowed by the central dogma account for DNA replication and for protein synthesis.

A.22.1 SEQUENCE CHANGES

We'll describe operations that can change a DNA sequence. The first three we can illustrate for single sequences; the fourth requires two sequences.

1. A *point mutation* is a change of a single base in a DNA or RNA sequence. (Some authors consider the insertion or a deletion of a single base to be a point mutation.) For example,

$$ACTGGAT\underline{A}TCCAC \rightarrow ACTGGAT\underline{C}TCCAC$$

where the changed base is underlined. A *transition* is the replacement of one purine with another (interchange A and G) or of one pyrimidine with another (interchange C and T). A *transversion* is the replacement of a purine with a pyrimidine, or of a pyrimidine with a purine.

Functionally, point mutations fall into three classes: silent, nonsense, and missense.

1a. A *silent mutation* replaces one base by another so both the mutated and the unmutated codons code for the same amino acid. For example, A A U can mutate to A A C, both of which code for asparagine. Silent mutations do not affect the protein coded by that portion of the DNA.

1b. A *nonsense mutation* adds or deletes a STOP or START codon. For example, the codon CGA, which codes for arginine, can point mutate to TGA, a STOP codon. This will shorten the protein, altering, or destroying, its function. A stop codon can be changed to code for an amino acid. This lengthens the protein, which also affects function. Deletion or insertion of a START codon can have analogous effects.

1c. A *missense mutation* replaces one base by another so the mutated and unmutated codon code for different amino acids. For example, in the beta hemoglobin gene the point mutation from the GAG codon to the GUG codon replaces the amino acid glutamic acid with valine. This single mutation causes sickle cell anemia.

But missense mutations also are important drivers of evolution. They're why you are a *Homo sapiens* rather than a bacterium. Okay, that's a bit of an oversimplification, but not as much as you might think.

2. An *insertion mutation* is the addition of a single base pair or a sequence of base pairs into a chromosome. For example,

$$ACTGGATATCCAC \rightarrow ACTGGATA\underline{CGGTC}TCCAC$$

where the inserted sequence is underlined on the right.

This can cause troubles anywhere in the genome, but we'll focus on what can happen if the insertion occurs in a sequence that codes for a protein. Recall that the sequence is organized into groups of three nucleotides called codons, and the sequence of codons determines the sequence of amino acids in a protein. This organization is called the *reading frame* of the sequence. A sequence has three reading frames. For example,

AGT GGA CTA CAA ATG CTG CCT AGA AGT
A GTG GAC TAC AAA TGC TGC CTA GAA GT
AG TGG ACT ACA AAT GCT GCC TAG AAG T

The reading frame begins when the START codon AUG is encountered, and it extends until a STOP codon is encountered.

If the number of inserted bases is divisible by 3, this is called an *in-frame insertion.* Usually this changes the number of amino acids in the protein and can change which acids occur in what order. In-frame insertions can modify, or destroy, the functionality of the protein.

If the number of inserted bases is not divisible by 3, this is called a *frameshift mutation* because it alters the reading frame in all positions after the insertion. In addition to changing the amino acid sequence, this reading frame shift can insert a STOP codon earlier in the sequence and so produce a truncated protein sequence. For example, if a sequence of length $3k+1$ is inserted at the start of a codon and downstream from the insertion the original DNA sequence included GGA TCT CCG, then the insertion and transcription will give

original DNA reading frame... *GGA TCT CCG* ...
shifted DNA reading frame... *GG ATC TCC G*...
transcribed mRNA... *CC UAG AGG C*...

and we see the STOP codon UAG occurs early in the mRNA transcribed from the shifted DNA. This can cause all manner of mischief.

3. A *deletion mutation* is the removal of a single base pair or a sequence of base pairs from a chromosome. For example,

$$ACT\underline{GG}ATATCCAC \rightarrow ACTATATCCAC$$

where the deleted sequence is underlined on the left.

As with insertions, deletions can result in frameshift mutations. Deletions of large sequences usually are fatal, but this is no surprise. Deletions are responsible for some cases of Duchenne muscular dystrophy and for some cases of cystic fibrosis, for example.

4. A *recombination* is shuffling bits of two strings. For example

$$ACTGGATATCCAC \rightarrow ACTGGA\underline{CAGCTTA}$$
$$\underline{GGTCACCAGCTTA} \rightarrow \underline{GGTCAC}TATCCAC$$

To distinguish the strings, one is underlined.

When a recombination occurs in a protein coding region, then a frameshift mutation results if the recombination site is not the beginning of a codon.

But remember the local maximum problem in fitness landscapes: selection pressure moves the genotype in the direction determined by the local gradient. This will find a local maximum, but a higher local maximum may be at some other location and to get there the fitness must endure a temporary fitness reduction. Reproductive success is the measure of fitness, so lower fitness means that the part of the population that remains at the local max will out-reproduce everything nearby. Survival involves individuals, not populations; anything that reduces individual fitness is selected against. With recombination an individual's offspring can travel great distances across the fitness landscape. Their descendants can locate another local fitness maximum, perhaps higher that the previous one.

A.22.2 SEQUENCE CORRELATIONS

As mentioned earlier, the human genome consists of about 3.2 billion base pairs, organized in 23 pairs of chromosomes that include about 19,000 genes. Another way to partition the sequence is by activity. About 1% of the human genome consists of *exons*, carriers of the codons for protein formation. Typically, a gene includes several exons. About 24% of our genome is *introns*, important for the regulation of gene expression, but they do not contribute to protein formation. In the formation of mRNA to produce a protein, the introns are removed from the copy and the exons are spliced together. The rest of our DNA is *intergenic*, regions between genes. At the moment, these still are mysterious, though promoters and enhancers have been identified in these regions. Do exon regions exhibit any long-range correlations? Do intron regions? We'll discuss a bit of the evidence.

Richard Voss [368, 369] is a pioneer in the mathematical study of correlations in DNA sequences, an active field now home to many researchers and many approaches. Voss and others studied the correlations of these sequences. Graphical approaches are tricky, because as Voss pointed out, any plot in a space of fewer than 4 dimensions will introduce spurious correlations. Think of the shadow of a 3-dimensional wire-frame model. The shadows of two wires can cross even though the wires don't touch.

The points of a plot in 4 dimensions are easy to generate, but any graphical representation must be a projection into 3 dimensions—2, really, if it's a picture on a page. Better is to compute the *correlation function* in order to find the strength of the relationship between nucleotides z_i positioned a distance w apart in the sequence:

$$C(w) = \langle z_i z_{i+w} \rangle$$

But how do we multiply nucleotides? What is the product of A and C? Voss defined the product this way:

$$z_i z_{i+w} = \begin{cases} 1 & \text{if } z_i \text{ and } z_{i+w} \text{ are the same nucleotide} \\ 0 & \text{if } z_i \text{ and } z_{i+w} \text{ are different nucleotides} \end{cases}$$

To illustrate this, we'll use the DNA sequence for amylase, an enzyme found in saliva that catalyzes the hydrolysis of starches into sugars. The sequence consists of 3,957 bases and begins T,G,A,A,T,T,C, ... Then

$$\begin{aligned} z_1 z_2 &= 0 \quad \text{because } z_1 = T \text{ and } z_2 = G \\ z_2 z_3 &= 0 \quad \text{because } z_2 = G \text{ and } z_3 = A \\ z_3 z_4 &= 1 \quad \text{because } z_3 = A \text{ and } z_4 = A \end{aligned}$$

and so on.

1 2 3 4

Figure A.14. Graphical representations of the first 200 products $z_i z_{i+w}$, $w = 1, 2, 3, 4$, for amylase.

In Fig. A.14 we plot the first 200 $z_i z_{i+w}$ for $w = 1, 2, 3$, and 4. The correlation functions for each of these are

$$\begin{aligned} C(1) &= \frac{1}{3956}(z_1 z_2 + z_2 z_3 + \cdots + z_{3956} z_{3957}) \approx 0.3079 \\ C(2) &= \frac{1}{3955}(z_1 z_3 + z_2 z_4 + \cdots + z_{3955} z_{3957}) \approx 0.3016 \\ C(3) &= \frac{1}{3954}(z_1 z_4 + z_2 z_5 + \cdots + z_{3954} z_{3957}) \approx 0.3002 \\ C(4) &= \frac{1}{3953}(z_1 z_5 + z_2 z_6 + \cdots + z_{3953} z_{3957}) \approx 0.3018 \end{aligned}$$

and so on.

Voss applied this method to the complete genome of human Cytomegalovirus strain AD 169, which has 229,354 bases. He found that $C(w)$ depends on w in roughly the same way as in fractal time series, with similar correlations across many scales. For individual genomes, $C(w)$ depends on w in a noisy way. To obtain smoother graphs, Voss averaged over GenBank classifications and found similar fractal dependencies for bacteria, viruses, plants, invertebrates, and vertebrates. With annotated GenBank entries, Voss established power law scalings (fractal correlations) within introns and within exons over large ensembles.

Across all forms of life, DNA exhibits fractal correlations. At the moment, we don't know why.

A.22.3 DNA GLOBULES

Finally, we'll discuss how 2 meters of unwound DNA can be curled up into the tiny nucleus of a cell so that any bit of it can be unwound for copying or for transcription in protein production.

Unwound and pulled straight, the DNA from one of our cells would stretch to nearly two meters, longer than some of us are tall. Placed end-to-end, the unwound DNA of each of our ten trillion cells would span the diameter of Pluto's orbit. The difficult task of folding this DNA into the cell nucleus, which has a diameter of about five millionths of a meter, is even more challenging than folding the 130-square-meter lungs into the 5-liter chest cavity, but it turns out that both folding problems have something in common. For DNA, this folding is accomplished by *histones*, protein components of *chromatin*, the combination of proteins and DNA making up cell nuclei. Chromatin compacts DNA into a tiny volume, helps prevent damage to DNA, allows DNA replication during cell division, and controls gene expression. Chromatin has two forms: more tightly folded *heterochromatin* and more loosely folded *euchromatin*. The looser folds of euchromatin allow gene regulatory proteins to bind with DNA in order to initiate gene transcription. Heterochromatins also work in gene regulation, and promote chromosome integrity by preventing the deterioration of chromosome ends and also by preventing the combination of chromosomes.

In statistical mechanics, a model of polymer folding is the equilibrium globule. (When polymer folding was introduced as a topic in the stat mech course I took as an undergrad physics major, one of my classmates complained to the teacher that he took physics courses to avoid "all that ook" of chemistry. He was not happy to encounter "ook" in a physics course. I found these connections fascinating.) This was an early model for DNA folding, and while it can fold

DNA into the right size package, in the process it introduces many knots. Knotted DNA does not unfold readily, yet parts of it must unfold to copy DNA for cell replication, and for gene regulation in protein production.

In 1988 Alexander Grosberg and coworkers [146] proposed the fractal globule, another folding conformation, and in 1993 [147] they proposed it as the solution to the DNA folding problem. Leonid Mirny's paper [237] gives a good description of the equilibrium and fractal globule models. Typically fractals produce complex structures by the repeated application of the same process across many size scales. An iterative process of folds on folds on folds is a natural choice. Recent molecular dynamics simulations by Dusan Racko and coworkers [302] show that an experimentally observed process called chromatin loop extrusion can actively unknot DNA in the formation of a fractal globule. Consequently, in the fractal globule, portions of the DNA strand can unfold and refold during gene activation and repression without unfolding the whole globule. The fractal globule hypothesis is supported by experiments of Erez Lieberman-Aiden and coworkers [209]. Job Dekker, a coauthor of this study, described a fractal globule as a globule made of globules made of globules made of globules.

Aurélien Bancaud and coworkers [27] report their experiment that offers strong evidence for how the nucleus maintains two distinct regions with different levels of gene activity. Most genes are inactive in regions that contain heterochromatin; most are active in regions that contain euchromatin. The structure of chromatin was thought to resemble a sponge: small molecules can pass through while large molecules are blocked. Their experiment tracked the movements of fluorescent molecules of different sizes. Contrary to expectation, all molecules, regardless of size, were obstructed to the same degree. This suggests that the cell nucleus has a similar structure across a range of scales. In this sense, the cell nucleus is fractal.

With its large crinkled surface euchromatin more fully occupies 3-dimensional space than does a less-complex surface. We can quantify this by saying that euchromatin has a higher fractal dimension (Recall Sect. 6.4). Heterochromatin has a lower fractal dimension and a smoother surface. Proteins activate a gene by binding to specific, sparsely distributed DNA sequences, so the rougher surface of euchromatins could cause proteins to explore larger portions of the DNA sequence, and this could help in locating the target. Proteins that inactivate genes bind to histones, so the smoother fractal surface of heterochromatin would encourage shorter movement. Alterations of the chromatin fractal structure could change the behavior of different areas of the DNA.

In 2012, Bancaud and another team [28] suggested reasons that evolution has given chromatin a fractal structure. They suggested additional experiments to test the fractal globule hypothesis.

That in some fundamental way our DNA may have fractal properties, not just in the long-range correlations of its sequence but also in its 3-dimensional structure, gives still more evidence that evolution discovers and uses the principles of geometry, including fractal geometry.

Genetics is in a very interesting circumstance now. So very much information is available that a new field (bioinformatics) has appeared just for the purpose of storing and analyzing this data. But so far we don't have something like Maxwell's equations for electromagnetism, a few elegant ideas with which all manner of circuits and fields can be studied. Maybe genetics is too complicated to be described by a small number of underlying rules. Every result appears to have special cases and exceptions, all greatly beloved by geneticists. Still, I hope so far we are looking at the trees and not the forest.

Appendix B Some Mathematica code

Here we'll give examples of Mathematica code to generate some of the figures in the text. Code that can be copied and pasted into Mathematica are at https://gauss.math.yale.edu/~frame/BiomathMma.html

B.11 EIGENVALUES IN HIGHER DIMENSIONS

The Mathematica code to find the eigenvalues of

$$m = \begin{bmatrix} 1 & 0 & 2 \\ 0 & -1 & 1 \\ 0 & -1 & 2 \end{bmatrix} \quad \text{that is,} \quad m = \{\{1,0,2\},\{0,-1,1\},\{0,-1,2\}\}$$

is again

```
Eigenvalues[m]
```

Mathematica will try to solve this exactly. For this matrix that's easy; it will give the eigenvalues we computed by hand in Sect. 13.1, that is, 1 and $(1 \pm \sqrt{5})/2$. For many other matrices the Eigenvalues command will give something much more complicated, or signal that it can't find the exact solutions.

The command to find numerical eigenvalues, which always works, is

```
N[Eigenvalues[m]]
```

For the matrix m above, this returns 1.61803, 1.0, and -0.618034.

This generalizes in the obvious way to $n \times n$ matrices for $n > 3$.

B.12 SIR CALCULATIONS

Here we'll present the Mathematica code to generate the curves in Fig. 13.3 of Sect. 13.2. In earlier sections we've given the code to plot curves similar to those in Figs. 13.1 and 13.2. First, generate the solution curves by

```
b=2; e=3; d=1.0; g=0.5; n=0.75; tmax=45;
sol = NDSolve[{
x'[t]==b*x[t] - e*x[t]*z[t] - d*x[t] + g*z[t],
y'[t]==e*x[t]*z[t] - n*y[t] - d*y[t],
z'[t]==n*y[t] - d*z[t] - g*z[t],
x[0]==0.5,y[0]==0.25,z[0]==0.25},{x,y,z},{t,0,tmax}]
```

To plot the three population curves on the same graph, use this command.

```
Plot[{Evaluate[x[t] /. sol], Evaluate[y[t] /. sol], Evaluate[z[t] /. sol]},
  {t, 0, tmax}]
```

To plot the trajectory of the right side of Fig. 13.3, use this command.

```
ParametricPlot3D[Evaluate[{x[t], y[t], z[t]} /. sol], {t, 0, tmax}]
```

Note the 3D after the ParametricPlot. The ParametricPlot command plots a trajectory in the plane; to plot a trajectory in 3 dimensions, use the ParametricPlot3D command.

B.13 MICHAELIS-MENTEN CALCULATIONS

Here we show code to generate the figures in Sect. 13.3, and to perform the experiments of the exercises of that section. This is a straightforward generalization of the approach of Sect. B.12.

```
a = 0.2; b = 0.03; c = 0.05; tmax = 100;
sol=NDSolve[{
x'[t]==-a*x[t]*y[t] + b*z[t],
y'[t]==-a*x[t]*y[t] + (b+c)*z[t],
z'[t]==a*x[t]*y[t] - (b+c)*z[t],
w'[t]==c*z[t],
x[0]==1,y[0]==.5,z[0]==0.0,w[0]==0},{x,y,z,w},{t,0,tmax}]
```

Use these commands to plot the four population curves.

```
Plot[Evaluate[x[t] /. sol], {t, 0, tmax}]
Plot[Evaluate[y[t] /. sol], {t, 0, tmax}]
Plot[Evaluate[z[t] /. sol], {t, 0, tmax}]
Plot[Evaluate[w[t] /. sol], {t, 0, tmax}]
```

Use this command to plot the four population curves on the same graph.

```
Plot[{Evaluate[w[t] /. sol], Evaluate[x[t] /. sol], Evaluate[y[t] /. sol], Evaluate[z[t]
/. sol]}, {t, 0, tmax}]
```

B.14 VIRUS DYNAMICS

Here we present Mathematica code to plot the curves of Fig. 13.11 of Sect. 13.4. We choose this graph because it is generated by a 4-dimensional system, the dynamics of HIV under protease inhibitor treatment modeled by Eq. (13.27). Generate the solution curves by

```
l=100000; d=0.1; a=0.5; b=.0000002; k=100; u=5; tmax=20;
sol = NDSolve[{
x'[t] == l - d*x[t] - b*x[t]*z[t],
y'[t] == b*x[t]*z[t] - a*y[t],
z'[t] == k*y[t] - u*z[t],
w'[t] == -u*w[t],
x[0] == 1000, y[0] == 1000, z[0] == 0, w[0] == 5000000}, {x, y, z, w}, {t, 0,
tmax}]
```

Here z stands for the variable z_n and w for the variable z_i in Eq. (13.27). To plot the curves z_i and z_n we use

```
Plot[{Evaluate[z[t] /. sol], Evaluate[w[t] /. sol]},
   {t, 0, tmax}]
```

B.15 IMMUNE SYSTEM DYNAMICS

Here is Mathematica code to generate the curves of Fig. 13.13 of Sect. 13.5. That is, we plot the CTL response to the infected cell population. First, generate the solution curves of Eq. (13.29) by

```
b=100000; dx=0.1; g=0.0000002; dy=0.5; e=0.002; k=80; dz=5;
   h=0.002; dw=50; tmax=10;
sol = NDSolve[{
x'[t] == b - dx*x[t]- g*x[t]*z[t],
y'[t] == g*x[t]*z[t] - dy*y[t] - e*y[t]*w[t],
z'[t] == k*y[t] - dz*z[t],
```

```
w'[t] == h*y[t]*w[t] - dw*w[t],
x[0] == 1000000, y[0] == 0, z[0] == 2000, w[0] == 0.005},
{x, y, z, w}, {t, 0, tmax},MaxSteps->2000]
```

Then this command plots the *y* and *w* curves in the same graph.

```
Plot[{Evaluate[y[t] /. sol], Evaluate[w[t] /. sol],},{t, 0, tmax}]
```

B.16 THE HODGKIN-HUXLEY EQUATIONS

Here we'll present code to solution curves $n(t)$, $m(t)$, $h(t)$, and $V(t)$, as well as trajectories $(V(t), m(t))$ for the Hodgkin-Huxley equations (13.33)–(13.36) of Sect. 13.6. Because there are some tricks involved in drawing the graphs of Fig. 13.17, we'll present the code we used to generate that figure. We'll recall how to generate rectangular pulses. And finally, because the Hodgkin-Huxley equations are so highly nonlinear, we'll show code to find the coordinates of fixed points.

Hodgkin-Huxley solution curves and trajectories

The Hodgkin-Huxley equations (13.33)–(13.36) with an imposed current $i(t)$are

$$\frac{dn}{dt} = \alpha_n(V)(1-n) - \beta_n(V)n$$

$$\frac{dm}{dt} = \alpha_m(V)(1-m) - \beta_m(V)m$$

$$\frac{dh}{dt} = \alpha_h(V)(1-h) - \beta_h(V)h$$

$$\frac{dV}{dt} = -\frac{1}{C}\left(g_{Na}(V)(V - V_{Na}) + g_K(V)(V - V_K) + g_L(V - V_L) - i\right)$$

where

$$\alpha_n(V) = \frac{0.01(V+10)}{e^{(V+10)/10} - 1} \qquad \beta_n(V) = 0.125e^{V/80}$$

$$\alpha_m(V) = \frac{0.1(V+25)}{e^{(V+25)/10} - 1} \qquad \beta_m(V) = 4e^{V/18}$$

$$\alpha_h(V) = 0.07e^{V/20} \qquad \beta_h(V) = \frac{1}{e^{(V+30)/10} + 1}$$

$$g_{Na}(V) = \bar{g}_{Na} m^3 h \qquad g_K(V) = \bar{g}_K n^4$$

Because it introduces a new feature—ignoring the beginning of the plot to allow the graph to settle down to its long-term behavior—we'll present the code to generate the third graph of Fig. 13.18. Run NDSolve for {t,0,tmax} and run ParametricPlot for {t,tmin,tmax}.

```
tmin=12; tmax=40;
```

```
c=1.9; vna=-115; vk=12; vl=-10.6; cgna=120; cgk=36; cgl=0.3;
i[t_]:=Sin[Pi*t] + Sin[Pi*t/Sqrt[2]];
gna[m_,h_]:=cgna*(m^3)*h;
gk[n_]:=cgk*n^4;
an[v_]:=0.01*(v + 10)/(Exp[(v + 10)/10] - 1);
bn[v_]:=0.125*Exp[v/80];
am[v_]:=0.1*(v + 25)/(Exp[(v + 25)/10] - 1);
bm[v_]:=4*Exp[v/18];
ah[v_]:=0.07*Exp[v/20];
bh[v_]:=1/(Exp[(v + 30)/10] + 1);
sol1 = NDSolve[{
n'[t]==an[v[t]]*(1 - n[t]) - bn[v[t]]*n[t],
m'[t]==am[v[t]]*(1 - m[t]) - bm[v[t]]*m[t],
h'[t]==ah[v[t]]*(1 - h[t]) - bh[v[t]]*h[t],
v'[t]==(-1/c)*(gna[m[t],h[t]]*(v[t] - vna) + gk[n[t]]*(v[t] - vk)
      + cgl*(v[t] - vl) + i[t]),
n[0]==0.577, m[0]==0.046, h[0]==0.874, v[0]==0.78},
{n, m, h, v}, {t, 0, tmax},MaxSteps- >6000]
sol2 = NDSolve[{
n'[t]==an[v[t]]*(1 - n[t]) - bn[v[t]]*n[t],
m'[t]==am[v[t]]*(1 - m[t]) - bm[v[t]]*m[t],
h'[t]==ah[v[t]]*(1 - h[t]) - bh[v[t]]*h[t],
v'[t]==(-1/c)*(gna[m[t],h[t]]*(v[t] - vna) + gk[n[t]]*(v[t] - vk)
      + cgl*(v[t] - vl) + i[t]),
n[0]==0.577, m[0]==0.045, h[0]==0.874, v[0]==0.78},
{n, m, h, v}, {t, 0, tmax},MaxSteps- >6000]
```

To plot the $(m(t), V(t))$ trajectories for both systems use this code.

```
ParametricPlot[{Evaluate[{m[t], v[t]} /. sol1],
   Evaluate[{m[t], v[t]} /. sol2]}, {t, tmin, tmax}, PlotRange -> All,
   PlotPoints -> 100, AspectRatio -> 1]
```

Plot of the nullclines of Fig. 13.17

We find an expression for the m-nullcline and the V-nullcline, both as functions of V. Then we plot both curves on the same graph.

```
gk0 = 36; ninf = 0.318; vk = 12; gl = 0.3; vl; = -10.6; gna = 120;
   hinf = 0.596; vna = -115;
am[v_]:=0.1*(v + 25)/(Exp[(v + 25)/10] - 1);
bm[v_]:=4*Exp[v/18];
mncl[v_]:=am[v]/(am[v] + bm[v])
```

```
vncl[v_]:=((-gko*(ninf^4)*(v - vk) - gl*(v - vl))/
        (gna*hinf*(v - vna)))^(1/3)
Plot[{mncl[v],vncl[v]}, {v,-122,2}]
```

Code to generate periodic rectangular pulses

```
i[t_]:=If[t<1,1,If[t<2,0,i[t-2]]]
```

Finding coordinates of fixed points

Points of intersection of the m- and V-nullclines.

```
FindRoot[mncl[v]==vncl[v],{v,-100},MaxIterations->30]
```

Because the FindRoot algorithm uses Newton's method, it must have an initial guess. Here $v = -100$ is that initial guess.

B.17 CHAOS IN PREDATOR-PREY MODELS

Here we'll see Mathematica code to plot trajectories of Gilpin's model (13.39) and (13.40) seen in Fig. 13.23 of Sect. 13.7. First, generate the solution curves by

```
r1=1; r2=1; r3=-1; a11=0.001; a12=0.001; a13=0.01; a21=0.0016;
a22=0.001; a23=0.001; a31=-0.005; a32=-0.0005; a33=0; tmax=2000;
sol = NDSolve[{
x'[t] == r1*x[t] - x[t]*(a11*x[t] + a12*y[t] + a13*z[t]),
y'[t] == r2*y[t] - y[t]*(a21*x[t] + a22*y[t] + a23*z[t]),
z'[t] == r3*z[t] - z[t]*(a31*x[t] + a32*y[t] + a33*z[t]),
x[0]==0.1, y[0]==0.1, z[0]==0.1},{x, y, z}, {t, 0, tmax},MaxSteps->20000]
```

Then to plot the trajectory $(x(t), y(t), z(t))$ use this command.

```
ParametricPlot3D[Evaluate[{x[t], y[t], z[t]} /. sol], {t, 0, tmax},
    PlotPoints->12000, BoxRatios->{1,1,1}, PlotRange->All]
```

B.18 THE DUFFING EQUATION

Here we'll present some code to investigate the forced Duffing equation (13.44) of Sect. 13.8. Because we already have seen code for the analogous figures for other systems, here we'll show code to generate the right image of Fig. 13.34. To plot the $x(t)$-curves for two different sets of initial values, we'll generate two solutions, sol1 and sol2, one for each set of initial values, and then plot the $x(t)$-curve of both solutions on the same graph.

```
a=0.005; b=-0.5; c=0.5; d=0.1; w=1; tmax=100;
sol1 = NDSolve[{
x'[t] == y[t],
```

```
y'[t] == -a*y[t] - b*x[t] - c*x[t]^3 + d*Cos[w*z[t]],
z'[t] == 1,
x[0]==1.0, y[0]==0, z[0]==0}, {x, y, z}, {t, 0, tmax}, MaxSteps -> 20000]
sol2 = NDSolve[{
x'[t] == y[t],
y'[t] == -a*y[t] - b*x[t] - c*x[t]^3 + d*Cos[w*z[t]],
z'[t] == 1,
x[0]==1.1, y[0]==0, z[0]==0}, {x, y, z}, {t, 0, tmax}, MaxSteps -> 20000]
```

This command plots the $x(t)$ curves of both systems.

```
Plot[{Evaluate[x[t] /. sol1], Evaluate[x[t] /. sol2]},{t, 0, tmax}]
```

To plot the differences of the $x(t)$-curves use this code.

```
Plot[Evaluate[x[t] /. sol1] - Evaluate[x[t] /. sol2], {t, tmin, tmax}]
```

B.19 CONTROL OF CHAOS

Here we present Mathematica code to generate the plots of Figs. 13.43 and 13.44 of Sect. 13.9. First, the left side of Fig. 13.43.

```
l[s_, x_] := s*x*(1 - x)
SeedRandom[1234]; s = 4; num = 200; del = 0.02; rd = 0.000001;
x = Random[Real, {0, 1}];
xlst = {x};
Do[{If[Abs[x - 0.25] < del, s = Solve[0.25 == t*x*(1 - x), t][[1, 1, 2]]
   + Random[Real, {-rd, rd}]],
x = l[s, x], AppendTo[xlst, x], s = 4, x = l[s, x], AppendTo[xlst, x]},
   {i, 1, num}]
ptlst = {}; Do[AppendTo[ptlst, Point[{i, xlst[[i]]}]], {i, 1, Length[xlst]}];
Show[Graphics[{Line[{{0, 1}, {0, 0}, {Length[xlst], 0}}], ptlst}],
   PlotRange -> {0, 1}, AspectRatio -> 1/2]
```

The initial point is a random real number between 0 and 1. The SeedRandom command selects the same random number each time the program is run. Why do that? If one run produces an especially suitable graph, better than those from subsequent runs with different SeedRandoms, the program can be rerun with the earlier random number seed and the desired graph is reproduced.

One more point before we describe the logic of the program: what is the [[1,1,2]] after the Solve command? The first x-value that lies within $\delta = 0.02$ of 0.25 is $x_{17} = 0.25589$. Then for $x = x_{17}$,

```
Solve[0.25 == t*x*(1 - x), t]             returns {{ t → 1.31295 }}
Solve[0.25 == t*x*(1 - x), t][[1]]        returns { t → 1.31295 }
Solve[0.25 == t*x*(1 - x), t][[1,1]]      returns t → 1.31295
Solve[0.25 == t*x*(1 - x), t][[1,1,2]]    returns 1.31295
```

Here we've used Mathematica's protocol for selecting elements of a list: list[[n]] selects the *n*th element of the list. For a list of lists, this process can be iterated. So for example, {1,{2,3,4},5}[[2,3]] = 4.

Now for the program mechanics. So long as $|x_i - 0.25| \geq \delta$, $x_{i+1} = l(4, x_i)$. Whenever $|x_i - 0.25| < \delta$, the logistic parameter s is adjusted to s' that gives $x_{i+1} = l(s', x_i) = 0.25$. The small random perturbation to the value of s' represents measurement error in any real experiment. Then s is reset to 4 so $x_{i+2} = l(s, x_{i+1}) = 0.75$. Because $x = 0.75$ is a fixed point for $l(4, x)$, if x_{i+1} were *exactly* 0.25, then $x_{i+2} = x_{i+3} + \cdots$. But of course the solution for s' is not exact, so x_{i+1} differs from 0.25 by a tiny amount. Because the fixed point $x = 0.75$ is unstable, further iterates wander away from $x = 0.75$. When an iterate comes within δ of 0.25, the s adjustment is reapplied. The left plot of Fig. 13.43 shows that sometimes many iterates are required in order for an x-value to be close enough to 0.25.

An attempt to speed this process is the reason for the program to generate Fig. 13.44.

```
l[s_, x_] := s*x*(1 - x)
SeedRandom[3142]; s = 4; num = 200; del = 0.02; eps = 0.05;
  rd = 0.000001;
x = Random[Real,{0,1}];
xlst = {x};
Do[{If[Abs[x - 0.25] < del, {s = Solve[0.25 == t*x*(1 - x), t][[1, 1, 2]]
  + Random[Real, {-rd, rd}], x = l[s, x], AppendTo[xlst, x]}],
  s = 4, x = l[s, x], AppendTo[xlst, x],
If[Abs[x - 0.75] > eps, {s = Solve[0.75 == t*x*(1 - x), t][[1, 1, 2]]
  + Random[Real, {-rd, rd}], x = l[s, x], AppendTo[xlst, x]}],
  s = 4, x = l[s, x], AppendTo[xlst, x]}, {i, 1, num}]
ptlst = {}; Do[AppendTo[ptlst, Point[{i, xlst[[i]]}]], {i, 1, Length[xlst]}];
Show[Graphics[{Line[{{0, 1}, {0, 0}, {Length[xlst], 0}}], ptlst}],
  PlotRange -> {0, 1}, AspectRatio -> 1/2]
```

The difference between this program and the previous one is that here the s adjustment occurs if either $|x_i - 0.25| < \delta$ or $|x_i - 0.75| > \epsilon$. Each adjustment occurs for only one iteration, after which s is reset to 4 and the next iteration performed.

B.20 SENSITIVITY ANALYSIS

Computer algebra systems such as Mathematica usually find right eigenvectors. In Sect. 14.4 we use left eigenvectors. Here's a way to find left eigenvectors with Mathematica. Recall the *transpose* M^{tr} of a matrix M is obtained by interchanging the rows and columns of M. For example,

$$M^{tr} = \begin{bmatrix} 3/4 & 0 & 1/4 \\ 1/4 & 1/2 & 0 \\ 0 & 1/2 & 3/4 \end{bmatrix}^{tr} = \begin{bmatrix} 3/4 & 1/4 & 0 \\ 0 & 1/2 & 1/2 \\ 1/4 & 0 & 3/4 \end{bmatrix}$$

and the transpose of a column vector is a row vector.

The Mathematica command to find the transpose of a matrix, written as a list of lists, is

```
Transpose[{{1,2},{3,4}}]
```

To view this as a matrix, instead of a list of lists, use

```
MatrixForm[Transpose[{{1,2},{3,4}}]]
```

For M^{tr} Mathematica gives this eigenvector of the eigenvalue 1:

$$\begin{bmatrix} 1 \\ 1 \\ 1 \end{bmatrix} \text{ Dividing by the sum of the entries gives } \begin{bmatrix} 1/3 \\ 1/3 \\ 1/3 \end{bmatrix}$$

Transposing to a row vector gives our original calculation of the left unit eigenvector.

B.21 THE CLANCY-RUDY MODEL

Here's the code we used to generate the solution of Practice Problem 16.4.1(c) of Sect. 16.4.

```
p = {{0.8, 0.5, 0, 0, 0}, {0.2, 0.2, 0.2, 0.2, 0}, {0, 0.2, 0.7, 0.1, 0},
{0, 0.1, 0.1, 0.6, 0.1}, {0, 0, 0, 0.1, 0.9}};
```

Next we initialize and generate the cumulative probability matrix.

```
cp = {{}, {}, {}, {}, {}};
cp[[1]] = p[[1]];
cp[[2]] = cp[[1]] + p[[2]];
cp[[3]] = cp[[2]] + p[[3]];
cp[[4]] = cp[[3]] + p[[4]];
cp[[5]] = cp[[4]] + p[[5]];
```

Start with a random state. Initialize the list of states, and initialize the state counters to 0.

```
s = Random[Integer, {1, 5}];
slst = {s}; ct[1] = 0; ct[2] = 0; ct[3] = 0; ct[4] = 0; ct[5] = 0;
num = 100000;
```

Next we generate a sequence of random numbers, and based on the current state and the cumulative probability matrix, we determine the next state and augment the appropriate state counter.

```
Do[{x = Random[Real, {0, 1}],
  If[x < cp[[1, s]], t = 1,
    If[x < cp[[2, s]], t = 2,
      If[x < cp[[3, s]], t = 3,
        If[x < cp[[4, s]], t = 4, t = 5]]]],
          s = t, ct[s] = ct[s] + 1, AppendTo[slst, s]}, {i, 1, num}]
```

Successive indents have been placed to help unpack the sequence of nested If statements. Finally, to compare the counts with the fractions calculated in Practice Problem 16.4.1 (b) use this command.

```
N[{ct[1]/num, ct[2]/num, ct[3]/num, ct[4]/num, ct[5]/num}]
```

B.22 A GENETIC TOGGLE SWITCH

Here is code to generate solution plots of the genetic toggle switch of Sect. 18.1. For reference, here are the equations, presented as Eq. (18.1).

$$\frac{dx}{dt} = -x + \frac{a_1}{1+y^{b_1}} \qquad \frac{dy}{dt} = -y + \frac{a_2}{1+x^{b_2}}$$

This code plots the curves $(t,x(t))$, $(t,y(t))$, and the trajectory $(x(t),y(t))$, for the genetic toggle switch equations with sinusoidal perturbations.

```
a1 = 2.5; a2 = 2.5; b1 = 2; b2 = 2; c = 0.08; tmax = 300;
sol = NDSolve[{
  x'[t] == -x[t] + a1/(1 + y[t]∧b1) + c*Sin[t/10],
  y'[t] == -y[t] + a2/(1 + x[t]∧b2)+ c*Sin[t/15],
  x[0] == 0.5, y[0] == 1.5},{x, y}, {t, 0, tmax},MaxSteps->8000]
Plot[{Evaluate[x[t] /. sol],Evaluate[y[t]/.sol]}, {t, 0, tmax},
  PlotRange->All]
ParametricPlot[Evaluate[{x[t],y[t]}/.sol],{t,0,tmax}, PlotRange->All,
  PlotPoints->100]
```

B.23 TRANSCRIPTION NETWORKS

Here is the code to generate graph (b) of Practice Problem 18.2.2 of Sect. 18.2.

```
ay=1; by=1; az=1; bz=2; kxy=0.5; kxz=1.0; kyz=0.75; tmax=10;
x[t_]:=If[t<5,1.25,0.55];
sol = NDSolve[{y'[t]==If[x[t]>kxy, by - ay*y[t],-ay*y[t]],
   z'[t]==If[x[t]>kxz && y[t]>kyz,bz - az*z[t],-az*z[t]],
   y[0]==0,z[0]==0}, {y,z}, {t, 0, tmax},MaxSteps->8000,
   Method->{"DiscontinuityProcessing"->False}]
Plot[{Evaluate[y[t] /. sol], Evaluate[z[t] /. sol], x[t]}, {t, 0, tmax},
   PlotRange -> All]
```

In the If statement that defines z′ the condition x[t]>kxz && y[t]>kyz means that both x[t]>kxz and y[t]>kyz must be true in order for z′[t] to equal bz - az*z[t].

The reason for the DiscontinuityProcessing command is to allow NDSolve to deal with the discontinuities in the definitions of y′ and z′. Modifications of the step function x[t] produce the other graphs in this practice problem. In earlier sections, for example the Norton-Simon model code in Appendix B.4, adding a discontinuous term to the right-hand side of a differential equation doesn't require the DiscontinuityProcessing modification. The transcription network model of this section is different: the entire right-hand side of the differential equation is discontinuous.

B.24 TOPOLOGICAL METHODS

Here we give the Mathematica code to generate the driven IFS of Sect. 19.3. First, paste the genome, lower-case and comma separated, between the brackets of z = {}. For a sequence of capital C, A, T, and G, change the function definitions from tr[c,{x_,y_}] to tr[C,{x_,y_}], etc.

```
dlst={g,g,g,g, ..., g,t,t,a,c,g};
tr[c,{x_,y_}]:={0.5*x,0.5*y};
tr[a,{x_,y_}]:={0.5*x + 0.5,0.5*y};
tr[t,{x_,y_}]:={0.5*x,0.5*y + 0.5};
tr[g,{x_,y_}]:={0.5*x + 0.5,0.5*y + 0.5};
ptlst={};
{x,y}={0.5,0.5};
Do[{{nx,ny}=tr[dlst[[i]],{x,y}],{x,y}={nx,ny},
     AppendTo[ptlst,Point[{x,y}]]},{i,1,Length[dlst]}]
Show[Graphics[{Line[{{0,0},{1,0},{1,1},{0,1},{0,0}}],PointSize[0.01],
```

```
ptlst}], PlotRange->{{-.01,1.01},{-.01,1.01}},
AspectRatio->Automatic]
```

Here is code to bin numerical data, intervals between successive heartbeats for example, entered in as comma-separated numbers in dlst.

```
dlst = { };
```

For equal size bins the bin boundaries are

```
rng = Max[dlst] - Min[dlst];
B1 = Min[dlst] + 0.25*rng;
B2 = Min[dlst] + 0.50*rng;
B3 = Min[dlst] + 0.75*rng;
```

For equal weight bins we find the bin boundaries here.

```
qlst = Quartiles[dlst];
B1=qlst[[1]];
B2=qlst[[2]];
B3=qlst[[3]];
```

For median centered bins with δ = d.

```
rng = Max[dlst] - Min[dlst];
d = ;
B2=Median[dlst];
B1=B2 - d*rng;
B3=B2 + d*rng;
```

To convert a number from dlst into a choice of which function to apply, use this nested If statement.

```
ind[z_]:=If[z < B1,1,If[z < B2,2,If[z < B3,3,4]]]
```

With the choice of bin boundaries, this code will generate the driven IFS plot and count the address occupancies. The number of length-1 addresses visited equals the length of the data set. The number of length-2 addresses visited is the length of the data set minus 1; the number of length-3 addresses visited is the length of the data set minus 2. This is why we occupy the first two length-1 addresses and the first length-2 address before we begin the loop to occupy the length -1, -2, and -3 addresses.

```
tr[1,{x_,y_}]:={0.5*x, 0.5*y};
tr[2,{x_,y_}]:={0.5*x + 0.5, 0.5*y};
tr[3,{x_,y_}]:={0.5*x, 0.5*y + 0.5};
tr[4,{x_,y_}]:={0.5*x + 0.5, 0.5*y + 0.5};
ind[z_]:=If[z < B1,1,If[z < B2,2,If[z < B3,3,4]]];
```

```
Do[ad1[i]=0.,{i,1,4}];
Do[Do[ad2[i,j]=0,{i,1,4}],{j,1,4}];
Do[Do[Do[ad3[i,j,k]=0,{i,1,4}],{j,1,4}],{k,1,4}];
{x,y}={.5,.5};
drlst={Point[{x,y}]};
w=ind[dlst[[1]]]; {x,y}=tr[w,{x,y}];
AppendTo[drlst,Point[{x,y}]];ad1[w]=ad1[w]+1;
v=ind[dlst[[2]]]; {x,y}=tr[v,{x,y}];
AppendTo[drlst,Point[{x,y}]]; ad1[v]=ad1[v]+1; adr[v,w]=ad2[v,w]+1;
Do[{u=ind[dlst[[i]]], {x,y}=tr[u,{x,y}],
AppendTo[drlst,Point[{x,y}]], ad1[u]=ad1[u]+1,
ad2[u,v]=ad2[u,v]+1, ad3[u,v,w]=ad3[u,v,w]+1, w=v,v=u},{i,3,Length[dlst]}];
```

To plot the driven IFS along with the length 2 address squares,

```
lnlst1={}; Do[{AppendTo[lnlst1,Line[{{x,0},{x,1}}]],
AppendTo[lnlst1,Line[{{0,x},{1,x}}]]},{x,0,1,1/4}]
Show[Graphics[{{RGBColor[.6,.6,.6],lnlst1},PointSize[.015],drlst}],
PlotRange->{{-.01,1.01},{-.01,1.01}},AspectRatio->1]
```

To find the number of points with address 124, after the program has run use the command

```
Print[ad3[1,2,4]]
```

Appendix C Some useful integrals and hints

Double angle formulas $\sin^2(x) = \frac{1-\cos(2x)}{2} \qquad \cos^2(x) = \frac{1+\cos(2x)}{2}$

1. $\int \sin^2(x)\, dx = \frac{1}{2}x - \frac{1}{4}\sin(2x) + C$
2. $\int \cos^2(x)\, dx = \frac{1}{2}x + \frac{1}{4}\sin(2x) + C$
3. $\int \tan^2(x)\, dx = \tan(x) - x + C$
4. $\int \cot^2(x)\, dx = -\cot(x) - x + C$
5. $\int \sec^2(x)\, dx = \tan(x) + C$
6. $\int \csc^2(x)\, dx = -\cot(x) + C$
7. $\int \sin^n(x)\, dx = -\frac{1}{n}\sin^{n-1}(x)\cos(x) + \frac{n-1}{n}\int \sin^{n-2}(x)\, dx$
8. $\int \cos^n(x)\, dx = \frac{1}{n}\cos^{n-1}(x)\sin(x) + \frac{n-1}{n}\int \cos^{n-2}(x)\, dx$

9. $\int \tan^n(x)\,dx = \frac{1}{n-1}\tan^{n-1}(x) - \int \tan^{n-2}(x)\,dx$

10. $\int \cot^n(x)\,dx = -\frac{1}{n-1}\cot^{n-1}(x) - \int \cot^{n-2}(x)\,dx$

11. $\int \sec^n(x)\,dx = \frac{1}{n-1}\sec^{n-2}(x)\tan(x) + \frac{n-2}{n-1}\int \sec^{n-2}(x)\,dx$

12. $\int \csc^n(x)\,dx = -\frac{1}{n-1}\csc^{n-2}(x)\cot(x) + \frac{n-2}{n-1}\int \csc^{n-2}(x)\,dx$

13. $\int \sin^m(x)\cos^{2k+1}(x)\,dx = \int \sin^m(x)(1-\sin^2(x))^k\cos(x)\,dx$ and substitute $u = \sin(x)$

14. $\int \sin^{2k+1}(x)\cos^m(x)\,dx = \int (1-\cos^2(x))^k\sin(x)\cos^m(x)\,dx$ and substitute $u = \cos(x)$

15. $\int \sin^{2m}(x)\cos^{2n}(x)\,dx$ use both double angle formulas

For integrals involving $\sqrt{a^2-x^2}$ substitute $x = a\sin(\theta)$.
For integrals involving $\sqrt{a^2+x^2}$ substitute $x = a\tan(\theta)$.
For integrals involving $\sqrt{x^2-a^2}$ substitute $x = a\sec(\theta)$.

16. $\int \sqrt{a^2+x^2}\,dx = \frac{x}{2}\sqrt{a^2+x^2} + \frac{a^2}{2}\ln\left|x+\sqrt{a^2+x^2}\right| + C$

17. $\int x^2\sqrt{a^2+x^2}\,dx = \frac{x}{8}(a^2+2x^2)\sqrt{a^2+x^2} - \frac{a^4}{8}\ln\left|x+\sqrt{a^2+x^2}\right| + C$

18. $\int \frac{\sqrt{a^2+x^2}}{x}\,dx = \sqrt{a^2+x^2} - a\ln\left|\frac{a+\sqrt{a^2+x^2}}{x}\right| + C$

19. $\int \frac{\sqrt{a^2+x^2}}{x^2}\,dx = -\frac{\sqrt{a^2+x^2}}{x} + \ln\left|x+\sqrt{a^2+x^2}\right| + C$

20. $\int \frac{dx}{\sqrt{a^2+x^2}} = \ln\left|x+\sqrt{a^2+x^2}\right| + C$

21. $\int \frac{x^2\,dx}{\sqrt{a^2+x^2}} = \frac{x}{2}\sqrt{a^2+x^2} - \frac{a^2}{2}\ln\left|x+\sqrt{a^2+x^2}\right| + C$

22. $\int \frac{dx}{x\sqrt{a^2+x^2}} = -\frac{1}{a}\ln\left|\frac{\sqrt{a^2+x^2}+a}{x}\right| + C$

23. $\int \frac{dx}{x^2\sqrt{a^2+x^2}} = -\frac{\sqrt{a^2+x^2}}{a^2x} + C$

24. $\int \sqrt{a^2-x^2}\,dx = \frac{x}{2}\sqrt{a^2-x^2} + \frac{a^2}{2}\sin^{-1}\left(\frac{x}{a}\right) + C$

25. $\int x^2\sqrt{a^2-x^2}\,dx = \frac{x}{8}(2x^2-a^2)\sqrt{a^2-x^2} + \frac{a^4}{8}\sin^{-1}\left(\frac{x}{a}\right) + C$

26. $\displaystyle\int \frac{\sqrt{a^2-x^2}}{x}\,dx = \sqrt{a^2-x^2} - a\ln\left|\frac{a+\sqrt{a^2-x^2}}{x}\right| + C$

27. $\displaystyle\int \frac{\sqrt{a^2-x^2}}{x^2}\,dx = -\frac{\sqrt{a^2-x^2}}{x} - \sin^{-1}\left(\frac{x}{a}\right) + C$

28. $\displaystyle\int \frac{dx}{\sqrt{a^2-x^2}} = \sin^{-1}\left(\frac{x}{a}\right) + C$

29. $\displaystyle\int \frac{x^2\,dx}{\sqrt{a^2-x^2}} = -\frac{x}{2}\sqrt{a^2-x^2} + \frac{a^2}{2}\sin^{-1}\left(\frac{x}{a}\right) + C$

30. $\displaystyle\int \frac{dx}{x\sqrt{a^2-x^2}} = -\frac{1}{a}\ln\left|\frac{\sqrt{a^2-x^2}+a}{x}\right| + C$

31. $\displaystyle\int \frac{dx}{x^2\sqrt{a^2-x^2}} = -\frac{\sqrt{a^2-x^2}}{a^2x} + C$

32. $\displaystyle\int \sqrt{x^2-a^2}\,dx = \frac{x}{2}\sqrt{x^2-a^2} - \frac{a^2}{2}\ln\left|x+\sqrt{x^2-a^2}\right| + C$

33. $\displaystyle\int x^2\sqrt{x^2-a^2}\,dx = \frac{x}{8}(2x^2-a^2)\sqrt{x^2-a^2} - \frac{a^4}{8}\ln\left|x+\sqrt{x^2-a^2}\right| + C$

34. $\displaystyle\int \frac{\sqrt{x^2-a^2}}{x}\,dx = \sqrt{x^2-a^2} - a\cos^{-1}\left(\frac{a}{|x|}\right) + C$

35. $\displaystyle\int \frac{\sqrt{x^2-a^2}}{x^2}\,dx = -\frac{\sqrt{x^2-a^2}}{x} + \ln\left|x+\sqrt{x^2-a^2}\right| + C$

36. $\displaystyle\int \frac{dx}{\sqrt{x^2-a^2}} = \ln\left|x+\sqrt{x^2-a^2}\right| + C$

37. $\displaystyle\int \frac{x^2\,dx}{\sqrt{x^2-a^2}} = \frac{x}{2}\sqrt{x^2-a^2} + \frac{a^2}{2}\ln\left|x+\sqrt{x^2-a^2}\right| + C$

38. $\displaystyle\int \frac{dx}{x\sqrt{x^2-a^2}} = \frac{1}{a}\sec^{-1}\left(\frac{x}{a}\right) + C$

39. $\displaystyle\int \frac{dx}{x^2\sqrt{x^2-a^2}} = \frac{\sqrt{x^2-a^2}}{a^2x} + C$

40. $\displaystyle\int \frac{Ax+B}{(ax+b)(cx+d)}\,dx = \frac{E}{a}\ln|ax+b| + \frac{F}{c}\ln|cx+d| + C$, where E and F are the solutions of $Ax+B = E(cx+d)+F(ax+b)$.

41. $\displaystyle\int \frac{1}{x^2+1}\,dx = \tan^{-1}(x) + C$

42. $\displaystyle\int \sec(x)\,dx = \ln|\sec(x)+\tan(x)| + C$

43. $\displaystyle\int \tan(x)\,dx = \ln|\sec(x)| + C$

44. $\int \csc(x)\,dx = \ln|\csc(x) - \cot(x)| + C$

45. $\int \cot(x)\,dx = \ln|\sin(x)| + C$

46. $\int \sin^{-1}(x)\,dx = x\sin^{-1}(x) + \sqrt{1-x^2} + C$

47. $\int \cos^{-1}(x)\,dx = x\cos^{-1}(x) - \sqrt{1-x^2} + C$

48. $\int \tan^{-1}(x)\,dx = x\tan^{-1}(x) - \frac{1}{2}\ln(1+x^2) + C$

49. $\int \ln(x)\,dx = -x + x\ln(x) + C$

Some differentiation formulas

1. $\frac{d}{dx}\sin^{-1}(x) = \frac{1}{\sqrt{1-x^2}}$

2. $\frac{d}{dx}\cos^{-1}(x) = -\frac{1}{\sqrt{1-x^2}}$

3. $\frac{d}{dx}\tan^{-1}(x) = \frac{1}{1+x^2}$

4. $\frac{d}{dx}\csc^{-1}(x) = -\frac{1}{x\sqrt{x^2-1}}$

5. $\frac{d}{dx}\sec^{-1}(x) = \frac{1}{x\sqrt{x^2-1}}$

6. $\frac{d}{dx}\cot^{-1}(x) = -\frac{1}{1+x^2}$

References

[1] C. Adami, *Introduction to Artificial Life*, Springer-Verlag, New York, 1998.

[2] K. Aihara, G. Matsumoto, "Chaotic oscillations and bifurcations in the squid giant axon," pgs. 257–269 of [176].

[3] L. Alberghina, H. Westerhoff, eds., *Systems Biology: Definitions and Perspectives*, Springer, New York, 2005.

[4] R. Albin, "The pleiotropic gene theory of senescence: supportive evidence from human genetic disease," *Ethology and Sociobiology* **9** (1988), 371–382.

[5] D. Allen, N. Bekken, K. Crisfulla, M. Espinoza, K. Hill, C. Kabeli, K. Knight-Frank, M. Mikalaitis, C. Rader, L. Trujillo, R. Unruh, O. Wilson, *ECG Interpretation Made Incredibly Easy*, 5th ed., Lippincott Williams & Wilkins, Philadelphia, 2011.

[6] E. Allman, J. Rhodes, *Mathematical Models in Biology: An Introduction*, Cambridge Univ. Pr., Cambridge, 2004.

[7] J. Almeida, "Sequence analysis by iterated maps, a review," *Briefings in Bioinformatics* **15** (2013), 369–375.

[8] J. Almeida, J. Carriço, A. Maretzek, P. Noble, M. Fletcher, "Analysis of genomic sequences by chaos game representation," *Bioinformatics* **17** (2001), 429–437.

[9] U. Alon, *An Introduction to Systems Biology: Design Principles of Biological Circuits*, Chapman & Hall, Boca Raton, 2007.

[10] S. Altschul, W. Gish, W. Miller, E. Myers, D. Lipman, "Basic Local Alignment Search Tool," *J. Molecular Biol.* **215** (1990), 403–410.

[11] C. Anderson, C. Stevens, "Voltage clamp analysis of acetylcholine produced end-plate current fluctuations at frog neuromuscular junction," *J. Physiol.* **235** (1973), 655–691.

[12] R. Anderson, R. May, *Infectious Diseases of Humans: Dynamics and Control*, Oxford Univ. Pr., Oxford, 1991.

[13] M. Antolin et al., "Evolution and medicine in undergraduate education: a prescription for all biology students," *Evolution* **66** (2012), 1991–2006.

[14] O. Avery, C. MacLeod, M. McCarty, "Studies on the chemical nature of the substance inducing transformation of pneumococcal types: Induction of transformation by a desoxyribonucleic acid fraction isolated from pneumococcus type III," *J. Experimental Medicine* **79** (1944), 137–158.

[15] J. Baedke, "The epigenetic landscape in the course of time: Conrad Hal Waddington's methodological impact on the life sciences," *Stud. Hist. Philos. Biol. Biomed. Sci.* **44** (2013), 756–773.

[16] S. Bahar, *The Essential Tension: Competition, Cooperation, and Multilevel Selection in Evolution*, Springer, New York, 2018.

[17] S. Bahar, F. Moss, "The nonlinear dynamics of the crayfish mechanoreceptor system," *Int. J. Bif. Chaos* **13** (2003), 2013–2034.

[18] S. Bahar, F. Moss, "Stochastic resonance and synchronization in the crayfish caudal photoreceptor," *Math. Biosci.* **188** (2004), 81–97.

[19] S. Bahar, A. Neiman, L. Wilkens, F. Moss, "Phase synchronization and stochastic resonance effects in the crayfish caudal photoreceptor," *Phys. Rev. E* **65** (2002), 050901-1–050901-4.

[20] P. Bak, *How Nature Works: The Science of Self-Organized Criticality*, Springer, New York, 1996.

[21] P. Bak, K. Chen, "Self-organized criticality," *Sci. Am.* (Jan. 1991), 46–53.

[22] P. Bak, H. Flyvbjerg, B. Lautrup, "Coevolution in a rugged fitness landscape," *Phys. Rev. A* **46** (1992), 6724–6730.

[23] P. Bak, C. Tang, K. Wiesenfeld, "Self-organized criticality," *Phys. Rev. A* **38** (1988), 364–374.

[24] J. Baladron, D. Fasoli, O. Faugeras, J. Touboul, "Mean-field description and propagation of chaos in networks of Hodgkin-Huxley and Fitzhugh-Nagumo neurons," *J. Math. Neuroscience* **2** (2012), DOI 10.1186/2190-8567-2-10.

[25] G. Balázsi, L. Kish, F. Moss, "Spatiotemporal stochastic resonance and its consequences in neural model systems," *Chaos* **11** (2001), 563–569.

[26] E. Ballestar, M. Esteller, B. Richardson, "The epigenetic face of systemic lupus erthematosus," *J. Immunology* **176** (2006), 7143–7147.

[27] A. Bancaud, S. Huet, N. Daigle, J. Mozziconacci, J. Beaudouin, J. Ellenberg, "Molecular crowding affects diffusion and binding of nuclear proteins in heterochromatin and reveals the fractal organization of chromatin," *EMBO J.* **28** (2009), 3785–3798.

[28] A. Bancaud, C. Lavelle, S. Huet, J. Ellenberg, "A fractal model for nuclear organization: current evidence and biological implications," *Nucleic Acids Res.* **40** (2012), 8783–8792.

[29] E. Barillot, L. Calzone, P. Hupe, J.-P. Vert, A. Zinovyev, *Computational Systems Biology of Cancer*, Chapman & Hall, Boca Raton, 2012.

[30] M. Barnsley, S. Demko, "Iterated function systems and the global construction of fractals," *Proc. Roy. Soc. London A* **399** (1985), 243–275.

[31] M. Barnsley, A. Sloan, "A better way to compress images," *Byte* **13** (1988), 215–223.

[32] A. Barrat, M. Bartélemy, A. Vespignani, *Dynamical Processes on Complex Networks*, Cambridge Univ. Pr., Cambridge, 2008.

[33] D. Basanta, J. Scott, M. Fishman, G. Ayala, S. Hayward, A. Anderson, "Investigating prostate cancer tumor-stroma interactions: clinical and biological insights from an evolutionary game," *British J. Cancer* **106** (2012), 174–181.

[34] S. Baskaran, P. Stadler, P. Schuster, "Approximate scaling properties of RNA free energy landscapes," *J. theor. Biol.* **181** (1996), 299–310.

[35] A. Baxevanis, B. Ouellette, *Bioinformatics: A Practical Guide to the Analysis of Genes and Proteins*, 2nd ed., Wiley, New York, 2001.

[36] G. Benettin, L. Galgani, A. Giorgilli, J.-M. Strelcyn, "Lyapunov characteristic exponents for smooth dynamical systems and for Hamiltonian systems; a method for computing all of them. Part 1: theory," *Meccanica* **15** (1980), 9–20.

[37] G. Benettin, L. Galgani, A. Giorgilli, J.-M. Strelcyn, "Lyapunov characteristic exponents for smooth dynamical systems and for Hamiltonian systems; a method for computing all of them. Part 2: numerical application," *Meccanica* **15** (1980), 21–30.

[38] S. Bezrukov, I. Vodyanoy, "Signal transduction across voltage-dependent ion channels of alamethicin in the presence of external noise," *Biophys. J.* textbd68 (1995), A368.

[39] S. Bezrukov, I. Vodyanoy, "Noise-induced enhancement of signal transduction across voltage-dependent ion channels," *Nature* textbd378 (1995), 362–364.

[40] S. Bezrukov, I. Vodyanoy, "Stochastic resonance and small-amplitude signal transduction in voltage-gated ion channels," pgs. 257–280 of [372].

[41] G. Birkhoff, "Extensions of Jentzsch's theorem," *Trans. Amer. Math. Soc.* **85** (1957), 219–227.

[42] J. Birman, R. Williams, "Knotted periodic orbits in dynamical systems—I: Lorenz's equations," *Topology* **22** (1983), 47–82.

[43] J. Bissonette, ed., *Wildlife and Landscape Ecology: Effects of Pattern and Scale*, Springer, New York, 1997.

[44] https://blast.ncbi.nlm.nih.gov/Blast.cgi

[45] S. Bonhoeffer, R. May, G. Shaw, M. Nowak, "Virus dynamics and drug therapy," *Proc. Nat. Acad. Sci.* **94** (1997), 6971–6976.

[46] D. Bonnet, J. Dick, "Human acute myeloid leukemia is organized as a hierarchy that originates from a primitive hematopoietic cell," *Nature Medicine* **3** (1997), 730–737.

[47] J. L. Borges, *A Universal History of Infamy*, Dutton, New York, 1979.

[48] C. Brack, M. Hirama, R. Lenhard-Schuller, S. Tonegawa, "A complete immunoglobulin gene is created by somatic recombination," *Cell* **15** (1978), 1–14.

[49] J. Bull, L. Meyers, M. Lachmann, "Quasispecies made simple," *PLoS Computational Biology* **1** (2005), 450–460.

[50] https://www.cancer.gov/about-nci/organization/ccg/research/-structural-genomics/tcga.

[51] http://cancergenome.nih.gov.

[52] J. Carlson, J. Doyle, "Highly optimized tolerance: a mechanism for power laws in designed systems," *Phys. Rev. E* **60** (1999), 1412–1427.

[53] L. Cartwright, J. Littlewood, "On non-linear differential equations of the second order. I. The equation $y'' + k(1-y^2) + y = b\lambda k \cos(\lambda t + a)$, k large," *J. London Math. Soc.* **20** (1942), 180–189.

[54] J. del Castillo, B. Katz, "Interaction at end-plate receptors between different choline derivatives," *Proc. Roy. Soc. B* **146** (1957), 369–381.

[55] A. Chatterjee, P. Smith, A. Perelson, "Hepatitis C viral kinetics: the past, present and future," *Clinics in Liver Disease* **17** (2013), 13–26.

[56] H. Chen, D. Maduranga, P. Mundra, J. Zheng, "Integrating epigenetic prior in dynamic Bayesian network for gene regulatory network inference," 2013 IEEE Symposium on Computational Intelligence in Bioinformatics and Computational Biology, pgs. 76–82.

[57] R. Chen, Y. Rubanova, J. Bettencourt, D. Duvenaud, "Neural ordinary differential equations," *32nd Conference of Neural Information Processing Systems*, Montréal, 2018, arXiv:1806:07366v3.

[58] C. Cheng, K. Yan, K. Yip, J. Rozowsky, R. Alexander, C. Shou, M. Gerstein, "A statistical framework for modeling gene expression using chromatin features and application to modENCODE datasets," *Genome Biol.* **12** (2011), R15.

[59] R. Chiu, K. Chan, Y. Gao, et al., "Noninvasive prenatal diagnosis of fetal chromosomal aneuploidy by massively parallel genomic sequencing of DNA in maternal plasma," *Proc. Nat. Acad. Sci. USA* **105** (2008), 20458–20463.

[60] D. Christini, J. Collins, "Using chaos control and tracking to suppress a pathological nonchaotic rhythm in a cardiac model," *Phys. Rev. E* **53** (1996), R49–R52.

[61] D. Christini, J. Collins, "Real-time adaptive, model-independent control of low-dimensional chaotic and nonchaotic dynamical systems," *IEEE Trans. Circ. Syst.* **44** (1997), 1027–1030.

[62] D. Christini, V. In, M. Spano, W. Ditto, J. Collins, "Real-time experimental control of a system in its chaotic and nonchaotic regimes," *Phys. Rev. E* **56** (1997), R3749–R3752.

[63] D. Christini, K. Stein, S. Markowitz, S. Mittal, D. Slotwiner, M. Scheiner, S. Iwai, B. Lerman, "Nonlinear-dynamical arrhythmia control in humans," *Proc. Nat. Acad. Sci.* **98** (2001), 5827–5832.

[64] S. Ciupe, R. Ribeiro, P. Nelson, A. Peterson, "Modeling the mechanisms of acute hepatitis B virus infection," *J. theor. Biol.* **247** (2007), 23–35.

[65] C. Clancy, R. Kass, "Defective cardiac ion channels: from mutations to clinical syndromes," *J. Clinical Invest.* **110** (2002), 1075–1077.

[66] C. Clancy, R. Kass, "Inherited and acquired vulnerability to ventricular arrhythmias: cardiac Na^+ and K^+ channels," *Physiol. Rev.* **85** (2005), 33–47.

[67] C. Clancy, R. Kass, "Theoretical investigation of the neuronal Na^+ channel SCN1A: abnormal gating and epilepsy," *Biophys. J.* **86** (2004), 2606–2614.

[68] C. Clancy, Y. Rudy, "Linking a genetic defect to its cellular phenotype in a cardiac arrhythmia," *Nature* **400** (1999), 566–569.

[69] C. Clancy, Y. Rudy, "Cellular consequences of HERG mutations in the long QT syndrome: precursors to sudden cardiac death," *Cardiovascular Res.* **50** (2001), 301–313.

[70] C. Clancy, Y. Rudy, "Na+ channel mutation that causes both Brugada and long QT syndrome phenotypes: a simulation study of mechanism," *Circulation* **105** (2002), 1208–1213.

[71] C. Clancy, M. Tateyama, H. Liu, X. Wehrens, R. Kass, "Non-equilibrium gating in cardiac Na^+ channels: an original mechanism of arrhythmia," *Circulation* **107** (2003), 2233–2237.

[72] C. Clancy, Z. Zhu, Y. Rudy, "Pharmacogenetics and anti-arrhythmic drug therapy: a theoretical investigation," *Am. J. Physiol. Heart Circ. Physiol.* *292* (2007), H66–H75.

[73] L. Clegg, F. Mac Gabhann, "Molecular mechanism matters: benefits of mechanistic computational models for drug development," *Pharmacol. Res.* **99** (2015), 149–155.

[74] K. Cole, H. Curtis, "Electric impedance of the squid giant axon during activity," *J. Gen. Physiol.* **22** (1939), 649–670.

[75] J. Collins, C. Chow, T. Imhoff, "Aperiodic stochastic resonance in excitable systems," *Phys. Rev. E* **52** (1995), R3321–R3324.

[76] J. Collins, T. Imhoff, P. Grigg, "Noise-mediated enhancements and decrements in human tactile sensation," *Phys. Rev. E* **56** (1997), 923–926.

[77] D. Colquhoun, A. Hawkes, "The principles of the stochastic interpretation of ion-channel mechanisms," pgs. 397–482 of [318].

[78] D. Colquhoun, B. Sakmann, "Fast events in single-channel currents activated by acetylcholine and its analogues at the frog muscle end-plate," *J. Physiol.* **369** (1985), 501–557.

[79] F. Crick, "On protein synthesis," pgs. 138–163 of [319].

[80] J. Cuevas, R. Geller, R. Garijo, J. López-Aldeguer, R. Sanjuán, "Extremely high mutation rate of HIV-1 in vivo," *PLoS Biol.* textbf13 (2015), e1002251.

[81] A. Dalgleish, P. Beverley, P. Clapham, D. Crawford, M. Greaves, R. Weiss, "The CD4(T4) antigen is an essential component of the receptor for the AIDS retrovirus," *Nature* **312** (1984), 763–767.

[82] N. Dees, S. Bahar, R. Garcia, F. Moss, "Patch exploitation in two dimensions: from *Daphnia* to simulated foragers," *J. theor. Biol.* **252** (2008), 69–76.

[83] R. Devaney, *An Introduction to Chaotic Dynamical Systems*, 2nd ed., Addison-Wesley, Redwood City, 1989.

[84] E. Diala, M. Cheah, D. Rowitch, R. Hoffman, "Extent of DNA methylation in human tumor cells," *J. Nat. Cancer Inst.* **71** (1983), 755–764.

[85] O. Diekmann, P. Heesterbeek, *Mathematical Epidemiology of Infectious Diseases*, Wiley, New York, 2000.

[86] D. Dingli, F. Chalub, F. Santos, S. Van Segbroeck, J. Pacheco, "Cancer phenotype as the outcome of an evolutionary game between normal and malignant cells," *British J. Cancer* **101** (2009), 1130–1136.

[87] D. Dingli, F. Chalub, F. Santos, S. Van Segbroeck, J. Pacheco, "Reply: Evolutionary game theory: lessons and limitations, a cancer perspective," *British J. Cancer* **101** (2009), 2062–2063.

[88] W. Ditto, S. Rauseo, M. Spano, "Experimental control of chaos," *Phys. Rev. Lett.* **65** (1990), 3211–3214.

[89] W. Ditto, M. Spano, V. In, J. Neff, B. Meadows, J. Langberg, A. Bolmann, K. McTeague, "Control of human atrial fibrillation," *Int. J. Bifurcation Chaos* **10** (2000), 593–601.

[90] E. Domingo, J. Sheldon, C. Perales, "Viral quasispecies evolution," *Microbiology and Molecular Biology Reviews* **76** (2012), 159–216.

[91] E. Domingo, P. Schuster, eds., *Quasispecies: From Theory to Experimental Systems*, Springer, Berlin, 2016.

[92] E. Domingo, P. Schuster, "What is a quasispecies? Historical origins and current scope," pgs. 1–22 of [91].

[93] J. Douglass, F. Moss, A. Longtin, "Statistical and dynamical interpretation of ISIH data from periodically stimulated sensory neurons," pgs. 993–1000 of [158].

[94] J. Douglass, L. Wilkens, E. Pantazelou, F. Moss, "Noise enhancement of information transfer in crayfish mechanoreceptors by stochastic resonance," *Nature* **365** (1993), 337–338.

[95] C. N. Driver, "A dictionary of driven iterated function systems to characterize chaotic time series," applied mathematics senior thesis, Yale University, 2014.

[96] R. Drmanac et al., "Human genome sequencing using unchained base reads on self-assembling DNA nanoarrays," *Science* **327** (2009), 78–81. 8888

[97] D. Easley, J. Kleinberg, *Networks, Crowds, and Markets: Reasoning about a Highly Connected World*, Cambridge Univ. Pr., Cambridge, 2010.

[98] G. Edelman, *Neural Darwinism: The Theory of Neuronal Group Selection*, Basic Books, New York, 1987.

[99] G. Edelman, *The Remembered Present: A Biological Theory of Consciousness*, Basic Books, New York, 1989.

[100] G. Edelman, *Bright Air, Brilliant Fire: On the Matter of the Mind*, Basic Books, New York, 1992.

[101] G. Edelman, *Wider than the Sky: The Phenomenal Gift of Consciousness*, Yale Univ. Pr., New Haven, 2005.

[102] G. Edelman, *Second Nature: Brain Science and Human Knowledge*, Yale Univ. Pr. New Haven, 2007.

[103] G. Edelman ,G. Tononi, *A Universe of Consciousness: How Matter Becomes Imagination*, Basic Books, New York, 2001.

[104] L. Edelstein-Keshet, *Mathematical Models in Biology*, Random House, New York, 1988.

[105] M. Eigen, J. McCaskill, P. Schuster, "Molecular quasi-species," *J. Phys. Chem.* **92** (1988), 6881–6891.

[106] M. Eigen, P. Schuster, "Hypercycles and quasi-species," *Naturwissensch.* **64** (1977), 541–565.

[107] S. Ellner, J. Guckenheimer, *Dynamic Models in Biology*, Princeton Univ. Pr., Princeton, 2006.

[108] P. Erdös, A. Rényi, "On random graphs. I," *Publ. Math. (Debrecenians)* **6** (1959), 290–297.

[109] P. Erdös, A. Rényi, "On the evolution of random graphs," *Publ. Math. Inst. Hungarian Acad. Sci.* **5** (1960), 17–60.

[110] *The Evolution & Medicine Review*, https://evmedreview.com.

[111] W. Ewens, G. Grant, *Statistical Methods in Bioinformatics: An Introduction*, 2nd ed., Springer, New York, 2005.

[112] V. Fantin, J. St-Pierre, P. Leder, "Attenuation of LDH-A expression uncovers a link between glycolysis, mitochondrial physiology, and tumor maintenance," *Cancer Cell* **9** (2006), 425–434.

[113] A. Feinberg, R. Ohlsson, S. Henikoff, "The epigenetic progenitor origin of human cancer," *Nature Reviews Genetics* **7** (2006), 21–33.

[114] W. Feller, *An Introduction to Probability Theory and Its Applications*, vol. 1, 2nd ed., Wiley, New York, 1957.

[115] B. Ferreira, M. Savi, A. de Paula, "Chaos control applied to cardiac rhythms represented by ECG signals," *Phys. Scr.* **89** (2014), 105203.

[116] R. Fisher, *The Genetical Theory of Natural Selection*, Clarendon Press, Oxford, 1930.

[117] R. Fitzhugh, "Thresholds and plateaus in the Hodgkin-Huxley nerve equation," *J. Gen. Physiol.* **43** (1960), 867–896.

[118] H. Flyvbjerg, B. Lautrup, "Evolution in a rugged fitness landscape," *Phys. Rev. A* **46** (1992), 6714–6723.

[119] W. Fontana et al., "RNA folding and combinatory landscapes," *Phys. Rev. E* **47** (1993), 2083–2099.

[120] M. Frame, A. Urry, *Fractal Worlds: Grown, Built, and Imagined*, Yale Univ. Pr., New Haven, 2016.

[121] S. Frank, "George Price's contributions to evolutionary genetics," *J. theor. Biol.* **175** (1995), 373–388.

[122] P. Fredrickson, J. Kaplan, E. Yorke, J. Yorke, "The Lyapunov dimension of strange attractors," *J. Diff. Eq.* **49** (1983), 185–207.

[123] W. Freeman, "The physiology of perception," *Sci. Amer.* **264** (February 1991), 78–85.

[124] J. Freund, L. Schimansky-Geier, B. Beisner, A. Neiman, D. Russell, T. Yakusheva, F. Moss, "Behavioral stochastic resonance: how the noise from a *Daphnia* swarm enhances individual prey capture by juvenile paddlefish," *J. theor. Biol.* **214** (2002), 71–83.

[125] G. Frobenius, "Über Matrizen aus negativen Elementen," *Sitz. Preuss. Akad. Wiss. Berlin* (1912), 456–477.

[126] P. Gailey, A. Neiman, J. Collins, F. Moss, "Stochastic resonance in ensembles of nondynamical elements: the role of internal noise," *Phys. Rev. Lett.* **79** (1997), 4701–4704.

[127] H. Gangal, G. Dar, "Mode locking, chaos and bifurcations in Hodgkin-Huxley neuron forced by sinusoidal current," *Chaotic Modeling and Simulation* **3** (2014), 287–294.

[128] B. Garcia, S. Busby, J. Shabanowitz, D. Hunt, N. Mishra, "Resetting the epigenetic histone code in the MRL-lpr/lpr mouse model of lupus by histone deacetylase inhibition," *J. Proteome Res.* **4** (2005), 2032–2042.

[129] T. Gardner, C. Cantor, J. Collins, "Construction of a genetic toggle switch in *Escherichia coli*," *Nature* **403** (2000), 339–342.

[130] A. Garfinkel, M. Spano, W. Ditto, J. Weiss, "Controlling cardiac chaos," *Science* **257** (1992), 1230–1235.

[131] R. Gatenby, T. Vincent, "Application of quantitative models from population biology and evolutionary game theory to tumor therapeutic strategies," *Cancer Research* **63** (2003), 919–927.

[132] S. Gavrilets, *Fitness Landscapes and the Origin of Species*, Princeton Univ. Pr., Princeton, 2004.

[133] The genome for the Enterobacteria phage lambda is at
https://www.ncbi.nlm.nih.gov/nuccore/215104
and the genome for the bacteriophage T7 is at
https://www.ncbi.nlm.nih.gov/nuccore/V01146.

[134] Genetics Home Reference, https://ghr.nlm.nih.gov/gene/SCN5A.

[135] N. Gerald, D. Dutta, R. Brajesh, S. Saini, "Mathematical modeling of movement on fitness landscapes," *BMC Systems Biology* (2019), 13:25.

[136] Z. Gills, C. Iwata, R. Roy, I. Schwartz, I. Triandaf, "Tracking unstable steady states: Extending the stability regime of a multimode laser system," *Phys. Rev. Lett.* **69** (1992), 3169–3172.

[137] M. Gilpin, "Spiral chaos in a predator-prey model," *Amer. Naturalist* **113** (1979), 306–308.

[138] L. Glass, M. Mackey, *From Clocks to Chaos: The Rhythms of Life*, Princeton Univ. Pr., Princeton, 1988.

[139] W. Glass et al., "CCR5 deficiency increases risk of symptomatic West Nile virus infection," *J. Experimental Medicine* **203** (2006), 35–40.

[140] J. Gleick, *Chaos: Making a New Science*, Viking, New York, 1987.

[141] P. Gluckman, F. Low, T. Buklijas, M. Hanson, A. Beedle, "How evolutionary principles improve the understanding of human health and disease," *Evol. Appl.* **4** (2011), 249–263.

[142] A. Goldberger, D. Rigney, B. West, "Chaos and fractals in human physiology," *Sci. Am.* (Feb. 1990), 42–49.

[143] E. Gorter, F. Grendel, "On bimolecular layers of lipoids on the chromocytes of the blood," *J. Exp. Medicine* **41** (1925), 439–443.

[144] L. Greenemeier, "Virtual ventricle: computer predicts dangers of arrhythmia drugs better than animal testing," *Sci. Am.* (Sept. 2011). https://www.scientificamerican.com/article/computer-heart-simulation-arrhythmia.

[145] P. Greenwood, L. Ward, D. Russell, A. Nieman, F. Moss, "Stochastic resonance enhances the electrosensory information available to paddlefish for prey capture," *Phys. Rev. Lett.* **84** (2000), 4773–4776.

[146] A. Grosberg, S. Nechaev, E. Shakhnovich, "The role of topological constraints in the kinetics of collapse of macromolecules," *J. Phys. France* **49** (1988), 2095–2100.

[147] A. Grosberg, Y. Rabin, S. Havlin, A. Neer, "Crumpled globule model of the three-dimensional structure of DNA," *Europhys. Lett.* **23** (1993), 373–378.

[148] J. Guckenheimer, R. Oliva, "Chaos in the Hodgkin-Huxley model," *SIAM J. Appl. Dyn. Syst.* **1** (2002), 105–114.

[149] T. Guillard, S. Pons, D. Roux, G. Pier, D. Skurnik, "Antibiotic resistance and virulence: Understanding the link and its consequences for prophylaxis and therapy," *Bioessays* **38** (2016), 682–693.

[150] J. Gunawardena, "Time-scale separation—Michaelis and Menten's old idea, still bearing fruit," *FEBS J.* **281** (2014), 473–488.

[151] M. Ha, D. Ng, W.-H. Li, Z. Chen, "Coordinated histone modifications are associated with gene expression variation within and between species," *Genome Research* **21** (2011), 590–598.

[152] J. Hadamard, "Les surfaces à courbures opposées et leur lignes geodesics," *J. de Math.* **4** (1898), 27–73.

[153] A. Haldane, M. Manhart, A. Morozov, "Biophysical fitness landscapes for transcription factor binding sites," *PLOS Computational Biology* **10** (2014), e1003683.

[154] J. B. S. Haldane, *New Paths in Genetics*, Harper, New York, 1942.

[155] K. Hall, D. Christini, M. Tremblay, J. Collins, L. Glass, J. Billette, "Dynamic control of cardiac alternans," *Phys. Rev. Lett.* **78** (1997), 4518–4521.

[156] W. Hamilton, "The genetical evolution of social behavior," *J. theor. Biol.* **7** (1064), 1–16.

[157] P. Hänggi, P. Jung, C. Zerbe, F. Moss, "Can colored noise improve stochastic resonance?" *J. Stat. Phys.* **70** (1993), 25–47.

[158] S. Hanson, J. Cowan, L. Giles, eds., *Proceedings of the Fifth Neural Information Processing Systems Conference*, Morgan Kaufmann, San Francisco, 1993.

[159] O. Harman, *The Price of Altruism: George Price and the Search for the Origins of Kindness*, Norton, New York, 2010.

[160] D. Hebb, *The Organization of Behavior*, Wiley, New York, 1949.

[161] C. Hedrich, G. Tsokos, "Epigenetic mechanisms in systemic lupus erythematosus and other autoimmune diseases," *Trends Mol. Med.* **17** (2011), 714–724.

[162] C. Heneghan, C. Chow, J. Collins, T. Imhoff, S. Lowen, M. Teich, "Information measures quantifying aperiodic stochastic resonance," *Phys. Rev. E* **54**, (1996), R2228–R2231.

[163] S. Herculano-Houzel, "The remarkable, yet not extraordinary, human brain as a scaled-up primate brain and its associated costs," *Proc. Nat. Acad. Sci., Suppl. 1* **109** (2012), 10661–10668.

[164] S. Herculano-Houzel, R. Lent, "Isotropic fractionator: a simple, rapid method for the quantification of total cell and neuron numbers in the brain," *J. Neurosci.* **25** (2005), 2518–2521.

[165] P. Hilton, *General Cohomology Theory and K-Theory*, Cambridge Univ. Pr., Cambridge, 1971.

[166] M. Hirsch, S. Smale, *Differential Equations, Dynamical Systems, and Linear Algebra*, Academic Press, New York, 1974.

[167] M. Hirsch, S. Smale, R. Devaney, *Differential Equations, Dynamical Systems, and an Introduction to Chaos*, Academic Press, Waltham, 2013.

[168] D. Ho, A. Neumann, A. Perelson, W. Chen, J. Leonard, M. Markowitz, "Rapid turnover of plasma virions and CD4 lymphocytes in HIV-1 infection," *Nature* **373** (1995), 123–126.

[169] A. Hodgkin, A. Huxley, "Currents carried by sodium and potassium ions through the membrane of the giant axon of *Loligo*," *J. Physiol.* **116** (1952), 449–472.

[170] A. Hodgkin, A. Huxley, "The components of membrane conductance in the giant axon of *Loligo*," *J. Physiol.* **116** (1952), 473–496.

[171] A. Hodgkin, A. Huxley, "The dual effect of membrane potential on sodium conductance in the giant axon of *Loligo*," *J. Physiol.* **116** (1952), 497–506.

[172] A. Hodgkin, A. Huxley, "A quantitative description of membrane current and its application to conduction and excitation in nerve," *J. Physiol.* **117** (1952), 500–544.

[173] A. Hodgkin, A. Huxley, B. Katz, "Measurement of the current-voltage relations in the membrane of the giant axon of *Loligo*," *J. Physiol.* **109** (1949), 424–448.

[174] J. Hofbauer, K. Sigmund, *The Theory of Evolution and Dynamical Systems*, Cambridge Univ. Pr., Cambridge, 1998.

[175] J. Hofbauer, K. Sigmund, *Evolutionary Games and Population Dynamics*, Cambridge Univ. Pr., Cambridge, 1988.

[176] A. Holden, *Chaos*, Princeton Univ. Pr., Princeton, 1986.

[177] R. Horn, C. Johnson, *Matrix Analysis*, Cambridge Univ. Pr., Cambridge, 1985.

[178] S. Hummert, K. Bohl, D. Basanta, A. Deutsch, S. Werner, G. Theißen, A. Schroeter, S. Schuster, "Evolutionary game theory: cells as players," *Mol. Biosyst.* **10** (2014), 3044–3065.

[179] E. Hunt, "Stabilizing high-period orbits in a chaotic system: The diode resonator, *Phys. Rev. Lett.* **67** (1991), 1953–1955.

[180] J. Hutchinson, "Fractals and self-similarity," *Indiana Univ. Math. J.* **30** (1981), 713–747.

[181] P. Iyidogan, K. Anderson, "Current perspectives on HIV-1 antiretroviral drug resistance," *Viruses* **6** (2014), 4095–4139.

[182] H. Jeffrey, "Chaos game representation of gene structure," *Nucl. Acid Res.* **18** (1990), 2163–2170.

[183] H. Jeffrey, "Chaos game visualization of sequences," *Comp. & Graphics* **16** (1992), 25–33.

[184] P. Jones, S. Baylin, "The fundamental role of epigenetic events in cancer," *Nature Reviews Genetics* **3** (2002), 415–428.

[185] J. Jungck, S. Donovan, A. Weisstein, N. Khiripet, S. Everse, "Bioinformatics education dissemination with an evolutionary problem solving perspective," *Briefings in Bioinformatics* **11** (2010), 570–581.

[186] J. Jungck, A. Weisstein, "Mathematics and evolutionary biology makes bioinformatics education comprehensible," *Briefings in Bioinformatics* **14** (2013), 599–609.

[187] K. Kaneko, I. Tsuda, *Complex Systems: Chaos and Beyond. A Constructive Approach with Applications in Life Sciences*, Springer-Verlag, Berlin, 1996.

[188] D. Kaplan, L. Glass, *Understanding Nonlinear Dynamics*, Springer-Verlag, New York, 1995.

[189] J. Kaplan, J. Yorke, "Chaotic behavior of multidimensional difference equations," pgs. 204–227 of [272].

[190] A. Karlin, "On the application of 'a plausible model' of allosteric proteins to the receptor for acetylcholine," *J. theor. Biol.* **16** (1967), 306–320.

[191] E. Karlsson, D. Kwiatkowski, P. Sabeti, "Natural selection and infectious disease in human populations," *Nat. Rev. Genet.* **15** (2014), 379–393.

[192] B. Katz, S. Thesleff, "A study of the 'desensitization' produced by acetylcholine at the motor end-plate," *J. Physiol.* **138** (1957), 63–80.

[193] T. Kautz, N. Forrester, "RNA virus fidelity mutants: a useful tool for evolutionary biology or a complex challenge," *Viruses* **10** (2018), 600–616.

[194] A. Kaznatcheev, J. Peacock, D. Basanta, A. Marusyk, J. Scott, "Fibroblasts and alectinib switch the evolutionary games played by non-small cell lung cancer," *Nature Ecology & Evolution* **3** (2019), 450–456.

[195] J. Kemeny, L. Snell, *Finite Markov Chains*, Springer, New York, 1960.

[196] M. Kimura, "Evolutionary rate at the molecular level," *Nature* **217** (1968), 624–626.

[197] J. King, L. Jukes, "Non-Darwinian evolution," *Science* **164** (1969), 788–798.

[198] T. Kirkwood, "Why can't we live forever?" *Sci. Am.* **303** (Sept. 2010), 42–49.

[199] A. Knudson, "Mutation and cancer: statistical study of retinoblastoma," *Proc. Nat. Acad. Sci. USA* **68** (1971), 820–823.

[200] S. Krishnamurthy, K. Warner, Z. Dong, A. Imai, et al., "Endothelial Interleukin-6 defines the tumorigenic potential of primary human cancer stem cells," *Stem Cells* **32** (2014), 2845–2857.

[201] A. Lander, "A calculus of purpose," *PLoS Biology* **2** (2004), 712–714.

[202] E. Lander, et al., "Initial sequencing and analysis of the human genome," *Nature* **409** (2001), 860–921.

[203] C. Lengauer, W. Kinzler, B. Vogelstein, "Genetic instability in colorectal cancers," *Nature* **386** (1997), 623–627.

[204] C. Lengauer, W. Kinzler, B. Vogelstein, "Genetic instability in human cancers," *Nature* **396** (1998), 643–649.

[205] A. Lesk, *Introduction to Bioinformatics*, Oxford Univ. Pr., Oxford, 2002.

[206] P. Leslie, "On the use of matrices in certain population mathematics," *Biometrika* **33** (1945), 183–212.

[207] P. Leslie, "Some further notes on the use of matrices in certain population mathematics," *Biometrika* **35** (1948), 213–245.

[208] M. Liberti, J. Locasale, "The Warburg effect: how does it benefit cancer cells?" *Trends Biochem. Sci.* **41** (2016), 211–218.

[209] E. Lieberman-Aiden et al., "Comprehensive mapping of long-range interactions reveals folding principles of the human genome," *Science* **326** (2009), 289–293.

[210] K. Lin, "Entrainment and chaos in a pulse-driven Hodgkin-Huxley oscillator," arXiv:math/0505161v4.

[211] D. Lind, B. Marcus, *Introduction to Symbolic Dynamics and Coding*, Cambridge Univ. Pr., Cambridge, 1995.

[212] S. Little, A. McLean, C. Spina, D. Richman, D. Havlir, "Viral dynamics of acute HIV-1 infection," *J. Ex. Med.* **190**, 841–850.

[213] H. Liu, M. Tateyama, C. Clancy, H. Abriel, R. Kass, "Channel openings are necessary but not sufficient for use-dependent block of cardiac Na^+ channels by flecainide: evidence from the analysis of disease-linked mutations," *J. Gen. Physiol.* **120** (2002), 39–51.

[214] E. Londin, C. Barash, "What is translational bioinformatics?" *Appl. & Translational Genomics* **6** (2015), 1–2.

[215] A. Longtin, A. Bulsara, D. Pierson, F. Moss, "Bistability and the dynamics of periodically forced sensory neurons," *Biol. Cybern.* **70** (1994), 569–578.

[216] M. Loog, M. Reinders, D. de Ridder, L. Wessels, eds., *Pattern Recognition in Bioinformatics*, Springer, Berlin, 2011).

[217] E. Lorenz, "Deterministic non-periodic flows," *J. Atmos. Sci.* **20** (1963), 130–141.

[218] C. Luo, Y. Rudy, "A dynamic model of the cardiac ventricular action potential. I. Simulations of ionic currents and concentration changes," *Circ. Res.* **74** (1994), 1071–1096.

[219] C. Luo, Y. Rudy, "A dynamic model of the cardiac ventricular action potential. II. Afterdepolarizations, triggered activity, and potentiation," *Circ. Res.* **74** (1994), 1097–1113.

[220] C. MacCluer, "The many proofs and applications of Perron's theorem," *SIAM Review* **42** (2000), 487–498.

[221] B. Maddox, "The double helix and the 'wronged heroine,' " *Nature* **421** (2003), 407–408.

[222] L. Mansky, H. Temin, "Lower in vivo mutation rate of human immunodeficiency virus type 1 than that predicted from the fidelity of purified reverse transcriptase," *J. Virol.* **29** (1995), 5087–5094.

[223] M. Margulies et al., "Genome sequencing in microfabricated high-density picolitre reactors," *Nature* **437** (2005), 376–380, and **441** (2006), 120.

[224] M. Markowitz, M. Louie, A. Hurley, E. Sun, M. Mascio, A. Perelson, D. Ho, "A novel antiviral intervention results in more accurate assessment of human immunodeficiency virus type 1 replication dynamics and T cell decay in vivo," *J. Virology* **77** (2003), 5037–5038.

[225] R. Marois, J. Ivanoff, "Capacity limits of information processing in the brain," *Trends in Cognitive Sciences* **9** (2005), 296–305.

[226] J. Marsden, A. Tromba, *Vector Calculus*, 3rd ed., Freeman, New York, 1988.

[227] R. May, "Simple mathematical models with very complicated dynamics," *Nature* **261** (1976), 459–467.

[228] J. Maynard Smith, "The theory of games and the evolution of animal conflicts," *J. theor. Biol.* **47** (1974), 209–221.

[229] J. Maynard Smith, G. Price, "The logic of animal conflict," *Nature* **246** (1973), 15–18.

[230] D. McCandlish, "Visualizing fitness landscapes," *Evolution* **65** (2011), 1544–1558.

[231] W. McCullouch, W. Pitts, "A logical calculus of ideas immanent in nervous activity," *Bull. Math. Biophysics* **5** (1943), 115–133.

[232] J. McEvoy, "Evolutionary game theory: lessons and limitations, a cancer perspective," *British J. Cancer* **101** (2009), 2060–2061.

[233] L. Merlo, J. Pepper, B. Reid, C. Maley, "Cancer as an evolutionary and ecological process," *Nature Reviews Cancer* **6** (2006), 924–935.

[234] L. Michaelis, M. Menten, "Die Kinetik der Invertinwirkung," *Biochem. Z.* **49** (1913), 333–369.

[235] R. Michod, *Darwinian Dynamics: Evolutionary Transitions in Fitness and Individuality*, Princeton Univ. Pr., Princeton, 1999.

[236] M. Minsky, S. Papert, *Perceptrons: An Introduction to Computational Geometry*, MIT Press, Cambridge, 1969.

[237] L. Mirny, "The fractal globule as a model of chromatin architecture in the cell," *Chromosome Res.* **19** (2011), 37–51.

[238] S. Moolgavkar, A. Knudson, "Mutation and cancer: a model for human carcinogenesis," *J. Nat. Cancer Inst.* **66** (1981), 1037–1052.

[239] J. Moreno, Z. Zhu, P.-C. Yang, J. Brakston, M.-T. Jeng, C. Kang, L. Wang, J. Bayer, D. Christini, N. Trayanova, C. Ripplinger, R. Kass, C. Clancy, "A computational model to predict the effects of class I anti-arrhythmic drugs on ventricular rhythms," *Sci. Transl. Med.* **3**, 98ra83 (2011).

[240] E. Mosekilde, Y. Maistrenko, D. Postnov, *Chaotic Synchronization: Applications to Living Systems*, World Scientific, Singapore, 2002.

[241] F. Moss, "Stochastic resonance at the molecular level," *Biophys. J.* **73** (1997), 2249–2250.

[242] F. Moss, "Noisy waves," *Nature* **391** (1998), 743–744.

[243] F. Moss, "Stochastic resonance: looking forward," pgs. 236–256 of [372].

[244] F. Moss, X. Pei, "Neurons in parallel," *Nature* **376** (1995), 211–212.

[245] D. Mount, *Bioinformatics: Sequence and Genome Analysis*, 2nd ed., Cold Spring Harbor Laboratory Press, Cold Spring Harbor, 2004.

[246] J. Murray, *Mathematical Biology I: An Introduction*, 3rd ed., Springer, New York, 2002.

[247] K. Murphy, S. Mian, "Modelling gene expression data using dynamic Bayesian networks,"Technical Reports, Computer Sci. Div., Univ. Calif., Berkeley **104** (1999).

[248] F. Nazari, A. Pearson, J. Nör, T. Jackson, "A mathematical model for IL-6-mediated, stem cell driven tumor growth and targeted treatment," *PLOS Computational Biology* **14** (2018), e1005920.

[249] E. Neher, B. Sakmann, "Single channel currents recorded from membrane of denervated frog muscle fibres," *Nature* **260** (1976), 799–802.

[250] A. Neiman, D. Russell, F. Moss, L. Schimansky-Geier, "Stochastic synchronization: applications to oscillatory electroreceptors," pgs. 239–248 of [323].

[251] A. Neiman, L. Schimansky-Geier, A. Cornell-Bell, F. Moss, "Noise-enhanced phase synchronization in excitable media," *Phys. Rev. Lett.* **83** (1999), 4896–4899.

[252] A. Neiman, L. Schimansky-Geier, F. Moss, B. Shulgin, J. Collins, "Synchronization of noisy systems by stochastic signals," *Phys. Rev. E* **60** (1999), 284–292.

[253] R. Nesse et al., "Making evolutionary biology a basic science for medicine," *Proc. Nat. Acad. Sci. USA* **107** (2010), 1800–1807.

[254] R. Nesse, S. Stearns, "The great opportunity: Evolutionary applications to medicine and public health," *Evolutionary Applications* **1** (2008), 28–48.

[255] R. Nesse, G. Williams, *Why We Get Sick: The New Science of Darwinian Medicine*, Vintage, New York, 1994.

[256] A. Neumann, N. Lam, H. Dahari, D. Gretch, T. Wiley, T. Layden, A. Perelson, "Hepatitis C viral dynamics in vivo and the antiviral efficacy of interferon-α therapy," *Science* **282** (1998), 103–107.

[257] M. Newman, A.-L. Barabási, D. Watts, *The Structure and Dynamics of Networks*, Princeton Univ. Pr., Princeton, 2006.

[258] M. Nowak, *Evolutionary Dynamics: Exploring the Equations of Life*, Harvard Univ. Pr., Cambridge, 2006.

[259] M. Nowak, S. Bonhoeffer, A. Hill, R. Boehme, H. Thomas, H. McDade, "Viral dynamics in hepatitis B virus infection," *Proc. Nat. Acad. Sci.* **93** (1996), 4398–4402.

[260] M. Nowak, R. Highfield, *SuperCooperators: Altruism, Evolution, and Why We Need Each Other to Succeed*, Free Press, New York, 2011.

[261] M. Nowak, R. May, *Virus Dynamics: Mathematical Principles of Immunology and Virology*, Oxford Univ. Pr., Oxford, 2000.

[262] D. Nozaki, J. Collins, Y. Yamamoto, "Mechanism of stochastic resonance enhancement in neuronal models driven by $1/f$ noise," *Phys. Rev. E* **60** (1999), 4637–4644.

[263] D. Nozaki, D. Mar, P. Grigg, J. Collins, "Effects of colored noise on stochastic resonance in sensory neurons," *Phys. Rev. Lett.* **82** (1999), 2402–2405.

[264] P. Nyren, B. Pettersson, M. Uhlen, "Solid phase DNA minisequencing by an enzymatic luminometric inorganic pyrophospate detection assay," *Analytical Biochem.* **208** (1993), 171–175.

[265] P. Orlando, R. Gatenby, J. Brown, "Cancer treatment as a game: integrating evolutionary game theory into the optimal control of chemotherapy," *Phys. Biol.* **9** (2012), doi:10.1088/1478-3975/9/6/065007.

[266] V. Oseledec, "A multiplicative ergodic theorem. Lyapunov characteristic numbers for dynamical systems," *Trans. Moscow Math. Soc.* **19** (1978), 197–231.

[267] E. Ott, C. Grebogi, J. Yorke, "Controlling Chaos," *Phys. Rev. Lett.* **64** (1990), 1196–1199.

[268] J. Pacheco, F. Santos, D. Dingli, "The ecology of cancer from an evolutionary game theory perspective," *Interface Focus* (2014), doi.org/10.1098/rsfs.2014.0019.

[269] L. Page, S. Brin, R. Motwani, T. Winograd, "The PageRank citation ranking: bringing order to the web," Technical Report 1999-66, Stanford InfoLab, Nov. 1999.

[270] D. Peak, "Taming chaos in the wild: A model-free technique for wildlife population control," pgs. 70–100 of [43].

[271] D. Peak, M. Frame, *Chaos Under Control: The Art and Science of Complexity*, Freeman, New York, 1994.

[272] H.-O. Peitgen, H.-O. Walther, *Functional Differential Equations and the Approximation of Fixed Points*, Springer, Berlin, 1979.

[273] A. Perelson, "Modelling viral and immune system dynamics," *Nature Reviews Immunology* **2** (2002), 28–36.

[274] A. Perelson, P. Essunger, Y. Cao, M. Vesanen, A. Hurley, K. Saksela, M. Markowitz, D. Ho, "Decay characteristics of HIV-1-infected compartments during combination therapy," *Nature* **387** (1997), 188–191.

[275] A. Perelson, P. Essunger, D. Ho, "Dynamics of HIV-1 and CD4+ lymphocytes in vivo," *AIDS* **11** (1997), S17–S24.

[276] A. Perelson, J. Guedj, "Modelling hepatitis C therapy—predicting effects of treatment," *Nature Reviews Gastroenterology and Hepatology* **12** (2015), 437–445.

[277] A. Perelson, S. Kauffman, eds., *Molecular Evolution on Rugged Landscapes: Proteins, RNA, and the Immune System*, Addison-Wesley, Redwood City, 1991.

[278] A. Perelson, A. Neumann, M. Markowitz, J. Leonard, D. Ho, "HIV-1 dynamics in vivo: virion clearance rate, infected cell life-span, and viral generation time," *Science* **271** (1996), 1582–1586.

[279] A. Perelson, R. Ribeiro, "Modeling the within-host dynamics of HIV infection," *BMC Biology* **11** (2013), 96–105.

[280] A. Pérez-Rivera et al., "The congenital long QT syndrome type 3: an update," *Indian Pacing and Electrophysiology J.* **18** (2018), 25–35.

[281] L. Perko, *Differential Equations and Dynamical Systems*, Springer, New York, 1991.

[282] O. Perron, "Zur Theorie zur Matrices," *Math. Annalen* **64** (1907), 248–263.

[283] A. Petronis, "The origin of schizophrenia: genetic thesis, epigenetic antithesis, and resolving synthesis," *Biological Psychiatry* **55** (2004), 965–970.

[284] V. Petrov, V. Gáspár, J. Masere, K. Showalter, "Controlling chaos in the Belousov-Zhabotinsky reaction," *Nature* **361** (1993), 240–243.

[285] https://collections.plos.org/translational-bioinformatics.

[286] H. Poincaré, *New Methods in Celestial Mechanics*, ed. D. Goroff, Amer. Inst. Physics, 1993.

[287] D. Pokholok, et al., "Genome-wide map of nucleosome acetylation and methylation in yeast," *Cell* **122** (2005), 517–527.

[288] G. Porreca, "Genome sequencing on nanoballs," *Nature Biotech.* **28** (2010), 43–44.

[289] W. Poundstone, *Prisoner's Dilemma: John von Neumann, Game Theory, and the Puzzle of the Bomb*, Doubleday, New York, 1992.

[290] R. Powers, *The Gold Bug Variations*, Morrow, New York, 1991.

[291] R. Powers, *The Overstory*, Norton, New York, 2018.

[292] R. Powers, *Prisoner's Dilemma*, HarperCollins, New York, 1988.

[293] G. Price, "Extension of covariance selection mathematics," *Ann. Hum. Genet. London* **35** (1972), 485–490.

[294] G. Price, "Fisher's 'fundamental theorem' made clear," *Ann. Hum. Genet. London* **36** (1972), 129–140.

[295] G. Price, "Selection and covariance," *Nature* **227** (1970), 520–521.

[296] M. Prince, R. Sivanandan, A. Kaczorowski, G. Wolf, M. Kaplan, P. Dalerba, I. Weissman, M. Clarke, L. Ailles, "Identification of a subpopulation of cells with cancer stem cell properties in head and neck squamous cell carcinoma," *Proc. Nat. Acad. Sci.* **104** (2007), 973 –n978.

[297] A. Priplata, J. Niemi, J. Harry, L. Lipsitz, J. Collins, "Vibrating insoles and balance control in elderly people," *Lancet* **362** (2003), 1123–1124.

[298] A. Priplata, J. Niemi, M. Salen, J. Harry, L. Lipsitz, J. Collins, "Noise-enhanced human balance control," *Phys. Rev.Lett.* **23** (2002), 238101-1–238101-4.

[299] A. Priplata, B. Patritti, J. Niemi, R. Hughes, D. Gravelle, L. Lipsitz, A. Veves, J. Stein, P. Bonato, J. Collins, "Noise-enhanced balance control in patients with diabetes and patients with stroke," *Ann. Neurol.* **59** (2006), 4–12.

[300] R. Prum, *The Evolution of Beauty: How Darwin's Forgotten Theory of Mate Choice Shapes the Animal World—and Us*, Doubleday, New York, 2017.

[301] D. Queller, "A general model for kin selection," *Evolution* **46** (1992), 376–380.

[302] D. Racko, F. Benedetti, D. Goundaroulis, A. Stasiak, "Chromatin loop extrusion and chromatin unknotting," *Polymers* **10** (2018), 1126–1137.

[303] V. Ramakrishnan, *Gene Machine: The Race to Decipher the Secrets of the Ribosome*, Basic Books, New York, 2018.

[304] A. Rapoport, A. Chammah, C. Orwant, *Prisoner's Dilemma*, Univ. of Michigan Pr., Ann Arbor, 1965.

[305] S. Reid, "Disease evolution: how new illnesses emerge when we change how we live," https://theconversation.com/disease-evolution-how-new-illnesses-emerge-when-we-change-how-we-live-54570

[306] C. Reidys, P. Stadler, "Combinatorial landscapes," *SIAM Review* **44** (2002), 3–54.

[307] T. Reya, S. Morrison, M. Clarke, I. Weissman, "Stem cells, cancer, and cancer stem cells," *Nature* **414** (2001), 105–111.

[308] R. Ribeiro, L. Qin, L. Chavez, D. Li, S. Self, A. Perelson, "Estimation of the initial viral growth rate and basic reproductive number during acute HIV-1 infection," *J. Virology* **84** (2010), 6096–6102.

[309] H. Richter, "Evolutionary optimization and dynamic fitness landscapes: from reaction-diffusion systems to chaotic CML," pgs. 409–446 of [398].

[310] H. Richter, A. Engelbrecht, eds., *Recent Advances in the Theory and Application of Fitness Landscapes*, Springer, Berlin, 2014.

[311] H. Richter, "Scale-invariance of ruggedness measures in fractal fitness landscapes," *Int. J. Parallel, Emergent, and Distributed Systems* **33** (2018), 460–473.

[312] B. Roberts, P.-C. Yang, S. Behrens, J. Moreno, C. Clancy, "Computational approaches to understand cardiac electrophysiology and arrhythmias," *Am. J. Physiol. Heart Circ. Physiol.* **303** (2012), H766–H783.

[313] F. Rosenblatt, "The perceptron: a probabilistic model for information storage and organization in the brain," *Psych. Rev.* **65** (1958), 386–408.

[314] T. Roth, F. Lubin, M. Sodhi, J. Kleinman, "Epigenetic mechanisms in schizophrenia," *Biochim. Biophys. Acta* **1790** (2009), 869–877.

[315] N. Rusk, "Torrents of sequence," *Nature Methods* **8** (2011), 44.

[316] G. Ryland et al., "Loss of heterozygosity: what is it good for?" *BMC Medical Genomics* **8** (2015), 45–56.

[317] B. Sakmann, E. Neher, "Patch clamp techniques for studying ionic channels in excitable membranes," *Ann. Rev. Physiology* **46** (1984), 455–472.

[318] B. Sakmann, E. Neher, eds., *Single Channel Recording*, 2nd ed., Springer, New York, 1995.

[319] F. Sanders, *Biological Replication of Macromolecules*, Cambridge Univ. Pr., Cambridge, 1958.

[320] F. Sanger, G. Air, B. Barrell, N. Brown, A. Coulson, C. Fiddes, C. Hutchinson, P. Slocombe, M. Smith, "Nucleotide sequence of bacteriophage phi X174 DNA," *Nature* **265** (1977), 687–695.

[321] M. Sarangdhar, C. Kambhampati, "Chaotic oscillations in a Hodgkin-Huxley neuron—quantifying similarity estimation of neural responses," *Proc. World Congr. Engineering, 2010*, ISSN:2078-0958.

[322] S. Schiff, K. Jerger, D. Duong, T. Chang, M. Spano, W. Ditto, "Controlling chaos in the brain," *Nature* **370** (1994), 615–620.

[323] L. Schimansky-Geier, D. Abbott, A. Neiman, C. Van der Broeck, eds., *Noise in Complex Systems and Stochastic Dynamics, Proc. SPIE* **5114** (2003).

[324] Systems biology markup language website, http://sbml.org/index.psp

[325] P. Schuster, "Mathematical modeling of evolution. Solved and open problems," *Theoret. Biosci.* **130** (2011), 71–89.

[326] P. Schuster, "Quasispecies on fitness landscapes," pgs. 61–120 of [91].

[327] T. Shen, S. Pajaro-Van de Stadt, N. Yeat, J. Lin, "Clinical applications of next generation sequencing in cancer: from panels, to exomes, to genomes," *Frontiers in Genetics* **6** (2015), art. 215.

[328] H. Shim, Y. Chun, B. Lewis, C. Dang, "A unique glucose-dependent apoptotic pathway induced by c-Myc," *Proc. Nat. Acad. Sci.* **95** (1998), 1511–1516.

[329] I. Shimada, T. Nagashima, "A numerical approach to ergodic problem of dissipative dynamical systems," *Prog. Theor. Phys.* **61** (1979), 1605–1616.

[330] T. Shirahata, "Numerical simulation of bistability between regular bursting and chaotic spiking in a mathematical model of snail neurons," *Int. J. Theor. and Math. Physics* **5** (2015), 145–150.

[331] K. Sigmund, *Games of Life: Exploration in Ecology, Evolution, and Behavior*, Oxford Univ. Pr., Oxford, 1993.

[332] E. Simonotto, M. Riani, C. Seife, M. Roberts, J. Twitty, F. Moss, "Visual perception of stochastic resonance," *Phys. Rev. Lett.* **78** (1997), 1186–1189.

[333] E. Simonotto et al., "fMRI studies of visual cortical activity during noise stimulation," *Neurocomputing* **26–27** (1999), 511–516.

[334] J. Simpore, S. Pignatelli, S. Barlati, S. Musumeci, "Modification in the frequency of Hb S and Hb C in Burkina Faso: an influence of migratory fluxes and improvement of patient health care," *Hemoglobin* **26** (2002), 113–120.

[335] J. Simpore, S. Pignatelli, S. Musumeci, "Anthropological consideration on prevalence and fitness of β C and β S genotypes in Burkina Faso," *International J. Anthropology* **17** (2002), 77–89.

[336] L. Smith, *Linear Algebra* 3rd Ed., Springer, New York, 1998.

[337] T. Smith, M. Waterman, "Identification of common molecular subsequences," *J. Mol. Biol.* **147** (1981), 195–197.

[338] L. Sompayrac, *How the Immune System Works*, 5th ed., Wiley, Chichester, 2016.

[339] G. Sorkin, "Combinatorial optimization, simulated annealing, and fractals," IBM Research Report RC 13674 (No. 61253).

[340] G. Sorkin, "Efficient simulated annealing on fractal energy landscapes," *Algorithmica* **6** (1991), 367–418.

[341] I. Splawski et al., "Variant of SCN5A sodium channel implicated in risk of cardiac arrhythmia," *Science* **297** (2002), 1333–1336.

[342] M. Stafford, L. Corey, Y. Cao, E. Daar, D. Ho, A. Perelson, "Modeling plasma virus concentration during primary HIV infection," *J. theor. Biol.* **203** (2000), 285–301.

[343] S. Stearns, *The Evolution of Life Histories*, Oxford Univ. Pr., Oxford, 1992.

[344] S. Stearns, J. Koella, *Evolution in Health and Disease*, 2nd ed., Oxford Univ. Pr., Oxford, 2009.

[345] S. Stearns, R. Medzhitov, *Evolutionary Medicine*, Sinauer Assoc., Sunderland, MA, 2016.

[346] S. Stearns, R. Nesse, D. Govindaraju, P. Ellison, "Evolutionary perspective on health and medicine," *Proc. Nat. Acad. Sci. USA* **107** (2010), 1691–1695.

[347] S. Stearns, R. Nesse, D. Haig, "Introducing evolutionary thinking for medicine," pgs. 3–15 of [344].

[348] D. Stehelin, H. Varmus, M. Bishop, P. Vogt, "DNA related to the transforming gene(s) of avian sarcoma viruses is present in normal avian DNA," *Nature* **260** (1976), 170–173.

[349] S. Sternberg, *Dynamical Systems*, Dover, Mineola, 2010.

[350] http://explore.pediatriccancergenomeproject.org

[351] S. Strogatz, *Nonlinear Dynamics and Chaos with Applications to Physics, Biology, Chemistry, and Engineering*, 2nd ed., Westview Press, Philadelphia, 2015.

[352] S. Strogatz, *Infinite Powers: How Calculus Reveals the Secrets of the Universe*, Houghton Mifflin Harcourt, Boston, 2019.

[353] S. Strogatz, "Outsmarting a virus with math," *Sci. Am.* **320** (April 2019), 70–73.

[354] J. Su, Y. Qi, S. Liu, X. Wu, J. Lv, H. Liu, R. Zhang, Y. Zhang, "Revealing epigenetic patterns in gene regulation through integrative analysis of epigenetic interaction network," *Mol. Biol. Rep.* **39** (2012), 1701–1712.

[355] C. Terrenoire, C. Clancy, J. Cormier, K. Sampson, R. Kass, "Autonomic control of cardiac action potentials: role of potassium channel kinetics in response to sympathetic stimulation," *Circ. Res.* **96** (2005), e25–e34.

[356] S. Thompson, S. Bakhoum, D. Compton, "Mechanisms of chromosomal instability," *Curr. Biol.* **20** (2010), R285–R295.

[357] H. Tijms, *Understanding Probability: Chance Rules in Everyday Life*, 2nd ed., Cambridge Univ. Pr., Cambridge, 2007.

[358] S. Tonegawa, "Somatic generation of antibody diversity," *Nature* **302** (1983), 575–581.

[359] S. Tonegawa, N. Hozumi, G. Matthyssens, R. Schuller, "Somatic changes in the content and context of immunoglobulin genes," *Cold Spring Harbor Symposia on Quantitative Biology* **41** (1977), 877–889.

[360] A. Trigos, R. Pearson, A. Papenfuss, D. Goode, "Altered interactions between unicellular and multicellular genes drive hallmarks of transformation in a diverse range of solid tumors," *Proc. Nat. Acad. Sci. USA* **114** (2017), 6406–6411.

[361] N. Vargas-Rondón, V. Villegas, M. Rondón-Lagos, "The role of chromosomal instability in cancer and therapeutic responses," *Cancers* **10** (2018), doi:10.3390.

[362] V. Vassilev, T. Fogarty, J. Miller, "Information characteristics and the structure of landscapes," *Evolutionary Computation* **8** (2000), 31–60.

[363] J. Venter et al., "The sequence of the human genome," *Science* **291** (2001), 1304–1305.

[364] B. Vogelstein, K. Kinzler, eds., *The Genetic Basis of Human Cancer*, 2nd ed., McGraw-Hill, Toronto, 2002.

[365] E. Voit, *A First Course in Systems Biology*, Garland Science, 2012.

[366] J. von Neumann, "Zur theorie der Gesellschaftsspiele," *Math. Ann.* **100** (1928), 295–300.

[367] J. von Neumann, O. Morgenstern, *Theory of Games and Economic Behavior*, Princeton Univ. Pr., Princeton, 1944.

[368] R. Voss, "Evolution of long-range fractal correlations and $1/f$ noise in DNA base sequences," *Phys. Rev. Lett.* **68** (1992), 3805–3808.

[369] R. Voss, "Long-range fractal correlations in DNA introns and exons," *Fractals* **2** (1994), 1–6.

[370] R. Voss, J. Clarke, "$1/f$ noise in music and speech," *Nature* **258** (1975), 317–318.

[371] R. Wallace, "A fractal model of HIV transmission on complex sociogeographic networks: towards analysis of large data sets," *Environment and Planning A* **25** (1993), 137–148.

[372] J. Walleczek, *Self-Organized Biological Dynamics and Nonlinear Control*, Cambridge Univ. Pr., Cambridge, 2000.

[373] O. Warburg, K. Posener, E. Negelein, "The metabolism of cancer cells," *Biochem. Z.* **152** (1924), 314–344.

[374] O. Warburg, "The metabolism of carcinoma cells," *J. Cancer Research* **9** (1925), 148–163.

[375] O. Warburg, "On the origin of cancer cells," *Science* **123** (1956), 309–314.

[376] X. Wei, S. Ghosh, M. Taylor, V. Johnson, E. Emini, P. Deutsch, J. Lifson, S. Bonhoeffer, M. Nowak, B. Hahn, M. Saag, G. Shaw, "Viral dynamics in human immunodeficiency virus type 1 infection," *Nature* **373** (1995), 117–122.

[377] E. Weinberger, "Fourier and Taylor series on fitness landscapes," *Biol. Cybernetics* **65** (1991), 321–330.

[378] E. Weinberger, P. Stadler, "Why *some* fitness landscapes are fractal," *J. theor. Biol.* **163** (1993), 255–275.

[379] J. Weiss, A. Garfinkel, M. Spano, W. Ditto, "Chaos and chaos control in biology," *J. Clin. Invest.* **93** (1994), 1355–1360.

[380] J. Wells, R. Nesse, R. Sear, R. Johnstone, S. Stearns, "Evolutionary public health: Introducing the concept," *Lancet* **390** (2017), 500–509.

[381] P. Werbos, "Beyond regression: new tools for prediction and analysis in the behavioral sciences," Ph. D. thesis, Harvard Univ., 1974.

[382] P. Werbos, *The Roots of Backpropagation: From Ordered Derivatives to Neural Networks and Political Forecasting*, Wiley, New York, 1994.

[383] K. Wiesenfeld, F. Moss, "Stochastic resonance and the benefits of noise: from ice ages to crayfish and SQUIDs," *Nature* **373** (1995), 33–36.

[384] C. Wilke, "Quasispecies theory in the context of population genetics," *BMC Evolutionary Biology* **5** (2005), 44–51.

[385] G. Williams, "Pleiotropy, natural selection, and the evolution of senescence," *Evolution* **11** (1957), 398–411.

[386] G. Williams, R. Nesse, "The dawn of Darwinian medicine," *Quart. Rev. Biology* **66** (1991), 1–22.

[387] A. Winfree, *The Geometry of Biological Time*, Springer-Verlag, Berlin, 1990.

[388] A. Winfree, *When Time Breaks Down: The Three-Dimensional Dynamics of Electrochemical Waves and Cardian Arrhythmias*, Princeton Univ. Pr., Princeton, 1987.

[389] A. Wolf, "Quantifying chaos with Lyapunov exponents," pgs. 273–290 of [176].

[390] M. Woolhouse, S. Gowtage-Sequeria, "Host range and emerging and reemerging pathogens," *Emerging Infectious Diseases* **11** (2005), 1842–1847.

[391] S. Wright, "Evolution in Mendelian populations," *Genetics* **16** (1931), 97–159.

[392] S. Wright, "The role of mutation, inbreeding, crossbreeding, and selection in evolution," *Proc. Sixth Int. Cong. Genetics* **1** (1932), 356–366.

[393] J. Xiong, *Essential Bioinformatics*, Cambridge Univ. Pr., Cambridge, 2006.

[394] X. Xu, S. Hoang, M. Mayo, S. Bekiranov, "Application of machine learning methods to histone methylation ChIP-Seq data reveals H4R3me2 globally represses gene expression," *BMC Bioinformatics* **11** (2010), 396.

[395] L. You, J. Brown, F. Thuijsman, J. Cunningham, R. Gatenby, J. Zhang, K. Staňková, "Spatial vs. non-spatial eco-evolutionary dynamics in a tumor growth model," *J. theor. Biol.* **435** (2017), 78–97.

[396] H. Yu, S. Zhu, B. Zhou, H. Xue, J. Han, "Inferring causal relationships among different histone modifications and gene expression," *Genome Research* **18** (2008), 1314–1324.

[397] T. Zaidi, J. Lyczak, M. Preston, G. Pier, "Cystic fibrosis transmembrane conductance regulator-mediated corneal epithelial cell ingestion of *Pseudomonas aeruginosa* is a key component in the pathogenesis of experimental murine keratitis," *Infection and Immunity* **67** (1999), 1481–1492.

[398] I. Zelinka et al., *Evolutionary Algorithms and Chaotic Systems*, Springer, Berlin, 2010.

[399] J. Zeng, K. Laurita, D. Rosenbaum, Y. Rudy, "Two components of the delayed rectifier K^+ current in ventricular myocytes of the guinea pig type: Theoretical formulation and their role in repolarization," *Circ. Res.* **77** (1995), 140–152.

[400] J. Zhang, J. Cunningham, J. Brown, R. Gatenby, "Integrating evolutionary dynamics into treatment of metastatic castrate-resistant prostate cancer," *Nature Commun.* **8** (2017), doi:10.1038/s41467-017-01968-5.

[401] J. Zheng, I. Chaturvedi, J. Rajapakse, "Integration of epigenetic data in Bayesian network modeling of gene regulatory networks," pgs. 87–96 of [216].

[402] N. Zinder, J. Lederberg, "Genetic exchange in Salmonella," *J. Bacteriol.* **64** (1952), 679–699.

Index

abiraterone acetate, 334
action potential, 45, 208, 209
adaptation, 157–161, 331, 332
adaptive control, 82
Agarwal, Divyansh, xvi
agonist, 198–200
AIDS, 24, 30, 374
Aihara, Kazuyuki, 51
AIS cell
  experienced, 376, 383, 385
  naive, 376, 379, 380, 382–385, 387, 388
algebraic multiplicity, 356–359
allele, 136–141, 143, 145, 146, 172, 180–185, 187, 215, 216, 220, 332
  dominant, 143, 144
  fixation, 182–189
  frequency, 137–140, 143–145, 180, 182
  heterogeneous, 140
  pair, 138, 145
  recessive, 143, 144
  sex-linked, 138
  TSG, 215–220, 222
Almeida, Jonas, 344
Alon, Uri, 314, 316, 324
Altschul, Stephen, 164
amniotic fluid, 351
amylase, 17, 164, 343, 412
Anderson, C., 201
Anderson, Karen, 34
androgen deprivation therapy (ADT), 333
anergized, 388
anneal, 337, 338
Antaeus, 373
antibiotic, 330, 331
antibiotic-resistant, 330, 331
antibody, 23, 171, 173, 177, 376, 379–383, 385, 386, 388
  F_{ab} region, 380–383
  F_c region, 380–383
  heavy chain (Hc), 380, 381

light chain (Lc), 380, 381
neutralizing, 380
anticodon, 407
anticodon loop, 407
anticontrol, 81
antiderivative, 242, 246, 249, 294, 389
antigen, 173, 177, 334, 375, 379–385, 387, 388
cognate, 380, 381, 383, 385, 386
self, 387, 388
antigen presenting cells (APC), 375, 383, 385
antigen recognition protein
α, 383
β, 383
antisense therapy, 171
apoptosis, 220, 378, 384, 386, 387
Archimedes, 281
arclength, 232, 233, 247, 281
by integration, 231
Arnold, Monique, xvi
ARQ 531, 177
arrhythmia, 81, 82, 86, 206–215
Artz, Steven, xvii
asymptotic
behavior, 44
value, 318, 321–323
atrial fibrillation (AFib), 81, 82
atrioventricular (AV) node, 207
autoimmune disease, 173, 174, 376, 387
autoimmune reaction, 382
autoregulator, 323
average selective value, 139
Avery, Oswald, 336
AZT, 30, 31

B cell, 23, 173, 177, 375, 376, 379–384, 386–388
class switching, 382
experienced, 385
memory, 383, 385, 386
naive, 380, 382, 384, 385, 388
plasma, 383, 386
proliferation, 382, 384
receptor (BCR), 381–384, 388
somatic hypermutation, 383, 388
B cell activation
T cell dependent, 382
T cell independent, 382
B7, 379, 383, 385
bacteria, 23, 293, 330, 331, 333, 373, 374, 380, 413
Bahar, Sonya, xvii, 149, 401
Bancaud, Aurélien, 414, 415
Barnsley, Michael, 341
Barre-Sinoussi, Francoise, 24
basic reproductive ratio (R_0), 26–30, 34–39, 41, 45
basin of attraction, 66, 68, 69
basis, 96, 125, 356–358, 396–398
Bayes' theorem, xxi, 175
Bayesian inference, 175
Bendixson's criterion, x, xiii, 263–266, 273
Benettin, Giancarlo, 395
Bézout's identity, 107–109
Bick, Ted, xv
bifurcation, 308
binomial
coefficient, 180
distribution, 180
expansion, 354
expected value, 180
variance, 181
bioinformatics, xiii, xvii, 153, 163–167, 340, 341, 415
translational, xiii, 336–341
biological altruism, 146, 149
biologics, 176
Birman, Joan, 57
birth rate, 8, 141
bistable system, 307
BLAST, xiii, 164–167, 337
blind spot, 332
BLOSUM, 165
Bobrownicki, Aiyana, xvi
Borges, Jorge Luis, 328
Breitenbach, T. E., xvii
Brownian motion, 89
Bruton's tyrosine kinase (BKT) inhibitors, 177
Bubinak, John, xvii

burst size, 26, 35, 374
Byrd, John, xvii

Calamia, Joseph, xvi
cancer, 24, 172, 173, 175–178, 215, 216, 219, 220, 325, 333, 341, 373, 376
  breast, 172, 331, 340
  cervical, 351
  CLL, 177
  colon, 215, 224
  Hodgkin's lymphoma, 176
  liver, 373
  prostate, 331, 333
    castrate-resistant (mCRPC), 333, 334
    metastatic (mPC), ix, 333
  skin, 172
cancer genome atlas, 341
cancer stem cell (CSC), 177, 178
cancer stem cell hypothesis, 177
Cantor, Charles, 307, 310
capsid, 23, 374
cardiac
  dynamics, 81
  SA pacemaker, 207
Cauchy, Augustin, 244
causal diagram, 175
*CD*28 receptor, 383, 385
*CD*4 receptor, 23, 24, 383
*CD*8 receptor, 23, 24, 383
cell surface receptors, 23, 40, 171, 177, 198, 199, 371, 374–378, 380–385, 387, 388
  activation, 378
  inhibition, 378, 379
chain-terminating nucleotide, 338, 339
chain-termination method, 336, 337
Chang, Amy, xvii
Chang, Sandy, xvi
chaos, xii, 1, 50, 51, 54–62, 65, 67, 68, 70, 73, 74, 78, 80–82, 86, 351, 421, 422
  control of, 1, 54, 55, 73–86, 359–361, 397, 422
  in cardiac dynamics, 81, 82, 86
  in neurological dynamics, 81
  in population dynamics, 55–59, 61–63, 80
chaotic dynamics, 51, 60, 61
Chapman-Kolmogorov equations, 91
characteristic equation, 2, 5, 115–117, 119, 121, 123, 124, 393, 394
chemokine, 384
chemotherapy, 171, 176, 178
Chen, Haifen, 175, 176
Christini, David, 82
chromatin, 172, 174, 413–415
chromatin loop extrusion, 414
chromosomal instability (CIN), 173, 220, 224
circulatory system, 378, 379
Clairaut's theorem, 231, 237, 287, 288
Clancy, Colleen, xvi, 178, 207–209, 211
Clancy-Rudy model, x, xiii, 207–209, 211
clearance rate, 25–27, 32–34, 37–39
Clegg, Lindsay, 209
closed curve, 9, 225, 236, 237, 250, 251, 256, 257, 260, 263, 267, 270, 271, 275, 288
  simple, 57, 236, 253, 254, 269–271
co-receptor, 382, 383
co-stimulation, 379, 383–385, 388
codon, 407–411
  START, 409, 410
  STOP, 409, 410
coevolution, 142, 152, 153
Cole, Kenneth, 372
Collins, James, 307, 310, 401
colon crypt, 215, 216, 218
Colquhoun, David, 199
Colquhoun-Sakmann model, 199, 200
comparison test for infinite series, 354
complement molecule
  *C*3, 376, 377, 382
  *C*3*a*, 376, 377
  *C*3*b*, 376, 377, 382
  *iC*3*b*, 377
complement system (CS), 23, 375–378, 380–382, 384
  alternative activation pathway, 376
  *C*1 complex, 381, 382
  classical activation pathway, 376, 380–382
  lectin activation pathway, 376, 377
complementary bases, 406
complex number, 14, 78, 117, 118, 391

argument, 391
imaginary part, 118
modulus, 391
polar representation, 117, 391
real part, 14, 30, 40, 43, 61, 64, 117
component (vector projection), 232, 233, 281, 282, 284, 397
conditional probability, xx, 88, 89, 175
connected space, 235, 236, 253
continuous control, 82
contraction, 341
contraction mapping principle (CMP), 390
convex combination, 110, 367
cooperative behavior, 168, 169
correlation
address, 349
function, 412
long-range, 415
Costner, Kevin, 397
covariance, 147, 148
COVID-19, 330
Coyne, Christopher, xvi
CpG, 172
islands, 172
Crick, Francis, 336, 407
critical point, 140, 155, 156
cross product, 280
crosslinked
BCR, 379, 382, 384
TCR, 383, 388
CTL, 37–41, 44
response, 37–45, 418
Cuevas, José, 34
curl, xiii, 225, 226, 285–291, 296, 299–301
cycle, 55, 74, 75, 78, 80, 83–85, 93, 97, 98, 106, 109, 111–113, 315
cylindrical coordinates, 293–295, 300
cystic fibrosis, 332, 410
cytokine, 23, 177, 375, 377, 378, 382, 384–386
profile, 384

Darwin, Charles, 142
death rate, 8, 141, 178
degree
in-, 314, 316
of freedom, 279, 280
of homogeneity, 129, 130
out-, 314, 316
Dekker, Job, 414
denature, 337–339
dendritic cells, 375, 379, 383, 384
follicular (FDC), 384, 388
Dengue virus, 330
dense, 55, 57, 74
dense overlapping regulon (DOR), 324
depolarization, 206–208
derivative
directional, 151, 152, 227–230
matrix, xix, 8–13, 15, 19, 21, 28, 29, 35, 36, 39, 40, 42, 43, 45, 51–53, 58, 59, 61, 170, 264, 275, 309, 398
partial, 127–130, 133, 152, 155, 156, 203, 226–228, 230, 231, 234–240, 242, 243, 250–265, 278, 285–288, 296–298, 300–303, 396, 403, 404
Descartes, René, 87
deterministic, 82
Devaney, Robert, 54
diagonalizable, 357, 359
dictionary of coupled maps, 349–351
Diekmann, Odo, 114
differential equation
autonomous, 54, 67
non-autonomous, 32, 68
differentiation
chain rule, 70, 130, 155, 227, 228, 234, 286, 287, 404
product rule, 126, 287, 355
quotient rule, 141
diffusion, 226, 302, 303
equation, 301
dimension
box-counting, 279, 398, 414
Euclidean, ix, xii, 1, 3, 9–11, 35, 36, 50, 51, 54–58, 64, 67, 68, 70, 82, 89, 110, 151, 152, 155, 166, 170, 182, 226, 227, 232, 277–280, 285, 293, 296, 299, 395, 398, 411, 412, 414–418
Dingli, David, 170
diploid, 180

directed graph, 175, 305
discrete dynamics, 88–104, 114–123, 180–189, 216–224, 342, 344–351
Ditto, William, 80, 81
div, 301
divergence, xiii, 225, 226, 296–303
DNA, 23, 24, 30, 31, 164, 165, 172–174, 304, 312–314, 331, 336–340, 343, 344, 374, 381, 383, 384, 406–415
  3′-end, 337, 407
  5′-end, 172, 337, 407
  polymerase, 337–340
  primer, 337–339
  sequence, 337, 338
  template, 337–340
DNA sequencing, 336–340
Do Carmo, Mariana, xvi
dominant, 136, 137, 143, 144
Donaldson, Simon, xiii
Donnally, Mike, xvii
dot product, 77, 86, 125, 152, 227, 229, 233, 279, 280, 282, 290, 397
Driver, Noelle, xvi, 348
Drmanac, Radoje, 340
Duchenne muscular dystrophy (DMD), 410
Duffing equation, 1, 54, 63, 64, 66, 67, 69, 70, 72, 73, 398–401, 421
  forced, 64, 65
Duffing system, 64, 65, 70, 71, 307
  forced, 70, 73
DUST, 164
dynamic control, 82
dynamical Bayesian network, 175

early after-depolarization, 208, 209
Ebola virus, 330
Eigen, Manfred, 150, 158
eigenvalue, xix, 1–16, 19, 21, 28–30, 35, 36, 40, 42, 43, 45, 52, 59, 61, 62, 64, 66, 75, 83, 88, 94–96, 98, 100, 102, 105, 106, 110, 115–120, 122–127, 130, 132, 159, 162, 170, 202–205, 210, 211, 214, 264, 271, 273, 275, 309, 310, 356–361, 367, 368, 390–393, 398, 416, 424
  complex, 14, 30, 39, 40, 43, 61, 64, 116–118, 123, 356, 358
  dominant, 96, 98, 102, 105, 106, 110, 118, 120, 122, 123, 125–129, 131, 132, 202, 212, 366, 368, 370, 393
  geometric multiplicity, 356–358
eigenvector, 1, 3–6, 28, 29, 35, 61, 66, 75–77, 83, 88, 94–96, 98–102, 105, 110, 116, 118–120, 123–128, 130–133, 159, 162, 202–205, 210–215, 356–361, 366–368, 370, 390–393, 424
  component equation, 95
  equation, 3, 100, 116, 118, 119, 126, 130
  generalized, 357, 358
  left, 110, 123–128, 131–133, 424
  stable, 75–78
  unit, 94–96, 98–102, 105, 118, 119, 124–128, 131–133, 159, 162, 202–205, 210–214, 366, 368, 424
  unstable, 75
Einstein, Albert, xiii, 180, 226
Einthoven, Willem, 206
elasticity, 129, 131–133
electrocardiogram (ECG), 206, 207
Ellner, Stephen, 201, 328
endoplasmic reticulum (ER), 375
endosome, 375, 376
epigenetic landscape, 172
epigenetics, 172, 174–176
epilepsy, 209
epitope, 380, 382
Erdös, Paul, 315
error threshold, 158, 160, 161, 163
euchromatin, 413, 414
Euclid's algorithm, 107, 108
Euler characteristic, 272
Euler's formula, 117, 272
Euler's theorem, 129, 130
Euler, Leonhard, 272
event, 90, 131
events
  disjoint, 97–99
  independent, 137, 181, 198, 217
  mutually exclusive, 90

evolutionary arms race, 331, 332, 378, 379, 388
evolutionary game theory, 168–170, 172, 335
evolutionary medicine, xiii, 329–336
evolutionary phase transition, 142
exon, 411, 413
expected number, 365
expected value, xx, 97, 100, 139, 146–149, 180, 181, 200–202, 205, 216–218, 316

Fairchild, William, xvi
Fas, 386
Fas ligand (FasL), 384, 386
FASTA, 166
feed-forward loop (FFL), 315–317, 319–321, 323
  coherent, 316, 317, 320, 321
  incoherent, 316, 317, 320, 321, 323
Fernandez, Rafael, xvi
Ferreira, Bianca, 82
fever, 333
Feynman, Richard, 328
Fibonacci numbers, 194, 196
final state sensitivity, 1, 54, 63, 65, 307
Fisher's theorem, 146–149
Fisher, Ronald, 146, 148
fitness, 140–142, 150–152, 156–158, 162, 168, 169, 331
  average, 139–142, 145, 146, 150–152, 158, 160–163, 170
fitness function, 151, 153, 155–157, 227
fitness landscape, xii, 150–159, 172, 226, 227, 331, 411
  fractal (rugged), 152, 153
Fitzhugh, Richard, 48
Fitzhugh-Nagumo equations, 401
fixed point, xiii, 7–13, 15, 16, 19, 22, 28, 29, 35, 36, 38, 40–43, 45, 49, 52–55, 58–62, 64–66, 69–76, 78–81, 84–86, 93, 98, 139, 140, 142, 159–161, 170, 171, 226, 250, 264, 266, 267, 269–273, 275, 276, 307–312, 320, 398, 402, 403, 419, 421, 423
  asymptotically stable, xix, 4, 7, 8, 12–14, 16, 19, 21, 22, 28, 30, 40, 43, 61, 64, 66, 68, 72, 270, 275, 309, 310, 398, 401
  center, 270
  equation, 42
  index, x, xiii, 226, 250, 270, 272, 273, 275–277
  node, xix, 14, 19, 64, 270, 275, 309, 310
  saddle, xix
  saddle point, 28, 52, 56, 61, 64, 66, 75, 78, 155, 171, 270, 271, 275, 276, 309, 361, 398
  spiral, xix, 4, 14, 30, 40, 64, 66, 72, 271, 398
  stability, xix, 7–13, 15, 19, 29, 36, 45, 49, 51, 53, 58, 62, 72, 160, 170, 171, 271, 309
  stable, 31, 93, 136, 171, 307, 310
  type, xix, 3, 15, 49, 51, 53, 72, 170
  unstable, xix, 4, 10, 12, 13, 28, 43, 59, 61, 79, 81, 98, 171, 270, 423
flux, 282, 302
fractal, xiii, 152, 153, 335, 341, 342, 375, 413–415
  correlation, 413
Frame, Kim, xvii
Frame, Mary, xvii
Frame, Ruth, 176
Frame, Steve, x, xi, xvii, 325
Frame, Walter, xvii, 176
Franklin, Rosalind, 336
Frobenius, Ferdinand, 105, 106, 390
Fubini's theorem, 244
Fubini, Guido, 244
fundamental theorem of calculus (FTC), 159, 234, 236
fundamental theorem of line integrals (FTLI), xiii, 225, 226, 234, 236, 239, 242, 243, 298, 403

Gallo, Robert, 24
Gardner, Timothy, 307, 310
Garfinkel, Alan, 81
gating
  background, 207, 208, 212–214
  burst, 207, 208, 212–214
Gauss' theorem, xiii, 226, 296, 298, 300, 302
Geisser, Daniel, xvii
gel electrophoresis, 339

Gelfand's formula, 392, 393
gene, 136, 137, 151, 163, 165, 172, 176, 209, 310, 312–314, 317, 319, 321, 323–325, 332, 333, 336, 337, 341, 374, 387, 411, 414
  activation, 174, 316, 341, 414
  activator, 313, 316, 317, 319–321
  beta hemoglobin, 409
  expression, 173, 175, 307, 313, 319–321, 341, 411, 413
  inactivation, 215, 216, 219, 220, 414
  pair, 136, 137, 141, 145
  pleiotropic, 152
  product, 313, 316
  regulation, 175, 313, 413, 414
  regulatory network, 157, 175, 176, 305, 307, 310, 323
  regulatory protein, 413
  repression, 316, 341, 414
  repressor, 264, 307, 309, 313, 316, 317, 320, 321
  segment, 381, 383
  silenced, 172
  therapy, 135
  transcription, 312, 413
  transcription network, ix, x, xiii, 305, 306, 312–317, 319, 321
  tumor suppressor (TSG), 172, 215–224, 325
genetic, 135, 340
genetic algorithm, 306
genetic drift, 180–189, 303
genetic toggle switch, 263, 264, 307, 310, 311, 425
genetic transduction, 330
genetic variation, 144, 145, 158, 169, 182
genetics, xii, 23, 135–140, 143–145, 150, 151, 154–162, 164, 166–168, 170–174, 177, 179, 336, 352, 406, 415
  molecular, x, 23, 37, 167, 341, 406–415
genome, 30, 34, 151, 158, 161–164, 166, 173, 314, 330–332, 336, 337, 339–341, 345, 381, 387, 409, 411, 413, 426
genotype, 136–138, 143–145, 150–153, 158, 171, 411
  frequency, 136–138, 143, 145
Gilpin model, 56–59, 61, 62, 421
Gilpin, Michael, 56
Gish, Warren, 164
Gleick, James, 74
globule
  equilibrium, 413
  fractal, 414, 415
glycolysis, 324, 325
Gorter, Evert, 371
gradient, xiii, 151, 152, 154, 155, 159, 225–231, 234–239, 241, 242, 250, 253, 285, 288, 296, 299, 301–303, 306, 371, 403, 404, 411
  descent, 306
Gram-Schmidt orthogonalization, 397, 398
graph
  connected, 105
  global regulator, 314
  sparse, 314
  strongly connected, 105, 106, 109, 113
graphical iteration, 78
greatest common divisor, 106–108, 113
Grebogi, Celso, 55, 74, 77
Green's theorem, xiii, 225, 241, 250–263, 271, 285–287, 291, 298, 299
Grendel, François, 371
Grosberg, Alexander, 414
Grossmann, Marcel, 180
growth rate, 34, 114, 121–123, 127, 128, 163, 307, 320, 335, 395, 396
  eventual, 96, 122, 132–134
Guckenheimer, John, 201, 328
Gumbel extreme value distribution, 166
Gurbatri, Candice, xvi

Hagan, Liz, xvi
Haldane, J. B. S., 331
half-life, 33, 34
  and mean lifetime, 34
Hall, John, xvi
Hamilton's altruism equation, 147, 149
Hamilton, William, 146
Hamming distance, 150, 151, 153, 154, 156, 160
hantavirus, 330
haploid, 180

Hardy, G. H., 136
Hardy-Weinberg equilibrium, 141, 143
Hardy-Weinberg law, xii, 138
Harris, Fran, xvi
Hartman-Grobman theorem, xix, 1
Havlickova, Miki, xvi
heartbeat, 74, 81, 207, 347, 427
heat equation, 302, 303
Hebb, Donald, 325
Heesterbeek, Hans, 114
hemochromatosis, 331
Henson, Taraji P., 397
hepatitis B, 22, 330, 373
hepatitis C, 22
Hercules, 373
Hesse, Hermann, 155
Hesse, Ludwig, 155
Hessian, 155, 156
heterochromatin, 413, 414
heterozygosity, 181, 182, 332
  loss of (LOH), 215, 216, 220
heterozygous, 136, 144, 146, 208
Hilbert, David, 390
Hill coefficient, 307, 313
Hill function, 313
hill-climbing problem, 152
Hirsch, Morris, 357
histone, 413, 414
  modification, 172–175
HIV, 22–26, 29–38, 158, 330, 333, 374, 418
  initial acute phase, 24, 27, 34
  second phase, 24, 25, 27, 30
  shoulder phase, 32
  third phase, 24, 25, 30
Ho, David, 25, 32
Hodgkin, Alan, 45–48, 51, 372
Hodgkin-Huxley
  equations, ix, 1, 47–54, 419–421
homeostasis, 324
homologues, 166
homozygosity, 332
homozygous, 136, 143, 144, 146
Hopf, Eberhard, 272
Hopf, Heinz, 273
host cell, 24, 31, 373, 374
human endogenous retroviral elements (HERV), 173
human T cell leukemia virus (HTLV), 24
Hutchinson, John, 341
Huxley, Andrew, 45–48, 51, 372
hypermethylation, 172
hypomethylated, 173
hypomethylation, 172, 173

ibrutinib, 177
IgA, 380, 381, 385
IgD, 380
IgE, 380
IgG, 380, 385, 386
IgM, 380–382
immune response dynamics, ix, 1, 11, 37, 371, 385, 387
immune system, x, 22–25, 37, 114, 145, 169, 177, 220, 331, 374–382, 384, 385, 387, 389, 418
  adaptive (AIS), 23, 375, 376, 379, 385–387
  innate (IIS), 23, 375–379, 387
immunology, 167
index, 226, 266, 267, 273
influenza, 153, 330, 374
insulin growth factor-1, 331
integral
  definite, 231, 242
  double, 243–250, 254–256, 258, 260–263, 281, 285, 292, 389, 390
  line, xiii, 225, 226, 229, 232–237, 239–243, 250–258, 260–263, 266, 271, 273–275, 281, 283, 285–292, 298, 403
  surface, xiii, 225, 281–286, 288–291, 296–302
  triple, 292, 296, 297, 302
integrase, 24, 30
integrating factors, 32, 33
integration
  substitution, 31, 233, 240, 247, 282, 389, 390, 430
intron, 411, 413
invariant chain (IC), 375, 376
inverse function theorem, 308

invertible, 76, 94–96, 110, 124, 184, 196, 197, 357, 364
ion channel, x, 46, 47, 179, 190, 207–211, 373, 374, 401
  blocker, 47
  calcium, 206
  dynamics, xii, xiii, 198–215
  potassium, 206
  sodium, 206–209
ion pump, 46, 81, 206, 371
ion semiconductor sequencing, 339, 340
iterated function system (IFS), 341, 342
  address, 342, 343
  bin boundaries, 346
  driven, x, 343–345, 347–351, 426–428
  equal size bins, 346–348
  equal weight bins, 346–348, 351
  median centered bins, 346–348
  random algorithm, 342, 343
Iyidogan, Pinar, 34

Jeffrey, H. Joel, 344
jellyfish, 335
Jenkins, Megan, xvi
Jordan curve theorem, 236
Jordan, Camille, 236
junctional diversity, 381
Jung, Misun, xvi
Jungck, John, xvii

Kankel, Douglas, xvi
Kaplan-Yorke conjecture, 398
Kass, Robert, 209
Kemeny, John, 362
Kinzler, Kenneth, 215, 220
knot, 57, 414
Knudson's two-hit hypothesis, 172, 215
Knudson, Alfred, 215
Koella, Jacob, 330

Laine, Mary, xvii
Lander, Arthur, 328
Laplacian, 303
Lauter, Miriam, xvi
law of conditioned probabilities, xxi
law of cosines, 227
Lawson, Terry, 273
Lederberg, Joshua, 330
Lengauer, Christoph, 220
Leonardo of Pisa, 194
Leslie matrix, 89, 110, 115, 116, 118–123, 127, 129, 393
Leslie, Patrick, 114
level curve, 226, 227, 229
Liapunov exponent, 60, 68, 395–398
Liapunov function, xx, 1, 141, 155, 403
Lieberman-Aiden, Erez, 414
Lief, Regina, xvi
Lienard
  limit cycle criterion, 402
  system, 402
  theorem, x, 402
life-cycle graph, 115
ligand dependence, 371
limit cycle, xx, 10, 53, 55, 57, 266, 267, 273, 398, 402, 406
  stable, 402, 406
limit point, 342
linear combination, 96, 396
linear differential equation, 31, 35, 159, 354, 355
linear independence, 358, 359, 396, 397
linear regression, 75
linear span, 278, 280, 293, 396
Lipman, David, 164
Little, Susan, 29
logistic, 56, 63, 423
  map, 55, 60, 78, 79, 83, 346, 348, 350, 351
  susceptible limitations, 11
long QT syndrome type 3 (LQT3), 206, 207, 209
  mutation, 207
loop, 109, 175, 203, 210–212, 236
Lorenz
  attractor, 57
  weather model, 57
Lothes, Scott, xvii, 325
Lotka-Euler equation, 116, 117
Lotka-Volterra
  equations, 264

population curves, 9
system, 40, 58, 264, 303
luciferase, 339
Luo-Rudy model, 208
lymph node, 293, 379, 381, 383, 384, 386, 388
lymphatic system, 376, 379, 387
lymphocyte, 23, 37, 293, 373, 384
lymphoid follicles, 384
lysosome, 377

Möbius band, 283
Maatta, Jean, xvii, xviii
Mac Gabhann, Feilim, 209
MacLeod, Colin, 336
macrophage, 23, 171, 375, 377–380, 384, 386
Magliula, Richard, xvii, xviii
major histocompatibility complex (MHC), 375, 383, 387, 388
class I, 375, 379, 383, 388, 389
class II, 375–377, 379, 383, 384
malaria, 145, 332
Mandelbrot, Benoit, 373
mannose, 377
mannose-binding lectin (MBL), 377
Mansky, Louis, 34
Markov chain, x, xiii, 88, 91–94, 96–100, 102–104, 110, 179, 180, 183, 184, 197, 198, 200, 207, 209, 215, 216, 362–366, 368, 373, 393
absorbing class, 365
absorbing states, 96, 97, 99, 183–186, 204, 217, 218, 220, 222
aperiodic, 97–99
closed states, 99
equilibrium distribution, 91–94, 97–100, 102–104, 106
ergodic class, 96, 97, 99, 362, 363, 365, 366, 368
ergodic states, 362–364, 366, 368
irreducible, 368–370
mean first passage time, 368
mean recurrence time, 369, 370
periodic, 97, 98
periodic states, 96, 97
primitive, 368
recurrent states, 97, 99
transient class, 96, 362, 363, 365, 366
transient states, 99, 183, 184, 204, 362–366
transition graph, 89, 103, 104, 207
transition matrix, 90, 91, 97–104, 106, 113, 200, 205, 206, 210–212, 214, 216, 217, 220, 222, 223, 363, 366, 368
Markov, Andrey, 99
Marmount, George, 372
Marquez, Jonathan, xvi
matrix
cofactor, 2
determinant, 1, 2, 4, 6, 196, 364, 394, 395
differential equations, xiii, 190–197
exponentiation, 190–197, 353, 354, 356
inverse, 83, 111
irreducible, 105, 106, 368, 390
lower triangular, 394, 395
multiplication, 76, 124, 192
nilpotent, 357, 358
norm, 392
power positive, 105, 106, 115
primitive, 105, 106, 109–114, 124, 366, 368, 390, 391
symmetric, 78, 360, 362
trace, 196
transpose, 94, 197
triangular, 12
upper triangular, 394, 395
Matsumoto, Gen, 51
May, Robert, 22, 29, 31, 32
Maynard Smith, John, 169
McCarty, Maclyn, 336
McCullouch, Warren, 325
mean lifetime, 26, 34, 44
membrane, 47, 81, 198, 200, 201, 206, 371
cell, 23, 24, 46, 370–372, 378, 384, 390
current, 199–202, 205
lipid, 23, 374
mucous, 374, 380
potential, 46, 47, 199, 201, 206, 371, 372
membrane attack complex (MAC), 377, 382
Mendel, Gregor, 136
Menten, Maud, 17, 19
metabolic network, 307, 324, 325

methylation, 172–174
Michaelis constant, 17
Michaelis, Leonor, 17, 19
Michaelis-Menten differential equations, 17
Michaelis-Menten kinetics, 1, 17, 20, 22, 417
Michaelis-Menten rate law, 17, 19, 20
Michod, Richard, 142, 149
Miller, Webb, 164
Mohamed, Aala, xvi
moment, second, xx
Montagnier, Luc, 24
Moolgavkar, Suresh, 215
Moore's law, 166
Morgenstern, Oskar, 168
Moss, Frank, 401
MRSA, 341
Muldoon, Maureen, xvii
multiple myeloma, 169, 171
mutation, 34, 151, 153, 157, 158, 160–162, 164, 172–174, 177, 180, 182, 206–209, 213, 215–217, 219, 220, 273, 306, 313, 314, 325, 330, 337, 378, 383, 409
  back, 160
  deletion, 151, 166, 177, 207, 332, 381, 409, 410
  frameshift, 410, 411
  in-frame, 410
  insertion, 151, 166, 381, 409, 410
  missense, 409
  nonsense, 409
  point, 30, 34, 151, 158, 160–163, 215, 216, 220, 409
  rate, 33, 34, 158–163, 216, 383
  recombination, 411
  silent, 409
  somatic, 341
  transition, 409
  transversion, 409
mutation-selection matrix, 159, 161–163
Myers, Eugene, 164

nAChR (nicotinic acetylcholine receptor), 198
Nagashima, Tomomasa, 395
nanoball, 340
  ligation, 339, 340
Nash equilibrium, 168
natural killer (NK) cells, 375, 378–380, 386, 388, 389
negative definite, 155, 156
Neher, Erwin, 372
neighborhood, 400
Nernst equation, 46, 47
Nesse, Randolph, 330, 331
network motif, 314, 316, 323, 324, 407
neural net, x, 209, 305, 307, 325–328
  back propagation, 326
  hidden layer, 325–327
  input layer, 325
  output layer, 325
  training set, 326, 327
  weight, 326, 327
neuron, 45, 46, 51, 81, 174, 209, 293, 324–327, 401
neutral mutation hypothesis, 145
neutrophil, 23, 375, 378–380, 386
Newton's method, 421
Newton, Isaac, 87
next-generation sequencing (NGS), 336, 337, 339, 340
noise, 63, 145, 179, 306, 310, 398–401
  $1/f$, 401
nonlinear differential equations, 54
normal distribution, 279
normal vector, 279, 280, 282–286, 289–291, 296, 297, 299–301
normalization factor, 279, 390
Norton-Simon model, 426
Nowak, Martin, 22, 29, 31, 32, 216
nullcline, 19, 21, 22, 28, 38, 48, 49, 53, 64, 71, 274, 275, 308, 420, 421
Nyren, Pål, 339

odd function, 402
Odell, Garrett, xv
OGY
  formula, 77–81, 83, 362
  method, 75, 80–82, 361
oncogene, 173, 220, 325
  activation, 173

open set, 235, 236, 253, 342
opsonize, 377, 378, 380, 384, 388
orbit
  closed (periodic), 40
  continuous, 40
orientable, 283
orientation, 251, 267, 285
oriented, 258
  clockwise, 258, 259, 261
  counterclockwise, 258, 260, 261, 273, 289–292
  positively, 251–256, 258, 262, 285–290
Oseledec, Valery, 396
Ott, Edward, 55, 74, 77

parameterization
  curve, 231–234, 240–242, 256, 261, 275, 277, 278, 281, 283, 286, 289, 290, 292
    tangent line, 278
    tangent vector, 231, 241, 289, 290
  graph, 277, 282, 283, 285–287, 291, 292, 297, 299
  sphere, 277, 281, 293
  surface, 277, 280, 281, 283–285
    tangent plane, 278
  torus, 278, 284
Park, Susie, xvi
partition, 88, 98, 347, 411
patch clamp, 197–199, 201, 305, 372
path, 66, 68, 70, 71, 90, 105, 106, 109, 110, 112, 115, 152–157, 171, 187, 198, 203, 211, 212, 228, 231–237, 239–241, 255, 257, 261, 269, 271, 288, 289, 292, 316, 317, 319, 363, 369, 405
  closed, 236, 237
  independence, 234–237, 288
path-connected, 235
pattern-recognition receptors (PRR), 377–379, 384
payoff matrix, 168, 169
Peak, David, xvi
pediatric cancer genome project, 341
pendulum
  frictionless, 237
  total energy, 237
peptide, 23, 375, 376, 379, 383
per capita birth rate, 8, 14
per capita death rate, 8, 26, 35
per capita growth rate, 141, 200
per capita reproduction rate, 170
Perelson, Alan, 25, 29
perforin–granzyme B, 384
Perko, Lawrence, 402
Perron, Oskar, 105, 106, 390
Perron-Frobenius theorem, xiii, 110, 116, 118, 124, 203–205, 212, 368, 390, 393
Pettersson, Bertil, 339
Peyer's patches, 384
phagosome, 376, 377
phenotype, 136–138, 144, 150, 169, 175, 334, 335, 340
phosphorylation, 172, 324
phylostratigraphy, 341
planimeter, 250, 253, 301
Poincaré section, 69
Poincaré, Henri, 272
Poincaré-Bendixson theorem, xx, 1, 9, 10, 54, 55, 67, 269, 273
polar coordinates, 247–249, 293, 300, 389
Pollard, David, xvi
polymerase chain reaction (PCR), 337, 338, 340
population curve, 9–16, 56, 59, 60, 62, 70, 417, 418
population dynamics, 9, 22
Porreca, Gregory, 340
positive definite, 155
potential
  electrical, 46, 47, 206, 371
  energy, 237
  function, 63, 65, 234, 235, 237, 239, 288, 400
power law, 314, 316, 413
Powers, Richard, 406
predator-prey, 1, 22, 40, 169
  equations, 50, 264
  system, 55–59, 61–63, 421, 422
Price's equation, xvii, 147–149
Price, George, 149, 169

principle of mass action, xxi, 10, 11, 17, 18, 26, 38, 179, 381, 384
prior distribution, 175
prisoner's dilemma, 168, 169
probability
  multiplication rule, 137, 181, 198
probability density function (pdf), xx
production rate, 25–27, 38, 39, 313
programmed death-1, 385, 386
promoter, 172, 307, 313, 314, 316, 317, 319, 320, 411
proportional perturbation feedback (PPF), 81, 82
proportional sensitivity, 129, 132
prostate specific antigen (PSA), 334
protease inhibitor, 25, 30–34, 36, 418
proteasome, 375
Prum, Richard, xvi, 142
pyrosequencing, 339, 340

quasi-steady state hypothesis, 19
quasispecies, 158, 159, 161–163
  distribution, 161, 163
  equation, xiii, 159, 160, 162, 170

Rényi, Alfréd, 315
Racko, Dusan, 414
Ramakrishnan, Venki, 407
random mating, 138
random unions, 137, 138
random variable(s), 364
  independent, 389
Rauseo, S. N., 80
reading frame, 410
recessive, 136, 137, 143, 144
Red Queen Principle, 332
relatively prime, 107, 109, 111, 112
repolarization, 206, 207, 209
reproductive rate, 133, 158, 179, 215
reproductive success, 142
reproductive value, 126
resting potential, 46, 47, 206
retrovirus, 23, 24, 173
return map, 361, 362
  first, 74, 75, 81
  second, 74, 75, 80
reverse transcriptase, 24, 30
  gene, 30, 34
  inhibitor, 30–34, 36
Riffle, David, xvii
Riffle, Linda, xvii, 25, 325
RNA, 23, 24, 30, 31, 153, 164, 374, 384, 406
  mRNA, 264, 313, 407, 410, 411
  polymerase, 312, 407
  RNAp, 312
  tRNA, 407
Rock, Miriam, xvi
Rosenblatt, Frank, 325
Rothberg, Jonathan, 339
rowhammer, 87
Rudy, Yoram, 207–209, 211
Rusk, Nicole, 340

Sakmann, Bert, 199, 372
Samelson, Hans, 273
Sanger, Frederick, 336–339
SARS-CoV, 330
scalar triple product, 293
scaling, 142, 413
schizophrenia, 174, 175
Schuster, Peter, 150, 158
Schwarzer, Ashley, xvi
second-derivative test, 155
secondary lymphoid organs (SLO), 384–388
SEG, 164
Segraves, William, xvi
Seiken, Arnold, xvi
SEIS model, 9
selectin, 378
selection, 138, 142
  aesthetic, 142
  clonal, 382, 383
  equation
    continuous, 140–142
    discrete, 139, 140, 142
  fundamental theorem, 139, 140
  kin, 146
  natural, 142, 146, 152, 158, 169–171, 314, 328, 330–332, 334–336, 383, 411
  negative, 387

positive, 387
senescence, 335, 336
sensitivity analysis, xiii, 114, 126–129, 131–133
sensitivity to initial conditions, 1, 50, 54, 55, 58, 65, 73, 74, 396
separation of variables, 31
separatrices, 66, 67
sequence space, 156, 159, 161, 162, 383
series
convergent, 190, 344, 353, 354
absolutely, 354
geometric, 212
set point, 24, 25
Shaw, George, 32
Shelley, Mary, 87
Shen, Tony, 340
Shimada, Ippei, 395
sickle cell, 145, 332, 409
sickle cell anemia, 332
Sigmund, Karl, 136
simply-connected, 250, 251, 253, 254, 256–259, 263–265, 271, 288
simulated annealing, 306
single-input module (SIM), 324
sinoatrial (SA) node, 207
SIR model, ix, 1, 7–9, 11–17, 121, 179, 417
SIS model, 26
Smale, Stephen, 357
Smith-Waterman algorithm, 166
Snell, J. Laurie, 362
solution curve, 321, 417–419, 421
Sompayrac, Lauren, xvi, 374
Sorkin, Gregory, 152, 153
Spano, Mark, 80
spherical coordinates, 293–295, 300
spleen, 384, 386
stable age distribution, 118, 120–123
Stankey, Christina, xvi
state space, 88, 99
Stearns, Stephen, xiv, xvi, 329
Steinman, Ralph, 379
stem cell, 171, 173, 177, 178, 215
Stephen, Bijan, xvi
Sternberg, Shlomo, 390
Stevens, C., 201
stochastic
matrix, 91, 110, 183, 202, 368
process, 88
resonance, xvii, 63, 398–402
Stokes' theorem, xiii, 226, 285–292, 298–300
strategy
mixed, 169
pure, 168
Strogatz, Steven, 26
submultiplicative norm, 392
substitution matrix, 165
surface area, 281
surface independence, 288–290
Suzuki, Daisetsu Teitaro, 178
symbolic dynamics, 341, 347, 351
systemic lupus erythematosus (SLE), 173, 175
systems biology, xiii, 304–328
Szymkowiak, Andy, xvi

T cell, 23–25, 37, 173, 379, 382–387
class switching, 384
effector, 386
experienced, 383, 385
helper, 23–27, 29–35, 173, 375, 377, 379, 382–385, 388
killer, 24, 37, 40, 375, 379, 383–386, 388
membrane, 24, 383
memory, 386
naive, 379, 383–385, 387, 388
natural regulatory, 388
receptor (TCR), 383, 384, 388
somatic hypermutation, 384
tangent plane, 278
tangent vector, 155, 231, 232, 237, 241, 250, 270, 278, 280–283, 286, 289, 290, 396
Taq polymerase, 337, 338
targeted therapy, 176
Taylor series, xx, 161, 190, 193, 246, 247
Taylor's theorem, 155
Temin, Howard, 34
tent map, 83, 86, 346
The Essential Tension: Competition, Cooperation and Multilevel Selection in Evolution, 149

Thomas, Taylor, xvi
Tijms, Henk, 362
time series, 74, 80, 89, 347–350, 413
tit-for-tat, 169
tocilizumab (TCZ), 177, 178
toll gene, 377
toll-like receptor, 377, 382
Tonegawa, Susumu, 381
Toskas, Paschalis, xvi
trace-determinant (tr-det) plane, xix, 72, 264
trajectory, 9–11, 28, 30, 35, 53–58, 60, 61, 65–74, 170, 250, 263, 264, 270, 273, 275, 276, 311, 357, 395, 396, 398, 402–406, 417, 419–421, 425
  closed, xiii, xx, 9, 70, 226, 263–266, 270, 271, 273, 275–277, 403
  periodic, 1, 9, 55, 57, 58, 70, 72, 74, 263
  spiral, 4, 55, 57, 61, 357, 406
transcription, 407, 410, 413
transcription factor, 172, 312–314, 317, 324
transcription input function, 313
transition graph, 89, 102, 103, 105, 106, 109, 111, 113, 115, 188, 189, 199, 202–204, 210–212, 362
transmembrane channel, 47, 88, 197, 226, 303, 371, 372
transmembrane protein
  $Ig\alpha$, 381, 382
  $Ig\beta$, 381, 382
transporter protein, 375
trapping region, 9, 264
tumor necrosis factor (TNF), 377, 379
tumor suppressor gene (TSG), x, xiii

Uhlen, Mathias, 339
uncoupled
  maps, 351
  modes, 212, 213
  systems, 16

variance, xx, 146, 181, 201
vector field, 21, 22, 225, 226, 229, 232–243, 250, 252, 253, 256, 266–277, 281–283, 285, 288–292, 296, 299–301, 402–404
  conservative, 234–237, 240, 250, 253, 254, 288, 289
vector projection (component), 232, 293
vector space
  dimension, 358, 393, 397
viral envelope, 23, 24, 374
viral load, 24, 25, 27, 29–31, 34
virion, 24–40, 42, 45, 374
virus, 22–25, 29, 32, 33, 36, 37, 43, 114, 153, 158, 161, 169, 330, 373–375, 377–380, 384, 388, 413
  core, 23, 374
  hepatitis, 374
  naked, 374
  RNA, 23, 158, 161, 163
virus dynamics, ix, 1, 11, 22–37, 114, 418
Vogelstein, Bert, 215, 220
voltage clamp, 46, 201, 205, 372
voltage dependence, 371
Volterra's trick, 55
von Neumann, John, 168
Voss, Richard, 411–413

Waddington, Conrad, 172
Wagner, Günter, xvi
Warburg effect, 325
Warburg, Otto, 325
Watson, James, 336
Weinberg, Wilhelm, 136
Weiss, Robin, 24
Werbos, Paul, 325
West Nile virus, 333
wild type, 145, 158, 160, 161, 173, 207, 208, 210
Williams, George, 330, 331
Williams, Robert, 57
Wright, Sewall, 150
Wright-Fisher model, 180–182, 184, 185, 187, 188
Wyman, Robert, xvi

x-nullcline, xix, 70, 71, 73, 274, 275, 307, 308, 402, 403 *y*-nullcline, xix, 70, 71, 73, 274, 275, 307, 308, 402

Yorke, James, 55, 74, 77

Zayyad, Zaina, xvi
Zhang, Jingsong, 333
Zhu, Zheng, 209
zoonotic pathogen, 331